Applied Mathematics-I

As per UBTER, Uttarakhand and AICTE Syllabus

for

First Semester Diploma Students

by

Er. M.K. Kanyal

KHANNA PUBLISHERS
Operational Office
4575/15, Onkar House, Ground Floor,
Darya Ganj, New Delhi 110 002
Phones : 011-23243042, 011-23243043 & Mob. 09811541460
email : contactus@khannapublishers.in

***Published by* :**

Romesh Chander Khanna & Vineet Khanna
for **KHANNA PUBLISHERS**
2-B, Nath Market, Nai Sarak,
Delhi-110006 (India)

Visit us at : www.khannapublishers.in

Hologram & Description

To all readers of our books, from yourself if being defrauded by pirates to prevent, please make sure that there is an Hologram on the cover of our books with the below specifications. If you find any book without Hologram and Description, please mail us at **contactus@khannapublishers.in**

Thanking you

ISBN No. : 978-93-87394-61-2

***First Edition* : 2019**

***Price* : ₹ 299.00**

Preface

It give us a great pleasure in presenting the new edition for Diploma level students of UBTER, Uttarakhand as per AICTE syllabus, for their **First Semester Diploma Course**. The aim of writing this book is to be provided the students a clear understanding of the basic concepts.

The subject matter of the book is arranged and presented such a way that the entire syllabus has been covered in the articles form in same sequence.

The book is divided into ***Four Units*** and ***Seventeen Chapters***. A large number of solved examples, exercises with answers given at the end of each chapter. This should be of a great help to understand the relevant concept and in developing problems solving skills.

Inspite of our best efforts, it is possible that some errors might have crept in. We shall acknowledge with gratitude, if any such error is brought in our notice.

Also, any suggestions and comments from students and teachers for improvement of the book are welcome.

We hope the book will be found useful by the readers.

—Publishers.

Contents

Unit—III Complex Number 205—236

Unit—IV Differential Calculus 237—366

Part 1 : Function and Limits

Part 2 : Continuity and Differentiation and its Applications

Unit—I

Algebra

1.1. VALUE OF ^np_r AND ^nc_r (WITHOUT PROOF)

• Factorial Notation

Let us consider n be a positive integer, the continued product of first n natural number is called factorial n, be denoted by n ! or $\lfloor n$.

Do you know?

$0 ! = 1$

Thus,

$$n ! = \lfloor n = n (n-1) (n-2) \ldots 3 . 2 . 1$$
$$= n (n-1) (n-2) (n-3) \ldots 3 . 2 . 1$$
$$= n [(n-1) (n-2) (n-3)] \ldots 3 . 2 . 1]$$
$$= n [(n-1) !]$$

Theorem:

(*i*) $n (n-1) (n-2) \ldots (n-r+1) = \dfrac{n!}{(n-r)!}$

$n (n-1) (n-2) \ldots (n-r+1)$

$$= n (n-1) (n-2) \ldots (n-r+1)$$

Multiplying numerator and denominator by $(n-r)$!

$$= \frac{n!}{(n-r)!}$$

For example, $(2n)! = (2n) (2n-1) (2n-2) (2n-3) \ldots 4 . 3 . 2 . 1$

$$= [(2n) (2n-2) (2n-4) \ldots 4 \ldots 2] [(2n-1) (2n-3) \ldots 3 \ldots 1]$$
$$= 2^n \times [n (n-1) (n-2) \ldots 2 . 1] \times [(2n-1) (2n-3) \ldots 3 \ldots 1]$$
$$= 2^n . (n !) . [1 . 3 . 5 \ldots (2n-1)]$$

Example 1. *Prove that inequality,*

$$(n!)^2 \le n^n . n ! < (2n) ! \text{ for all positive integer } n.$$

Sol. $(2n) ! = (1 . 2 . 3 \ldots n) (n+1) (n+2) \ldots (2n-1) (2n) > (n !) n^n$

[$\because$ $(n+r) > n$ far $r = 1, 2 \ldots n$]

$(n !) . n^n < (2n) !$ ...(*i*)

$(n !)^2 = (1 . 2 . 3 \ldots n) (n !) \le n^n . (n !)$ ...(*ii*)

[$\because$ $r \le h$ for each $r = 1, 2 .. n$]

From (*i*) and (*ii*), we get

$$(n\,!)^2 \leq n^n\,.\,n\,! < (2n)\,!$$

Example 2. *LCM [4 ! 5 ! 6 !]*

Sol. We have

$$4\,! = (4 \times 3 \times 2 \times 1) = 2^3 \times 3$$
$$5\,! = (5 \times 4 \times 3 \times 2 \times 1) = 2^3 \times 3 \times 5$$
$$6\,! = (6 \times 5 \times 4 \times 3 \times 2 \times 1) = 2^4 \times 3^2 \times 5$$

$\therefore$ LCM [4 !, 5 !, 6 !] = $(2^4 \times 3^2 \times 5) = 720$.

Example 3. *Find the value of* $\dfrac{8!}{4!\,5!}$

Sol. Given, $\dfrac{8!}{4!\,5!}$

First, we have 8!

$$8! = 8 \times 7 \times 6 \times 5 \times 4 \times 3 \times 2 \times 1$$
$$= 40320$$

Now consider 4!5!

First $4! = 4 \times 3 \times 2 \times 1$
$= 24$

and $5! = 5 \times 4 \times 3 \times 2 \times 1$
$= 120$

$\therefore$ $4!\,5! = 2880$

Hence, $\dfrac{8!}{4!\,5!} = \dfrac{40320}{2880} = 14$

Example 4. *Show that (n ! + 1) is not divisible by any natural number between 2 and n*

Sol. $n\,! = n\,(n-1)\,(n-2)\,...\,3\,.\,2\,.\,1$

$n!$ is divisible by every natural number between 2 and n.

If $(n\,!+1)$ is divided by any natural number between 2 and n, leaves 1 as remainder.

So, $(n\,!+1)$ is not divisible by any natural number between 2 and n.

Example 5. *There are 4 routes between Allahabad and Ghaziabad. In how many different ways can a man go from Allahabad to Ghaziabad and return, if for returning.*

(*i*) *any of the routes is taken*

(*ii*) *the same route is taken*

(*iii*) *the same route is not taken?*

Sol. (*i*) If the man may take route for going from Allahabad to Ghaziabad.

When he return by any routes, in 4 different ways.

Total number of ways on going to Ghaziabad and returning back to Allahabad = $4 \times 4 = 16$

(*ii*) Now, 4 ways of going to Ghaziabad and in this case only 1 way of returning

$\therefore$ total number of ways = $4 \times 1 = 4$

(*iii*) 4 ways of going to Ghaziabad,

3 ways of returning back to Ghaziabad

Example 6. *How many words (with or without meaning) of three distinct letters of the English alphabet are there?*

Sol. There are 26 letters of the English alphabet

There are 25 ways of filling up the second place.

$\therefore$ the required number of words

$$= 26 \times 25 \times 24 = 15600.$$

Example 7. *How many numbers are there between 100 and 1000 such that 7 is the units place.*

Sol. Hundred's place can be filled with any of the digits from 1 to 9.

So, the hundred's place can be filled with any of the digits from 1 to 9.

The units place can be filled only 1 way.

$\therefore$ Required numbers = $9 \times 10 \times 1 = 90$.

1.1.1. ^np_r (Permutations)

The different arrangement which can be made out of a given number of things by taking sum or all 0 at a time, are called permutations.

The number of all permutations of n dissimilar things taken r at a time is given by

$$^np_r = n(n-1)(n-2) \ldots (n-r+1)$$

$$= \frac{n!}{(n-r)!}.$$

Example 8. *Calculate :*

(*i*) $^8p_8$ (*ii*) $^{75}p_2$ (*iii*) $^{12}p_4$.

Sol. We have (*i*) $^8p_8 = 8! = 8 \times 7 \times 6 \times 5 \times 4 \times 3 \times 2 \times 1 = 40320$

$$(ii)\ ^{75}p_2 = \frac{75!}{(75-2)!} = \frac{75!}{73!}$$

$$= \frac{75 \times 74 \times 73!}{73!} = 5550$$

$$(iii)\ ^{12}p_4 = \frac{12!}{(12-4!)} = \frac{12 \times 11 \times 10 \times 9 \times 8!}{8!}$$

$$= 11880.$$

Example 9. *Prove that* (*i*) $^np_n = {}^np_{n-1}$

(*ii*) $^np_r = n \,.\, ^{n-1}p_{r-1}$

(*iii*) $^np_r = {}^{n-1}p_r + r.\, ^{n-1}p_{r-1}$.

Sol. We have (*i*) $^np_{n-1} = \dfrac{n!}{[n-(n-1)]!} = \dfrac{n!}{1!} = n! = {}^np_n$

$$(ii) \qquad ^np_r = \frac{n!}{(n-r)!} = \frac{n(n-1)!}{[(n-1)-(r-1)]!} = n \,.\, ^{n-1}p_{r-1}$$

(iii) $\quad {}^{n-1}p_r + r \,.\, {}^{p-1}p_{r-1} = \left\{\dfrac{(n-1)!}{(n-1-r)!} + r \,.\, \dfrac{(n-1)!}{[(n-1)-(r-1)]!}\right\}$

$$= \left\{\frac{(n-1)!}{(n-1-r)!} + r \,.\, \frac{(n-1)!}{[(n-r)!]}\right\}$$

$$= \left\{\frac{(n-1)!}{(n-r-1)!} + r \,.\, \frac{(n-1)!}{[(n-r)(n-r-1)!]}\right\}$$

$$= \frac{(n-1)!}{(n-r-1)!}\left\{1 + \frac{r}{(n-r)}\right\} = \frac{n \,.\, (n-1)!}{(n-r)(n-r-1)!}$$

$$= n\,!/(n-r)\,! = {}^{n}p_{r}.$$

Example 10. *In how many ways can the letters of the word 'PENCIL' be arranged so that N is always next to E.*

Sol. Putting E and N together and considering them as one letter, we have to arrange 5 letters at 5 places.

This is ${}^5p_5 = 5! = 120$ ways.

Example 11. *(i) How many words can be formed by using all the letters of the word 'ALLAHABAD'.*

(ii) How many of these words will not contain both L together?

Sol. (*i*) There are 9 letters out of which 'A' is repeated 4 times, 'L' is repeated 2 times, and the rest are all different.

$\therefore \qquad$ No. of words $= \dfrac{9!}{4!\,2!} = 7560$

(*ii*) When both 'L' are together, we can letter. So, we will have to arranged 8 letters out of which 'A' is repeated 4 times.

No. of arrangement with both 'L' together $= \dfrac{8!}{4!} = 1680$

So the no. of arrangement with both 'L' not occurring together = 7560 – 1680 = 5880.

Example 12. *How many 6 rings of different types had in 4 fingers.*

Sol. First ring can be worn in any of 4 fingers, there are 4 ways of wearing it.

Thus, rings may be worn in 4 ways.

$\therefore$ number of ways $= 4^6 = 4096$.

Theorem 1. *The number of circular permutations of n different objects is (n – 1) !.*

Proof. Fixing the position of an object can be done in *n* ways, as the position of anyone of them may be fixed.

So, each circular permutation corresponds to *n* linear permutations, depending upon where from we start.

There are n ! linear permutations, it follows that there are $\dfrac{n!}{n}$, *i.e.*, $(n-1)$! circular permutations.

Theorem 2. *The number of ways in which n persons can be seated round a table is (n – 1)!.*

Proof. Let us consider fix the position of one person and then arrange the remaining $(n-1)$ persons in all possible ways. Clearly, this can be done in $(n-1)$! ways.

So, the required number of ways $= (n-1)$!.

Theorem 3. *Show that the number of ways in which n different beads can be arranged to form a necklace is 1/2 (n – 1) !.*

Proof. Fixing the position of one bead, the remaining $(n - 1)$ beads can be arranged in $(n - 1)$! ways.

In case of arranging the beads, there is no distinction between the clockwise and the anticlockwise arrangements.

Thus, the required number of ways = $1/2\ (n - 1)$!.

1.1.2. ^nC_r (Combinations)

Each different groups which can be formed by taking some or all of a number of objects of their arrangements is called a combination.

- If number of all combinations of n distinct objects, taken or at a time, is given by

$$^nC_r = \frac{n!}{(r!) \times (n-r)!}$$

- If $0 \le r \le n$, then $^nC_r = {}^nC_{n-r}$ • If $1 \le r \le n$, then $^nC_r + {}^nC_{r-1} = {}^{n+1}C_r$
- If $1 \le r \le n$, then $n \times {}^{n-1}C_{r-1} = (n - r + 1) \times {}^nC_{r-1}$
- If n and r are positive integers such that $1 \le r \le n$, then $\dfrac{^nC_r}{^nC_{r-1}} = \dfrac{n-r+1}{r}$
- If $1 \le r \le n$, then $^nC_r + {}^nC_{r+1} = {}^{n+1}C_{r+n}$
- If $^nC_p = {}^nC_q \Rightarrow p = q$ or $p + q = n$.

1.1.3. Binomial Theorem (without proof) for Positive Integral Index

$$(x + y)^n = {}^nC_0 + x^n + {}^nC_1\, x^{n-1}y + {}^nC_2\, x^{n-2}\, y^2 + \ldots + {}^nC_{n-1}\, xy^{n-1} + {}^nC_n\, y^n$$

Here, $(x + y)^0 = 1$

$$(x + y)^1 = (x + y) = {}^1C_0\, x + {}^1C_1\, y$$
$$(x + y)^2 = x^2 + 2xy + y^2 = {}^2C_0\, x^2 + {}^2C_1\, xy + {}^2C_2\, y^2$$
$$(x + y)^3 = x^3 + 3x^2y + 3xy^2 + y^3 = {}^3C_0\, x^3 + {}^3C_1\, x^2y + {}^3C_2\, xy^2 + {}^3C_3\, y^3$$
$$(x + y)^4 = x^4 + 4x^3y + 6x^2y^2 + 4xy^3 + y^4$$
$$= {}^4C_0\, x^4 + {}^4C_1\, x^3y + {}^4C_2\, x^2y^2 + {}^4C_3\, xy^3 + {}^4C_4\, y^4 \text{ and so on.}$$

Pascal's Triangle

The pascal's triangle showing the coefficients of various terms in a binomical expansion.

Index	*Coefficients*
0	1
1	1 ∇ 1
2	1 ∇ 2 ∇ 1
3	1 ∇ 3 ∇ 3 ∇ 1
4	1 ∇ 4 ∇ 6 ∇ 4 ∇ 1
5	1 ∇ 5 ∇ 10 ∇ 10 ∇ 5 ∇ 1
6	1 ∇ 6 ∇ 15 ∇ 20 ∇ 15 ∇ 6 ∇ 1
7	1 ∇ 7 ∇ 21 ∇ 35 ∇ 35 ∇ 21 ∇ 7 ∇ 1
8	1 8 28 56 70 56 28 8 1

We observe that:

- every row starts with 1 and ends with 1;
- every coefficient (except the first and the last) in a row is the sun of two coefficients in the preceeding row, one just before it and the other just after it.

Look at the following pattern:

For $n = 6$, we have the coefficients ${}^6C_0, {}^6C_1, {}^6C_2, {}^6C_3, {}^6C_4, {}^6C_5, {}^6C_6$

For $n = 7$, we have the coefficients ${}^7C_0, {}^7C_1, {}^7C_2, {}^7C_3, {}^7C_4, {}^7C_5, {}^7C_6, {}^7C_7,$

For $n = 8$, we have the coefficients ${}^8C_0, {}^8C_1, {}^8C_2, {}^8C_3, {}^8C_4, {}^8C_5, {}^8C_6, {}^8C_7, {}^8C_8$ and so on.

1.1.3.1. *Binomial Theorem for Expansion Terms*

Theorem 4. *If x and y are real numbers then for all n ∈ N,*

$$(x + y)^n = {}^nC_0x^n + {}^nC_1\, x^{n-1}\, y + {}^nC_2x^{n-2}\, y^2 + ... + {}^nC_{n-1}xy^{n-1} + {}^nC_ny^n,$$

i.e.,

$$(x + y)^n = \sum_{r=0}^{n} C_r\ .\ x^{n-r}\, y^r.$$

Proof. We shall prove the theorem by using the principle of mathematical induction.

Let P (n) be the statement:

$$(x + y)^n = {}^nC_0x^n + {}^nC_1\, x^{n-1}\, y + {}^nC_2x^{n-2}\, y^2 + ... + {}^nC_{n-1}xy^{n-1} + {}^nC_ny^n.$$

Clearly, the statement P (1) is

$$(x + y)^1 = {}^1C_0\, x^1 + {}^1C_1\, x^0y = (x + y).$$

This shows that P (1) is true.

Let P (m) be true. Then,

$$(x + y)^m = {}^mC_0x^m + {}^mC_1\, x^{m-1}\, y + {}^mC_2x^{m-2}\, y^2 + ...\ {}^mC_{m-1} + {}^mC_my^m \qquad ...(i)$$

Multiplying both sides of (i) by $(x + y)$, we get

$$\begin{aligned}(x + y)^{m+1} &= {}^mC_0x^{m+1} + {}^mC_0x^my + {}^mC_1x^my \\ &\quad + {}^mC_1x^{m-1}\, y^2 + {}^mC_2x^{m-1}\, y^2 + {}^mC_2x^{m-2}\, y^3 + ... \\ &\quad + {}^mC_{m-1}x^2y^{m-1} + {}^mC_{m-1}\, xy^m + {}^mC_mxy^m + {}^mC_m{}^{m+1} \\ &= {}^mC_0x^{m+1}\, [{}^mC_0 + {}^mC_1]\, x^my + [{}^mC_1 + {}^nC_2]\, x^{m-1}\, y^2 + ... \\ &\quad + [{}^mC_{m-1} + {}^mC_m]\, xy^m + {}^mC_m\, y^{m+1} \\ &= {}^{m+1}C_0x^{m+1} + {}^{m+1}C_1x^my + {}^{m+1}\, C_2x^{m-1}\, y^2 + ... \\ &\quad + {}^{m+1}C_mxy^m + {}^{m+1}\, C_{m+1}\, y^{m+1} \\ &\quad [\because {}^{m+1}C_0 = 1 = {}^mC_0;\ {}^mC_m = 1 = {}^{m+1}C_{m+1} \text{ and } {}^mC_{r-1} + {}^mC_r = {}^{m+1}C_r]\end{aligned}$$

This shows that P $(m + 1)$ is true, whenever P(m) is true.

Thus, by the principle of mathematical induction, the theorem for all $n \in$ N.

Remark. ${}^nC_0, {}^nC_1, {}^nC_2, ..., {}^nC_n$ are known as binomial coefficients.

1.1.3.2. *General Term*

The expansion of $(x + y)^n$ has $(n + 1)$ terms. Since ${}^nC_r = {}^nC_{n-r}$, we have ${}^nC_0 = {}^nC_n$; ${}^nC_1 = {}^nC_{n-1}$, and so on.

Thus, the coefficient of the terms equidistant from the beginning and the end in a binomial expansion are equal.

General term in the expansion of $(x + y)^n$ and general term = $(r + 1)$th term,

$$t_{r+1} = {}^nC_rx^{n-r}\, y^r.$$

1.1.3.3. *Middle Term*

The middle terms in the expansion of $(x + y)^n$ *:*

We know that the expansion of $(x + y)$ has $(n + 1)$ terms.

(*i*) *If n is even, the middle term* = $\left(\frac{n}{2}+1\right)$ *th term.*

For example, in the expansion of $(x + 2y)^8$, the middle term is $\left(\frac{8}{2}+1\right)^{\text{th}}$ *i.e.* 5th term.

(*ii*) *If n is odd, the middle term*

$$\frac{1}{2}\ (n + 1)th\ term\ and\ \left\{\frac{1}{2}(n+1)+1\right\} th\ term.$$

For example, is the expansion of $(2x - y)^7$, the middle term is $\left(\frac{7+1}{2}\right)^{\text{th}} = 4^{\text{th}}$ and $\left(\frac{7+1}{2}\right)^{\text{th}}$ *i.e.*, 5th terns.

pth term from the end in $(x + y)^n$

= $(n + 1 - p + 1)$th term from the beginning.

= $(n - p + 2)$th term from the beginning.

Deductions

Replacing x by 1 and y by x in binomial expansion of $(x + y)^n$, we have

$$(1 + x)^n = {}^nC_0 + {}^nC_1x + {}^nC_2x^2 + \ldots + {}^nC_nx^n.$$

In this expansion, $(r + 1)$th term is the general term, given by

$$t_{r+1} = {}^nC_rx^r.$$

Replacing y by $-y$ in the binomial expansion of $(x + y)^n$, we get

$$(x - y)^n = {}^nC_0x^n - {}^nC_1x^{n-1}y + {}^nC_2x^{n-2}y^2 + \ldots$$

$$\Rightarrow \quad (x - y)^n = \sum_{r=0}^{n} (-1)^n \,.\, {}^nC_rx^{n-r}\, y^r$$

$$\Rightarrow \quad t_{r+1} = (-1)^r \,.\, {}^nC_rx^{n-r}\, y^r.$$

In this expansion, $t_{r+q} = (-1)^r \,.\, {}^nC_rx^{n-r}y^r.$

$$(1 - x)^n = {}^nC_0 - {}^nC_1x + {}^nC_2x^2 + \ldots + (-1)^n \,.\, {}^nC_nx^n.$$

$$\Rightarrow \quad (1 - x)^n = \sum_{r=0}^{n} (-1)^n \,.\, {}^nC_rx^r$$

Hence, in the expression of $(1 - x)^n$, we have $t_{r+1} = (-1)^r \,.\, {}^nC_rx^r$.

Example 13. *Expand* $(x^2 + 2y)^5$ *by the binomial theorem.*

Sol. Using the binomial theorem, we have

$$(x^2 + 2y)^5 = {}^5C_0 \,.\, (x^2)^5 + {}^5C_1 \,.\, (x^2)^4\,(2y) + {}^5C_2 \,.\, (x^2)^3, (2y)^2$$
$$+ {}^5C_3 \,.\, (x^2)^2\,(2y)^3 \,.\, {}^5C_4 \,.\, x^2\,(2y)^4 + {}^5C_5 \,.\, (2y)^5$$
$$= x^{10} + 5x^8\,(2y) + 10x^6\,(4y^2) + 10x^4\,(8y^3) + 5x^2\,(16y^4) + 32y^5$$
$$= x^{10} + 10x^8y + 40x^6y^2 + 80x^4y^3 + 32y^5$$

Example 14. *Expand* $\left(2x - \frac{3}{y}\right)^5$ *by the binomial theorem.*

Sol. Using the binomial theorem, we have

$$\left(2x - \frac{3}{y}\right)^5 = {}^5C_0 \,.\, (2x)^5 - {}^5C_1 \,.\, (2x)^4 \,.\, \left(\frac{3}{y}\right) + {}^5C_2 \,.\, (2x)^3 \,.\, \left(\frac{3}{y}\right)^2$$

$$- {}^5C_3 \,.\, (2x)^2 \left(\frac{3}{y}\right)^3 + {}^5C_4 \,.\, (2x) \left(\frac{3}{y}\right)^4 - {}^5C_5 \,.\, \left(\frac{3}{y}\right)^5$$

$$= 32x^5 - \left(5 \times 16x^4 \times \frac{3}{y}\right) + \left(10 \times 8x^3 \times \frac{9}{y^2}\right) - \left(10 \times 4x^2 \times \frac{27}{y^3}\right)$$

$$+ \left(5 \times 2x \times \frac{81}{y^4}\right) - \frac{243}{y^5}$$

$$= 32x^5 - \frac{240x^4}{y} + \frac{720x^3}{y^2} - \frac{1080x^2}{y^3} + \frac{810x}{y^4} - \frac{243}{y^5}.$$

Example 15. *Using the binomial theorem, expand* $[(x + y)^5 + (x - y)^5]$ *and hence find the value of* $[(\sqrt{2} + 1)^5 + (\sqrt{2} - 1)^5]$.

Sol. We have, $(x + y)^5 + (x - y)^5$

$$= [{}^5C_0x^5 + {}^5C_1x^4y + {}^5C_2x^3y^2 + {}^5C_3x^2y^3 + {}^5C_4xy^4 + {}^5C_5y^5]$$
$$+[{}^5C_0x^5 - {}^5C_1x^4y + {}^5C_2x^3y^2 - {}^5C_3x^2y^3 + {}^5C_4xy^4 - {}^5C_5y^5]$$
$$= 2\, [{}^5C_0x^5 + {}^5C_2x^3y^2 + {}^5C_4\, xy^4]$$
$$= 2\, [x^5 + 10x^3y^2 + 5xy^4]$$

Putting, $x = \sqrt{2}$ and $y = 1$

$$(\sqrt{2} + 1)^5 + (\sqrt{2} - 1)^5 = 2[(\sqrt{2}) + 10(\sqrt{2})^3 \,.\, 1^2 + 5(\sqrt{2})\, 1^4]$$
$$= 2\, [4\sqrt{2} + 20\sqrt{2} + 5\sqrt{2}\,] = 58\sqrt{2}\,.$$

Coefficient of x and Term Independent of x in Binomial Theorem

The coefficients of each expansion are the entries in Row n of Pascal's Triangle. So, the coefficient of each term r of the expansion of $(x + y)^n$ is given by $C(n, r - 1)$. The exponents of x descend, starting with n, and the exponents of y ascend, starting with 0, so the rth term of the expansion of $(x + y)^2$ contains $x_{n-(r-1)_{y_{r-1}}}$

In the expansion of $\left(x + \frac{1}{x}\right)^{2n}$, where, $n \neq 0$, middle term is $\left(\frac{2n+1+1}{2}\right)^{\text{th}}$. $(n + 1)^{\text{th}}$ term as $2n$ is even.

This term is called the term independent of x or the constant term.

Example 16. Expand $\left(x^2 + \dfrac{3}{x}\right)^4$, $x \neq 0$.

Sol. $$\left(x^2 + \frac{3}{x}\right)^4 = {}^4C_0\,(x^2)^4 + {}^4C_1\,(x^2)^3\left(\frac{3}{x}\right) + {}^4C_2\,(x^2)^2\left(\frac{3}{x}\right)^2 + {}^4C_3\,(x^2)\left(\frac{3}{x}\right)^3 + {}^4C_1\left(\frac{3}{x}\right)^4$$

$$= x^8 + 4x^6 \cdot \frac{3}{x} + 6x^4 \cdot \frac{9}{x^2} + 4x^2 \cdot \frac{27}{x^3} + \frac{81}{x^4}$$

$$= x^8 + 12x^5 + 54x^2 + \frac{108}{x} + \frac{81}{x^4}.$$

Example 17. *Expand* $(98)^5$.

Sol. $(98)^5 = (100 - 2)^5$

$$= {}^5C_0\,(100)^5 - {}^5C_1\,(100)^4 \cdot 2 + {}^5C_2\,(100)^3 2^2 - {}^5C_3\,(100)^2\,(2)^3 + {}^5C_4\,(100)\,(2)^4 - {}^5C_3\,(2)^5$$

$$= 10000000000 - 500000000 \times 2 + 10 \times 1000000 \times 4 - 10 \times 10000 \times 8 \times 5 \times 100 \times 16 - 32$$

$$= 9039207968.$$

Example 18. *Find the 10th term of* $(2x^2 + 1/x)^{12}$.

Sol. We know that $(a + b)^n$

$(r + 1)^{\text{th}}$ term, $\quad t_{r+1} = {}^nC_r = {}^nC_r\, a^{n-r}\, b^r$

$\therefore$ 10th term, $\quad t_{10} = t_{9+1} = {}^{12}C_9\,(2x^2)^{12-9} \cdot \left(\dfrac{1}{x}\right)^9 \quad \left[\because a = 2x^2, b = \dfrac{1}{x}, n = 12, r = 9\right]$

$$= -{}^9C_4\left(\frac{4x}{5}\right)^4\left(\frac{5}{2x}\right)^5 \qquad [\because {}^9C_5 = {}^9C_4]$$

$$= -\left[\frac{9 \times 8 \times 7 \times 6}{4 \times 3 \times 2 \times 1}\,\frac{2^8\, x^4}{5^4} \cdot \frac{5^5}{2^5\, x^5}\right].$$

$$= -\frac{5040}{x}.$$

Example 19. *Find the* 5th *term from the end in the expansion* $\left(\dfrac{x^3}{2} - \dfrac{2}{x^2}\right)^7$.

Sol. pth term from the end $= (n - p + 2)^{\text{th}}$ term from the beginning

$\therefore$ In the expression of $\left(\dfrac{x^3}{2} - \dfrac{2}{x^2}\right)^n$, we have

5th term from the end $= (9 - 5 + 2)^{\text{th}}$ term $= 6^{\text{th}}$ term $= t_6 = t_{5+1}$

$$= (-1)^5 \cdot {}^9C_5\left(\frac{x^3}{2}\right)^{9-5} \cdot \left(\frac{2}{x^2}\right)^5 = -{}^9C_4 \cdot \left(\frac{x^3}{2}\right)^4\left(\frac{2}{x^2}\right)^5$$

$$= -\left(\frac{9 \times 8 \times 7 \times 6}{4 \times 3 \times 2 \times 1} \times \frac{x^{12}}{16} \cdot \frac{32}{x^{10}}\right) = -252x^2.$$

Example 20. *Find the middle term in the expansion of* $\left(x - \frac{1}{2y}\right)^{10}$.

Sol. Total number of terms

$$\text{Middle term} = 6^{\text{th}} \text{ term} = t_{5+1}$$

$$= (-1)^5 \cdot {}^{10}C_5\, x^{(10-5)} \cdot \left(\frac{1}{2y}\right)^5 = \frac{-63x^5}{8y^5}.$$

Example 21. *Find the coefficients of* x^{32} *and* x^{-17} *in the expansion of* $\left(x^4 - \frac{1}{x^3}\right)^{15}$.

Sol. Let the term containing x^{32} be the $(r + 1)^{\text{th}}$ term

$$t_{r+1} = (-1)^r\, {}^{15}C_r\, (x^4)^{15-r} \left(\frac{1}{x^3}\right)^r$$

$$= (-1)^r\, {}^{15}C_r\, x^{(60-7r)} \qquad \ldots(i)$$

Putting $60 - 7r = 32$, we get $r = 4$.

$\therefore$ $(4 + 1)^{\text{th}}$, *i.e.*, the 5th term contains x^{32}

Putting $r = 4$ in (*i*), we get

$$t^5 = (-1)^4 \cdot {}^{15}C_4\, x^{32} = \left(\frac{15 \times 14 \times 13 \times 12}{4 \times 3 \times 2 \times 1}\right) x^{32} = 1365x^{32}.$$

$\therefore$ coefficient of x^{32} in the given expansion is 1365.

Again, let the term containing x^{-17} be the $(s + 1)^{\text{th}}$ term.

We have, $$t_{s+1} = (-1)^s\, {}^{15}C_s\, (x^4)^{(15-s)} \left(\frac{1}{x^3}\right)^5$$

$$= (-1)^s\, {}^{15}C_s\, x^{(60-7s)} \qquad \ldots(ii)$$

Putting $60 - 7s = -17$, we get $s = 11$.

Hence, $(11 + 1)$th, *i.e.*, 12th term contains x^{-17}.

Putting $s = 11$ in (*ii*), we get

$$t_{12} = (-1)^{11}\, {}^{15}C_{17}\, x^{-17}$$

$$= -{}^{15}C_4\, x^{-17} = \left(\frac{15 \times 14 \times 13 \times 12}{4 \times 3 \times 2 \times 1}\right) x^{-17}$$

$$= -1365x^{-17}$$

Thus, the coefficient of x^{-17} in the given expansion is -1365.

Example 22. *Find the term independent of x in the expansion of*

(*i*) $\left(x^2 + \frac{1}{x}\right)^9$ (*ii*) $\left(2x + \frac{1}{x}\right)^{10}$.

Sol. (*i*) Let the $(r + 1)$th term be independent of x.

In the expansion of $\left(x^2 + \frac{1}{x}\right)^9$, we have

$$t_{r+1} = {}^9C_r \cdot (x^2)^{9-r} \cdot \left(\frac{1}{x}\right)^r = {}^9C_r\, x^{18-3r} \qquad \ldots(i)$$

Putting $18 - 3r = 0$, we get $r = 6$.

$\therefore$ $(6 + 1)$th term contains x^0, *i.e.*, it is independent of x.

Putting $r = 6$ in (i), we get

$$t_{6+1} = {}^9C_6 x^0 = {}^9C_3 = \left(\frac{9 \times 8 \times 7}{3 \times 2 \times 1}\right) = 84.$$

(ii) Let the $(r + 1)$th term be independent of x.

In the expansion of $\left(2x - \frac{1}{x}\right)^{10}$, we have

$$t_{r+1} = (-1)^r \cdot {}^{10}C_r \cdot (2x)^{(10-r)} \cdot \left(\frac{1}{x}\right)^r$$

$$= (-1)^r\, {}^{10}C_r\, 2^{(10-r)}\, x^{(10-2r)} \qquad ...(ii)$$

Putting, $10 - 2r = 0$, we get $r = 5$.

$\therefore$ $(5 + 1)$th term contains x^0, *i.e.*, it is independent of x.

Putting $r = 5$ in (ii), we get

$$t_{5+1} = (-1)^5\, {}^{10}C_5 \cdot 2^{(10-5)} \cdot x^0$$

$$= -{}^{10}C_5 \cdot 2^5 = -32 \times \left(\frac{10 \times 9 \times 8 \times 7 \times 6}{5 \times 4 \times 3 \times 2 \times 1}\right)$$

$$= -8064.$$

Example 23. *Prove that there is no term involving x^6 in the expansion of $\left(2x^2 - \frac{3}{x}\right)^{11}$, where $x \neq 0$.*

Sol. Let the $(r + 1)$th term involve x^6.

Now, $$t_{r+1} = (-1)^r\, {}^{11}C_r \cdot (2x^2)^{11-r} \cdot \left(\frac{3}{x}\right)^r$$

$$= (-1)^r\, {}^{11}C_r \cdot 2^{(11-r)} \cdot 3r \cdot x^{(22-3r)}.$$

Putting $22 - 3r = 6$, we get $r = \frac{16}{3}$, which is a fraction.

Since, the value of r cannot be a fraction, it follows that there is no term containing x^6.

Example 24. *Find the coefficient of x^5 in the expansion of the product $(1 + 2x)^5 (1 - x)^7$.*

Sol. We have, $(1 + 2x)^6 (1 - x)^7$

$$= [1 + {}^6C_1 \cdot (2x) + {}^6C_2 \cdot (2x)^2 + {}^6C_3 \cdot (2x)^3 + {}^6C_4 \cdot (2x)^4 + {}^6C_5 \cdot (2x)^5 + ...]$$
$$\times [1 - {}^7C_1 \cdot x + {}^7C_2 \cdot x^2 - {}^7C_3 \cdot x^3 + {}^7C_4 \cdot x^4 - {}^7C_5 \cdot x^5 + ...]$$
$$= [1 + 12x + 60x^2 + 160x^3 + 240x^4 + 192x^5 + ...]$$
$$\times [1 - 7x + 21x^2 - 35x^3 + 35x^4 - 21x^5 + ...].$$

$\therefore$ Coefficient of x^5 in the product

$$= [1 \times (-21) + (12 \times 35) + 60 \times (-35) + 160 \times 21 + 240 \times (-7) + 192 \times 1]$$
$$= (-21 + 420 - 2100 + 3360 - 1680 + 192) = 171.$$

Example 25. *Using the binomial theorem, find the value of*

(i) $(102)^6$ *(ii) $(99)^4$.*

Sol. (*i*) $(102)^6 = (100 + 2)^6$

$$= {}^6C_0 \times (100)^6 + {}^6C_1 \times (100)^5 \times 2 + {}^6C_2 \times (100)^4 \times (2)^2 + {}^6C_3 \times (100)^3 \times (2)^3 + {}^6C_4 \times (100)^2 \times (2)^4 + {}^6C_5 \times 100 \times (2)^5 + {}^6C_6 \times (2)^6$$

$$= (100)^6 + 12 \times (100)^5 + 60 \times (100)^4 + 160 \times (100)^3 + 240 \times (100)^2 + 19200 + 64$$

$$= 1000000000000 + 12000000000 + 600000000 + 160000000 + 2400000 + 19200 + 64$$

$$= 1126162419264.$$

(*ii*) $(99)^4 = (100 - 1)^4$

$$= {}^4C_0 \times (100)^4 - {}^4C_1 \times (100)^3 \times 1 + {}^4C_2 \times (100)^2 \times (1)^2 - {}^4C_3 \times 100 \times (1)^3 + {}^4C_4 \times (1)^4$$

$$= (100)^4 - 4 \times (100)^3 + 6 \times (100)^2 - 400 + 1$$

$$= (100000000 - 4000000 + 60000 - 400 + 1)$$

$$= 96059601.$$

Example 26. *Using the binomial theorem, find the values of (i) $(10.1)^5$ and (ii) $(0.99)^{15}$ up to 4 places of decimals.*

Sol. (*i*) $(10.1)^5 = (10 + 0.1)^5$

$$= {}^5C_0 \times (10)^5 + {}^5C_1 \times (10)^4 \times (0.1) + {}^5C_2 \times (10)^3 \times (0.1)^2 + {}^5C_3 \times (10)^2 \times (0.1)^3 + {}^5C_4 \times 10 \times (0.1)^4 + {}^5C_5 \times (0.1)^5$$

$$= (10)^5 + 5 \times (10)^4 \times (0.1) + (10)^4 \times (0.01) + (10)^3 \times (0.001) + 50 \times (0.0001) + (0.00001)$$

$$= 100000 + 50000 \times 0.1 + 10000 \times 0.01 + 1000 \times 0.001 + 50 \times 0.0001 + 0.00001$$

$$= 100000 + 5000 + 100 + 1 + 0.0050 + 0.00001$$

$$= 105101.00501.$$

(*ii*) $(0.99)^2 = (1 - 0.01)^{15}$

$$= 1 - {}^{15}C_1 \times (0.01) + {}^{15}C_2 \times (0.01)^2 + {}^{15}C_3 \times (0.01)^3 + \ldots$$

[leaving higher powers of 0.01]

$$= (1 - 15 \times 0.01 + 105 \times 0.0001 - 455 \times 0.000001)$$

$$= (1 - 0.15 + 0.0105 - 0.000455) = 0.860045$$

Example 27. *Show that the middle term in the expansion of*

$$(1 + x)^{2n} \text{ is } \frac{1 \cdot 3 \cdot 5 \ldots (2n - 1)}{n!} \cdot 2^n \cdot x^n, \text{ where } n \in N.$$

Sol. Clearly, the number of terms in the expansion of $(1 + x)^{2n}$ is $(2n + 1)$.

So, middle term = $(n + 1)$th term = t_{n+1}

$$= {}^{2n}C_n \cdot x^n = \frac{(2n)!}{(n!) \times (n!)} \cdot x^n$$

$$= \frac{1 \cdot 2 \cdot 3 \cdot 4 \ldots (2n - 2)(2n - 1)(2n)}{(n!) \times (n!)} \cdot x^n$$

$$= \frac{1\cdot 3\cdot 5 \ldots (2n-3)(2n-1)] \times [2\cdot 4\,.\,6 \ldots (2n-2)(2n)]}{(n\,!)(n\,!)} \cdot x^n$$

$$= \frac{1\cdot 3\cdot 5 \ldots (2n-1) \times 2^n \times [1\cdot 2\,.\,3 \ldots (n-1)\cdot n]}{(n\,!)(n\,!)} \cdot x^n$$

$$= \frac{1\cdot 3\cdot 5 \ldots (2n-1)] \times 2^n \times x^n}{(n\,!)}.$$

Example 28. *Show that the coefficients of the middle term in the expansion of $(1 + x)^{2n}$ is the sum of the coefficients of two middle terms in the expansion of $(1 + x)^{2n-1}$.*

Sol. The expansion of $(1 + x)^{2n}$ contains $(2n + 1)$ terms and therefore, the $(n + 1)$th term is the middle term.

∴ the middle term in the expansion of $(1 + x)^{2n}$ is given by

$$t_{n+1} = {}^{2n}C_n\, x^n = \frac{(2n)!}{(n\,!)\cdot(n\,!)} x^n = \frac{(2n)!}{(n\,!)^2} x^n.$$

Hence, the coefficient of the middle term of $(1 + x)^{2n}$ is $\dfrac{(2n)!}{(n\,!)^2}$.

Again, the expansion of $(1 + x)^{2n-1}$ contains $2n$ terms.

Thus, the middle terms are the nth and the $(n + 1)$th terms.

Now, $t_n = t_{(n-1)+1} = {}^{2n-1}C_{n-1}\, x^{n-1}$ and $t_{n+1} = {}^{2n-1}C_n x^n$.

∴ the sum of the coefficients of the middle terms of $(1 + x)^{2n-1}$

$$= {}^{2n-1}C_{n-1} + {}^{2n-1}C_n$$

$$= \frac{(2n-1)!}{(n-1)! \times (n\,!)} + \frac{(2n-1)!}{(n\,!) \times (n-1)!}$$

$$= \frac{2 \times (2n-1)! \times n}{(n\,!) \times n \times (n-1)!}$$

[multiplying numerator and denominator by n]

$$= \frac{(2n)\cdot(2n-1)!}{(n\,!)\cdot(n\,!)} = \frac{(2n)!}{(n\,!)^2}$$

= The coefficient of the middle term of $(1 + x)^{2n}$.

Example 29. *Find the term independent of x in the expansion of* $\left(\dfrac{1}{3x} - \dfrac{3}{2}x^2\right)^6$.

Sol. We have $\quad T_{r+1} = {}^6C_r \left(\dfrac{1}{3x}\right)^{6-r} \left(-\dfrac{3}{2}x^2\right)^r = {}^6C_r \left(\dfrac{1}{3x}\right)^{6-r} \left(-\dfrac{3}{2}\right)^r . x^{3r-6}$

$3r - 6 = 0$ *i.e.*, $r = 2$,

$$x = T_3 = {}^6C_2 \left(\frac{1}{3}\right)^4 \left(-\frac{3}{2}\right)^2 = \frac{5}{12}.$$

Example 30. *Show that the middle term in the expansion of $(1 + x)^{2n}$ is $\dfrac{1\cdot 3\cdot 5\,.\,(2n-1)}{n\,!}$ $2nx^n$, where n is +ve integer.*

Sol. Here, $2n$ is even, middle term of the expansion $(1 + x)^{2n}$ is $\left(\frac{2n}{2}+1\right)^{\text{th}}$ *i.e.*, $(n + 1)^{\text{th}}$ term

$$T_{n+1} = {}^{2n}C_n (1)^{2n-n} (x)^n = {}^{2n}C_n x^n = \frac{(2n)!}{n!\,n!} x^n$$

$$= \frac{2n(n-1)(2n-2)\ldots 4.3.2.1}{n!\,n!} x^n$$

$$= \frac{1.2.3.4\,(2n-2)(2n-1)(2n)}{n!\,n!} x^n$$

$$= \frac{[1.3.5\,(2n-1)\,[2.4.6\ldots(2n)]}{n!\,n!} x^n$$

$$= \frac{[1.3.5\,(2n-1)]2^n\,[1.2.3\ldots n]}{n!\,n!} x^n$$

$$= \frac{[1.3.5\,(2n-1)]\,n!}{n!\,n!} 2^n . x^n$$

$$= \frac{1.3.5\ldots(2n-1)\,n!}{n!} 2^n x^n$$

Example 31. *2nd, 3rd and 4th terms in the expansion of $(x + y)^n$ are 240, 720 and 1080 respectively. Find the value of x, y and n.*

Sol. $(x + y)^n = {}^nC_0 x^n + {}^nC_1 x^{n-1}y + {}^nC_2 x^{n-2}y^2 + {}^nC_3 x^{n-3}y^3 + \ldots$

$$(x + y)^n = x^n + nx^{n-1}y + \frac{n(n-1)}{2} . x^{n-2}y^2 + \frac{n(n-1)(n-2)}{6} x^{n-3}y^3 + \ldots$$

$$\therefore \quad nx^{n-1}y = 240 \quad \ldots(i)$$

$$\frac{n(n-1)}{2} . x^{n-2}y^2 = 720 \quad \ldots(ii)$$

$$\frac{n(n-1)(n-2)}{6} x^{n-3}y^3 = 1080 \quad \ldots(iii)$$

Dividing (*ii*) by (*i*) and (*iii*) by (*ii*), we get

$$\frac{(n-1).y}{2x} = 3 \text{ and } \frac{(n-2).y}{3x} = \frac{3}{2}$$

$$\Rightarrow \quad \frac{6}{n-1} = \frac{9}{2(n-2)} \Rightarrow \quad n = 5.$$

Put $n = 5$ in (*i*) and (*ii*), we get

$$x^4y = 48$$

$$x^3y^2 = 72$$

$$\Rightarrow \quad y = \frac{48}{x^4} \text{ and } y^2 = \frac{72}{x^3}$$

$$\Rightarrow \quad \left(\frac{48}{x^4}\right)^2 = \frac{72}{x^3} \Rightarrow x^5 = \frac{48 \times 48}{72} = 32 = (2)^5$$

$$\Rightarrow \quad x = 2$$

$$y = \frac{48}{x^4} + \frac{48}{16} = 3$$

Hence, $x = 2, y = 3$ and $n = 5$.

Example 32. *If x^p occurs in the expansion of $\left(x^2 + \frac{1}{x}\right)^{2n}$, show that its coefficient is*

$$\frac{(2n)!}{\left[\frac{1}{3}(4n - p)!\right] \cdot \left[\frac{1}{3}(2n + p)!\right]}.$$

Sol. Expansion of $\left(x^2 + \frac{1}{x}\right)^{2n}$, we have

$$t_{r+1} = {}^{2n}C_r \cdot (x^2)^{2n-r} \cdot \left(\frac{1}{x}\right)^r = {}^{2n}C_r \cdot x^{(4n-3r)}$$

Putting $(4n - 3r) = p$, we get $r = \frac{1}{3}(4n - p)$

$\therefore$ Coefficient of $x^p = {}^{2n}C_{1/3(4n-p)}$

$$= \frac{(2n)!}{\left[\frac{1}{3}(4n - p)!\right] \times \left[2n - \frac{1}{3}(4n - p)\right]!}$$

$$= \frac{(2n)!}{\left[\frac{1}{3}(3n - p)!\right] \cdot \left[\frac{1}{3}(2n + p)!\right]}.$$

Example 33. *Calculate the value of r, if the coefficients of (2r + 4)th and (r – 2)th terms in the expansion of $(1 + x)^{18}$ are equal.*

Sol. We have

$$t_{2r+4} = t_{(2r+3)+1} = {}^{18}C_{2r+3}x^{2r+3}$$

and $$t_{r-2} = t_{(r-3)+1} = {}^{18}C_{r-3}\, x^{r-3}$$

$\therefore$ Coefficient of $t_{(2r+4)}$ = coefficient of $t_{(r-2)}$

$$\Rightarrow \quad {}^{18}C_{2r+3} = {}^{18}C_{r-3}$$

$$\Rightarrow \quad 2r + 3 = r - 3 \quad \text{or} \quad 2r + 3 + r - 3 = 18$$

$$\Rightarrow \quad r = -6 \quad \text{or} \quad r = 6$$

But, the negative value of r is not possible, so $r = 6$.

Example 34. *If the coefficients of a^{r-1}, a^r, a^{r+1} in the binomial expansion $(1 + a)^n$ are in arithmetic progression, prove that $n^2 - n(4r + 1) + 4r^2 - 2 = 0$.*

Sol. The general term in the expansion of $(1 + a)^n$ is given by

$$t_{r+1} = {}^nC_r\, a^r.$$

$\therefore$ Coefficients of a^{r-1}, a^r and a^{r+1} in the expansion are ${}^nC_{r-1}$, ${}^nC_r$ and ${}^nC_{r+1}$ respectively.

Now, ${}^nC_{r-1}$, ${}^nC_r$ and ${}^nC_{r+1}$ are in A.P.

$$\Rightarrow \quad {}^{2n}C_r = {}^nC_{r-1} + {}^nC_{r+1}$$

$$\Rightarrow \quad \frac{{}^nC_{r-1}}{{}^nC_r} + \frac{{}^nC_{r+1}}{{}^nC_r} = 2$$

$$\Rightarrow \quad \frac{n!}{(r-1)!.(n-r+1)!} \times \frac{(r!).(n-r)!}{n!} + \frac{n!}{(r-1)!.(n-r+1)!} \times \frac{(r!).(n-r)!}{n!} = 2$$

$$\Rightarrow \quad \frac{r}{(n-r+1)} + \frac{n-r}{r+1} = 2$$

$[\because \ (n-r+1)! = (n-r+1).(n-r)!$ and $(r!) = r.(r-1)!]$

$\Rightarrow \quad r(r+1) + (n-r)(n-r+1) = 2(r+1)(n-r+1)$

$\Rightarrow \quad n^2 - n(4r+1) + 4r^2 - 2 = 0.$

Example 35. *If the coefficients of (r – 1)th, rth and (r + 1)th terms in the expansion of $(x+1)^n$ are in the ratio 1 : 3 : 5, find n and r.*

Sol. We know that, $(x+1)^n = (1+x)^n$

Now, $\quad t_{r-1} = t_{(r-2)+1} = {}^nC_{r-2}\, x^{r-2};\ t_r = t_{(r-1)+1} = {}^nC_{r-1}\, x^{r-1}$

and $\quad t_{r+1} = {}^nC_r\, x^r.$

$\therefore \quad$ (coeff. of t_{r-1}) : (coeff. of t_r) ; (coeff. of t_{r+1}) = 1 : 3 : 5

$\Rightarrow \quad {}^nC_{r-2} : {}^nC_{r-1} : {}^nC_r = 1 : 3 : 5.$

$$\Rightarrow \quad \frac{{}^nC_{r-2}}{{}^nC_{r-1}} = \frac{1}{3} \text{ and } \frac{{}^nC_{r-1}}{{}^nC_r} = \frac{3}{5}.$$

But, $$\frac{{}^nC_{r-2}}{{}^nC_{r-1}} = \frac{1}{3} \Rightarrow \frac{n!}{(r-2)!.(n-r+2)!} \times \frac{(r-1)!.(n-r+1)!}{n!} = \frac{1}{3}$$

$$\Rightarrow \quad \frac{(r-1)}{(n-r+2)} = \frac{1}{3}$$

$\Rightarrow \quad n - 4r = -5 \qquad \ldots(i)$

And, $$\frac{{}^nC_{r-1}}{{}^nC_r} = \frac{3}{5} \Rightarrow \frac{n}{(n-r+1)} = \frac{3}{5}$$

$\Rightarrow \quad 3n - 8r = -3 \qquad \ldots(ii)$

Solving (*i*) and (*ii*), we get $n = 7$, and $r = 3$.

Example 36. *If P be the sum of odd terms and Q that of even terms in the expansion of $(x+a)^n$, prove that*

(i) $(P^2 - Q^2) = (x^2 - a^2)^n$;

(ii) $4PQ = [(x+a)^{2n} - (x-a)^{2n}]$;

(iii) $2(P^2 + Q^2) = [(x+a)^{2n} + (x-a)^{2n}]$.

Sol. We have know that

$$(x+a)^n = {}^nC_0\, x^n + {}^nC_1\, x^{n-1}a + {}^nC_2\, x^{n-2}a^2 - \ldots + {}^nC_n\, a^n$$
$$= t_1 - t_2 + t_3 + \ldots + t_n + t_{n+1}$$
$$= (t_1 + t_3 + t_5 + \ldots) + (t_2 + t_4 + t_6 + \ldots)$$
$$= P + Q. \qquad \ldots(i)$$

And, $$(x - a)^n = {}^nC_0\, x^n + {}^nC_1\, x^{n-1} a + {}^nC_2\, x^{n-2} a^2 - \ldots + (-1)^n\, {}^nC_n\, a^n$$
$$= (t_1 - t_2 + t_3 - t_4 + \ldots)$$
$$= (t_1 + t_3 + t_5 + \ldots) - (t_2 + t_4 + t_6 + \ldots)$$
$$= P - Q \qquad \ldots(ii)$$

(*i*) Multiplying (*i*) and (*ii*), we get

$$(x + a)^n \,.\, (x - a)^n = (P^2 - Q^2) \text{ or } (x^2 - a^2)^n = (P^2 - Q^2)$$

(*ii*) Squaring (*i*) and (*ii*) and subtracting, we get

$$[(x + a)^{2n} - (x - a)^{2n}] = [(P + Q)^2 - (P - Q)^2] = 4\,PQ$$

(*iii*) Squaring (*i*) and (*ii*) and adding

$$\Rightarrow \quad [(x + a)^{2n} + (x - a)^{2n}] = [(P + Q)^2 + (P - Q)^2] + 2\,(P^2 + Q^2).$$

Example 37. *Evaluate* $(x^2 - \sqrt{1 - x^2}\,)^4 + (x^2 + \sqrt{1 - x^4}\,)^4$.

Sol. Putting $\sqrt{1 - x^2} = y$, we get

The given expression $= (x^2 - y)^4 + (x^2 + y)^4 = 2\,[x^8 + {}^4C_2\, x^4 y^2 + {}^4C_4\, y^4]$

$$= 2\left[x^8 + \frac{4 \times 3}{2 \times 1} x^4 \cdot (1 - x^2) + (1 - x^2)^2\right]$$
$$= 2\,[x^8 + 6x^4\,(1 - x^2) + (1 - 2x^2 + x^4)]$$
$$= 2x^8 - 12x^6 + 14x^4 - 4x^2 + 2.$$

Example 38. *Find the coefficient of* x^{11} *in the expansion of* $\left(x^3 - \dfrac{2}{x^2}\right)^{12}$.

Sol. Let the general term, *i.e.*, $(r + 1)^{\text{th}}$ contain x^{11}.

We have, $$T_{r+1} = {}^{12}C_r\, (x^3)^{12-r} \left(-\frac{2}{x^2}\right)^r$$
$$= {}^{12}C_r\, x^{36 - 3r - 2r}\, (-1)^r\, 2^r$$
$$= {}^{12}C_r\, (-1)^r\, 2^r\, x^{36 - 5r}$$

For x^{11}, we observe that

$$36 - 5r = 11, \textit{ i.e.}, r = 5$$

Hence, the coefficient of x^{11} is

$${}^{12}C_5\, (-1)^5\, 2^5 = \frac{12 \times 11 \times 10 \times 9 \times 8}{5 \times 4 \times 3 \times 2} \times 32 = -25344.$$

Example 39. *Find whether the expansion of* $\left(x^2 - \dfrac{2}{x}\right)^{18}$ *will contain a term containing* x^{10}?

Sol. Let T_{r+1} contain x^{10}. Then,

$$T_{r+1} = {}^{18}C_r\, (x^2)^{18-r} \left(\frac{-2}{x}\right)^r$$
$$= {}^{18}C_r\, x^{36 - 2r}\, (-1)^r \,.\, 2^r\, x^{-r}$$
$$= (-1)^r\, 2^r \,.\, {}^{18}C_r\, x^{36 - 3r}$$

Thus, $$36 - 3r = 10, \textit{ i.e.}, r = \frac{26}{3}$$

Since r is a fraction, the given expansion cannot have a term containing x^{10}.

Example 40. *Find the term independent of x in the expansion of* $\left(\frac{\sqrt{x}}{\sqrt{3}}+\frac{\sqrt{3}}{2x^2}\right)^{10}$.

Sol. Let $(r+1)^{\text{th}}$ term be independent of x which is given by

$$T_{r+1} = {}^{10}C_r \left(\sqrt{\frac{x}{3}}\right)^{10-r} \left(\frac{\sqrt{3}}{2x^2}\right)^r$$

$$= \left(\frac{x}{3}\right)^{\frac{10-r}{2}} 3^{r/2}\left(\frac{1}{2^r\, x^{2r}}\right)$$

$$= {}^{10}C_r\, 3^{\frac{r}{2}-\frac{10-r}{2}}\, 2^{-r}\, x^{\frac{10-r}{2}-2r}$$

Since the term is independent of x, we have

$$\frac{10-r}{2} - 2r = 0 \;\Rightarrow\; r = 2$$

Thus, 3rd term is independent of x and its value is given by

$$T_3 = {}^{10}C_2\, \frac{3^{-3}}{4} = \frac{10\times 9}{2\times 1}\times\frac{1}{9\times 12} = \frac{5}{12}$$

Example 41. *Find the middle term in the expansion of* $\left(2ax - \frac{b}{x^2}\right)^{12}$.

Sol. Since the power of binomial is even, it has one middle term which is the $\left(\frac{12+2}{2}\right)^{\text{th}}$ term and it is given by

$$T_7 = {}^{12}C_6\,(2\,ax)^6\left(\frac{-b}{x^2}\right)^6$$

$$= {}^{12}C_6\,\frac{2^6\, a^6\, x^6\,.(-b)^6}{x^{12}}$$

$$= {}^{12}C_6\,\frac{2^6\, a^6\, b^6}{x^6} = \frac{59136\, a^6\, b^6}{x^6}.$$

Example 42. *Find the middle term (terms) in the expansion of* $\left(\frac{p}{x}+\frac{x}{p}\right)^9$.

Sol. Since the power of binomial is odd. Therefore, we have two middle terms which are 5th and 6th terms. These are given by

$$T_5 = {}^9C_4\left(\frac{p}{x}\right)^5\left(\frac{x}{p}\right)^4 = {}^9C_4\,\frac{p}{x} = \frac{126\,p}{x}$$

and

$$T_6 = {}^9C_5\left(\frac{p}{x}\right)^4\left(\frac{x}{p}\right)^5 = {}^9C_5\,\frac{x}{p} = \frac{126\,x}{p}.$$

Example 43. *Show that $2^{4n+4} - 15n - 16$, where $n \in N$ is divisible by 225.*

Sol. We have

$$\begin{aligned} 2^{4n+4} - 15n - 16 &= 2^{4(n+1)} - 15n - 16 \\ &= 16^{n+1} - 15n - 16 \\ &= (1 + 15)^{n+1} - 15n - 16 \\ &= {}^{n+1}C_0\, 15^0 + {}^{n+1}C_1\, 15^1 + {}^{n+1}C_2\, 15^2 + {}^{n+1}C_3\, 15^3 + \ldots \\ &\quad + {}^{n+1}C_{n+1} - (15)^{n+1}\, 15n - 16 \\ &= 1 + (n + 1)\, 15 + {}^{n+1}C_2\, 15^2 + {}^{n+1}C_3\, 15^3 + \ldots \\ &\quad + {}^{n+1}C_{n+1}\, (15)^{n+1} - 15n - 16 \\ &= 1 + 15n + 15 + {}^{n+1}C_2\, 15^2 + {}^{n+1}C_3\, 15^3 + \ldots \\ &\quad + {}^{n+1}C_{n+1}\, (15)^{n+1} - 15n - 16 \\ &= 15^2\, [{}^{n+1}C_2 + {}^{n+1}C_3\, 15 + \ldots \text{ so on}] \end{aligned}$$

Thus, $2^{4n+4} - 15n - 16$ is divisible by 225.

Example 44. *Find numerically the greatest term in the expansion of $(2 + 3x)^9$, where $x = \frac{3}{2}$.*

Sol. We have $(2 + 3x)^9 = 2^9 \left(1 + \frac{3x}{2}\right)^9$

Now, $$\frac{T_{r+1}}{T_r} = \frac{2^9 \left[{}^9C_r \left(\frac{3x}{2}\right)^r\right]}{2^9 \left[{}^9C_{r-1} \left(\frac{3x}{2}\right)^{r-1}\right]}$$

$$= \frac{{}^9C_r}{{}^9C_{r-1}} \left|\frac{3x}{2}\right| = \frac{\lfloor 9}{\lfloor r \lfloor 9-r} \cdot \frac{\lfloor r-1 \lfloor 10-r}{\lfloor 9} \left|\frac{3x}{2}\right|$$

$$= \frac{10-r}{r} \left|\frac{3x}{2}\right| = \frac{10-r}{4}\left(\frac{9}{4}\right). \qquad \left(\because \text{ Since } x = \frac{3}{2}\right)$$

Therefore, $$\frac{T_{r+1}}{T_r} \geq 1 \Rightarrow \frac{90-9r}{4r} \geq 1$$

$\Rightarrow$ $90 - 9r \geq 4r$

$\Rightarrow$ $r \leq \frac{90}{13}$

$\Rightarrow$ $r \leq 6\frac{12}{13}$

Thus, the maximum value of r is 6. Therefore, the greatest term is $T_{r+1} = T_7$.

Hence, $$T_7 = 2^9 \left[{}^9C_6 \left(\frac{3x}{2}\right)^6\right], \text{ where } x = \frac{3}{2}$$

$$= 2^9 \cdot {}^9C_6 \left(\frac{9}{4}\right)^6 = 2^9 \cdot \frac{9 \times 8 \times 7}{3 \times 2 \times 1}\left(\frac{3^{12}}{2^{12}}\right) = \frac{7 \times 3^{13}}{2}.$$

Example 45. *If n is a positive integer, find the coefficient of x^{-1} in the expansion of $(1 + x)^n$ $\left(1+\frac{1}{x}\right)^n$.*

Sol. We have

$$(1 + x)^n \left(1+\frac{1}{x}\right)^n = (1 + x)^n \left(\frac{x+1}{x}\right)^n = \frac{(1+x)^{2n}}{x^n}$$

Now to find the coefficient of x^{-1} in $(1 + x)^n \left(1+\frac{1}{x}\right)^n$, it is equivalent to finding coeffcient of x^{-1} in $\frac{(1+x)^{2n}}{x^n}$ which turn is equal to the coefficient of x^{n-1} in the expansion.

$$(1 + x)^{2n} = {}^{2n}C_0\, x^0 + {}^{2n}C_1 x^1 + {}^{2n}C_2\, x^2 + \ldots\ {}^{2n}C_{n-1}\, x^{n-1} + \ldots\ {}^{2n}C_{2n}\, x^{2n}$$

$\therefore$ Coefficient of x^{n-1} is ${}^{2n}C_{n-1} = \frac{\lfloor 2n}{\lfloor n - \lfloor 2n-n+1} = \frac{\lfloor 2n}{\lfloor n-1 \lfloor n+1}$.

EXERCISE 1.1.1.

1. Expansion of each of the following:
   (*i*) $(3x + 2y)^8$ (*ii*) $(p - q)^8$
   (*iii*) $(1 - x)^8$
2. Find the value of each of following:
   (*i*) $(1001)^5$ **[Ans.** 1005010010005001]
   (*ii*) $(1.01)^5$ **[Ans.** 1.0510100501]
   (*iii*) $(999)^4$ **[Ans.** 996005996001
   (*iv*) $(\sqrt{3} + 1)^5 - (\sqrt{3} - 1)^5$ **[Ans.** 152]
   (*v*) $(2 + \sqrt{3})^7 + (2 - \sqrt{3})^7$ **[Ans.** 9884]
3. Expand by using binomial theorem:
   (*i*) $(3x + 2y)^4$ **[Ans.** $81x^4 + 216\, x^3y + 216\, x^2y^2 + 96\, xy^3 + 16y^4$]
   (*ii*) $(1 + 2x - 3x^2)^5$ **[Ans.** $1 + 10x + 25x^2 - 10x^3 - 190\, x^4 + 9x^5 + 570\, x^6 - 360x^7 - 675x^8 + 810x^9 - 243x^{10}$]
   (*iii*) $\left(\sqrt[3]{x} - \sqrt[3]{y}\right)^6$ **[Ans.** $x^2 - 6x^{5/3}\, y^{1/3} + 15x^{4/3}\, y^{2/3} - 20xy + 15x^{2/3}\, y^{4/3} - 6x^{1/3}\, y^{5/3} + y^2$]
   (*iv*) $\left(x^2 - \frac{2}{x}\right)^7$ $\left[\textbf{Ans. } x^{14} - 14x^{11} + 84x^8 - 280\,x^5 + 560\,x^2 - \frac{672}{x} + \frac{448}{x^2} - \frac{128}{x^7}\right]$
4. (*i*) Find the 16th term in the expansion of $(\sqrt{x} - \sqrt{y})^{17}$. **[Ans.** $136\, xy^{15/2}$]
   (*ii*) Find the 10th term in the expansion of $\left(\frac{a}{b} - \frac{2b}{a^2}\right)^{12}$ $\left[\textbf{Ans. } \frac{(-366080)b^5}{a^{14}}\right]$
   (*iii*) Find the 7th term in the expansion of $\left(\frac{4x}{5} + \frac{5}{2x}\right)^8$ $\left[\textbf{Ans. } \frac{4375}{x^4}\right]$
5. Write the general term in the expansion of $(x^2 - y)^6$ **[Ans.** $(-1)^r\ {}^6C_r \cdot x^{(12-2r)} \cdot y^r$.]

6. (*i*) Find the 4th term from the end in the expansion of $(4x/5 - 5/2x)^9$ [**Ans.** (*i*) $10500/x^3$]
(*ii*) Find the 5th term from the end in the expansion $(x - 1/x)^{12}$ [**Ans.** $495x^4$]

7. Expand the following:

(*i*) $\left(1+\frac{2}{3}x\right)^6$ (*ii*) $(3x - y^2)^5$ (*iii*) $\left(\frac{x^4}{4}+\frac{2}{x}\right)^4$

(*iv*) $\left(x^2+\frac{1}{x^3}\right)^7$ (*v*) $\left(x^3+\frac{1}{x^2}\right)^5$ (*vi*) $\left(\frac{1}{x^2}-x^3\right)^4$

8. Find the specified terms in the expansion of each of the following:

(*i*) $(1 - 2x)^7$: 3rd term [**Hint:** Here $r = 2$] (*ii*) $\left(x-\frac{1}{2x}\right)^6$: middle term(s)

(*iii*) $(3x - 4y)^6$: 4th term (*iv*) $\left(y^2-\frac{1}{y}\right)^{11}$: middle term(s)

(*v*) $(x^3 - y^3)^{12}$: 4th term (*vi*) $(1 - 3x^2)^{10}$: middle term(s)

(*vii*) $(-3x - 4y)^6$: 5th term (*viii*) Write the rth term in the expansion of $(x - 2y)^6$.

9. Write the $(r + 1)$th term in the expansion of each of the following, where $n \in N$:

(*i*) $(3x - y^2)^n$ (*ii*) $\left(x^3+\frac{1}{x}\right)^n$

10. Write the $(r - 1)$th term in the expansion of $(1 + 2x)^8$.

11. If T_r, denotes the rth term in the expansion of $(1 + x)^n$ in ascending powers of x (n being a natural number), prove that

$$r(r + 1)\, T_{r+2} = (n - r + 1)(n - r)\, x^2\, T_r$$

[**Hint:** $T_r = {}^nC_{r-1}\, x^{r-1}$ and $T_{r+2} = {}^nC_{r+1}\, x^{r+1}$]

12. k_r is the coefficient of x^{r-1} in the expansion of $(1 + 2x)^{10}$ in ascending powers of x and $k_{r+2} = 4k_r$. Find the value of r.
[**Hint:** $k_r = {}^{10}C_{r-1}\, 2^{r-1}$ and $k^{r+2} = {}^{10}C_{r+1}\, 2^{r+1}$]

13. The coefficients of the 5th, 6th and 7th terms in the expansion of $(1 + a)^n$ (n being a natural number) are in A.P. Find n.
[**Hint:** ${}^nC_5 - {}^nC_4 = {}^nC_6 - {}^nC_5$]

14. Expand $(1 + y + y^2)^4$.
[**Hint:** $(1 + y + y^2)^4 = \{(1 + y) + y^2\}^4$]

15. Write the expansion of each of the following:

(*i*) $(1 - x)^{-4}, |x| < 1$ (*ii*) $\frac{1}{(1+x)^3}, |x| < 1$

(*iii*) $(3 - z)^{-4}, |z| < 3$ (*iv*) $\frac{1}{(1+3x)^{3/2}}, |x| < \frac{1}{3}$

16. State the condition under which the expansion of $(x - 2y)^{-3}$ will be valid in ascending powers of y. Also, write the expansion.

17. State the condition under which the expansion of $(x-3y)^{\frac{-1}{2}}$ will be valid in ascending powers of x. Also, write the expansion.

18. Expand the following, stating the condition of y under which the expansion will be valid:

(*i*) $\frac{1}{(2+y)^4}$ (*ii*) $(3-y)^{\frac{-2}{3}}$

19. Find the value of each of the following up to three decimal places, using the necessary number of terms in the expansion:

(*i*) $(0.99)^{-4}$ (*ii*) $(1.03)^{-3}$

(*iii*) $\sqrt[3]{26}$ **[Hint:** $(26)^{\frac{1}{3}} = (27-1)^{\frac{1}{3}}$]

(*iv*) $\sqrt[7]{127}$ **[Hint:** $127^{\frac{1}{7}} = (128-1)^{\frac{1}{7}}$]

(*v*) $\sqrt[5]{35}$ **[Hint:** $(35)^{\frac{1}{5}} = \{32+3\}^{\frac{1}{5}}$]

(*vi*) $\sqrt[5]{31}$ **[Hint:** $(31)^{\frac{1}{5}} = \{32-1\}^{\frac{1}{5}}$]

(*vii*) $\sqrt[3]{1001}$ **[Hint:** $(100)^{\frac{1}{3}} = \{1000+1\}^{\frac{1}{3}}$]

20. Assuming y to be so small that y^2 and higher powers of y can be neglected, find the value of each of the following:

(*i*) $(1+5y)^{-2}\,(1+2y)^{\frac{-3}{2}}$ (*ii*) $\dfrac{(1-4)y^{-3}\,(1-2y^2)^{\frac{1}{2}}}{(4-y)^{3/2}}$

(*iii*) $\dfrac{\sqrt{1-3y}+(1-y)^{\frac{5}{3}}}{\sqrt{4-y}}$

21. Find the coefficients of x^4 in the expansion of $(1+x)^n\,(1-x)^n$ deduce that $C_2 = C_0C_4 - C_1C_3 + C_2C_2 - C_3\,C_1 + C_4\,C_0$. **[Ans.** $C_0\,C_4 - C_1C_3 + C_2C_2 - C_3\,C_1 + C_4C_0$]

22. Using binomial theorem prove that $6^n - 5n$, when divided by 25, always leaves a remainder 1, where $n \in N$.

23. Using binomial theorem prove that $(3^{2n+2} - 8n - 9)$ is divisible by 64, where n is a positive integer.

24. If a and b are different integers, prove that $(a^n - b^n)$ is divisible by $(a - b)$, where ever n is a positive integer.

25. The first three terms in binomial expansion of $(x+y)^n$ are 1, 56 and 1372. Find the value of x and y. **[Ans.** $x = 1, y = 7$]

26. If the binomial expansion of $(a+b)^n$, the coefficeints of the 4th and 13th terms are equal to each other. Find n.

27. If the coefficients of 2nd, 3rd and 4th terms in the expansion of $(1+n)^{2n}$ in A.P. Show that $2n^2 - 9n + 7 = 0$.

28. Find the coefficient of x^5 in the expansion of $(1+x)^3\,(1-x)^6$. **[Ans.** 6]

29. Simplify the following expression.

(*i*) $(x + 1/x)^6 - (x - 1/x)^6$ $\left[\textbf{Ans. } (i)\ 12\left(x^4 + 1\frac{1}{x^4} + \frac{10}{3}\right)\right]$

(*ii*) $(2+\sqrt{x})^4 + (2-\sqrt{x})^4$ **[Ans.** (*ii*) $2\,(16 + 24x + x^2)$]

(*iii*) $[x + \sqrt{x^2-1}\,]^6 + [x - \sqrt{x^2-1}\,]^6$ **[Ans.** $64x^6 - 96x^4 + 36x^2 - 2$]

30. Show that the ratio of the coefficient of x^{10} in $(1-x^2)^{10}$ and term independent of x in $\left(x - \dfrac{2}{x}\right)^{10}$ is 1 : 32.

31. Find the coefficient of

(*i*) x^7 in the expansion of $(x^2 + 1/x)^{11}$ **[Ans.** 462]

(*ii*) x^{18} in the expansion of $\left(x^2 + \dfrac{39}{x}\right)^{15}$ **[Ans.** 110565 a^4]

(*iii*) x^6 in the expansion of $(3x^2 - 1/3x)^9$ **[Ans.** 378]

32. Find the middle term in the expansion of

(*i*) $(3 + x)^6$ **[Ans.** 540 x^3]

(*ii*) $\left(1 - \dfrac{x^2}{2}\right)^{14}$ **[Ans.** –429/16 x^{14}]

(*iii*) $\left(\dfrac{a}{3} + 9b\right)^8$ **[Ans.** 5670 $a^4\,b^4$]

(*iv*) $\left(x^2 - \dfrac{2}{x}\right)^{10}$ **[Ans.** –8064 x^5]

33. Find the two middle terms in the expansion

(*i*) $\left(3a - \dfrac{a^3}{6}\right)^9$ $\left[\textbf{Ans. } \dfrac{189}{8}a^{17}, -\dfrac{21}{16}a^{19}\right]$

(*ii*) $(x^4 - 1/x^3)^{11}$ **[Ans.** – 462 x^9, 462 x^2]

(*iii*) $\left(3x - \dfrac{2}{x^2}\right)^{15}$ $\left[\textbf{Ans. } -\dfrac{6435 \times 3^8 \times 2^7}{x^6} \text{ and } \dfrac{6435 \times 3^7 \times 2^9}{x^9}\right]$

(*iv*) $(x^2 + a^2)^5$ **[Ans.** 10 $x^6\,a^4$]

34. Show that the term containing x^3 does not exist in the expansion of $(3x - 1/2x)^8$.

35. Using binomial theorem, expand:

(*i*) $(2x - 1/x)^4$ $\left[\textbf{Ans. } 16x^4 - 32x^2 - 24 - \dfrac{12}{x^2} + \dfrac{1}{x^4}\right]$

(*ii*) $(1 + 2x - 3x^2)$

[Ans. $10x + 25x^2 - 40x^3 - 190x^4 + 92x^5 + 570\,x^6 - 360\,x^7 - 675\,x^8 + 810\,x^9 - 243\,x^{10}$]

36. Evaluate $(3 + \sqrt{2})^5 - (3 - \sqrt{2})^5$ **[Ans.** $1210\sqrt{2}$]

37. Find middle terms in the following:

(*i*) $(x/2 + 7y)^{10}$ **[Ans.** 132355.125 x^5y^5]

(*ii*) $(3x - x^2/6)^9$ **[Ans.** 23.625 x^{21}]

38. Find the term independent of x in the expansion of:

(*i*) $\left(\dfrac{1}{3x} - \dfrac{3}{2}x^2\right)^6$ **[Ans.** 5/12] (*ii*) $\left(\dfrac{3x^2}{2} - \dfrac{1}{3x}\right)^9$ $\left[\textbf{Ans. } T_7 = \dfrac{7}{18}\right]$

(*iii*) $\left(\dfrac{\sqrt{3}}{2x^2} + \sqrt{\dfrac{x}{3}}\right)^{10}$ **[Ans.** $T_3 = 5/12$]

39. Find:

(*i*) $(\sqrt{5} + 2)^5 (\sqrt{5} - 2)^6$ **[Ans.** 1364]

(*ii*) $(\sqrt{2} + \sqrt{3})^6 - (\sqrt{2} - \sqrt{3})^6$ **[Ans.** $396\sqrt{6}$]

40. If the coefficient of $(r - 1)^{\text{th}}$, r^{th} and $(r + 1)$th term in the expansion of $(x + 1)^n$ are in the ratio 1 : 3 : 5, find n and r. **[Ans.** $n = 7, r = 3$]

1.1.4. Binomial Theorem for any Index (Expansion)

In binomial theorem, we extend the same idea *for the expansion of* $(1 + x)^n$, *where n is a negative integer or a fraction and x is a real number such that* $|x| < 1$.

If n is a negative integer or a fraction, $C(n, r)$ is meaningless.

The statement of the thorem is the same except that we replace $C(n, r)$ by

$$\frac{n(n-1)(n-2)\dots(n-r+1)}{r!}.$$

Binomial Theorem for a Rational Index. *Let n be any rational number, positive or negative, integral or fractional, and x be any real number such that* $|x| < 1$. *Then,*

$$(1+x)^n = 1 + nx + \frac{n(n-1)}{2!}x^2 + \dots + \frac{n(n-1)\dots(n-r+1)}{r!}-. x^r + \dots \infty$$

- The proof of the theorem is beyond the scope of this course.
- The condition $|x| < 1$ is necessary for the validity of the theorem. Clearly, when $x = 1$ and $n = -1$, the above statement is not true, since $|x| < 1$ does not hold.

We know that the binomial theorem for any rational index n and a real number x such that $|x| < 1$ states that:

$$(1+x)^n = 1 + nx + \frac{n(n-1)}{2!}.x^2 + \dots + \frac{n(n-1)(n-2)\dots(n-r+1)}{r!}x^r + \dots \infty \quad \dots(i)$$

Now, replacing n by $-n$ in eqn. (*i*), we get

$$(1+x)^{-n} = 1 + (-n)x + \frac{(-n)(-n-1)}{2!}.x^2 + \dots + \frac{(-n)(-n-1)(-n-2)}{3!}.x^3 + \dots + \frac{(-n)(-n-1)\dots(-n-r+1)}{r!}.x^r + \dots$$

$$= 1 - nx + \frac{n(n+1)}{2!}.x^2 - \frac{n(n+1)(n+2)}{3!}.x^3 + \dots + (-1)r.\frac{n(n+1)\dots(n+r-1)}{r!}x^r + \dots$$

Do you know?

In a binomial expansion for a rational index:

(*i*) In $(1 + x)^n$, we have

$$t_{r+1} = \frac{n(n-1)(n-2)\dots(n-r+1)}{r!}.x^r.$$

(*ii*) In $(1 - x)^n$, we have t_{r+1}

$$= (-1)^r.\frac{n(n-1)(n-2)\dots(n-r+1)}{r!}.x^r.$$

(*iii*) In $(1 + x)^{-n}$, we have t_{r+1}

$$= (-1)^r.\frac{n(n-1)(n-2)\dots(n-r-1)}{r!}.x^r.$$

(*iv*) In $(1 - x)^{-n}$, we have t_{r+1}

$$= \frac{n(n-1)(n-2)\dots(n-r+1)}{r!}.x^r.$$

- **General term in the expansion of $(1 + x)^{-n}$:**

$$t_{r+1} = (-1)^r.\frac{n(n+1)(n+2)\dots(n+r-1)}{r!}.x^r.$$

Now, again replacing x by $-x$ and n by $-n$ in (i), we get

$$(1-x)^{-n} = 1 + (-n)(-x) + \frac{(-n)(-n-1)}{2!} \cdot (-x)^2$$

$$+ \frac{(-n)(-n-1)(-r+2)}{3!} \cdot (-x)^3 + \ldots$$

$$+ \frac{(-n)(-n-1)\ldots(-n-r+1)}{r!} \cdot (-x)^r + \ldots$$

$$= 1 + nx + \frac{n(n+1)}{2!} \cdot x^2 + \frac{n(n+1)(n+2)}{3!} \cdot x^3 + \ldots$$

$$+ \frac{n(n+1)\ldots(n+r-1)}{r!} \cdot x^r + \ldots$$

- **General term in the expansion of $(1-x)^{-n}$:**

$$t_{r+1} = \frac{n(n+1)(n+2)\ldots(n+r-1)}{r!} \cdot x^r.$$

Replacing x by $-x$ in (i), we get

$$(1-x)^n = 1 + n(-x) + \frac{n(n-1)}{2!} \cdot (-x)^2$$

$$+ \frac{n(n-1)(n-2)}{3!} \cdot (-x)^3 + \ldots$$

$$+ \frac{n(n-1)(n-2)\ldots(n-r+1)}{r!} \cdot (-x)^r + \ldots$$

$$= 1 - nx + \frac{n(n-1)}{2!} \cdot x^2 - \frac{n(n-1)(n-2)}{3!} x^3 + \ldots$$

$$+ (-1)^r \cdot \frac{n(n-1)(n-2)\ldots(n-r+1)}{r!} \cdot x^r + \ldots$$

- **General term in the expansion of $(1-x)^n$:**

$$t_{r+1} = (-1)^n \cdot \frac{n(n-1)(n-2)\ldots(n-r+1)}{r!} \cdot x^r.$$

- **Application of Binomial Theorem for Approximate Values**

The binomial expansions sometime have infinitely many terms. In such cases, for further calculations: an approximate value involving only the first few terms may be enough for us. Let us the approximate values.

Example 46. *Find the cube root of 1.03 up to three decimal places.*

Sol. We want to find $(1.03)^{1/3}$ up to three decimal places.

Now, $(1.03)^{1/3} = (1 + 0.03)^{1/3}$ ...(i)

Since $|0.03| < 1$, from (*i*), we have

$$(1+0.03)^{1/3} = 1 + \frac{1}{3}(0.03) + \frac{\frac{1}{3}\left(\frac{1}{3}-1\right)}{2!}(0.03)^2 + \ldots \qquad \ldots(ii)$$

Now, we need to approximate the value up to three decimal places. Since a non-zero digit in the fourth decimal place may affect the digit in the third place in the process of rounding off, we need to consider those terms in the expansion which produce a non-zero digit in the first, second, third or fourth decimal place.

Therefore, we can take the sum of the first three terms in the Expansion (*i*), and ignore the rest.

$$\therefore \quad (1.03)^{1/3} = 1 + 0.01 + \frac{\frac{1}{3}\left(-\frac{2}{3}\right)}{2!}(0.0009)$$

$$= 1 + 0.01 - 0.0001$$

$$= 1.0099$$

$$\approx 1.010, \text{ taking the value up to three decimal places.}$$

Now, the digit after the third decimal place is greater than 5, so we have increased the third decimal place by 1.

Expansion of $(x + y)^n$, for any rational n

Case I. When $y < x$, *i.e.*, $\frac{y}{x} < 1$

In this case, we have

$$(x+y)^n = \left\{x\left(1+\frac{y}{x}\right)\right\}^n = x^n\left(1+\frac{y}{x}\right)^n$$

$$= x^n\left\{1 + n\cdot\frac{y}{x} + \frac{n(n-1)}{2!}\cdot\left(\frac{y}{x}\right)^2 + \frac{n(n-1)(n-2)}{3!}\cdot\left(\frac{y}{x}\right)^3 + \ldots\right\}$$

Case II. When $x < y$, *i.e.*, $\frac{x}{y} < 1$

In this case, we have

$$(x+y)^n = \left[y\left(\frac{x}{y}+1\right)\right]^n = y^n\left(1+\frac{y}{x}\right)^n$$

$$= y^n\left\{1 + n\cdot\left(\frac{x}{y}\right) + \frac{n(n-1)}{2!}\cdot\left(\frac{x}{y}\right)^2 + \frac{n(n-1)(n-2)}{3!}\cdot\left(\frac{x}{y}\right)^3 + \ldots\right\}$$

Application:

Putting $n = 1, 2$ in the expansion of $(1 + x)^{-n}$, we get

(*i*) $(1 + x)^{-1} = 1 - x + x^2 - x^3 + x^4 - \ldots + (-1)^r x^r + \ldots \infty$.

(*ii*) $(1 + x)^{-2} = 1 - 2x + 3x^2 - \ldots + (-1)^r . (r + 1) . x^r + \ldots \infty$

Putting $n = 1, 2$ in the expansion of $(1 - x)^{-n}$, we get

(*i*) $(1 - x)^{-1} = 1 + x + x^2 + x^3 \ldots + x^r + \ldots \infty$.

(*ii*) $(1 - x)^{-2} = 1 + 2x + 3x^2 + 4x^3 + \ldots + (r + 1)x^r + \ldots + \infty$.

Example 47. *Find the value of $(0.98)^{-3}$ up to 2 decimal places.*

Sol. We have $(0.98)^{-3} = (1 - 0.02)^{-3}$

$$= [1 + (-3)(-0.02) + \frac{(-3)(-4)}{2!}(-0.02)^2 + ...]$$

$$= (1 + 0.06 + 0.0024) = 1.0624 \approx 1.06$$

Example 48. *Find $\sqrt[3]{126}$ up to 3 decimal places.*

Sol. $\sqrt[3]{126} = (126)^{1/3} = (125 + 1)^{1/3} = (125)^{1/3} = (1 + 1/125)^{1/3}$

$$= 5(1 + 1/5^3)^{1/3}$$

$$\Rightarrow \quad 5\left[1 + 1/3 + 1/5^3 + \frac{1/3\,(1/3 - 1)}{2!}\left(\frac{1}{5^3}\right)^2 + ...\right]$$

$$\Rightarrow \quad \left[5 + \frac{1}{75} - \frac{1}{28125}\right] = \left[5 + \frac{(575 - 1)}{28125}\right]$$

$$\Rightarrow \quad 5 + \frac{374}{28125} = 5 + 0.01329 = 5.01329 \approx 5.013$$

Example 49. *If p is nearly equal to q and n > 1, show that*

$$\frac{(n+1)p + (n-1)q}{(n-1)p + (n+1)q} = \left(\frac{p}{q}\right)^{1/n}.$$

Find the approximate value of $\left(\frac{99}{101}\right)^{1/6}$.

Sol. Let us consider $p = (q + h)$, where h is so small that its square and higher powers may be neglected. Then,

$$\frac{(n+1)p + (n-1)q}{(n-1)p + (n+1)q} = \frac{(n+1)(q+h) + (n-1)q}{(n-1)(q+h) + (n+1)q}$$

$$= \frac{2nq + (n+1)h}{2nq + (n-1)h} = \frac{\left[1 + \left(\frac{n+1}{2nq}\right)h\right]}{\left[1 + \left(\frac{n-1}{2nq}\right)h\right]}$$

[On dividing numerator and denominator by $2nq$]

$$= \left[1 + \left(\frac{n+1}{2nq}\right)h\right]\left[1 + \left(\frac{n-1}{2nq}\right)h\right]^{-1}$$

$$= \left[1 + \left(\frac{n+1}{2nq}\right)h\right]\left[1 - \left(\frac{n-1}{2nq}\right)h\right]$$

[Expanding the 2nd expression and neglecting h^2, h^3, etc.]

$$= \left[1 + \left(\frac{n+1}{2nq}\right)h - \left(\frac{n-1}{2nq}\right)h\right] \quad \text{[neglecting } h^2\text{]}$$

$$= \left(1 + \frac{h}{nq}\right).$$

Also, $$\left(\frac{p}{q}\right)^{1/n} = \left(\frac{q+h}{nq}\right)^{1/n} = \left(1 + \frac{h}{1}\right)^{1/n} = \left(1 + \frac{h}{nq}\right)$$

[expanding and neglecting h^2 and higher powers of h]

$$\therefore \quad \frac{(n+1)p + (n-1)q}{(n-1)p + (n+1)q} = \left(\frac{p}{q}\right)^{1/n}.$$

Deduction. Taking $p = 99$, $q = 101$ and $n = 6$ in the above result, we get

$$\left(\frac{99}{101}\right)^{1/6} = \frac{(6+1) \times 99 + (6-1) \times 101}{(6-1) \times 99 + (6+1) \times 101} = \frac{1198}{1202} = \frac{599}{601}.$$

Example 50. *Expand $(2 + 3x)^{-5}$ up to four terms in (i) ascending powers of x, (ii) descending powers of x. Mention the condition for validity of expansion in each case.*

Sol. (*i*) $(2 + 3x)^{-5} = \left[2\left(1 + \frac{3}{2}x\right)\right]^{-5}$

$$= 2^{-5}\left(1 + \frac{3}{2}x\right)^{-5}, \text{ expandable, when } \frac{3}{2}|x| < 1, \text{ i.e., } |x| < \frac{2}{3}$$

$$= \frac{1}{32}.\left[1 + (-5).\left(\frac{3}{2}x\right) + \frac{(-5)(-6)}{2!}.\left(\frac{3}{2}x\right)^2 + \frac{(-5)(-6)(-7)}{3!}.\left(\frac{3}{2}x\right)^3 + ...\right]$$

$$= \frac{1}{32}.\left[1 - \frac{15}{2}x + \frac{135}{4}x^2 - \frac{945}{8}x^3 + ...\right]$$

(*ii*) $(2 + 3x)^{-5} = \left[3x\left(1 + \frac{2}{3x}\right)\right]^{-5}$

$$= (3x)^{-5}\left(1 + \frac{2}{3x}\right)^{-5}, \text{ when } \left|\frac{2}{3x}\right| < 1, |x| > \frac{2}{3}$$

$$= \frac{1}{243\,x^5}\left[1 + (-5)\left(\frac{2}{3x}\right) + \frac{(-5)(-6)}{2!}\left(\frac{2}{3x}\right)^2 + \frac{(-5)(-6)(-7)}{3!}.\left(\frac{2}{3x}\right)^3 + ...\right]$$

$$= \frac{1}{243\,x^5} \times \left[1 - \frac{10}{3x} + \frac{20}{3x^2} - \frac{280}{27\,x^3} + ...\right]$$

$$= \frac{2}{243} \times \left[\frac{1}{x^5} - \frac{10}{3}.\frac{1}{x^6} + \frac{20}{3}.\frac{1}{x^7} - \frac{280}{27}.\frac{1}{x^8} + ...\right]$$

Example 51. *If $\frac{(1-3x)^{1/2} + (1-x)^{5/3}}{\sqrt{4-x}}$ is approximately equal to a + bx for all small values of x, find a and b.*

Sol. According to binomial theorem,

$$\frac{(1-3x)^{1/2}+(1-x)^{5/3}}{\sqrt{4-x}} = [(1-3x)^{1/2}+(1-x)^{5/3}] \times [(4-x)^{-1/2}]$$

$$= [(1-3x)^{1/2}+(1-x)^{5/3} \times 4^{-1/2}\,[1-x/4]^{-1/2}$$

$$= \left[\left\{1+\frac{1}{2}(-3x)\right\}+\left\{1+\frac{5}{3}(-x)\right\}\right]\times\frac{1}{2}\times\left[1+\left(\frac{-1}{2}\right)\left(-\frac{x}{4}\right)\right]$$

$$= \left[1-\frac{3x}{2}+1-\frac{5x}{3}\right]\times\left[\frac{1}{2}+\frac{x}{16}\right]$$

$$= \left(\frac{2-19x}{6}\right)\left(\frac{1}{2}+\frac{x}{16}\right) \Rightarrow \left(1+\frac{x}{8}-\frac{19x}{12}\right)=\left(1-\frac{35x}{24}\right)$$

Given expression = $(a + bx)$

$$\Rightarrow \quad 1-\frac{35x}{24} = a+bx$$

$$\therefore \quad a = 1 \text{ and } b = \frac{-35}{24}.$$

Example 52. *If x be nearly equal to 1, show that* $\dfrac{ax^b - bx^a}{x^b - x^a} = \dfrac{1}{(x-a)}$ *(approx).*

Sol. Putting $x = 1 + h$, we get

$$\frac{ax^b - bx^a}{x^b - x^a} = \frac{a(1+h)^b - b(1+h)^a}{(1+h)^b - (1+h)^a} = \frac{a[1+bh+\ldots]-b[1+ah+\ldots]}{[1+bh\ldots]-[1+ah+\ldots]}$$

[neglecting h^2]

$$= \frac{a(1+bh)-(b(1+ah)}{(1+bx)-(1+ax)} = \frac{a-b}{(b-a)h} = \frac{-1}{h} = \frac{1}{1-x}.$$

Example 53. *Prove that:* $(1 + x + x^2 + x^3 + \ldots)(1 - x + x^2 - x^3 + \ldots) = (1 + x^2 + x^4 + x^6 + \ldots)$.

Sol. $(1 + x + x^2 + x^3 + \ldots)(1 - x + x^2 - x^3 + \ldots)$

$$= (1-x)^{-1}(1+x)^{-1} = [(1-x)(1+x)]^{-1} = (1-x^2)^{-1}$$

$$= (1 + x^2 + x^4 + x^6 + \ldots) = \text{R.H.S.}$$

Example 54. *Find the coefficient of* x^6 *in the expansion of* $(1 - 2x)^{-5/2}$.

Sol. General term in the expansion of $(1 - y)^{-n}$ is

$$t_{r+1} = \frac{n(n+1)(n+2)\ldots(n+r-1)}{r!}\, y^r$$

$$\therefore \quad t_{r+1} = \frac{(5/2)\left(\frac{5}{2}+1\right)\left(\frac{5}{2}+2\right)+\ldots\left(\frac{5}{2}+r-1\right)}{r!}\,(2x)^r$$

$$= \frac{5.7.9\ldots(3+2r)}{2^r\,(r!)}\, 2^r . x^r$$

Putting $r = 6$, we get

$$t_7 = \frac{5.7.9.11.13.15}{2^6\,(6\,!)}\,2^6\,x^6 = \frac{15015}{16}\,x^6$$

Thus, the coefficient of x^6 in the expansion of $(1-2x)^{-5/2}$ is $\frac{15015}{16}$.

EXERCISE 1.1.2.

1. (*i*) Expand $(1+x)^{-3}$ to 5 terms and find the general term.

$$\left[\textbf{Ans. } (1-3x+6x^2-10x^3+15x^4+...),\ (-1)^r\,.\,\frac{3.4.5....(r+2)}{r\,!}\,x^r\right]$$

(*ii*) Expand $(1-x)^{-3}$ to 5 terms and find the general term.

$$\left[\textbf{Ans. } (1-3x+6x^2+10x^3+15x^4+...),\ \frac{3.4.5....(r+2)}{r\,!}\,x^r\right]$$

2. Expand each up to 4 terms:

(*i*) $\frac{1}{\sqrt{4-5x^2}}$ $\left[\textbf{Ans. } x < \frac{2}{\sqrt{5}}, \left(\frac{1}{2}+\frac{5}{16}x^2+\frac{75}{256}x^4+\frac{625}{2048}x^6\right)\right]$

(*ii*) $\frac{1}{3\sqrt{(27-6x)^2}}$ $\left[\textbf{Ans. } x < \frac{9}{2}, \frac{1}{9}\left(1+\frac{4}{27}x+\frac{20}{729}x^2+\frac{320}{59049}x^3+...\right)\right]$

(*iii*) $(9-6x)^{-3/2}$ $\left[\textbf{Ans. } x < \frac{3}{2}, \left(\frac{1}{27}+\frac{1}{27}x+\frac{5}{162}x^2+\frac{35}{1458}x^3+...\right)\right]$

(*iv*) $\sqrt{9-4x^2}$ $\left[\textbf{Ans. } x < \frac{3}{2}, \left(3-\frac{2}{3}x^2-\frac{2}{27}x^4-\frac{4}{243}x^6+...\right)\right]$

(*v*) $\sqrt{9+2x}$ $\left[\textbf{Ans. } x < \frac{9}{2} < \left(3+\frac{1}{3}x-\frac{1}{54}x^2+\frac{1}{486}x^3-...\right)\right]$

(*vi*) $(8+x)^{4/3}$ $\left[\textbf{Ans. } x < 8, \left(16+\frac{8}{3}x+\frac{1}{18}x^2-\frac{1}{618}x^3+...\right)\right]$

3. Find the coefficient of x^9 in the expansion of $(1-2x^2)^{-x/2}$. $\left[\textbf{Ans. } \frac{715}{2}\right]$

4. Find the coefficient of x^8 in the expansion of $(1-x^2)^{-2/3}$. $\left[\textbf{Ans. } \frac{110}{243}\right]$

5. Find the coefficient of x^7 in the expansion of $(x-2x^2)^{-3}$. [**Ans.** 67584]

6. Find the coefficient of x^3 in the expansion of $(1+x)^{-3}$. [**Ans.** $\sqrt{-10}$]

7. Find the 5th term in the expansion of $(1-2x)^{-1}$. [**Ans.** $16x^4$]

8. Find the 7th term in the expansion of $(1-x^2)^{-2/3}$. $\left[\textbf{Ans. } \frac{2618}{6561}x^{12}\right]$

9. Find the 4th term in the expansion of $(x-1)^{-3}$. [**Ans.** $-10x^3$]

10. Find the 5th term in the expansion of $(1-2x^3)^{11/2}$. $\left[\textbf{Ans. } \dfrac{1155}{8} x^{12}\right]$

11. Write the general term in each of the following expansion:

(i) $(2-3x^2)^{-2/3}$ $\left[\textbf{Ans. } \dfrac{2.5.8.\dots(3r+1)}{2^{2/3}\,(r\,!)} \cdot \dfrac{x^{2r}}{2^r}\right]$

(ii) $(3-x)^{-25}$ $\left[\textbf{Ans. } \dfrac{25.26\dots 24+r}{3^{(25+r)}.r\,!} x^r\right]$

(iii) $(1-x)^{-4}$ $\left[\textbf{Ans. } \dfrac{(r+1)(r+2)(r+3)}{1.2.3} x^r\right]$

(iv) $(1+2x)^{-1/2}$ $\left[\textbf{Ans. } (-1)^r . \dfrac{1.3.5\dots(2r-1)}{r\,!} x^r\right]$

12. Show that $(1+x+x^2+\dots\infty)(1+2x+5x^2+\dots\infty) = (1+5x+6x^2+\dots\infty)$.

13. Show that $(1+x+x^2+\dots\infty)^2 = (1+2x+3x^2+\dots\infty)$.

14. Show that $(1+x+x^2+\dots\infty)^3 = (1+3x+6x^2+\dots\infty)$.

15. Given x is nearly equal to 1, show that $\dfrac{mx^m - nx^n}{m-n} = x^{m+n}$.

16. Given x is very small, compared with a. Show that $\sqrt{\left(\dfrac{a}{a+x}\right)} + \sqrt{\left(\dfrac{a}{a-x}\right)} = \left(2+\dfrac{3x^2}{4a^2}\right)$.

17. If p and q be nearly equal, then prove that $\sqrt{\dfrac{p}{q}} = \dfrac{p}{p+q} + \dfrac{p+q}{4q}$.

18. If y is so small that its square and higher powers can be neglected, show that

(i) $\dfrac{(1+3/4\,y)^{-4}\,(16-3y)^{1/2}}{(8+y)^{2/3}} = \left(1-\dfrac{305}{96}x\right)$ nearly

(ii) $\dfrac{(16+5y)^{1/2}-(27-4y)^{2/3}}{5y+6} = \left(\dfrac{1}{6}-\dfrac{13}{1296}y\right)$ nearly

(iii) $\left(\dfrac{1+y}{1-y}\right)^2 = (1+4y)$

(iv) $\dfrac{1+y}{1-y} = 1+2y$.

19. Calculate:

(i) $\sqrt[4]{16.08}$ correct to four places of decimal. [**Ans.** 2.0025]

(ii) $(1.025)^{-1/3}$ correct to three places of decimal. [**Ans.** 0.091]

(iii) $3\sqrt{1010}$ correct to four places of decimal. [**Ans.** 10.0332] (*iv*) $1/\sqrt{47}$ correct to four places of decimal. [**Ans.** 0.1459]

(*v*) Cube root of 999 up to 3 places of decimal. [**Ans.** 9.996]

(*vi*) Cube root of 127 up to 3 places of decimal. [**Ans.** 5.0264]

20. Find the coefficient of x^{10} in the expansion of $\dfrac{3+2x}{(1-x)^2}$.

21. Prove that the coefficient of x^r in the expansion of $(1-2x)^{-1/2}$ is $\dfrac{(2r)!}{2^r.(r!)^2}$.

22. Expansion of $(2x^2 - 1/x)^{10}$, find the middle term. [**Ans.** $-8060\,x^5$]

23. Expansion of $(2x^2 - 1/x)^{10}$, find the term whose simplified from involves $1/x$. $\left[\textbf{Ans. } -\dfrac{960}{x}\right]$

24. Find the first negative term in the expansion of $(1+x)^{7/2}$.

$$\left[\textbf{Ans. } t_{r+1} = \frac{\frac{7}{2}\left(\frac{7}{2}-1\right)\left(\frac{7}{2}-2\right)\ldots\left(\frac{7}{2}-r+1\right)}{r!}x^r\right]$$

The term will be negative for the value of r for which $\left(\dfrac{7}{2}-r+1\right) < 0, r > 4\,\dfrac{1}{2}$, and $r = 5$.

25. Prove that $\sum\limits_{r=0}^{n} r^2\, {}^nC_r\, p^r q^{n-r} = npq + n^2p^2$, if $p + 1 = 1$.

26. If $\sum\limits_{r=0}^{n} (-1)^r . {}^nC_r \left[\dfrac{1}{2^r} + \dfrac{3^r}{2^{2r}} + \dfrac{7^r}{2^{3r}} + \ldots \text{ to } m \text{ terms}\right] = k\left(1 - \dfrac{1}{2^{mn}}\right)$, then find the value of k.

$\left[\textbf{Ans. } \dfrac{1}{2^n - 1}\right]$

27. Find the coefficient of x^{50} in the expression:

$(1+x)^{1000} + 2x\,.\,(1+x)^{999} + 3x^2\,(1+x)^{998} + \ldots + 1001\,x^{1000}$. [**Ans.** ${}^{1002}C_{50}$]

28. Given $s_n = 1 + q + q^2 + \ldots. + q^n$ and $S_n = 1 + \dfrac{q+1}{2} + \left(\dfrac{q+1}{2}\right)^2 + \ldots + \left(\dfrac{q+1}{2}\right)^n$, $q \neq 1$, prove that ${}^{n+1}C_1 + {}^{n+1}C_2 . s_1 + {}^{n+1}C_3 . s_2 + \ldots + {}^{n+1}C_{n+1} . s_n = 2^n\,.\,S_n$.

29. Write down the binomial expansion of $(1+x)^{n+1}$, when $x = 8$. Deduce that $9^{n+1} - 8n - 9$ is divisible by 64, whenever n is a positive integer.

30. Find the value of 'x' for which the fourth term in the expansion, $\left(5^{\frac{2}{5}\log_5 \sqrt{4^x + 44}} + \dfrac{1}{5^{\log_5 \sqrt[3]{2^{x-1} + 7}}}\right)^8$ is 336.

31. In the binomial expansion of $\left(\sqrt[3]{2} + \dfrac{1}{\sqrt[3]{3}}\right)^n$, the ratio of the 7th term from the beginning to the 7th term from the end is 1 : 6 ; find n.

32. Find the remainder, when $32^{32^{32}}$ is divided by 7. **[Ans.** 4]

33. Find the terms independent of 'x' in the expansion of the expression, $(1 + x + 2x^3)$ $\left(\frac{3}{2}x^2 - \frac{1}{3x}\right)^9$. **[Ans.** 17/54]

34. Show that the term independent of x in the expansion of $\left(1 + x + \frac{6}{x}\right)^{10}$ is, $1 + \sum_{r=1}^{5} {}^{10}C_{2r}\ {}^{2r}C_r\ 6^r$.

35. If $(1 + x)^n = C_0 + C_1x + C_2x^2 + ... + C_n x^n$, then show that:

$$C_1(1-x) - \frac{C_2}{2}(1-x)^2 + \frac{C_3}{3}(1-x)^3 - ..... + (-1)^{n-1}\frac{1}{n}(1-x)^n$$

$$= (1-x) + \frac{1}{2}(1-x^2) + \frac{1}{3}(1-x^3) + ... + \frac{1}{n}(1-x^n).$$

36. Prove that, $({}^{2n}C_1)^2 + 2.({}^{2n}C_2)^2 + 3.({}^{2n}C_3)^2 + ... + 2n.({}^{2n}C_{2n})^2 = \frac{(4n-1)!}{\{(2n-1)!\}^2}$.

37. If $(3\sqrt{3} + 5)^n = p + f$, where p is an integer and f is a proper fraction, then find the value of $(3\sqrt{3} + 5)^n$, $n \in N$. $1 - f$, if n is even and f, if n is odd.

38. Find the coefficient os $a^5 b^4 c^7$ in the expansion of $(bc + ca + ab)^8$. **[Ans.** 280]

39. If in the expansion of $(1 - x)^{2n+1}$, the co-efficient of x^r is denoted by a_r, then prove that $a_{r-1} + a_{2n-r} = 0$.

40. Show that if the greatest term in the expansion of $(1 + x)^{2n}$ has also the greatest co-efficient, then 'x' lies between, $\frac{n}{n+1}$ and $\frac{n+1}{n}$.

41. Prove that $53^{53}\quad 33^{33}$ is divisible by 10.

42. Assuming 'x' to be so small that x^2 and higher powers of 'x' can be neglected, show that,

$$\frac{\left(1 + \frac{3}{4}x\right)^{-4}(16 - 3x)^{1/2}}{(8 + x)^{2/3}} \text{ is approximately equal to, } 1 - \frac{305}{96}x.$$

43. If $(1 + x + x^2 + ... + x^p)^n = a_0 + a_1x + a_2x^2 + ... + a_{np} . x^{np}$, then find the value of :

$a_1 + 2a_2 + 3a_3 + ... + np . a_{np}$. $\left[\textbf{Ans. } \frac{np}{2}(p+1)^n\right]$

1.2. PARTIAL FRACTION

A fraction is a symbol indicating the division of integers. For example, $\frac{13}{9}, \frac{2}{3}$ are fractions and are called Common Fraction. The dividend (upper number) is called the numerator $N(x)$ and the divisor (lower number) is called the denominator, $D(x)$.

From the previous study of elementary algebra, we have learnt how the sum of different fractions can be found by taking L.C.M. and then add all the fractions. For example,

(*i*) $\frac{1}{x-1} + \frac{2}{x+2} = \frac{3x}{(x-1)(x+2)}$

(*ii*) $\frac{2}{x+1} + \frac{1}{(x+1)^2} + \frac{3}{x-2} = \frac{9x^2+5x-3}{(x+1)^2(x-2)}$

We split up a single fraction into a number of fractions whose denominators are the factors of denominator of that fraction. So these are called **Partial fractions.**

To express a single rational fraction into the sum of two or more single tational fractions is called **Partial fraction.**

$$\frac{2x+x^2-1}{x(x^2-1)} = \frac{1}{x} + \frac{1}{x-1} + \frac{1}{x+1}$$

$\frac{2x+x^2-1}{x(x^2-1)}$ is the resultant fraction and $\frac{1}{x} + \frac{1}{x-1} + \frac{1}{x+1}$ are its partial fractions.

Polynomial: Any expression of the form $P(x) = a_n x^n + a_{n-1} x^{n-1} + ... + a_2 x^2 + a_1 x + a_0$ where $a_n, a_{n-1}, ..., a_2, a_1, a_0$ are real constanats, if $a_n \neq 0$, then $P(x)$ is called polynomial of degree n.

• Definition

We know that $\frac{p}{q}, q \neq 0$ is called a rational number. Similarly the quotient of two polynomials $\frac{N(x)}{D(x)}$ where $D(x) \neq 0$, with no common factors, is called a rational fraction. A rational fraction is of two types:

(*a*) Proper Fraction

A rotional fraction $\frac{N(x)}{D(x)}$ is called a proper fraction, if the degree of numerator $N(x)$ is less than the degree of denominator $D(x)$.

For example,

(*i*) $\frac{9x^2-9x+6}{(x-1)(2x-1)(x+2)}$ (*ii*) $\frac{6x+27}{3x^2-9x}$

(*b*) Improper Fraction

A rational fraction $\frac{N(x)}{D(x)}$ is called an improper fraction, if the degree of the numerator $N(x)$ is greater than or equal to the degree of the denominator $D(x)$.

For example,

(i) $\dfrac{2x^3 - 5x^2 - 3x - 10}{x^2 - 1}$ (ii) $\dfrac{6x^2 - 5x^2 - 7}{3x^2 - 2x - 1}$.

Note: A improper fraction can be expressed, by division, as the sum of a polynomial and a proper fraction.

For example:

$$\frac{6x^2 + 5x^2 - 7}{3x^2 - 2x - 1} = (2x + 3) + \frac{8x - 4}{x^2 - 2x - 1}$$

Which is obtained as, divide $6x^2 + 5x^2 - 7$ by $3x^2 - 2x - 1$, then we get a polynomial $(2x + 3)$ and a proper fraction $\dfrac{8x - 4}{x^2 - 2x - 1}$.

- If in the denominator $D(x)$ a linear factor $(ax + b)$ occurs n times *i.e.*, $(ax + b)^n$, then there will be n partial fractions of the form $\dfrac{A_1}{ax + b} + \dfrac{A_2}{(ax + b)^2} + \dfrac{A_3}{(ax + b)^3} + \ldots + \dfrac{A_n}{(ax + b)^n}$, where $A_1, A_2, A_3 \ldots A_n$ are constants whose values are to be determined.
- If in the denominator $D(x)$, a quadratic factor $ax^2 + bx + c$ occurs and is non-repeating, its partial fraction will be of the form $\dfrac{Ax + B}{ax^2 + bx + c}$, where A and B are constants whose values are to be determined.
- If in the denominator a quadratic factor $ax^2 + bx + c$ occurs n times, *i.e.*, $(ax^2 + bx + c)^n$, then there will be n partial fractions of the form:

$$\frac{A_1x + B_1}{ax^2 + bx + c} + \frac{A_2x + B_2}{(ax^2 + bx + c)^2} + \ldots + \frac{A_nx + B_3}{(ax^2 + bx + c)^3} + \ldots + \frac{A_nx + B_n}{(ax^2 + bx + x)^n}$$

where $A_1, A_2, A_3, \ldots A_n$ and $B_1, B_2, B_3, \ldots B_n$ are constants whose values are to be determined.

(i) $\dfrac{1}{x + 2} + \dfrac{1}{x + 3} = \dfrac{2x^2 + x + 1}{(x + 3)\,(3 - 1)^2}$

(ii) $\dfrac{1}{x + 3} + \dfrac{1}{x - 1} + \dfrac{1}{(x + 1)^2}$

$= \dfrac{2x^2 + x + 1}{(x + 3)\,(x - 1)^2}$

Do you know?

If two polynomials are equal, for all values of the variables, then the coefficients having same degree on both sides are equal, for example, if $px^2 + qx + a = 2x^2 - 3x + 5 \;\forall\; x$, then $p = 2, q = -3$ and $a = 5$.

1.2.1. Linear Factors

The fraction having two or more factors in the denominator into simpler fraction. The denominators of the simple or partial fractions are given by the factors of the fractions while numerators are unknown and as such be assumed by us. Then by simple algebraic methods.

$$\frac{2x + 5}{(x + 2)(x + 3)} = \frac{A}{x + 2} + \frac{B}{x + 3}$$

In general, rational algebraic fraction is

$$\frac{a_0x^n + a_1x^{n-1} + a_2x^{n-2} + \dots a_n}{b_0x^m + b_1x^{m-1} + b_2x^{m-2} + \dots b_m}$$

here $a_0, a_1, a_2, \dots, a_n, b_1, b_2, \dots b_n$ are constants. Here, the degree of the numerator is n and the degree of the denominator is m. If $n \geq m$, the given fraction is called an improper fraction. In case $n > m$, the given fraction is a proper fraction. By improper fraction can be written as the sum of an integral part and a proper fraction.

$$\frac{x^3 + 3x + 5}{(x - 2)(x - 3)} = (x + 5) + \frac{22x - 25}{(x - 2)(x - 3)}$$

on the otherhand an improper fraction where $n = 3$ and $m = 2$ and after division, the R.H.S. consists of the integral part $(x + 5)$ and the proper fraction.

$$\frac{22x - 25}{(x - 2)(x - 3)}$$

Now we check that the given algebraic fraction in a proper fraction. Now we must divide the numerator, by the denominator to get a proper fraction.

$$\frac{x^3 + 2x + 5}{(x - 1)(x + 2)(x - 3)} = 1 + \frac{A}{(x - 1)} + \frac{B}{(x + 2)} + \frac{C}{(x - 3)}$$

actual division the quotient on integral part will be unity and the resulting proper fraction can then be resolved into three partial fraction.

Procedure to resolve a given rational fractions into partial fractions

(*i*) *Check whether the given fraction is a proper fraction.* If the degree of the numerator is higher or equal to that of the denominator, divide the numerator by the denominator till the remainder is of lower degree than the denominator.

(*ii*) *Frctorize the denomiator into real factors.* These will be either linear or quadratic, and some factors repeated.

(*iii*) *Resolve the proper fraction into sum of partial fractions such that*

- to a *non-repeated linear factor* $x - a$ in the denominator corresponds a pratial fraction of the form $A/(x - a)$;
- to a *repeated linear factor* $(x - a)^r$ in the denominator corresponds the sum of r partial fractions of the form $\frac{A_1}{x - a} + \frac{A_2}{(x - a)^2} + \frac{A_3}{(x - a)^3} + \dots + \frac{A_r}{(x + a)^r}$.
- to a *non-repeated quadratic factor* $(x^2 + ax + b)$ in the denominator, corresponds a partial fraction of the form $\frac{Ax + B}{x^2 + ax + b}$;
- to a *repeated quadratic factor* $(x^2 + ax + b)^r$ in the denominator, corresponds the sum of r partial fractions of the for $\frac{A_1x + B_1}{x^2 + ax + b} + \frac{A_2x + B_2}{(x^2 + ax + b)^2} + \dots + \frac{A_rx + B_r}{(x^2 + ax + b)^r}$.

Then we have to determine the unknown constants A, A_1, B_1, B_1 etc. as follows:

(iv) To obtain the partial fraction corresponding to the non-repeated linear factor $x - a$ in the denomiator, put $x = a$ everywhere in the given fraction expect in the factor $x - a$ itself.

In all other cases, equate the given fraction to a sum of suitable partial fractions in accordance with (*i*) and (*ii*) above, having found the partial fractions corresponding to the non-repeated linear factors by the above rule. Then multiply both sides by the denominator of the given fraction and equate the cofficients of like powers of x or substitute convenient numerical values of x on both sides. Finally solve the simplest of the resulting equations to find the unknown constants.

1.2.2. Repeated Linear Factors

In case of repeated first degree factors of the denominator, we have

$$\frac{N(x)}{D(x)} = \frac{A_1}{(x-a)} + \frac{A_2}{(x-a)^2} + \frac{A_3}{(x-a)^3} + \frac{B}{(x-b)} + \frac{C}{(x-c)} \qquad \ldots(i)$$

where $D(x) = (x - a)^3 (x - b)(x - c)$

Here, $(x - a)$ is repeated thrice. Now, equating numerators, we get

$$N(x) \equiv A_1(x-a)^2(x-b)(x-c) + A_2(x-a)(x-b)(x-c) + A_3(x-b)(x-c) + B(x-a)^3(x-c) + B(x-a)^3(x-b) \qquad \ldots(ii)$$

On putting $x = a$ in (*ii*), we get A_3. Similarly, on putting $x = b$ and $x = c$ in (*ii*), we get B and C respectively. To get A_1 and A_2, we equate any two of the co-efficients of x^4, x^3, x^2, x or the constant terms on both sides of (*ii*).

We also get A_1, A_2, etc. by putting $x = 0, 1, 2$ etc. in the indentity (*ii*) and then solving equations involving the unknown constants.

1.2.3. Non-repeated Linear Factors

In case of non-repeated first degree factors of the denominator, we have

$$\frac{N(x)}{D(x)} \equiv \frac{A}{(x-a)} + \frac{B}{(x-b)} + \frac{C}{(x-c)} \qquad \ldots(i)$$

where $D(x) = (x - a)(x - b)(x - c)$

Therefore, $N(x) \equiv A(x - b)(x - c) + B(x - a)(x - c) + C(x - a)(x - b) \qquad \ldots(ii)$

Now the values of A, B, C are found by putting $x = a$, $x = b$, $x = c$ respectively in the indentity (*ii*). They can also be obtained by equating coefficients of x^2, x and constant terms on both sides of (*ii*) and then solving the three equations involving A, B, C.

Short-Cut Method

Multiplying both sides of (*i*) by $(x - a)$, we get

$$(x-a)\frac{N(x)}{D(x)} = A + \frac{B(x-a)}{(x-b)} + \frac{C(x-a)}{(x-c)}$$

$$\Rightarrow \quad \frac{N(x)}{(x-b)(x-c)} = A + \frac{B(x-a)}{(x-b)} + \frac{C(x-a)}{(x-c)}$$

Now, let $x = a$, then

$$\frac{N(a)}{(a-b)(a-c)} = A.$$

Similarly, $\dfrac{N(b)}{(b-a)(b-c)} = B$

and $\qquad C = \dfrac{N(c)}{(c-a)(c-b)}$.

Example 1. *Resolve into partial fractions* $\dfrac{x^2 - 10x + 13}{(x-1)(x-2)(x-3)}$.

Sol. Let
$$\frac{N(x)}{D(x)} = \frac{x^2 - 10x + 13}{(x-1)(x-2)(x-3)}$$
$$= \frac{A}{(x-1)} + \frac{B}{(x-2)} + \frac{C}{(x-3)}$$
$$= \frac{A(x-2)(x-3) + B(x+1)(x-3) + C(x-1)(x-2)}{(x-1)(x-2)(x-3)}$$

Equating numerators, we have
$$x^2 - 10x + 13 = A(x-2)(x-3) + B(x-1)(x-3) + C(x-1)(x-2)$$

Put $x = 1$, $\quad 4 = A(1-2)(1-3) + 0 + 0 \quad \Rightarrow \quad A = 2$

Put $x = 2$, $\quad -3 = 0 + B(2-1)(2-3) + 0 \quad \Rightarrow \quad B = 3$

Put $x = 3$, $\quad -8 = 0 + 0 + C(3-1)(3-2) \quad \Rightarrow \quad C = -4$

Therefore, the partial fractions are
$$\frac{2}{(x-1)} + \frac{3}{(x-2)} - \frac{4}{(x-3)}.$$

Short-cut Method

$$A = \left[(x-1)\frac{N(x)}{D(x)}\right]_{x=1}$$
$$= \left[\frac{x^2 - 10x + 13}{(x-2)(x-3)}\right]_{x=1}$$
$$= \frac{1 - 10 + 13}{(1-2)(1-3)} = \frac{4}{2} = 2.$$
$$B = \left[(x-2) - \frac{N(x)}{D(x)}\right]_{x=2}$$
$$= \left[\frac{x^2 - 10x + 13}{(x-1)(x-3)}\right]_{x=2} = 3$$
$$C = \left[(x-3)\frac{N(x)}{D(x)}\right]_{x=3}$$
$$= \left[\frac{x^2 - 10x + 13}{(x-1)(x-2)}\right]_{x=3} = -4.$$

Thus, we get the unknown constants *A, B, C* easily and that leads to partial fractions.

Example 2. *Resolve into partial fractions* $\dfrac{3x+2}{(x+1)^2(x-2)}$.

Sol. Let $\dfrac{N(x)}{D(x)} = \dfrac{3x+2}{(x+1)^2\,(x-2)}$

$$\equiv \frac{A}{(x+1)} + \frac{B}{(x+1)^2} + \frac{C}{(x-2)}$$

or $$(3x+2) \equiv A(x+1)\,(x-2) + B(x-2) + C(x+1)^2 \qquad \ldots(i)$$

Let $x = -1$, then $\quad -1 = A(0) + B(-3) + C(0) \quad \Rightarrow \quad B = \dfrac{1}{3}$.

Let $x = 2$, then $\quad 8 = A(0) + B(0) + C(9) \quad \Rightarrow \quad C = \dfrac{8}{9}$.

To obtain A, substituting $x = 0$ in (i), we have

$$2 = A(1)\,(-2) + B\,(-2) + C(1)^2$$

or $$-2A - 2B + C = 2$$

or $$-2A - \frac{2}{3} + \frac{8}{9} = 2 \quad \text{or} \quad A = -\frac{8}{9}$$

Therefore, the partial fractions are

$$-\frac{8}{9(x+1)} + \frac{1}{3(x+1)^2} + \frac{8}{9\,(x-2)}.$$

Short Method

$$B = \left[(x+1)^2\,\frac{N(x)}{D(x)}\right]_{x=-1} = \left[\frac{3x+2}{(x-2)}\right]_{x=-1} = \frac{1}{3}$$

$$C = \left[(x-2)\,\frac{N(x)}{D(x)}\right]_{x=2} = \left[\frac{3x+2}{(x+1)}\right]_{x=2} = \frac{8}{9}$$

and we obtain A by the method given earlier, *i.e.*, y putting $x = 0$ in (i).

1.2.4. Non-Reducible Quadratic Factors

We shall introduce to the reverse process of breaking a given fraction into a sum of simple fractions. These simple fractions are called Denominator containing irreducible non-repeated quadratic factors of the fraction. Denominator containing second degree non-repeated factor.

Let $$\frac{N(x)}{D(x)} = \frac{Ax+B}{a^2+bx+c} + \frac{Cx+D}{dx^2+ex+f}$$

where $$D(x) = (ax^2+bx+c)\,(dx^2+ex+f)$$

Equating numerators both sides, we get

$$N(x) = (Ax+B)(dx^2+ex+f) + (Cx+D)(ax^2+bx+c) \quad \ldots(i)$$

Now, equating coefficients of x^3, x^2, x, the constants forms and solving the four equations, we get A, B, C and D sometimes, it is easier to put $x = 0$, 1 etc. (i) and then obtain the value of *ABCD*.

• Applications in Engineering Problems

In current years, the partial fractions, both fractional and integer orders have been documented as a dominant modelling procedure. To precisely reproduce the non-local, frequency and history-dependent properties of power law phenomena, some different modelling tools based on fractional operators have to be introduced.

In particular, the advantages of fractional calculus and fractional order models meaning differential systems involving fractional order integrodifferential operators and their applications have already been intensively studied during the last few decades with excellent results.

The long-range temporal spatial dependence phenomena inherent to the fractional order systems present unique pecularities not supported by their integer order counter part, which permit better models of the dynamics of complex processes. Although non-integer differentiation has become a popular tool for modelling and controlling the behaviours of physical systems from diverse applied branches of the science, many problems remain to be explored and solved.

While the investigation of the phenomena is described by the interaction of many organisms, the microsimulation plays an important role, and as a result the computers become more and more scientific instruments.

The objective of this special issue is to report and review the latest progress in the following areas of partial fractions.

Potential topics include, but are not limited to:

- Partial fractions and theii applications in science and engineering
- Modelling and simulation real world phenomena with partial fractions
- Analytical and numerical methods for partial fractions
- New applications of the iterations method
- Anomalous diffusion.

Example 3. *Resolve into partial fractions* $\dfrac{x^2 + 23x + 18}{(x^2 + x - 2)(2x - 1)}$.

Sol. Since it is a proper fraction, we write it is

$$\frac{x^2 + 23x + 18}{(x - 1)(x + 2)(2x - 1)} = \frac{A}{x - 1} + \frac{B}{x + 2} + \frac{C}{2x - 1}$$

To find A, put $x = 1$ in the L.H.S. except in '$x - 1$' itself. Then,

$$A = \frac{1 + 23 + 18}{(1 + 2)(2 - 1)} = 2.$$

To find B, put $x = -2$ in the L.H.S. except in factor '$x + 2$' itself. Then,

$$B = \frac{4 - 46 - 18}{(-2 - 1)(-4 - 1)} = -\frac{60}{15} = -4.$$

To find C, put $x = 1/2$ in the L.H.S. except in factor '$2x - 1$' Then,

$$C = \frac{\frac{1}{4} + 23 \cdot \frac{1}{2} - 18}{\left(\frac{1}{2} - 1\right)\left(\frac{1}{2} + 2\right)} = \frac{-\frac{25}{4}}{-\frac{5}{4}} = 5.$$

Thus,

$$\frac{x^2 + 23x - 18}{(x^2 + x - 2)(2x - 1)} = \frac{2}{x - 1} - \frac{4}{x + 2} + \frac{5}{2x - 1}.$$

Example 4. *Resolve into partial fractions* $\frac{x^2 - x + 1}{(x + 1)(x - 1)^2}$.

Sol. Since it is a proper fraction, we write

$$\frac{x^2 - x + 1}{(x + 1)(x - 1)^2} = \frac{A}{x + 1} + \frac{B}{x - 1} + \frac{C}{(x - 1)^2}$$

Multiply throughout $(x + 1)(x - 1)^2$. Then

$$x^2 - x + 1 = A(x - 1)^2 + B(x + 1)(x - 1) + C(x + 1)$$

Putting $x = -1$, we get $3 = 4A$, $\therefore$ $A = \frac{3}{4}$

Putting $x = 1$, we get $1 = 2C$, $\therefore$ $C = \frac{1}{2}$

Equating co-efficients of x^2, we have

$$1 = \frac{3}{4} + B \quad \therefore \quad B = \frac{1}{4}$$

Hence, $$\frac{x^2 - x + 1}{(x + 1)(x - 1)^2} = \frac{3}{4(x + 1)} + \frac{1}{4(x - 1)} + \frac{1}{2(x - 1)^2}.$$

Example 5. *Break into partial fractions* $\frac{x^4 - 5x^3 + 10x^2 - 8x - 1}{(x - 1)^3 (x - 2)}$.

Sol. Since the highest degree term in the numerator and denominator both is x^4, therefore on division, the first term of the quotient will be 1, let us assume that

$$\frac{x^4 - 5x^3 + 10x^2 - 8x - 1}{(x - 1)^3 (x - 2)} = 1 + \frac{A}{x - 2} + \frac{B}{x - 1} + \frac{C}{(x - 1)^2} + \frac{D}{(x - 1)^3}$$

Multiply throughout by $(x - 1)^3(x - 2)$.

Then $x^4 - 5x^3 + 10x^2 - 8x - 1 = (x - 1)^3 (x - 2) + A(x - 1)^3 + B(x - 1)^2 (x - 2)$
$+ C(x - 1)(x - 2) + D(x - 2)$

Putting $x = 2$, we get $-1 = A$ $\therefore$ $A = -1$

Putting $x = 1$, we get $-3 = D(-1)$ $\therefore$ $D = 3$

Equation constant terms, $-1 = 2 - A - 2B + 2C + 2D$

or $-B + C = 1$

Putting $x = -1$, we get

$$2B - C = 0$$

From (*i*) and (*ii*), we get

$$B = 1, C = 2$$

Thus, the given fraction

$$= 1 - \frac{1}{x-2} + \frac{1}{x-1} + \frac{2}{(x-1)^2} + \frac{3}{(x-1)^3}.$$

Example 6. *Decompose into partial fractions* $\dfrac{x+3}{(1-x)^2\,(1+x^2)}$.

Sol. Since it is a proper fraction, we write

$$\frac{x+3}{(1-x)^2\,(x^2+1)} = \frac{A}{x-1} + \frac{B}{(x-1)^2} + \frac{Cx+D}{x^2+1}$$

Multiply throughout by $(x - 1)^2 (x^2 + 1)$,

then $\quad x + 3 = A(x - 1)(x^2 + 1) + B(x^2 + 1) + (Cx + D)(x - 1)^2$

Putting $x = 1$, we get $4 = B(1 + 1)$ $\therefore$ $B = 2$

Equating coefficients of x^3,

$$0 = A + C \qquad ...(i)$$

Equating coefficients of x^2

$$0 = - A + B + D - 2C \qquad ...(ii)$$

Equating constant terms,

$$3 = - A + B = D \qquad ...(iii)$$

[(*ii*) – (*iii*)] given $- 3 = - 2C$.

i.e., $\quad C = 3/2$

From (*i*), $\quad A = - C = - 3/2$

From (*iii*), $\quad 3 = \frac{3}{2} + 2 + D.$

$\therefore$ $\quad D = -\frac{1}{2}$

Hence, $\dfrac{x+3}{(1-x)^2\,(x^2+1)} = \dfrac{-3}{2(x-1)} + \dfrac{2}{(x-1)^2} + \dfrac{3x-1}{2(x^2+1)}$.

Example 7. *Resolve into partial fractions* $\dfrac{x^4 + 4x^3 + 7}{(x-1)(x^2+1)^2}$.

Sol. We write

$$\frac{x^4 + 4x^3 + 7}{(x-1)(x^2+1)^2} = \frac{A}{x-1} + \frac{Bx+C}{x^2+1} + \frac{Dx+E}{(x^2+1)^2}$$

Multiply both sides by $(x - 1)^2 (x^2 + 1)^2$.

Then, $\quad x^4 + 4x^3 + 7 = A(x^2 + 1)^2 + (Bx + C)(x + 1)(x^2 + 1) + (Dx + E)(x - 1)$

Putting $x = 1$, we have $\quad 12 = A(1 + 1)^2 \quad \therefore \quad A = 3$

Equating coefficients of x^4, $1 = A + B \quad \therefore \quad B = - 2$

Equating coeffcients of x^3, $4 = C - B \quad \therefore \quad C = 4 + B = 2$

Equating coefficients of x^2, $0 = 2A + B - C + D$ $\therefore$ $D = C - B - 2A = -2$
Equating constant terms, $7 = A - C - E.$ $\therefore$ $E = A - C - 7 = -6$

Hence, $$\frac{x^4 + 4x^3 + 7}{(x-1)(x^2+1)^2} = \frac{3}{(x-1)} + \frac{2+2x}{(x^2+1)} + \frac{2x+6}{(x^2+1)^2}.$$

Example 8. *Resolve into partial fractions* $\dfrac{3x^2 - 13x + 16}{(x-2)^3}$.

Sol. Here, we put

$$x - 2 = t, \quad x = t + 2$$

$$= \frac{3(t+2) - 13(t+2) + 16}{t^3} = \frac{3t^2 - t - 2}{t^2}$$

$$= \frac{3}{t} - \frac{1}{t^2} + \frac{1}{t^3} = \frac{3}{x-2} - \frac{1}{(x-2)^2} + \frac{2}{(x-2)^3}.$$

Example 9. *Resolve into partial fraction:*

$$\frac{(x^2+1)(x^2+2)}{(x^2+3)(x^2+4)} x^2$$

Sol.

$$\frac{(x^2+1)(x^2+2)}{(x^2+3)(x^2+4)} \text{ Let } x^2 = t$$

$$\frac{N(t)}{D(t)} = \frac{(t+1)(t+2)}{(t+3)(t+4)} = 1 - \frac{A}{t+3} + \frac{B}{t+4}$$

$\Rightarrow$ $$(t+1)(t+2) = (t+3)(t+4) + A(t+4) + B(t+4)$$

Putting $t = -3$ and $t = -4$, we have

$$A = 2, \quad B = -6$$

$\Rightarrow$ $$1 + \frac{2}{x^2+3} - \frac{6}{x^2+4}.$$

Example 10. *Resolve the fractions* $\dfrac{x^2 + 1}{(x^3 - 1)}$.

Sol. Let $$\frac{N(x)}{D(x)} = \frac{x^2+1}{(x-1)(x^2+x+1)}$$

$$= \frac{A}{x-1} + \frac{Bx+C}{(x^2+x+1)} \quad \ldots(i)$$

$$A = \left[\frac{x^2+1}{x^2+x+1}\right]_{x=1} = \frac{2}{3}$$

Let $x = 0$, and $x = -1$

We put $-1 = -A + C$

$$\Rightarrow \qquad C = -\frac{1}{3}$$

$$-1 = -\frac{1}{2}A - B + C$$

$$\Rightarrow \qquad B = \frac{1}{3}.$$

Thus, the partial fractions

$$= \frac{3}{3\,(x-1)} + \frac{1}{3}\left(\frac{x-1}{x^2+2+1}\right).$$

Example 11. *Solve the partial fractions* $\frac{x^2+1}{x^4+x^2+1}$.

Sol.
$$\frac{N(x)}{D(x)} = \frac{x^2+1}{(x^2+x+1)(x^2-x+1)}$$

$$= \frac{Ax+B}{x^2+x+1} + \frac{Cx+D}{x^2-x+1}$$

$$(x^2+1) = (Ax+B)\,(x^2-x+1) + (Cx+D)\,(x^2+x+1)$$

Equating co-efficients of x^3, x^2, x.

$$0 = A + C$$
$$1 = -A + B + C + D$$
$$0 = A - B + C + D$$
$$1 = B + D$$

Solving these equations, we get

$$A = 0, \quad B = \frac{1}{2}, \quad C = 0, \quad D = \frac{1}{2}$$

Thus,
$$\frac{1}{2(x^2+1+1)} + \frac{1}{2(x^2-x+1)}.$$

Example 12. *Solve into partial fractions* $\frac{3x^2-13x+16}{(x-2)^3}$.

Sol. We put $\quad x - 2 = t, \quad \text{or} \quad x = t + 2$

$$= \frac{3(t+2)^2 - 13(t+2) + 16}{t^3}$$

$$= \frac{3t^2 - t + 2}{t^3} = \frac{3}{t} - \frac{1}{t^2} + \frac{2}{t^3}$$

$$= \frac{3}{x-2} - \frac{1}{(x-x)^2} + \frac{2}{(x-2)^3}.$$

Example 13. *Solve:* $\frac{9x+25}{(x+3)^2}$.

Sol. $$\frac{9x+25}{(x+3)^2} = \frac{A}{x+3} + \frac{B}{(x+3)^2}$$

Now, $$\frac{9x+25}{(x+3)^2} = \frac{A(x+3)+B}{(x+3)^2}$$

$$9x + 25 = A(x+3) + B$$

Put $x = -3$

$$9(-3) + 25 = A(-3+3) + B$$
$$B = -2$$

Again, $x = 0$

$$25 = A(3) - 2$$
$$\Rightarrow \quad 27 = 3A$$
$$A = 9$$

$$\therefore \quad \frac{9x+25}{(x+3)^2} = \frac{9}{x+3} - \frac{2}{(x+3)^2}.$$

Example 14. *Solve:* $\dfrac{4x^2}{(x-1)(x-2)^2}$.

Sol. $$\frac{4x^2}{(x-1)(x-2)^2} = \frac{A}{x-1} + \frac{B}{(x-2)} + \frac{C}{(x-2)^2}$$

$$\Rightarrow \quad \frac{4x^2}{(x-1)(x-2)^2} = \frac{A(x-2)^2 + B(x-1)(x+2) + C(x-1)}{(x-1)(x-2)^2}$$

$$\Rightarrow \quad 4x^2 = A(x-2)^2 + B(x-1)(x-2) + C(x-1)$$

Put $x = 1, \Rightarrow 4 = A(-1)^2 \Rightarrow A = 4$

Put $x = 2, \Rightarrow 16 = C(1) \Rightarrow C = 16$

Now, all value of x, that would be easy to work with $(x = 0)$.

$$4(0)^2 = 4(-2)^2 = B(-1)(-2) + 16(-1)$$
$$0 = 16 + 2B - 16$$
$$2B = 0 \Rightarrow B = 0$$

$$\therefore \quad \frac{4x^2}{(x-1)(x-2)^2} = \frac{4}{x-1} + \frac{16}{(x-2)^2}.$$

Example 15. *Solve:* $\dfrac{9-9x}{2x^2+7x-4}$.

Sol. $$\frac{9-9x}{2x^2+7x-4} = \frac{9-9x}{(2x-1)(x+4)} = \frac{A}{(2x-1)} + \frac{B}{x+4}$$

Taking LCM:

$$\frac{9-9x}{2x^2+7x-4} = \frac{9-9x}{(2x-1)(x+4)} = \frac{A}{(2x-1)} + \frac{B}{x+4}$$

$$\Rightarrow \qquad 9 - 9x = A(x + 4) + B\,(2x - 1)$$

When $\quad x = -4 \;\Rightarrow\; 45 = -9B \qquad \therefore \quad B = -5$

When $\quad x = -\dfrac{1}{2} \;\Rightarrow\; \dfrac{9}{2} = A\left(\dfrac{9}{2}\right) \;\Rightarrow\; A = 1$

$$\therefore \qquad \frac{9 - 9x}{2x^2 + 7x - 4} = \frac{1}{2x - 1} - \frac{5}{x + 4}.$$

Example 16. *Solve:* $\dfrac{8p^2 - 12}{p\,(p^2 + 2p - 6)}$.

Sol.
$$\frac{8p^2 - 12}{p\,(p^2 + 2p - 6)} = \frac{A}{p} + \frac{Bp + C}{p^2 + 2p - 6}$$

Now, LCM is $p\,(p^2 + 2p - 6)$

$$\frac{8p^2 - 12}{p\,(p^2 + 2p - 6)} = \frac{1}{2}\,\frac{A\,(p^2 + 2p - 6) + p(Bp + C)}{p\,(p^2 + 2p - 6)}$$

Now, $\quad 8p^2 - 12 = A(p^2 + 2p - 6) + p(Bp + C)$

$$\Rightarrow \qquad 8p^2 - 12 = Ap^2 + 2Ap - 6A + Bp^2 + Cp$$

$$\Rightarrow \qquad 8p^2 - 12 = (A + B)p^2 + (2A + C)p = 6A$$

$$A + B = 8$$

$$2A + C = 0$$

Now, $\quad -6A = -12$

$$A = 2$$

$$2 + B = 8 \;\Rightarrow\; B = 6$$

$$2(2) + C = 0 \;\Rightarrow\; C = -4$$

$$\therefore \qquad \frac{8p^2 - 12}{p\,(p^2 + 2p - 6)} = \frac{2}{p} + \frac{6p - 4}{p^2 + 2p - 6}.$$

Example 17. *Solve:* $\dfrac{3x^2 + 7x - 4}{(x^2 + 2)^2}$.

Sol.
$$\frac{3x^2 + 7x - 4}{(x^2 + 2)^2} = \frac{Ax + B}{x^2 + 2} + \frac{Cx + D}{(x^2 + 2)^2}$$

$$\Rightarrow \qquad \frac{3x^2 + 7x - 4}{(x^2 + 2)^2} = \frac{(Ax + B)(x^2 + 2) + Cx + D}{(x^2 + 2)^2}$$

$$\Rightarrow \qquad 3x^2 + 7x - 4 = (Ax + B)\,(x^2 + 2) + Cx + D$$

$$\Rightarrow \qquad 3x^3 + 7x - 4 = Ax^3 + 2Ax + Bx^2 + 2B + Cx + D$$

$$\Rightarrow \qquad 3x^3 + 7x - 4 = Ax^3 + Bx^2 + (2A + C)\,x + 2B + D$$

$$A = 3$$

$$B = 0$$

$$2A + C = 7$$

$$2B + D = -4$$
$$2(3) + C = 7 \Rightarrow 6 + C = 7 \Rightarrow C = 1$$
$$2(0) + D = -4 \qquad \Rightarrow D = -4$$

$$\therefore \qquad \frac{3x^2 + 7x - 4}{(x^2 + 2)^2} = \frac{3x}{x^2 + 2} + \frac{x - 4}{(x^2 + 2)^2}.$$

Example 18. *Solve into partial fractions* $\dfrac{x^2 + 1}{(x + 1)^3 (x - 2)}$.

Sol. Put, $\quad x + 1 = t$

$$x = t - 1$$

$$= \frac{(t-1)^2 + 1}{t^3 (t - 3)} = \frac{t^2 - 2t + 2}{t^2 (t - 3)}$$

$$= -\frac{1}{t^3}\left(\frac{2 - 2t + t^3}{3 - t}\right) = -\frac{1}{t^3}\left\{\frac{2}{3} - \frac{4t}{9} + \frac{5t^2}{27} + \frac{5t^3}{27(3 - t)}\right\}$$

$$= -\frac{2}{3t^3} + \frac{4}{9t^2} - \frac{5}{27t} - \frac{5}{27(3 - t)}$$

$$= -\frac{2}{3(x+1)^3} + \frac{4}{9(x+1)^2} - \frac{3}{27(x + 1)} + \frac{5}{27(x - 2)}.$$

Example 19. *Solve* $\dfrac{8x - 42}{x^2 + 3x - 18}$.

Sol. $$\frac{8x - 42}{x^2 - 3x - 18} = \frac{8x - 42}{(x + 6)(x - 3)}$$

$$\frac{8x - 42}{x^2 - 3x - 18} = -\frac{A(x - 3)}{(x + 6)(x - 3)} + \frac{B(x + 6)}{(x + 6)(x - 3)} = \frac{A(x - 3) + B(x + 6)}{(x + 6)(x - 3)}$$

$\Rightarrow \qquad 8x - 42 = A(x - 3) + B(x + 4)$

Put $\qquad x = 3$

$\therefore \qquad 8(3) - 42 = A(3 - 3) + B(3 + 6)$

$$-18 = 9B$$
$$-2 = B$$

Put $\qquad x = -6$

$\therefore \qquad 8(-6) - 42 = A(-6 - 3) + B(-6 + 6)$

$$-90 = 9A$$
$$A = 10$$

$$\therefore \qquad \frac{8x - 42}{x^2 + 3x - 18} = \frac{10}{x + 6} + \frac{-2}{x - 3} = \frac{10}{x + 6} - \frac{2}{x - 3}.$$

Example 20. *Solve* $\dfrac{3x + 5}{(1 - 2x)^2}$.

Sol. $$\frac{3x + 5}{(1 - 2x)^2} = \frac{A}{(1 - 2x)^2} + \frac{B}{1 - 2x}$$

Now, $3x + 5 = A + B\,(1 - 2x)$

$$5 = A + B$$

$$3x = -\,2Bx$$

Solving, $A = \dfrac{13}{2}$ and $B = -\dfrac{3}{2}$

$$\therefore \quad \frac{3x + 5}{(1 - 2x)^2} = \frac{13/2}{(1 - 2x)^2} + \frac{-\,3/2}{(1 - 2x)}.$$

Example 21. *Solve* $f(x) = \dfrac{1}{x^2 + 2x - 3}$.

Sol. Let $q(x) = x^2 + 2x - 3 = (x + 3)\,(x - 1)$

Now, $$= \frac{1}{x^2 + 2x - 3} = \frac{A}{x + 3} + \frac{B}{x - 1}$$

$\Rightarrow$ $1 = A(x - 1) + B(x + 3)$

Put $x = -\,3,$

$$A = -\,1/4$$

When $x = 1, \quad B = 1/4$

$$\therefore \quad \frac{1}{x^2 + 2x - 3} = \frac{1}{4}\left(\frac{-\,1}{x + 3} + \frac{1}{x - 1}\right).$$

Example 22. *Solve* $f(x) = \dfrac{x^3 + 16}{x^3 - 4x^2 + 8x}$.

Sol. $$f(x) = 1 + \frac{4x^2 - 8x + 16}{x^3 + 4x^2 + 8x}$$

$$= 1 + \frac{4x^2 - 8x + 16}{x\,(x^2 + 4x + 8)}$$

Now, $x^2 - 4x + 8$

Thus, $D = b^2 - 4ac = 0$

Thus, $(-\,4)^2 - 4 \times 8 = -\,16$ is negative

$$\frac{4x^2 - 8x + 16}{x\,(x^2 + 4x + 8)} = \frac{A}{x} + \frac{Bx + C}{x^2 - 4x + 8}$$

Multiply by $x^3 - 4x^2 + 8x$

Thus, $4x^2 - 8x + 16 = A(x^2 - 4x + 8) + (Bx + C)x$

Put $x = 0, \quad 16 = 8A$

$$A = 2,$$

Comparing the x^2 coefficients.

$$4 = A + B = 2 + B \;\Rightarrow\; B = 2$$

Thus, $-\,8 = -\,4A + C = -\,8 + C$

$\therefore$ $C = 0$

$$f(x) = 1 + 2\left(\frac{1}{x} + \frac{x}{x^2 - 4x + 8}\right)$$

$$\Rightarrow \quad \frac{x}{x^2 - 4x + 8} = \frac{D}{x - (2 + 2i)} + \frac{E}{x - (2 - 2i)}$$

Multiply by the denominator

$$x = D(x - (2 - 2i)) + E(x - (2 - 2i))$$

$$f(x) = \frac{x^3 + 16}{x^3 - 4x^2 + 8x} = 1 + \frac{2}{x} + \frac{1 - i}{x - (2 + 2i)} + \frac{1 + i}{x - (2 - 2i)}$$

$$D = \frac{1 + i}{2i} = \frac{1 - i}{2i},$$

$$E = \frac{1 - i}{-2i} = \frac{1 + i}{2}.$$

Example 23. *Solve* $\dfrac{x^2 - 3x + 1}{(x - 1)^2 (x - 2)}$.

Sol.

$$\frac{x^2 - 3x + 1}{(x - 1)^2 (x - 2)} = \frac{A}{x - 1} + \frac{B}{(x - 1)^2} + \frac{C}{x - 2}$$

$$x^2 - 3x + 1 = A(x - 1)(x - 2) + B(x - 2) + C(x - 1)^2 \quad \ldots(i)$$

Putting, $x - 1 = 0 \Rightarrow x = 1$

$$(1)^2 - 3(1) + 1 = B(1 - 2)$$

$$1 - 3 + 1 = -B$$

$$B = 1$$

Putting, $x - 2 = 0 \Rightarrow x = 2$ in (i), we get

$$(2)^2 - 3(2) + 1 = C(2 - 1)^2$$

$$4 - 6 + 1 = C(1)^2$$

$$-1 = C$$

Now, $x^2 - 3x + 1 = A(x^2 - 3x + 2) + B(x - 2) + C(x^2 - 2x + 1)$

Comparing the co-efficient of like

$$A + C = 1$$

$$A = 1 - C = 1 - (-1) = 1 + 1 = 2$$

$$A = 2$$

$$\therefore \quad \frac{x^2 - 3x + 1}{(x - 1)^2 (x - 2)} = \frac{2}{x - 1} + \frac{1}{(x + 1)^2} + \frac{1}{x - 2}.$$

Example 24. *Solve* $\dfrac{4 + 7x}{(2 + 3x)(1 + x^2)} = \dfrac{A}{2 + 3x} + \dfrac{B}{1 + x} + \dfrac{C}{(1 + x^2)}$.

Sol. $4 + 7x = A(1 + x^2) + B(2 + 3x)(1 + x) + C(2 + 3x) \quad \ldots(i)$

Put $2 + 3x = 0$

$$\Rightarrow \quad x = -2/3$$

$$4 + 7(-2/3) = A(1 - 2/3)^2$$

$$\Rightarrow \quad 4 - 14/3 = A(-1/3)^2$$

$$-2/3 = 1/9A$$

$$A = -2/3 \times \frac{9}{1} = -6 \quad \boxed{A = 6}$$

Putting $1 + x = 0$, $\Rightarrow$ $x = -1$ in eq. (*i*), we get

$$4 + 7(-1) = (2 - 3)$$

$$\Rightarrow \quad \boxed{C = 3}$$

$$4 + 7x = A(x^2 + 2x + 1) + B(2 + 5x + 3x^2) + C(2 + 3x)$$

$$A + 3B = 0$$

$$\Rightarrow \quad -6 + 3B = 0 \quad \boxed{B = 0}$$

$$\therefore \quad \frac{4 + 7x}{(2 + 3x)(1 + x^2)} = \frac{-6}{2 + 3x} + \frac{2}{1 + x} + \frac{3}{(1 + x)^2}.$$

Example 25. *Solve* $\dfrac{x^2 + x + 2}{x^2 (x^2 + 3)^2}$.

Sol.

$$\frac{x^2 + x + 2}{x^2 (x^2 + 3)^2} = \frac{A}{x} + \frac{B}{x^2} + \frac{Cx + D}{x^2 + 3} + \frac{Ex + F}{(x^2 + 3)^2} \quad \ldots(i)$$

Taking L.C.M. $x^2 + x + 2 = Ax(x^2 + 3)^2 + B(x^2 + 3)^2 + (Cx + D)x^2 (x^2 + 3) + (Ex + F)x^2$

Put $x = 0$ on both sides

$$2 = B(0 + B)^2$$

$$2 = 9B$$

$$\Rightarrow \quad \boxed{B = 2/9}$$

$$x^2 + x + 2 = A(x^4 + 6x^2 + 9) + B(x^4 + 6x^2 + 9) + C(x^5 + 3x^2) + D(x^4 + 3x^2) + E(x^3) + Fx^2$$

$$\therefore \quad x^2 + x + 2 = (A + C)x^5 + (B + D)x^4 + (6A + 3C + E)x^3 + (6B + 3D + F)x^2 + (x + 9B)$$

Comparing the co-efficient of like powers of x on both sides of Eq. (1), we have

Co-efficient of x^5 : $A + C = 0$ ...(*i*)

Co-efficient of x^4 : $B - D = 0$ ...(*ii*)

Co-efficient of x^3 : $6A + 3C + E = 0$ ...(*iii*)

Co-efficient of x^2 : $6B + 3D + F = 1$ ...(*iv*)

Co-efficient of x : $9A = 1$ ...(*v*)

Co-efficient term : $9B = 1$ ...(*vi*)

From (*v*) $9A = 1$

$$\therefore \quad \boxed{A = \frac{1}{9}}$$

From (*i*) $A + C = 0$

$$C = -A$$

$$\therefore \quad \boxed{C = -\frac{1}{9}}$$

From (i) $\quad B + D = 0$

$$D = -B$$

$$\therefore \quad \boxed{D = -\frac{2}{9}}$$

From (iii) $\quad 6A + 3C + E = 6\left(\frac{1}{9}\right) + 3\left(-\frac{1}{9}\right) + E = 0$

$$E = \frac{3}{9} - \frac{6}{9}$$

$$\therefore \quad \boxed{E = -\frac{1}{3}}$$

From (iv), $\quad 6B + 3D + F = 1$

$$F = 1 - 6B - 3D$$

$$= 1 - 6\left(\frac{2}{9}\right) - 3\left(\frac{2}{9}\right) \quad \Rightarrow \quad F = 1/3$$

$$= \frac{\frac{1}{9}}{x} + \frac{\frac{2}{9}}{x^2} + \frac{-\frac{1}{9x} - \frac{2}{9}}{x^2 + 3} + \frac{-\frac{1}{3x} + \frac{1}{3}}{(x^2 + 3)^2}$$

$$= \frac{1}{9x} + \frac{2}{9x^2} - \frac{x + 2}{9x^2 + 3} - \frac{x - 1}{3(x^2 + 3)^2}.$$

Example 26. *Solve* $\dfrac{x^2}{(1 - x)(1 + x^2)^2}$.

Sol.

$$\frac{x^2}{(1 - x)(1 + x^2)^2} = \frac{A}{1 - x} + \frac{Bx + C}{(1 + x^2)} + \frac{Dx + E}{(1 + x^2)^2}$$

$$\Rightarrow \quad x^2 = A(1 + x^2)^2 + (Bx + C)(1 - x)(1 + x^2) + (Dx + E)(1 - x)$$

$$\Rightarrow \quad x^2 = A(1 + 2x^2 + x^4) + (Bx + C)(1 - x + x^2 - x^3) + (Dx + E)(1 - x) \quad \ldots(i)$$

Put $\quad 1 - x = 0$

$$(1)^2 = A(1 + (1)^2)$$

$$1 = 4A \qquad \boxed{A = 1/4}$$

$$x^2 = A(1 + 2x^2 + x^4) + B(x - x^2 + x^3 - x^4) + C(1 - x + x^2 - x^3) + D(x - x^2) + E(1 - x) \quad \ldots(ii)$$

Co-efficient of x^4 : $\quad A - B = 0 \quad \ldots(i)$

Co-efficient of x^3 : $\quad B - C = 0 \quad \ldots(ii)$

Co-efficient of x^2 : $\quad 2A - B + C - D = 1 \quad \ldots(iii)$

Co-efficient of x : $B - C + D - E = 0$ ...(*iv*)

Co-efficient term : $A + C + E = 0$ ...(*v*)

$$B = 1/4, \quad A = 1/4$$

From (*i*), $C = 1/4$

From (*iii*), $D = 2A - B + C - 1 = 2\left(\frac{1}{4}\right) - \frac{1}{4} + \frac{1}{4} - 1$

$$D = -\frac{1}{2}$$

From (*v*), $E = -A - C \Rightarrow -\frac{1}{4} - \frac{1}{4} = -\frac{1}{2}$

$$\frac{1/4}{1 - x} + \frac{\frac{1}{4}x + \frac{1}{4}}{1 + x^2} + \frac{1/2x - 1/2}{(1 + x^2)^2}$$

$$\therefore \quad \frac{x^2}{(1 - x)(1 + x^2)^2} = \frac{1}{4(1 - x)} + \frac{x + 1}{4(1 + x^2)} - \frac{x + 1}{2(1 + x^2)^2}.$$

EXERCISE 1.1.3.

Solve:

1. $\dfrac{(1 + 2t)(1 + 3t)(1 + 4t)}{(1 - 2t)(1 - 2t)(1 - 4t)}$ $\left[\textbf{Ans. } -1 + \dfrac{30}{1 - 2t} - \dfrac{70}{1 - 3t} + \dfrac{42}{1 - 4t}\right]$

2. $\dfrac{x^3 - 6x^2 + 13x - 5}{(x - 1)(x - 2)(x - 3)}$ $\left[\textbf{Ans. } 1 + \dfrac{3}{2(x - 1)} - \dfrac{5}{x - 2} + \dfrac{7}{2(x - 3)}\right]$

3. $\dfrac{s(s + 1)(s + 2)}{(s + 3)(s + 4)(s + 5)}$ $\left[\textbf{Ans. } 1 - \dfrac{3}{s + 3} + \dfrac{24}{s + 4} - \dfrac{30}{s + 5}\right]$

4. $\dfrac{6t^3 + t^2 - 7}{3t^2 - 2t - 1}$ $\left[\textbf{Ans. } 2t + 3 + \dfrac{1}{t - 1} + \dfrac{5}{3t + 1}\right]$

5. $\dfrac{p - 4}{(p - 4)(p^2 - 3p + 2)}$ $\left[\textbf{Ans. } \dfrac{3}{5(p - 1)} - \dfrac{4}{15(p + 4)} - \dfrac{1}{3(p - 2)}\right]$

6. $\dfrac{(s - 1)(s - 2)}{(s - 3)(s - 4)}$ $\left[\textbf{Ans. } 1 - \dfrac{2}{s - 3} + \dfrac{6}{s - 4}\right]$

7. $\dfrac{6t^2 - t - 10}{(t - 1)(t - 2)(t + 4)}$ $\left[\textbf{Ans. } \dfrac{1}{t - 1} + \dfrac{2}{t - 2} + \dfrac{3}{t + 4}\right]$

8. $\dfrac{2x - 1}{(x - 1)(2x - 3)}$ $\left[\textbf{Ans. } \dfrac{4}{2x - 3} - \dfrac{1}{x - 1}\right]$

9. $\dfrac{t^2 + 1}{(t^2 + 2)(t^2 + 3)(t^2 + 4)}$ $\left[\textbf{Ans. } \dfrac{2}{t^2 - 3} - \dfrac{1}{2(t^2 + 2)} - \dfrac{3}{2(t^2 + 4)}\right]$

10. $\dfrac{t^6 + 3t^4 + 3t^2 + 5}{(t^2 + 1)(t^2 + 2)}$ $\left[\textbf{Ans. } t^2 + \dfrac{4}{t^2 + 1} - \dfrac{3}{t^2 + 2}\right]$

11. $\dfrac{y + 1}{(y^2 + 1)(y - 1)^2}$ $\left[\textbf{Ans. } \dfrac{y - 1}{2(y^2 + 1)} - \dfrac{1}{2(y - 1)} + \dfrac{1}{(y - 1)^2}\right]$

12. $\dfrac{x^4 + 1}{(x - 1)(x^2 + 1)}$ $\left[\textbf{Ans. } x + 1 + \dfrac{1}{x - 1} - \dfrac{x + 1}{x^2 + 1}\right]$

13. $\dfrac{s^2 + 1}{s^3 + 1}$ $\left[\textbf{Ans. } \dfrac{2}{3(s + 1)} + \dfrac{s + 1}{3(s^2 - s + 1)}\right]$

14. $\dfrac{s - 4}{(s^2 - 3s + 2)(s^2 + 4)}$ $\left[\textbf{Ans. } \dfrac{1}{20}\left(\dfrac{1^2}{s - 1} - \dfrac{5}{s - 2} - \dfrac{7s + 2}{s^2 + 4}\right)\right]$

15. $\dfrac{1}{(p^2 + p)(p^2 - 1)}$ $\left[\textbf{Ans. } \dfrac{1}{4(p - 1)} - \dfrac{1}{p} + \dfrac{3}{4(p + 1)} + \dfrac{1}{2(p + 1)^2}\right]$

16. $\dfrac{t^2}{(t - 1)^3 (t + 1)}$ $\left[\textbf{Ans. } \dfrac{1}{8(t - 1)} - \dfrac{1}{8(t + 1)} + \dfrac{3}{4(t - 1)^2} + \dfrac{1}{2(t - 1)^3}\right]$

17. $\dfrac{2t^2 + 3}{(t - 1)(t + 1)^2}$ $\left[\textbf{Ans. } \dfrac{5}{4(t - 1)} + \dfrac{3}{4(t + 1)} - \dfrac{1}{2(t + 1)^2}\right]$

18. $\dfrac{3k^2 - 8k^2 + 10}{(k - 1)^4}$ $\left[\textbf{Ans. } \dfrac{3}{k - 1} + \dfrac{1}{(k - 1)^2} - \dfrac{7}{(k - 1)^3} - \dfrac{5}{(k + 1)^4}\right]$

19. $\dfrac{1}{p^6 - 1}$

[**Hint:**

$$\frac{1}{p^6 - 1} = \frac{1}{(p^3 - 1)(p^3 + 1)} = \frac{1}{2}\left(\frac{1}{p^3 - 1} - \frac{1}{p^3 + 1}\right)$$

$$= \frac{1}{2(p - 1)(p^2 + p + 1)} - \frac{1}{2(p + 1)(p^2 - p + 1)}$$

Now, let

$$\frac{1}{(p - 1)(p^2 + p + 1)} = \frac{A}{p - 1} + \frac{Bp + C}{p^2 + p + 1}$$

and $$\frac{1}{(p + 1)(p^2 - p + 1)} = \frac{D}{p + 1} + \frac{Ep + F}{p^2 + p + 1}$$

$$\left[\textbf{Ans. } \frac{1}{2}\left[\frac{1}{3(p - 1)} - \frac{p + 2}{3(p^2 + p + 1)}\right] - \frac{1}{2}\left[\frac{1}{3(p + 1)} - \frac{p - 2}{3(p^2 - p + 1)}\right]\right]$$

20. $\dfrac{a(r^2 - 2a^2)}{r^4 + 4a^4}$ $\left[\textbf{Ans. } \dfrac{1}{2}\left(\dfrac{r-a}{r^2 - 2ar + 2a^2} - \dfrac{r}{r^2 + 2ar + 2a^2}\right)\right]$

21. $\dfrac{p}{p^4 + 4a^4}$ $\left[\textbf{Ans. } \dfrac{1}{4a}\left(\dfrac{1}{p^2 - 2ap + 2a^2} - \dfrac{1}{p^2 + 2a + 2a^2}\right)\right]$

22. $\dfrac{p}{p^4 + p^2 + 1}$ $\left[\textbf{Ans. } \dfrac{1}{2}\left(\dfrac{1}{p^2 - p + 1} - \dfrac{1}{p^2 + p + 1}\right)\right]$

23. $\dfrac{1}{x^4 - a^2}$ $\left[\textbf{Ans. } \dfrac{1}{2a^2}\left\{\dfrac{1}{2a}\left(\dfrac{1}{x-1} - \dfrac{1}{x-a}\right) - \dfrac{1}{x^2 + a^2}\right\}\right]$

24. $\dfrac{y^2 + 4y}{(1-y)(1+y^2)^2}$ $\left[\textbf{Ans. } \dfrac{5}{4(1-y)} + \dfrac{5(y+1)}{4(1+y^2)} + \dfrac{1}{2}\dfrac{(3y-5)}{(1+y^2)^2}\right]$

25. $\dfrac{1}{(p^2 + 1)^2 (p^2 - 3)}$ $\left[\textbf{Ans. } \dfrac{1}{2(p^2+1)^2} - \dfrac{1}{4(p^2+1)} + \dfrac{1}{4(p^2+3)}\right]$

26. $\dfrac{p^2 - p + 1}{(p-1)^2 (p-2)(p^2+1)}$ $\left[\textbf{Ans. } \dfrac{3}{5(p-2)} - \dfrac{1}{2(p-1)^2} - \dfrac{1}{2(x-1)} - \dfrac{x+2}{10(x^2+1)}\right]$

27. $\dfrac{2p^2 - 11p + 5}{(p-3)(p^2 + 2p + 5)}$ $\left[\textbf{Ans. } \dfrac{3p}{p^2 + 2p - 5} - \dfrac{1}{p-3}\right]$

28. $\dfrac{p^4}{p^3 + 1}$ $\left[\textbf{Ans. } p + \dfrac{1}{3(p+1)} - \dfrac{p+1}{3(p^2 - p + 1)}\right]$

29. $\dfrac{p^2 - 3}{(p+2)(p^2+1)}$ $\left[\textbf{Ans. } \dfrac{5}{5(p+2)} + \dfrac{4(p-2)}{5(p^2+1)}\right]$

30. $\dfrac{r^5}{(r^2 - 1)(r+2)^2}$ $\left[\textbf{Ans. } r - \dfrac{4}{3} + \dfrac{1}{2(r+1)} + \dfrac{1}{18(r-1)} - \dfrac{32}{9(r-2)} + \dfrac{32}{3(r+2)^2}\right]$

31. $\dfrac{1}{(1-2q)^2 (1-3q)}$ $\left[\textbf{Ans. } \dfrac{9}{1-3q} - \dfrac{6}{1-2q} - \dfrac{2}{(1-2q)^2}\right]$

32. $\dfrac{p-1}{(p+1)(p-2)^2}$ $\left[\textbf{Ans. } \dfrac{2}{9(p-2)} + \dfrac{1}{3(p-2)^2} - \dfrac{2}{9(p+1)}\right]$

33. $\dfrac{p^3 - 5}{p^2 - 3p + 2}$ $\left[\textbf{Ans. } p + 3 + \dfrac{3}{p-1} + \dfrac{3}{p-2}\right]$

34. $\dfrac{q+4}{(q^2 - 4)(q+1)}$ $\left[\textbf{Ans. } \dfrac{2}{2(q-2)} + \dfrac{1}{2(q+2)} - \dfrac{1}{q+1}\right]$

35. $\dfrac{1 - q + 6q^2}{q - q^3}$ $\left[\textbf{Ans. } \dfrac{1}{q} + \dfrac{3}{1-q} - \dfrac{4}{1+q}\right]$

36. $\dfrac{2x^3 + x^2 - x - 3}{x(x-1)(2x+3)}$ $\left[\textbf{Ans. } 1 + \dfrac{1}{x} - \dfrac{1}{5(x-1)} - \dfrac{8}{5(2x+3)}\right]$

37. $\dfrac{x^4}{(x-1)(x-2)(x-3)}$ $\left[\textbf{Ans. } x + 6 + \dfrac{1}{2(x-1)} - \dfrac{16}{x-2} + \dfrac{81}{2(x-3)}\right]$

38. $\dfrac{9x^2 - 9x + 6}{(x-1)(2x-1)(x+12)}$ $\left[\textbf{Ans. } \dfrac{2}{x-1} - \dfrac{3}{2x-1} + \dfrac{4}{x+12}\right]$

39. $\dfrac{6x + 27}{4x^3 - 9x}$ $\left[\textbf{Ans. } \dfrac{3}{x} + \dfrac{4}{2x-3} + \dfrac{2}{2x+3}\right]$

40. $\dfrac{1}{(1-x)(1-2x)(1-3x)}$ $\left[\textbf{Ans. } \dfrac{1}{2(1-x)} - \dfrac{4}{(1-2x)} + \dfrac{9}{2(1-3x)}\right]$

41. $\dfrac{2x^3 - x^2 + 1}{(x-3)(x-1)(x-3)}$ $\left[\textbf{Ans. } \dfrac{2+31}{4(x+3)} + \dfrac{1}{12(x-1)} - \dfrac{137}{6(x+5)}\right]$

42. $\dfrac{1}{(1-ax)(1-bx)(1-cx)}$

$$\left[\textbf{Ans. } \frac{a^2}{(a-b)(b-c)(1-ax)} + \frac{b^2}{(b-a)(b-c)(1-bx)} + \frac{c^2}{(c-b)(c-a)(1-cx)}\right]$$

43. $\dfrac{x}{(x-a)(x-b)(x-c)}$ $\left[\textbf{Ans. } \dfrac{a}{(a-b)(a-c)(x-a)}\right]$

44. $\dfrac{(x-1)(x-2)(x-3)}{(x-4)(x-5)(x-6)}$ $\left[\textbf{Ans. } 1 + \dfrac{3}{x-4} - \dfrac{24}{x-5} + \dfrac{30}{x-6}\right]$

45. $\dfrac{3x^2 - 2x - 5}{(x-2)(x+2)(x+3)}$ $\left[\textbf{Ans. } \dfrac{3}{20(x-2)} - \dfrac{11}{4(x-2)} + \dfrac{28}{5(x+3)}\right]$

46. $\dfrac{2x + 5}{x^2 + 5x + 6}$ $\left[\textbf{Ans. } \dfrac{1}{x+2} + \dfrac{1}{x+3}\right]$

47. $\dfrac{2x + 3}{(x-2)(x+5)}$ $\left[\textbf{Ans. } \dfrac{1}{x-2} + \dfrac{1}{x+5}\right]$

48. $\dfrac{x + 4}{(x-2)^2(x+1)}$ $\left[\textbf{Ans. } -\dfrac{1}{3(x-2)} + \dfrac{2}{(x-2)^2} + \dfrac{1}{3(x+1)}\right]$

49. $\dfrac{1}{(x+1)(x^2-1)}$ $\left[\textbf{Ans. } \dfrac{1}{4(x-1)} - \dfrac{1}{4(x+1)} - \dfrac{1}{2(x+1)^2}\right]$

50. $\dfrac{4x^3}{(x+1)^2(x^2-1)}$ $\left[\textbf{Ans. } \dfrac{1}{2(x-1)} + \dfrac{7}{2(x+1)} - \dfrac{5}{(x+1)^2} + \dfrac{2}{(x-1)^3}\right]$

51. $\dfrac{2x+1}{(x+3)(x-1)(x+2)^2}$ $\left[\textbf{Ans. } \dfrac{5}{4(x+3)} + \dfrac{1}{12(x-1)} - \dfrac{4}{3(x+2)} + \dfrac{1}{(x+2)^2}\right]$

52. $\dfrac{6x^2-11x-32}{(x+6)(x+1)^2}$ $\left[\textbf{Ans. } \dfrac{10}{x+6} - \dfrac{4}{x+1} - \dfrac{3}{(x-1)^2}\right]$

53. $\dfrac{x^2-x-3}{(x-1)^3}$ $\left[\textbf{Ans. } \dfrac{1}{x-1} + \dfrac{1}{(x-1)^2} - \dfrac{3}{(x-3)^2}\right]$

54. $\dfrac{5x^2+36x-27}{x^4-6x^3+9x^2}$ $\left[\textbf{Ans. } \dfrac{2}{x} - \dfrac{3}{x^2} - \dfrac{2}{(x-3)} + \dfrac{14}{(x-3)^2}\right]$

55. $\dfrac{4x^2-13x}{(x+3)(x-2)^2}$ $\left[\textbf{Ans. } \dfrac{3}{x+3} + \dfrac{1}{x-2} - \dfrac{2}{(x-2)^2}\right]$

56. $\dfrac{x^4+1}{x^2(x-1)}$ $\left[\textbf{Ans. } x+1-\dfrac{1}{x} - \dfrac{1}{x^2} + \dfrac{2}{x-1}\right]$

57. $\dfrac{x^3-8x^2+17x+1}{(x-3)^3}$ $\left[\textbf{Ans. } 1+\dfrac{1}{x-3} - \dfrac{4}{(x-3)^2} + \dfrac{7}{(x-3)^3}\right]$

58. $\dfrac{2x+1}{(x+1)(x-3)^2}$ $\left[\textbf{Ans. } \dfrac{-3}{25(x+2)} + \dfrac{3}{25(x-3)} + \dfrac{7}{5(x-3)^2}\right]$

59. $\dfrac{x^2+3x-1}{(x-2)(x^2+5)}$ $\left[\textbf{Ans. } \dfrac{1}{x-2} + \dfrac{3}{x^2+5}\right]$

60. $\dfrac{x^2-x+2}{(x+1)(x^2+3)}$ $\left[\textbf{Ans. } \dfrac{1}{x+1} - \dfrac{x}{x^2+3}\right]$

61. $\dfrac{3x+7}{(x+3)(x^2+1)}$ $\left[\textbf{Ans. } -\dfrac{1}{5(x+3)} + \dfrac{x+12}{5(x^2+1)}\right]$

62. $\dfrac{1}{(x^3+1)}$ $\left[\textbf{Ans. } \dfrac{1}{3(x+1)} - \dfrac{(x-2)}{3(x^2-x+1)}\right]$

63. $\dfrac{1}{(x+1)(x^2+1)}$ $\left[\textbf{Ans. } \dfrac{1}{2(x+1)} - \dfrac{x+1}{2(x^2+1)}\right]$

64. $\dfrac{3x+7}{(x^2+x+1)(x^2-4)}$ $\left[\textbf{Ans. } \dfrac{13}{28(x+2)} - \dfrac{1}{12(x+2)} - \dfrac{8x+31}{21(x^2+x+1)}\right]$

65. $\dfrac{3x^2-x+1}{(x+1)(x^2-x+3)}$ $\left[\textbf{Ans. } \dfrac{1}{x+1} + \dfrac{2x-2}{x^2-x+3}\right]$

66. $\dfrac{x+a}{x^2(x-a)(x^2+a^2)}$ $\left[\textbf{Ans. } \dfrac{1}{a^3}\left(\dfrac{1}{x-a} + \dfrac{x}{x^2 a^2} - \dfrac{2}{x} - \dfrac{a}{x^2}\right)\right]$

67. $\dfrac{x^5}{x^4-1}$ $\left[\textbf{Ans. } x+\dfrac{1}{4(x-1)}+\dfrac{1}{4(x+1)}-\dfrac{x}{2(x^2+1)}\right]$

68. $\dfrac{x^2+x+1}{(x^2-x-2)(x^2-2)}$ $\left[\textbf{Ans. } \dfrac{1}{3(x+1)}+\dfrac{7}{6(x-2)}-\dfrac{3x+2}{2(x^2-2)}\right]$

69. $\dfrac{1}{x^3-1}$ $\left[\textbf{Ans. } \dfrac{1}{3(x-1)}-\dfrac{x+2}{3(x^2+x+1)}\right]$

70. $\dfrac{x^2+3x+3}{(x^2-1)(x^2+4)}$ $\left[\textbf{Ans. } \dfrac{7}{10(x-1)}-\dfrac{1}{10(x+1)}-\dfrac{3x-1}{5(x^2+4)}\right]$

71. $\dfrac{7}{(x+1)(x^2+2)^2}$ $\left[\textbf{Ans. } \dfrac{7}{9(x+1)}-\dfrac{7x-7}{9(x^2+2)}-\dfrac{7x-7}{3(x^2+2)^2}\right]$

72. $\dfrac{x^2}{(1+x)(1+x^2)^2}$ $\left[\textbf{Ans. } \dfrac{1}{4(1+x)}-\dfrac{x-1}{4(1+x^2)}+\dfrac{x-1}{2(1+x^2)^2}\right]$

73. $\dfrac{5x^2+3x+9}{x(x^2+3)^2}$ $\left[\textbf{Ans. } \dfrac{1}{x}-\dfrac{x}{x^2+3}+\dfrac{2x+3}{(x^2+3)^2}\right]$

74. $\dfrac{4x^4+3x^3+6x^2+5x}{(x-1)(x^2+x+1)^2}$ $\left[\textbf{Ans. } \dfrac{2}{x-1}+\dfrac{2x-1}{x^2+x+1}+\dfrac{3}{(x^2+x+1)^2}\right]$

75. $\dfrac{2x^4-3x^2-4x}{(x+1)(x^2+2)^2}$ $\left[\textbf{Ans. } \dfrac{1}{3(x+1)}+\dfrac{5(x-1)}{3(x^2+2)}-\dfrac{2(3x-1)}{(x^2+1)^2}\right]$

76. $\dfrac{x^3-15x^2-8x-7}{(2x-5)(1+x^2)^2}$ $\left[\textbf{Ans. } -\dfrac{2}{2x-5}+\dfrac{x+3}{1+x^2}+\dfrac{x-2}{(1+x^2)^2}\right]$

77. $\dfrac{49}{(x-2)(x^2+3)^2}$ $\left[\textbf{Ans. } \dfrac{1}{x-2}-\dfrac{x+2}{x^2+3}-\dfrac{7x+14}{(x^2+3)^2}\right]$

78. $\dfrac{8x^2}{(1-x^2)(1+x^2)^2}$ $\left[\textbf{Ans. } \dfrac{1}{1-x}+\dfrac{1}{1+x}+\dfrac{2}{1+x^2}-\dfrac{4}{(1+x^2)^2}\right]$

79. $\dfrac{x^4+x^3+2x^2-7}{(x+2)(x^2+x+1)^2}$ $\left[\textbf{Ans. } \dfrac{1}{x+2}+\dfrac{2x-3}{(x^2+x+1)^2}-\dfrac{1}{x^2+x+1}\right]$

80. $\dfrac{x^2+2}{(x^2+1)(x^2+4)^2}$ $\left[\textbf{Ans. } \dfrac{1}{9(x^2+1)}-\dfrac{1}{9(x^2+4)}+\dfrac{2}{3(x^2+4)^2}\right]$

81. $\dfrac{1}{x^4+x^2+1}$ $\left[\textbf{Ans. } -\dfrac{(x-1)}{2(x^2-x+1)}+\dfrac{(x+1)}{2(x^2+x+1)}\right]$

1.3. DETERMINANTS

1.3.1. Definition

Let, Corresponding to each square matrix $A = \begin{vmatrix} a_{11} & a_{12} & a_{13} & ...a_{1n} \\ a_{21} & a_{22} & a_{23} & ...a_{2n} \\ ... & ... & ... & ... \\ ... & ... & ... & ... \\ a_{n-1} & a_{n-2} & a_{n-3} & ...a_{nm} \end{vmatrix}$

There is associated an expression called the determinant of A, denoted by det A or | A |, *i.e.*,

$$\det A = |A| = \begin{vmatrix} a_{11} & a_{12} & a_{13} & ...a_{1n} \\ a_{21} & a_{22} & a_{23} & ...a_{2n} \\ ... & ... & ... & ... \\ a_{n_1} & a_{n_2} & a_{n_3} & ...a_{nm} \end{vmatrix}$$

A determinant is an arrangement of numbers and thus each determinant has a fixed value. A determinant having n rows and n columns is called a determinant of order n.

- **Order of Determinant**

Order '1': The determinant of 'a' (1 × 1) matrix $[a]$ is given as $|\ a\ | = a$.

Order '2': We have, $\begin{vmatrix} a_{11} & a_{12} \\ a_{21} & a_{22} \end{vmatrix} = (a_{11}a_{22} - a_{21}a_{12})$

The expression $\begin{vmatrix} a_1 & b_1 \\ a_2 & b_2 \end{vmatrix}$ is called determinant of second order, when a_1, b_1, a_2, b_2 are called *elements*, which are arranged along two horizontal lines called *rows* and two vertical lines are called *columns*.

Order 3 or more: Value of a determinant of order 3 or more to solve the following method.

i.e., $\begin{vmatrix} a_1 & b_1 & c_1 \\ a_2 & b_2 & c_2 \\ a_3 & b_3 & c_3 \end{vmatrix}$ is called determinant of the *third order*.

1.3.2. Properties of Determinant

The following properties enable us to evaluate a determinant without expanding the same.

Property 1. *A determinant remains unaltered by changing its rows into columns and columns into rows.*

i.e.,
$$\Delta = \begin{vmatrix} a_1 & b_1 & c_1 \\ a_2 & b_2 & c_2 \\ a_3 & b_3 & c_3 \end{vmatrix} = \begin{vmatrix} a_1 & a_2 & a_3 \\ b_1 & b_2 & b_3 \\ c_1 & c_2 & c_3 \end{vmatrix}$$

Property 2. *If any two rows (columns) of a determinant are interchanged the determinant retains its numerical value but changes in sign.*

Let
$$\Delta = \begin{vmatrix} a_1 & b_1 & c_1 \\ a_2 & b_2 & c_2 \\ a_3 & b_3 & c_3 \end{vmatrix} \quad \text{[Expand by } R_1\text{]}$$

$$= a_1(b_2c_3 - b_3c_2) - b_1(a_2c_3 - a_3c_2) + c_1(a_2b_3 - a_3b_2)$$

Interchanging C_2 and C_3, we have

$$\Delta' = \begin{vmatrix} a_1 & c_1 & b_1 \\ a_2 & c_2 & b_2 \\ a_3 & c_3 & b_3 \end{vmatrix} \quad \text{[Expand by } R_1]$$

$$= a_1(c_2b_3 - c_3b_2) - c_1(a_2b_3 - a_3b_2) + b_1(a_2c_3 - a_3c_2)$$

$$= -\{a_1(b_2c_3 - b_3c_2) - b_1(a_2c_3 - a_3c_2) + c_1(a_2b_3 - a_3b_2)\}$$

$$= -\Delta.$$

Property 3. *A determinant vanishes, if two rows (or columns) are identical.* Consider a determinant Δ in which two parallel lines are identical. Interchange of the identical lines leaves the determinant unaltered yet by the previous property, the interchange of two parallel lines changes the sign of the determinant.

Hence, $\Delta = \Delta' = -\Delta$ or $\Delta = 0$.

Property 4. Factor Theorem. *If the elements of a determinant Δ are functions of x and two rows (or columns) become identical when $x = a$, then $x - a$ is a factor of Δ.*

Property 5. *If each element of a line be multiplied by the same factor, the whole determinant is multiplied by that factor.*

i.e.,
$$\begin{vmatrix} a_1 & pb_1 & c_1 \\ a_2 & pb_2 & c_2 \\ a_3 & pb_3 & c_3 \end{vmatrix} = p\begin{vmatrix} a_1 & b_1 & c_1 \\ a_2 & b_2 & c_2 \\ a_3 & b_3 & c_3 \end{vmatrix}$$

Similarly,
$$\begin{vmatrix} a_1 & b_1 & c_1 \\ ka_2 & kb_2 & kc_2 \\ a_3 & b_3 & c_3 \end{vmatrix} = k\begin{vmatrix} a_1 & b_1 & c_1 \\ a_2 & b_2 & c_2 \\ a_3 & b_3 & c_3 \end{vmatrix}.$$

Property 6. *If each element of a line consists of m terms, the determinant can be expressed as the sum of m determinants.*

e.g.,
$$\begin{vmatrix} a_1 & b_1 & c_1 + d_1 - e_1 \\ a_2 & b_2 & c_2 + d_2 - e_2 \\ a_3 & b_3 & c_3 + d_3 - e_3 \end{vmatrix}$$

$$= \begin{vmatrix} a_1 & b_1 & c_1 \\ a_2 & b_2 & c_2 \\ a_3 & b_3 & c_3 \end{vmatrix} + \begin{vmatrix} a_1 & b_1 & d_1 \\ a_2 & b_2 & d_2 \\ a_3 & b_3 & d_3 \end{vmatrix} - \begin{vmatrix} a_1 & b_1 & e_1 \\ a_2 & b_2 & e_2 \\ a_3 & b_3 & e_3 \end{vmatrix}.$$

Property 7. *If to each element of a line be added equi-multiples of the corresponding elements of one or more parallel lines, the determinant remains unaltered.*

Let
$$\Delta = \begin{vmatrix} a_1 & b_1 & c_1 \\ a_2 & b_2 & c_2 \\ a_3 & b_3 & c_3 \end{vmatrix}$$

Then
$$\Delta' = \begin{vmatrix} a_1 + pb_1 - qc_1 & b_1 & c_1 \\ a_2 + pb_2 - qc_2 & b_2 & c_2 \\ a_3 + pb_3 - qc_3 & b_3 & c_3 \end{vmatrix}$$

$$= \begin{vmatrix} a_1 & b_1 & c_1 \\ a_2 & b_2 & c_2 \\ a_3 & b_3 & c_3 \end{vmatrix} + \begin{vmatrix} pb_1 & b_1 & c_1 \\ pb_2 & b_2 & c_2 \\ pb_3 & b_3 & c_3 \end{vmatrix} + \begin{vmatrix} -qc_1 & b_1 & c_1 \\ -qc_2 & b_2 & c_2 \\ -qc_3 & b_3 & c_3 \end{vmatrix}$$

$$= \Delta + p \,.\, 0 - q \,.\, 0$$

$$= \Delta.$$

Obs. *This property is very useful for simplifying determinants.* To add equi-multiples of parallel lines, we shall employ the following notation:

Suppose to the elements of the second row, we add p times the elements of the first row and q times the element of the third row; then we say:

Operate $R_2 + pR_1 + qR_3$.

Similarly Operate '$C_3 + mC_1 - nC_2$'

means that to the elements of the third column add m times the elements of the first column and $-n$ times the elements of the second column.

1.3.3. Expansion of Determinants of Order 2 and 3

A determinant can be expanded in terms of any row (or column) as follows.

Multiply each elements of the row (or column) in terms of which we intend expanding the determinant, by its cofactor and then add up all these terms.

Expanding the determinant by Ist row, we get

$$= \begin{vmatrix} a_{11} & a_{12} & a_{13} \\ a_{21} & a_{22} & a_{23} \\ a_{31} & a_{32} & a_{33} \end{vmatrix}$$

$$= a_{11} \text{ (its cofactor)} + a_{12} \text{ (its cofactor)} + a_{13} \text{ (its cofactor)}$$

$$= a_{11} \begin{vmatrix} a_{22} & a_{23} \\ a_{32} & a_{33} \end{vmatrix} - a_{12} \begin{vmatrix} a_{21} & a_{23} \\ a_{31} & a_{33} \end{vmatrix} + a_{13} \begin{vmatrix} a_{21} & a_{22} \\ a_{31} & a_{32} \end{vmatrix}$$

$$= a_{11} . (a_{22}\, a_{33} - a_{32}\, a_{23}) - a_{12}\, (a_{21}\, a_{33} - a_{31}\, a_{23}) + a_{13}\, (a_{21}\, a_{32} - a_{31}\, a_{22})$$

We may expand it by any row or column.

Expansion along First Row (R_1)

Step 1. Multiply first element a_{11} of R_1 by $(-1)^{(1+1)}$ $[-1)^{\text{sum of suffixes in }} a_{11}]$ and with the second order determinant obtained by deleting the elements of first row (R_1) and first column (C_1) of $|A|$ as a_{11} lies in R_1 and C_1.

i.e., $$(-1)^{1+1}\, a_{11} \begin{bmatrix} a_{22} & a_{23} \\ a_{32} & a_{33} \end{bmatrix}$$

Step 2. Multiply 2nd element a_{12} by $(-1)^{1+2}$ $[(-1)^{\text{sum of suffixes in }} a_{12}]$ and the second or R_1 order determinant obtained by deleting elements of first row (R_1) and 2nd column (C_2) of $|A|$ as a_{12} lies in R_1 and C_2.

i.e., $$(-1)^{1+2}\, a_{12} \begin{bmatrix} a_{21} & a_{23} \\ a_{31} & a_{33} \end{bmatrix}$$

Step 3. Multiply third element a_{13} of R_1 by $(-1)^{1+3}$ $[(-1)^{\text{sum of suffixes in }} a_{13}]$ and the second order determinant obtained by deleting elements of first row (R_1) and third column (C_3) of $|A|$ as a_{13} lies in R_1 and C_3.

i.e., $$(-1)^{1+3}\, a_{13} \begin{bmatrix} a_{21} & a_{22} \\ a_{31} & a_{32} \end{bmatrix}$$

Step 4. Now the expansion of determinant of A, that is, $|A|$ written as sum of all three terms obtained in steps 1, 2, and 3 above is given by

$$\det A = |A| = (-1)^{1+1} a_{11} \begin{bmatrix} a_{22} & a_{23} \\ a_{32} & a_{33} \end{bmatrix} + (-1)^{1+2} a_{12} \begin{bmatrix} a_{21} & a_{23} \\ a_{31} & a_{33} \end{bmatrix} + (-1)^{1+3} a_{13} \begin{bmatrix} a_{21} & a_{22} \\ a_{31} & a_{32} \end{bmatrix}$$

$$|A| = a_{11}(a_{22}a_{33} - a_{32}a_{23}) - a_{12}(a_{21}a_{33} - a_{31}a_{23}) + a_{13}(a_{21}a_{32} - a_{31}a_{22})$$

$$= a_{11}a_{22}a_{33} - a_{11}a_{32}a_{23} - a_{12}a_{21}a_{33} - a_{12}a_{31}a_{23} + a_{13}a_{21}a_{32} - a_{13}a_{31}a_{22}$$

Note. We shall apply for four steps together.

Expansion along second row (R_2)

$$|A| = \begin{vmatrix} a_{11} & a_{12} & a_{13} \\ a_{21} & a_{22} & a_{23} \\ a_{31} & a_{32} & a_{33} \end{vmatrix}$$

Expanding along R_2, we get

$$|A| = (-1)^{21} a_{21} \begin{bmatrix} a_{12} & a_{13} \\ a_{32} & a_{33} \end{bmatrix} + (-1)^{2+2} a_{22} \begin{bmatrix} a_{11} & a_{13} \\ a_{31} & a_{33} \end{bmatrix} + (-1)^{2+3} a_{23} \begin{bmatrix} a_{11} & a_{12} \\ a_{31} & a_{32} \end{bmatrix}$$

$$= a_{21}(a_{12}a_{33} - a_{32}a_{13}) + (a_{11}a_{33} - 1\, a_{31}a_{13}) - a_{23}(a_{11}a_{32} - a_{31}a_{12})$$

$$|A| = a_{21}a_{12}a_{33} + a_{21}a_{32}a_{13} + a_{22}a_{11}a_{33} - a_{22}a_{31}a_{13} - a_{23}a_{11}a_{32} + a_{23}a_{31}a_{21}$$

$$= a_{11}a_{22}a_{33} - a_{11}a_{23}a_{32} - a_{12}a_{21}a_{33} + a_{12}a_{23}a_{31} + a_{13}a_{21}a_{32} - a_{13}a_{31}a_{22} \quad \text{...(2)}$$

Example 1. *Factorise* $\Delta = \begin{vmatrix} 1 & 1 & 1 \\ a & b & c \\ a^2 & b^2 & c^2 \end{vmatrix}$.

Sol. Putting $a = b$, the first two columns become identical and hence $\Delta = 0$.

$\therefore$ $a - b$ is a factor of Δ.

Similarly $(b - c)$ and $(c - a)$ are also factors of Δ.

Since each term of Δ is of the third degree, we have

$$\Delta = k(a - b)(b - c)(c - a)$$

where k is a numerical constant to be found.

The leading term in $\Delta = bc^2$

The corresponding term on R.H.S. $= kbc^2$

$\therefore$ $k = 1$

Hence $\Delta = (a - b)(b - c)(c - a)$.

Example 2. *If* $\begin{vmatrix} a & a^2 & a^3 - 1 \\ b & b^2 & b^3 - 1 \\ c & c^2 & c^3 - 1 \end{vmatrix} = 0$ *in which a, b, c are different, show that abc = 1.*

Sol. As each term of C_3 in the given determinant consists of two terms, we express it as a sum of two determinants.

$$\therefore \quad \begin{vmatrix} a & a^2 & a^3 - 1 \\ b & b^2 & b^3 - 1 \\ c & c^2 & c^3 - 1 \end{vmatrix} = \begin{vmatrix} a & a^2 & a^3 \\ b & b^2 & b^3 \\ c & c^2 & c^3 \end{vmatrix} + \begin{vmatrix} a & a^2 & -1 \\ b & b^2 & -1 \\ c & c^2 & -1 \end{vmatrix}$$

$$= abc \begin{vmatrix} 1 & a & a^2 \\ 1 & b & b^2 \\ 1 & c & c^2 \end{vmatrix} - \begin{vmatrix} a & a^2 & 1 \\ b & b^2 & 1 \\ c & c^2 & 1 \end{vmatrix}$$

[Taking common a, b, c from C_1, C_2, C_3 respectively of the first determinant and – 1 from C_3 of the second determinant].

$$= abc \begin{vmatrix} 1 & a & a^2 \\ 1 & b & b^2 \\ 1 & c & c^2 \end{vmatrix} - \begin{vmatrix} 1 & a & a^2 \\ 1 & b & b^2 \\ 1 & c & c^2 \end{vmatrix}$$

[Passing C_3 over C_2 and C_1 in the second determinant]

$$\therefore \quad \begin{vmatrix} 1 & a & a^2 \\ 1 & b & b^2 \\ 1 & c & c^2 \end{vmatrix} (abc - 1) = 0$$

Hence, $abc = 1$, since

$$\begin{vmatrix} 1 & a & a^2 \\ 1 & b & b^2 \\ 1 & c & c^2 \end{vmatrix} \neq 0 \text{ as } a, b, c \text{ are all different.}$$

Do you know?

1. In determinant by any row or column using minors, we keep in view the following symbols for determinant of order three:

$$\begin{vmatrix} + & - & + \\ - & + & - \\ + & - & + \end{vmatrix}.$$

2. If a row or a column of determinant consists of all zeros, the value of the determinant is zero.

Example 3. *Evaluate the determinant* Δ

$$= \begin{vmatrix} 1 & 2 & 4 \\ -1 & 3 & 0 \\ 4 & 1 & 0 \end{vmatrix}.$$

Sol. In third column, two entries are zero. So expanding along third column

$$\Delta = 4 \begin{vmatrix} -1 & 3 \\ 4 & 1 \end{vmatrix} - 0 \begin{vmatrix} 1 & 2 \\ 4 & 1 \end{vmatrix} + 0 \begin{vmatrix} 1 & 2 \\ -1 & 3 \end{vmatrix}$$

$$= 4\,(-1 - 12) - 0 + 0 = -52.$$

Example 4. *Evaluate* $\Delta = \begin{vmatrix} 0 & \sin\alpha & -\cos\alpha \\ -\sin\alpha & 0 & \sin\beta \\ \cos\alpha & -\sin\beta & 0 \end{vmatrix}.$

Sol. Expand along R_1, we get

$$\Delta = 0 \begin{vmatrix} 0 & \sin\beta \\ -\sin\beta & 0 \end{vmatrix} - \sin\alpha \begin{vmatrix} -\sin\alpha & \sin\beta \\ \cos\alpha & 0 \end{vmatrix} - \cos\alpha \begin{vmatrix} -\sin\alpha & 0 \\ \cos\alpha & -\sin\beta \end{vmatrix}$$

$$= 0 - \sin\alpha\,(0 - \sin\beta\cos\alpha) - \cos\alpha\,(\sin\alpha\sin\beta - 0)$$

$$= \sin\alpha\sin\beta\cos\alpha - \cos\alpha\sin\alpha\sin\beta = 0.$$

Example 5. *Evaluate* $\Delta = \begin{vmatrix} 2 & -3 & 5 \\ 6 & 0 & 4 \\ 1 & 5 & -7 \end{vmatrix}.$

Sol. Expanding the determinant along first row,

$$\Delta = 2\begin{vmatrix} 0 & 4 \\ 5 & -7 \end{vmatrix} - (-3)\begin{vmatrix} 6 & 4 \\ 1 & -7 \end{vmatrix} + 5\begin{vmatrix} 6 & 0 \\ 1 & 5 \end{vmatrix}$$

$$= 2\,(0 - 20) + 3\,(-42 - 4) + 5\,(30 - 0) = -28$$

By interchanging rows and columns, we get

$$\Delta_1 = \begin{vmatrix} 2 & 6 & 1 \\ -3 & 0 & 5 \\ 5 & 4 & -7 \end{vmatrix} \text{ (expanding along first column)}$$

$$= 2\begin{vmatrix} 0 & 5 \\ 4 & -7 \end{vmatrix} - (-3)\begin{vmatrix} 6 & 1 \\ 4 & -7 \end{vmatrix} + 5\begin{vmatrix} 6 & 1 \\ 0 & 5 \end{vmatrix}$$

$$= 2\,(0 - 20) + (-42 - 4) + 5\,(30 - 0) = -28$$

Example 6. *Prove that* $\begin{vmatrix} b+c & a & a \\ b & c+a & b \\ c & c & a+b \end{vmatrix} = 4\,abc.$

Sol. Let $\Delta = \begin{vmatrix} b+c & a & a \\ b & c+a & b \\ c & c & a+b \end{vmatrix}$

Applying $R_1 \to R_1 - R_2 - R_3$ to Δ, we get

$$\Delta = \begin{vmatrix} a & -2c & -2b \\ b & c+a & b \\ c & c & a+b \end{vmatrix}$$

Expanding along R_1, we obtain

$$\Delta = 0\begin{vmatrix} c+a & b \\ c & a+b \end{vmatrix} - (-2\,c)\begin{vmatrix} b & b \\ c & a+b \end{vmatrix} + (-2b)\begin{vmatrix} b & c+a \\ c & c \end{vmatrix}$$

$$= 2c\,(ab + b^2 - bc) - 2b\,(bc - c^2 - ac)$$
$$= 2abc + 2cb^2 - 2bc^2 - 2b^2c + 2bc^2 + 2abc$$
$$= 4abc.$$

Example 7. *If x, y, z are different and* $\Delta = \begin{vmatrix} x & x^2 & 1+x^3 \\ y & y^2 & 1+y^3 \\ z & x^2 & 1+z^3 \end{vmatrix} = 0$, *then show that* $1 + xyz = 0$.

Sol. $\Delta = \begin{vmatrix} x & x^2 & 1+x^3 \\ y & y^2 & 1+y^3 \\ z & x^2 & 1+z^3 \end{vmatrix}$

$$= \begin{vmatrix} x & x^2 & 1 \\ y & y^2 & 1 \\ z & x^2 & 1 \end{vmatrix} + \begin{vmatrix} x & x^2 & x^3 \\ y & y^2 & y^3 \\ z & z^2 & z^3 \end{vmatrix}$$

$$= (-1)^2 \begin{vmatrix} 1 & x & x^2 \\ 1 & y & y^2 \\ 1 & z & z^2 \end{vmatrix} + xyz \begin{vmatrix} 1 & x & x^2 \\ 1 & y & y^2 \\ 1 & z & z^2 \end{vmatrix} \quad \left[\begin{matrix}\text{Using} \\ \text{property}\end{matrix}\right]$$

$$= \begin{vmatrix} 1 & x & x^2 \\ 1 & y & y^2 \\ 1 & z & z^2 \end{vmatrix} (1 + xyz)$$

$$= (1 + xyz) \begin{vmatrix} 1 & x & x^2 \\ 0 & y - x & y^2 - x^2 \\ 0 & z - x & z^2 - x^2 \end{vmatrix}$$

Now, using $R_2 \to R_2 - R_1$ and $R_3 \to R_3 - R_1$

Taking out common factor $(y - x)$ from R_2 $(z - x)$ from R_3

$$\Delta = (1 + xyz)(y - x)(z - x) \begin{vmatrix} 1 & x & x^2 \\ 0 & 1 & y + x \\ 0 & 1 & z + x \end{vmatrix}$$

$$= (1 + xyz)(y - x)(z - x) \text{ on expanding}$$

$\Delta = 0$, and x, y, z are all different *i.e.*, $x - y \neq 0, y - x \neq 0, z - x \neq 0$.

We get $\quad 1 + xyz = 0.$

Example 8. *Show that* $\begin{vmatrix} 1+a & 1 & 1 \\ 1 & 1+b & 1 \\ 1 & 1 & 1+c \end{vmatrix} = abc\left(1 + \dfrac{1}{a} + \dfrac{1}{b} + \dfrac{1}{c}\right) = abc + bc + ca + ab.$

Sol. Taking out factors a, b, c common R_1, R_2 and R_3.

$$\text{L.H.S.} = abc \begin{vmatrix} \frac{1}{a} + 1 & \frac{1}{a} & \frac{1}{a} \\ \frac{1}{b} & \frac{1}{b} + 1 & \frac{1}{b} \\ \frac{1}{c} & \frac{1}{c} & \frac{1}{c} + 1 \end{vmatrix}$$

Applying $R_1 \to R_1 + R_2 + R_3$

$$\Delta = abc \begin{vmatrix} 1 + \frac{1}{a} + \frac{1}{b} + \frac{1}{c} & 1 + \frac{1}{a} + \frac{1}{b} + \frac{1}{c} & 1 + \frac{1}{a} + \frac{1}{b} + \frac{1}{c} \\ \frac{1}{b} & \frac{1}{b} + 1 & \frac{1}{b} \\ \frac{1}{c} & \frac{1}{c} & \frac{1}{c} + 1 \end{vmatrix}$$

$$= abc = \left(1 + \frac{1}{a} + \frac{1}{b} + \frac{1}{c}\right) \begin{vmatrix} 1 & 1 & 1 \\ \frac{1}{b} & \frac{1}{b}+1 & \frac{1}{b} \\ \frac{1}{c} & \frac{1}{c} & \frac{1}{c}+1 \end{vmatrix}$$

$C_2 \to C_2 - C_1, C_3 \to C_3 - C_1$

$$\Delta = abc\left(1 + \frac{1}{a} + \frac{1}{b} + \frac{1}{c}\right) \begin{vmatrix} 1 & 0 & 0 \\ \frac{1}{b} & 1 & 0 \\ \frac{1}{c} & 0 & 1 \end{vmatrix}$$

$= abc\,(1 + 1/a + 1/b + 1/c)\,[1\,(1 - 0)] = abc\,(1 + 1/a + 1/b + 1/c)$

$= abc + bc + ca + ab$ R.H.S.

Example 9. *Using properties of determinants, prove that*

$$\begin{vmatrix} a-b-c & 2a & 2a \\ 2b & b-c-a & 2b \\ 2c & 2c & c-a-b \end{vmatrix} = (a+b+c)^3.$$

Sol. Let the given determinant be Δ, then

$$\Delta = \begin{vmatrix} a-b-c & 2a & 2a \\ 2b & b-c-a & 2b \\ 2c & 2c & c-a-b \end{vmatrix}$$

$$= \begin{vmatrix} a+b+c & a+b+c & a+b+c \\ 2b & b-c-a & 2b \\ 2c & 2c & c-a-b \end{vmatrix} \quad [R_1 \to R_1 + R_2 + R_3]$$

$$= (a+b+c) \begin{vmatrix} 1 & 1 & 1 \\ 2b & b-c-a & 2b \\ 2c & 2c & c-a-b \end{vmatrix}$$

Taking $(a + b + c)$ common from R_1

$$= (a+b+c) \begin{vmatrix} 1 & 0 & 0 \\ 2b & -(a+b+c) & 0 \\ 2c & 0 & -(a+b+c) \end{vmatrix}$$

$C_2 \to C_2 - C_1$ and $C_3 \to C_3 - C_1$,

$$= (a+b+c)\,.\,1 \begin{vmatrix} -(a+b+c) & 0 \\ 0 & -(a+b+c) \end{vmatrix}$$

$= (a + b + c)\,(a + b + c)^2 = (a + b + c)^3$

Hence, $\Delta = (a + b + c)^3$.

Example 10. *If A + B + C = π, prove that*

$$= \begin{vmatrix} sin\,(A+B+C) & sin\,(A+C) & cos\,C \\ -sin\,B & 0 & tan\,A \\ cos\,(\pi - c) & tan(\pi - A) & 0 \end{vmatrix} = 0$$

Sol. $\begin{vmatrix} 0 & \sin B & \cos C \\ -\sin B & 0 & \tan A \\ -\cos C & -\tan A & 0 \end{vmatrix}$ $\left[\because \begin{array}{r} \sin \pi = 0, \sin(\pi - B) = \sin B \\ \cos(\pi - C) = -\cos C \\ \tan(\pi - A) = -\tan A \end{array}\right]$

$$= 0 + \sin B \begin{vmatrix} \sin B & \cos C \\ -\tan A & 0 \end{vmatrix} - \cos C \begin{vmatrix} \sin B & \cos C \\ 0 & \tan A \end{vmatrix}$$

$$= (\sin B\,.\,\tan A \cos C - \sin B \tan A\,.\,\cos C) = 0.$$

Example 11. *Using properties of determinants, prove that*

$$\begin{vmatrix} b+c & q+r & y+z \\ c+a & r+p & z+x \\ a+b & p+q & x+y \end{vmatrix} = 2 \begin{vmatrix} a & p & x \\ b & q & y \\ c & r & z \end{vmatrix}.$$

Sol. L.H.S. $\begin{vmatrix} b+c & q+r & y+z \\ c+a & r+p & z+x \\ a+b & p+q & x+y \end{vmatrix}$

$$= 2 \begin{vmatrix} a+b+c & p+q+r & x+y+z \\ c+a & r+p & z+x \\ a+b & p+q & x+y \end{vmatrix}$$

$R_1 \to (R_1 + R_2 + R_3)$ and taking 2 common

$$= 2 \begin{vmatrix} a+b+c & p+q+r & x+y+z \\ -b & -q & -y \\ -c & -r & -z \end{vmatrix}$$

$R_2 \to (R_2 - R_1)$ and $R_3 \to R_3 - R_1$

$$= 2 \begin{vmatrix} a & p & x \\ -b & -q & -y \\ -c & -r & -z \end{vmatrix}$$

$R_1 \to R_1 + R_2 + R_3$

$$= 2\,(-1)\,(-1) \begin{vmatrix} a & p & x \\ b & q & y \\ c & r & z \end{vmatrix}$$

taking (–1) common from each one of R_2 and R_3

$$= 2 \begin{vmatrix} a & p & x \\ b & q & y \\ c & r & z \end{vmatrix} = \text{R.H.S.}$$

Example 12. *Evaluate* $\begin{vmatrix} 9 & 9 & 12 \\ 1 & -3 & -4 \\ 1 & 9 & 12 \end{vmatrix}$.

Sol. Let the given determinant be Δ, then

$$\Delta = \begin{vmatrix} 9 & 9 & 12 \\ 1 & -3 & -4 \\ 1 & 9 & 12 \end{vmatrix} = \begin{vmatrix} 0 & 36 & 48 \\ 1 & -3 & -4 \\ 0 & 12 & 16 \end{vmatrix} = (12 \times 4) \begin{vmatrix} 0 & 3 & 4 \\ 1 & -3 & -4 \\ 0 & 3 & 4 \end{vmatrix}$$

{taking out 12 common from R_1 and 4 common from R_3}

$= 48 \times 0 = 0$ [R_1 and R_2 are identical]

$\Delta = 0$.

Example 13. *Without expanding. Prove that*

$$\begin{vmatrix} x+y & y+z & z+x \\ z & x & y \\ 1 & 1 & 1 \end{vmatrix} = 0.$$

Sol. Let the given determinant be Δ, then

$$\Delta = \begin{vmatrix} x+y & y+z & z+x \\ z & x & y \\ 1 & 1 & 1 \end{vmatrix}$$

$$= \begin{vmatrix} x+y+z & x+y+z & x+y+z \\ z & x & y \\ 1 & 1 & 1 \end{vmatrix}$$

$R_1 \to R_1 + R_2$

$$= (x+y+z) \begin{vmatrix} 1 & 1 & 1 \\ z & x & y \\ 1 & 1 & 1 \end{vmatrix}$$

taking $(x + y + z)$ common from R_1

$= (x + y + z) \times 0 = 0$ [R_1 and R_3 are identical]

Hence, $\Delta = 0$.

Example 14. *Solve the equation* $\begin{vmatrix} x+2 & 2x+3 & 3x+4 \\ 2x+3 & 3x+4 & 4x+5 \\ 3x+5 & 5x+8 & 10x+17 \end{vmatrix}$.

Sol. $R_3 - (R_1 + R_2)$, we get

$$= \begin{vmatrix} x+2 & 2x+3 & 3x+4 \\ 2x+3 & 3x+4 & 4x+5 \\ 0 & 1 & 3x+8 \end{vmatrix}$$

$R_2 \to R_2 - R_1$ and $R_1 \to R_1 + R_2$

$$= \begin{vmatrix} x+2 & 2x+3 & 3x+4 \\ 2x+3 & 3x+4 & 4x+5 \\ 0 & 1 & 3x+8 \end{vmatrix} = 0$$

$$= \begin{vmatrix} x+2 & 2x+4 & 6x+12 \\ x+1 & x+1 & x+1 \\ 0 & 1 & 3x+8 \end{vmatrix} = 0$$

$$= (x+1)(x+2) \begin{vmatrix} 1 & 2 & 6 \\ 1 & 1 & 1 \\ 0 & 1 & 3x+8 \end{vmatrix} = 0$$

To bring one more zero in C_1,

$$\therefore \quad (x+1)(x+2) \begin{vmatrix} 1 & 1 & 5 \\ 1 & 1 & 1 \\ 0 & 1 & 3x+8 \end{vmatrix} = 0$$

$$\Rightarrow \quad -(x+1)(x+2)(3x+8-5) = 0$$

$$3(x-1)(x+2)(x+1) = 0$$

$$x = -1, -1, -2.$$

Example 15. *Prove that* $\begin{vmatrix} a & b & c \\ a-b & b-c & c-a \\ b+c & c+a & a+b \end{vmatrix} = (a^3 + b^3 + c^3 - 3\,abc).$

Sol. $\Delta = \begin{vmatrix} a & b & c \\ a-b & b-c & c-a \\ b+c & c+a & a+b \end{vmatrix} = \begin{vmatrix} a+b+c & b & c \\ 0 & b-c & c-a \\ 2(a+b+c) & c+a & a+b \end{vmatrix}$

$C \to C_1 + C_2 + C_3$

$$= (a+b+c) \begin{vmatrix} 1 & b & c \\ 0 & b-c & c-a \\ 2 & c+a & a+b \end{vmatrix} \text{ taking } (a+b+c) \text{ common}$$

$$= a+b+c \begin{vmatrix} 1 & b & c \\ 0 & b-c & c-a \\ 0 & c+a-2b & a+b-2c \end{vmatrix}$$

$R_3 \to R_3 - 2R_1$

$$= (a+b+c) \begin{vmatrix} b-c & c-a \\ c+a-2b & a+b-2c \end{vmatrix} \qquad [\text{expanded by } C_1]$$

$$= (a+b+c)\begin{vmatrix} b-c & c-a \\ a-b & b-c \end{vmatrix} \quad R_2 \to R_2 + R_1$$

$$= (a+b+c) \cdot [(b-c)^2 - (a-b)(c-a)]$$

$$= (a+b+c)(a^2+b^2+c^2-ab-bc-ca)$$

$$= a^3+b^3+c^3-3abc \quad \textbf{Proved.}$$

Example 16. *Prove that* $\begin{vmatrix} 3a & -a+b & -a+c \\ a-b & 3b & c-b \\ a-c & b-c & 3c \end{vmatrix} = 3(a+b+c)(ab+bc+ca).$

Sol. Let the given determinant of Δ,

$$\Delta = \begin{vmatrix} 3a & -a+b & -a+c \\ a-b & 3b & c-b \\ a-c & b-c & 3c \end{vmatrix}$$

$C_1 \to C_1 + C_2 + C_3$

$$= \begin{vmatrix} a+b+c & -a+b & -a+c \\ a+b+c & 3b & c-b \\ a+b+c & b-c & 3c \end{vmatrix} = (a+b+c)\begin{vmatrix} 1 & -a+b & -a+c \\ 0 & 2b+a & a-b \\ 0 & a-c & 2c+a \end{vmatrix}$$

$R_2 \to R_2 - R_1$ and $R_2 \to R_3 - R_1$

$$= (a+b+c) \cdot 1 \begin{vmatrix} 2b+c & a-b \\ a-c & 2c+a \end{vmatrix}$$

$$= (a+b+c)[(2b+a)(2c+a) - (a-c)(a-b)]$$

$$= (a+b+c)[(4bc+2ab+2ac+a^2-(a^2-ab-ac+bc)]$$

$$= 3(a+b+c)(ab+bc+ca)$$

$$\Delta = 3(a+b+c)(ab+bc+ca). \quad \textbf{Hence Proved.}$$

Example 17. *Prove that* $\begin{vmatrix} 1 & a & a^3 \\ 1 & b & b^3 \\ 1 & c & c^3 \end{vmatrix} = (a-b)(b-c)(c-a)(a+b+c).$

Sol. The given determinant $\begin{vmatrix} 1 & a & a^3 \\ 1 & b & b^3 \\ 1 & c & c^3 \end{vmatrix}$

$R_2 \to R_2 - R_1$ and $R_2 \to R_3 - R_1$

$$= (b-a)(c-a)\begin{vmatrix} 1 & a & a^3 \\ 0 & 1 & b^2+ba+a^2 \\ 0 & 1 & c^2+ca+a^2 \end{vmatrix}$$

taking $(b-a)$ from R_2 and $(c-a)$ from R_3

$$= (b-a)(c-a) \cdot 1 \cdot \begin{vmatrix} 1 & b^2+ba+a^2 \\ 1 & c^2+ca+a^2 \end{vmatrix}$$

Expanding by C_1

$$= (b-a)(c-a)[(c^2+ca+a^2)-(b^2+ba+c^2)]$$
$$= (b-a)(c-a)[(c^2-b^2)+a(c-b)]$$
$$= (b-a)(c-a)(c-b)(a+b+c)$$
$$= (a-b)(b-c)(c-a)(a+b+c) \quad \textbf{Proved.}$$

Example 18. *Prove that* $\begin{vmatrix} 1+a & 1 & 1 & 1 \\ 1 & 1+b & 1 & 1 \\ 1 & 1 & 1+c & 1 \\ 1 & 1 & 1 & 1+d \end{vmatrix} = abcd\left(1+\dfrac{1}{a}+\dfrac{1}{b}+\dfrac{1}{c}+\dfrac{1}{d}\right).$

Sol. Let Δ be the given determinant. Taking $abcd$ common from R_1, R_2, R_3, R_4 respectively.

$$\Delta = abcd \begin{vmatrix} {}^{-1}+1 & a^{-1} & a^{-1} & a^{-1} \\ {}^{-1} & b^{-1}+1 & b^{-1} & b^{-1} \\ c & c^{-1} & c^{-1}+1 & c^{-1} \\ d^{-1} & d^{-1} & d^{-1} & d^{-1}+1 \end{vmatrix}$$

$R_1 \to R_1 + (R_2+R_3+R_4)$ and take out the common factor from R_1.

$$= abcd\,(1+a^{-1}+b^{-1}+c^{-1}+d^{-1}) \begin{vmatrix} 1 & 1 & 1 & 1 \\ b^{-1} & b^{-1}+1 & b^{-1} & b^{-1} \\ c^{-1} & c^{-1} & c^{-1}+1 & c^{-1} \\ d^{-1} & d^{-1} & d^{-1} & d^{-1}+1 \end{vmatrix}$$

$C_2 - C_1, C_3 - C_1, C_4 - C_1$ $$= abcd\left(1+\frac{1}{a}+\frac{1}{b}+\frac{1}{c}+\frac{1}{d}\right) \begin{vmatrix} 1 & 0 & 0 & 0 \\ b^{-1} & 1 & 0 & 0 \\ c^{-1} & 0 & 1 & 0 \\ d^{-1} & 0 & 0 & 1 \end{vmatrix}$$

$$= abcd\left(1+\frac{1}{a}+\frac{1}{b}+\frac{1}{c}+\frac{1}{d}\right).$$

Example 19. *Prove that* $\begin{vmatrix} \dfrac{1}{a} & a^2 & bc \\ \dfrac{1}{b} & b^2 & ca \\ \dfrac{1}{c} & c^2 & ab \end{vmatrix} = 0.$

Sol. Given determinant be Δ. Then,

$$\Delta = \begin{vmatrix} \dfrac{1}{a} & a^2 & bc \\ \dfrac{1}{b} & b^2 & ca \\ \dfrac{1}{c} & c^2 & ab \end{vmatrix} = \frac{1}{abc}\begin{vmatrix} 1 & a^3 & abc \\ 1 & b^3 & abc \\ 1 & c^3 & abc \end{vmatrix}$$

[Multiplying R_1, R_2, R_3 by a, b, c respectively and dividing Δ by abc]

$$= \frac{1}{abc} \cdot abc \cdot \begin{vmatrix} 1 & a^3 & 1 \\ 1 & b^3 & 1 \\ 1 & c^3 & 1 \end{vmatrix}$$

$= (1 \times 10) = 0$ [$\because$ C_1 and C_3 are identical]

Hence, $\Delta = 0$.

Example 20. *Prove that* $\begin{vmatrix} -bc & b^2+bc & c^2+bc \\ a^2+ac & -ac & c^2+ac \\ a^2+ab & b^2+ab & -ab \end{vmatrix} = (ab+bc+ac)^3$.

Sol. Given

$$\begin{vmatrix} -bc & b^2+bc & c^2+bc \\ a^2+ac & -ac & c^2+ac \\ a^2+ab & b^2+ab & -ab \end{vmatrix} = \frac{1}{abc}\begin{vmatrix} -abc & ab^2+abc & ac^2+abc \\ a^2b+abc & -abc & a^2b+abc \\ a^2c+abc & b^2c+abc & -abc \end{vmatrix}$$

[$R_1 \to aR_1$, $R_2 \to bR_2$ and $R_3 \to cR_3$ and so divide the det. by abc]

$$= \left(\frac{abc}{abc}\right)\begin{vmatrix} -bc & ab+ac & ac+ab \\ ab+bc & -ac & bc+ab \\ ac+bc & bc+ac & -ab \end{vmatrix}$$

[Taking out a, b, c common from C_1, C_2, C_3 respectively]

$$= \begin{vmatrix} ab+bc+ac & ab+bc+ac & ab+bc+ac \\ ab+bc & -ac & bc+ab \\ ac+bc & bc+ac & -ab \end{vmatrix} \quad [R_1 \to (R_1+R_2+R_3)]$$

$$= (ab+bc+ac) \cdot \begin{vmatrix} 1 & 1 & 1 \\ ab+bc & -ac & bc+ab \\ ac+bc & bc+ac & -ab \end{vmatrix}$$

$$= (ab+bc+ac) \cdot \begin{vmatrix} 0 & 0 & 1 \\ 0 & -(ab+bc+ac) & bc+ab \\ (ab+bc+ac) & (ab+bc+ac) & -ab \end{vmatrix}$$

[$C_1 \to (C_1 - C_3)$ and $C_2 \to (C_2 - C_3)$]

$= (ab+bc+ac)^3$ [Expanding by R_1]

1.3.4. Solution of Simultaneous Equations using Crammer's Rule (in 2 and 3 unknown)

Let the equations

$$a_1x + b_1y + c_1z = d_1$$
$$a_2x + b_2y + c_2z = d_2$$
$$a_3x + b_3y + c_3z = d_3$$

Solving this first, we calculate determinant of coefficients.

$$\Delta = \begin{vmatrix} a_1 & b_1 & c_1 \\ a_2 & b_2 & c_2 \\ a_3 & b_3 & c_3 \end{vmatrix}$$

then, $$x\Delta = \begin{vmatrix} xa_1 & b_1 & c_1 \\ xa_2 & b_2 & c_2 \\ xa_3 & b_3 & c_3 \end{vmatrix} \quad [\text{Operate } C_1 + yC_2 + zC_3]$$

$$= \begin{vmatrix} a_1x + b_1y + c_1z & b_1 & c_1 \\ a_2x + b_2y + c_2z & b_2 & c_2 \\ a_3x + b_3y + c_3z & b_3 & c_3 \end{vmatrix} = \begin{vmatrix} d_1 & b_1 & c_1 \\ d_2 & b_2 & c_2 \\ d_3 & b_3 & c_3 \end{vmatrix} \qquad [\text{by } (i)]$$

Thus, $$x = \begin{vmatrix} d_1 & b_1 & c_1 \\ d_2 & b_2 & c_2 \\ d_3 & b_3 & c_3 \end{vmatrix} + \begin{vmatrix} a_1 & b_1 & c_1 \\ a_2 & b_2 & c_2 \\ a_3 & b_3 & c_3 \end{vmatrix} \qquad ...(ii)$$

provided $\Delta \neq 0$.

Similarly, $$y = \begin{vmatrix} a_1 & d_1 & c_1 \\ a_2 & d_2 & c_2 \\ a_3 & d_3 & c_3 \end{vmatrix} + \begin{vmatrix} a_1 & b_1 & c_1 \\ a_2 & b_2 & c_2 \\ a_3 & b_3 & c_3 \end{vmatrix} \qquad ...(iii)$$

and $$z = \begin{vmatrix} a_1 & b_1 & d_1 \\ a_2 & b_2 & d_2 \\ a_3 & b_3 & d_3 \end{vmatrix} + \begin{vmatrix} a_1 & b_1 & c_1 \\ a_2 & b_2 & c_2 \\ a_3 & b_3 & c_3 \end{vmatrix} \qquad ...(iv)$$

Equations (ii), (iii) and (iv) giving the values of x, y, z constitute the **Cramer's rule.**

Example 21. *Apply Cramer's rule to solve the equations*

$$3x + y + 2z = 3,\ 2x - 3y - z = -3,\ x + 2y + z = 4.$$

Sol. Here, $$\Delta = \begin{vmatrix} 3 & 1 & 2 \\ 2 & -3 & -1 \\ 1 & 2 & 1 \end{vmatrix} \quad [\text{Expand by } C_1]$$

$$= 3(-3 + 2) - 2(1 - 4) + (-1 + 6) = 8$$

$\therefore$ $$x = \frac{1}{\Delta}\begin{vmatrix} 3 & 1 & 2 \\ -3 & -3 & -1 \\ 4 & 2 & 1 \end{vmatrix} \quad [\text{Expand by } C_1]$$

$$= \frac{1}{8}\,[3(-3 + 2) + 3(1 - 4) + 4(-1 + 6)] = 1$$

Similarly, $$y = \frac{1}{\Delta}\begin{vmatrix} 3 & 3 & 2 \\ 2 & -3 & -1 \\ 1 & 4 & 1 \end{vmatrix} = 2$$

and $$z = \frac{1}{\Delta}\begin{vmatrix} 3 & 1 & 3 \\ 2 & -3 & -3 \\ 1 & 2 & 4 \end{vmatrix} = -1$$

Hence, $x = 1, y = 2, z = -1.$

• Determinants and Cramer's Rule

Unique Solution of a 2 × 2 System

The 2 × 2 system

$$ax + by = e, \qquad \text{...(i)}$$
$$cx + dy = f,$$

has a unique solution provided $\Delta = ad - bc$ is non-zero, in which case the solution is given by

$$x = \frac{de - bf}{ad - bc}, \quad y = \frac{af - ce}{ad - bc}. \qquad \text{...(ii)}$$

This result, called **Cramer's Rule** for 2 × 2 systems, is usually learned in college algebra as part of determinant theory.

Determinants of Order 2

College algebra introduces matrix notation and determinant notation:

$$A = \begin{pmatrix} a & b \\ c & d \end{pmatrix}, \quad \det(A) = \begin{vmatrix} a & b \\ c & d \end{vmatrix}.$$

Evaluation of a 2 × 2 determinant is by **Cramer's Rule:**

$$\begin{vmatrix} \mathbf{a} & b \\ c & \mathbf{d} \end{vmatrix} = \mathbf{ad} - bc.$$

The boldface product **ad** is the product of the main diagonal entries and the other product bc is from the anti-diagonal.

Cramer's 2 × 2 rule in determinant notation is

$$x = \frac{\begin{vmatrix} e & b \\ f & d \end{vmatrix}}{\begin{vmatrix} a & b \\ b & d \end{vmatrix}}, \quad y = \frac{\begin{vmatrix} a & e \\ c & f \end{vmatrix}}{\begin{vmatrix} a & b \\ c & d \end{vmatrix}}. \qquad \text{...(iii)}$$

Unique Solution of an $n \times n$ System

Cramer's rule can be generalized to an $n \times n$ system of equations = $A\vec{x} = \vec{b}$

or

$$\begin{aligned} a_{11}x_1 + a_{12}x_2 + \ldots + a_{1n}x_n &= b_1, \\ a_{21}x_1 + a_{22}x_2 + \ldots + a_{2n}x_n &= b_2, \qquad \text{...(iv)} \\ \vdots \qquad \vdots \qquad \cdots \qquad \vdots \quad &\vdots \\ a_{n1}x_1 + a_{n2}x_2 + \ldots + a_{nn}x_n &= b_n. \end{aligned}$$

System (iv) has a unique solution provided the **determinant of coefficients** $\Delta = \det(A)$ is non-zero, in which case the solution is given by

$$x_1 = \frac{\Delta_1}{\Delta}, x_2 = \frac{\Delta_2}{\Delta}, \ldots, x_n = \frac{\Delta_n}{\Delta}. \qquad \text{...(v)}$$

The determinant Δ_j equals $\det(B_j)$ where matrix B_j is matrix A with column j replaced by $\vec{b} = (b_1, \ldots, b_n)$, which is the right side of system (iv). The result is called **Cramer's Rule** for $n \times n$ systems. Determinants will be defined shortly; intuition from the 2 × 2 case and Sarrus' rule should suffix for the moment.

Determinant Notation for Cramer's Rule. The **determinant of coefficients** for system $A\vec{x} = \vec{b}$ is denoted by

$$\Delta = \begin{vmatrix} a_{11} & a_{12} & \cdots & a_{1n} \\ a_{21} & a_{22} & \cdots & 1_{2n} \\ \vdots & \vdots & \cdots & \vdots \\ a_{n1} & a_{n2} & \cdots & a_{nn} \end{vmatrix} \quad ...(vi)$$

The other n determinants in Cramer's rule (v) are given by

$$\Delta_1 = \begin{vmatrix} b_1 & a_{12} & \cdots & a_{1n} \\ b_2 & a_{22} & \cdots & 1_{2n} \\ \vdots & \vdots & \cdots & \vdots \\ b_n & a_{n2} & \cdots & a_{nn} \end{vmatrix}, ..., \Delta_n = \begin{vmatrix} a_{11} & a_{12} & \cdots & b_1 \\ a_{21} & a_{22} & \cdots & b_2 \\ \vdots & \vdots & \cdots & \vdots \\ a_{n1} & a_{n2} & \cdots & b_n \end{vmatrix} \quad ...(vii)$$

The literature is filled with conflicting notations for matrices, vectors and determinants. The reader should take care to use vertical bars *only* for determinants and absolute values, *e.g.*, | A | makes sense for a matrix A or a constant A. For clarity, the notation det (A) is preferred, when A is a matrix. The notation | A | implies that *a determinant is a number,* computed by | A | $= \pm$ A when $n = 1$, and $|A| = a_{11}a_{22} - a_{12}a_{21}$ when $n = 2$. For $n \geq 3$, | A | is computed by similar but increasingly complicated formulas.

Advantages and Disadvantages of Cramer's Rule

Advantages: Cramer's Rule is that we find the values of x, y, or z without having to know any other values of x, y, or z. For example, we needed to find just the values of y, Cramer's Rule would work well. Another thing that Cramer Rule is that if any of the values of x, y, or z is are fractions, you do not have to plug in a fraction to find the other values. Each value can be found independently.

Disadvantages: Disadvantages to using Cramer's rule is if the values of D is zero, then Cramer's Rule will not work because you cannot divide by zero. However, if the value of D is zero then, you know that the solution is either "No Solution" or "Infinite Solutions". You will have to use a different technique such as Addition/Elimination to find out whether the answer is "No Solution" or "Infinite Solutions".

Cramer's Rule

$$ax + by = e$$
$$cx + dy = f$$

$$x = \frac{\Delta x}{\Delta}, y = \frac{\Delta y}{\Delta}$$

$$\Delta = \begin{vmatrix} a & b \\ c & d \end{vmatrix}, \Delta_x = \begin{vmatrix} e & b \\ f & d \end{vmatrix}, \Delta_y = \begin{vmatrix} a & e \\ c & f \end{vmatrix}.$$

- **Applications of Determinants to Co-ordinate Geometry and to Solve Engineering Problems**

Area of a Triangle: Area of a triangle whose vertices are $(x_1 y_1)$, $(x_2 y_2)$ and (x_3, y_3) is given by

$$\Delta = \frac{1}{2}[x_1(y_2 - y_3) + x_2(y_3 - y_1) + x_3(y_1 - y_2)]$$

Now this can be written in the form of a determinant

$$\Delta = \frac{1}{2}\begin{vmatrix} x_1 & y_1 & 1 \\ x_2 & y_2 & 1 \\ x_3 & y_3 & 1 \end{vmatrix} \quad ...(i)$$

- Since area is a positive quantity, we always take the absolute value of determinant in (Δ).
- The area is given both +ve and –ve values of the determinant for calculation.
- The area of triangle formed by three collinear points is zero.
- Application of determinants and matrices for solving the system of linear equations in two or three variables and checking the consistency of the system of linear equations.

Consistent System. A system of equation is said to be consistent, if its solution (one or more) exists.

Inconsistent System. A system of equation is said to be in consistent, if its solution does not exist.

Condition for collinearity of three points

Let A (x_1, y_1), B (x_2, y_2) and C (x_3, y_3) be three given points. Then, A, B, C are collinear.

$\Rightarrow$ area (ΔABC) = 0

$$\Leftrightarrow \quad \frac{1}{2}\begin{vmatrix} x_1 & y_1 & 1 \\ x_2 & y_2 & 1 \\ x_3 & y_3 & 1 \end{vmatrix} = 0 \Rightarrow \begin{vmatrix} x_1 & y_1 & 1 \\ x_2 & y_2 & 1 \\ x_3 & y_3 & 1 \end{vmatrix} = 0$$

$\Rightarrow$ $x_1 (y_2 - y_3) + x_2 (y_3 - y_1) + x_3 (y_1 - y_2) = 0$

$\therefore$ ABC are collinear.

Example 22. *Prove that, the points A (a, b + c),B (b, c + a) and C(c, a + b) are collinear.*

Sol. The points are collinear then area of ΔABC = 0

Now, area (ΔABC)

$$= \frac{1}{2}\begin{vmatrix} a & b+c & 1 \\ b & c+a & 1 \\ c & a+b & 1 \end{vmatrix}$$

$$= \frac{1}{2}\begin{vmatrix} a & a+b+c & 1 \\ b & a+b+c & 1 \\ c & a+b+c & 1 \end{vmatrix} \quad [C_2 \to C_2 + C_1]$$

$$= \frac{1}{2}(a+b+c)\begin{vmatrix} a & 1 & 1 \\ b & 1 & 1 \\ c & 1 & 1 \end{vmatrix}$$

$$= \frac{1}{2}(a+b+c) \times 0 = 0 \quad [\because C_2 \text{ and } C_3 \text{ are identical}]$$

Example 23. *Find the equation of the line joining the points A (1, 2) and B (3, 6) using determinants.*

Sol. P (x, y) be a point on AB.

Then, the points A, P and B are collinear.

area (ΔAPB) = 0

$$= \frac{1}{2}\begin{vmatrix} 1 & 2 & 1 \\ x & y & 1 \\ 3 & 6 & 1 \end{vmatrix} = 0 \Rightarrow \begin{vmatrix} 1 & 2 & 1 \\ x & y & 1 \\ 3 & 6 & 1 \end{vmatrix} = 0$$

$$\Rightarrow \begin{vmatrix} 1 & 2 & 1 \\ x & y & 1 \\ 0 & 0 & -2 \end{vmatrix} = 0 \; [R_3 \to R_3 - 3R_1]$$

$$\Rightarrow (-2)\begin{vmatrix} 1 & 2 \\ x & y \end{vmatrix} = 0 \Rightarrow (y - 2x) = 0, y = 2x$$

Thus, the required equation is $y = 2x$.

Example 24. *If A (1, 3), B (0, 0) and C (k, 0) be three points such that ar (ΔABC) = 3 sq. units. Find the value of k.*

Sol. Area (ΔABC) = 3 sq. units

$$\frac{1}{2}\begin{vmatrix} 1 & 3 & 1 \\ 0 & 0 & 1 \\ k & 0 & 1 \end{vmatrix} = \pm 3 \Leftrightarrow \begin{vmatrix} 1 & 3 & 1 \\ 0 & 0 & 1 \\ k & 0 & 1 \end{vmatrix} = \pm 6$$

$$(-1)\begin{vmatrix} 1 & 3 \\ k & 0 \end{vmatrix} = \pm 6 \Leftrightarrow 3k = \pm 6 \Leftrightarrow k = \pm 2.$$

Thus, $k = \pm 2$.

Example 25. *Find the value of k in order that the points (5, 5), (k, 1) and (10, 7) are collinear.*

Sol. The given points are collinear

$$\Rightarrow \begin{vmatrix} 5 & 5 & 1 \\ k & 1 & 1 \\ 10 & 7 & 1 \end{vmatrix} = 0$$

$$\Rightarrow \begin{vmatrix} 5 & 5 & 1 \\ k-5 & -4 & 0 \\ 5 & 2 & 0 \end{vmatrix} = 0 \qquad [R_2 \to R_2 - R_1, \text{ and } R_3 \to R_3 - R_1]$$

$$\Rightarrow 1 . [2(k-5) + 20] = 0$$

$$\Rightarrow 2k + 10 = 0$$

$$\Rightarrow k = -5.$$

Example 26. *If the points (a, b), (a', b') and (a – a', b – b') are collinear, show that ab' = a'b.*

Sol. The given points are collinear, we have $\begin{vmatrix} a & b & 1 \\ a' & b' & 1 \\ a-a' & b-b' & 1 \end{vmatrix} = 0$

$$\Rightarrow \begin{vmatrix} a & b & 1 \\ a'-a & b'-b & 0 \\ -a' & -b' & 0 \end{vmatrix} = 0 \qquad [R_2 \to R_2 - R_1 \text{ and } R_3 \to R_3 - R_1]$$

$$\Rightarrow 1.[(a'-a)(-b') + a'(b'-b)] \qquad \text{[expanding by } C_3]$$

$\Rightarrow \qquad ab' - a'b$

$\Rightarrow \qquad ab' = a'b.$

Example 27. *What is the area of the triangle whose vertices are (–2, –3), (3, 2) and (–1, –8)?*

Sol. Given, $(x_1 = -2, y_1 = -3)$ $(x_2 = 3, y_2 = 2)$ $(x_3 = -1, y_3 = -8)$

$$\therefore \qquad \text{Area} = \frac{1}{2}\begin{vmatrix} x_1 & y_1 & 1 \\ x_2 & y_2 & 1 \\ x_3 & y_3 & 1 \end{vmatrix} = \frac{1}{2}\begin{vmatrix} -2 & -3 & 1 \\ 3 & 2 & 1 \\ -1 & -8 & 1 \end{vmatrix}$$

$$\Rightarrow \qquad \frac{1}{2}\begin{vmatrix} -2 & -3 & 1 \\ 5 & 5 & 0 \\ 1 & -5 & 0 \end{vmatrix} \quad [R_2 \to (R_2 - R_1) \quad \text{and} \quad R_3 \to (R_3 - R_1)]$$

$$\Rightarrow \qquad \frac{1}{2}.1.\begin{bmatrix} 5 & 5 \\ 1 & -5 \end{bmatrix} = \frac{1}{2} \times (-25 - 5) = -15$$

Thus, area $(\Delta ABC) = |-15| = 15$ square units.

Example 28. *A* $\begin{vmatrix} 1 & \sin\theta & 1 \\ -\sin\theta & 1 & \sin\theta \\ -1 & -\sin\theta & 1 \end{vmatrix}$ *where* $0 \le \theta \le 2\pi$.

Sol. $A = \begin{vmatrix} 1 & \sin\theta & 1 \\ -\sin\theta & 1 & \sin\theta \\ -1 & -\sin\theta & 1 \end{vmatrix}$

$$|A| = 1(1 + \sin^2\theta) - \sin\theta(-\sin\theta + \sin\theta) + 1(\sin^2\theta + 1)$$

$$= 1 + \sin^2\theta + \sin^2\theta + 1 = 2 + 2\sin^2\theta = 2(1 + \sin^2\theta)$$

Now, $\qquad 0 \le \theta \le 2\pi$

$\Rightarrow \qquad 0 \le \sin\theta \le 1$

$\Rightarrow \qquad 1 \le 1 + \sin^2\theta \le 2$

$\Rightarrow \qquad 2 \le 2(1 + \sin^2\theta) \le 4$

$\qquad$ Det (A) $\le [2, 4]$.

Example 29. *If* $f(\theta) = \begin{vmatrix} \cos^2\theta & \cos\theta\sin\theta & -\sin\theta \\ \cos\theta\sin\theta & \sin^2\theta & \cos\theta \\ \sin\theta & -\cos\theta & 0 \end{vmatrix}$. *Then find the value of* $f(\pi/6)$.

Sol.

$$f(\theta) = \begin{vmatrix} \cos^2\theta & \cos\theta\sin\theta & -\sin\theta \\ \cos\theta\sin\theta & \sin^2\theta & \cos\theta \\ \sin\theta & -\cos\theta & 0 \end{vmatrix}$$

$$\Rightarrow \qquad f\left(\frac{\pi}{6}\right) = \begin{vmatrix} \cos^2\frac{\pi}{6} & \cos\frac{\pi}{6}\sin\frac{\pi}{6} & -\sin\frac{\pi}{6} \\ \cos\frac{\pi}{6}\sin\frac{\pi}{6} & \sin^2\frac{\pi}{6} & \cos\frac{\pi}{6} \\ \sin\frac{\pi}{6} & -\cos\frac{\pi}{6} & 0 \end{vmatrix}$$

$$= \begin{vmatrix} \frac{3}{4} & \frac{\sqrt{3}}{4} & -\frac{1}{2} \\ \frac{\sqrt{3}}{4} & \frac{1}{4} & \frac{\sqrt{3}}{2} \\ \frac{1}{2} & -\frac{\sqrt{3}}{2} & 0 \end{vmatrix}$$

$$= 3/4\ ((1/4)0 + \sqrt{3}/4\ (-\sqrt{3}/4) - 1/2\ (-3/8) - (1/8))$$

(Expanding along first row)

$$\Rightarrow \qquad f(\pi/6) = 9/16 + 3/16 + 4/16 = 1.$$

Example 30. *Show that the determinant* $\begin{vmatrix} a^2+x & ab & ac \\ ab & b^2+x & bc \\ ac & bc & c^2+x \end{vmatrix}$ *is divisible by* x^2.

Sol. Let,

$$\Delta = \begin{vmatrix} a^2+x & ab & ac \\ ab & b^2+x & bc \\ ac & bc & c^2+x \end{vmatrix}$$

Putting $x = 0$, we get

$$\Delta = \begin{vmatrix} a^2 & ab & ac \\ ab & b^2 & bc \\ ac & bc & c^2 \end{vmatrix} = abc \begin{vmatrix} a & a & a \\ b & b & b \\ c & c & c \end{vmatrix} = 0$$

At $x = 0$, $\Delta = 0$ with three columns identical. Therefore, $(x \to 0)$ is a repeated root of the equation $\Delta = 0 \Rightarrow x^2$ is a factor of determinant.

Thus, Δ is divisible by x^2. **Hence Proved**

Example 31. *Determine the value of 'a' for which the following system of equation has a non trival Sol.*

$$a^3x + (a+1)^3 y = (a+2)^3 z = 0$$

$$ax + (a+1)y + (a+2)z = 0$$

$$x + y + z = 0.$$

Sol. For this system of equations to have a non-trivial Sol.

$$\Delta = \begin{vmatrix} a^2 & (a+1)^2 & (a+b)^2 \\ a & b^2 & (a+2) \\ 1 & 1 & 1 \end{vmatrix} = 0, \left\{\begin{matrix} \text{by} \\ C_2 \to C_2 \to C_1 \\ C_3 \to C_3 \to C_1 \end{matrix}\right\}$$

$$\Delta = \begin{vmatrix} a^2 & 3a^2+3a+1 & 6a^2+12a+8 \\ a & 1 & 2 \\ 1 & 0 & 0 \end{vmatrix}$$

$$\Rightarrow \quad (3a^2 + 3a + 1)(2) + (6a^2 + 12a + 8)(-1) = 0$$

$$\Rightarrow \quad 6a^2 + 6a + 2 - 6a^2 - 12a - 8 = 0$$

$$\Rightarrow \quad -6a - 6 = 0$$

$$\Rightarrow \quad a = -1.$$

Example 32. *Without expanding at any stage show that*

$$\begin{vmatrix} x^3+x & x+1 & x-2 \\ 2x^2+3x-1 & 3x & 3x-3 \\ x^2+2x+3 & 2x-1 & 2x-1 \end{vmatrix} = x\,A + B.$$

Sol.

$$\Delta = \begin{vmatrix} x^3+x & x+1 & x-2 \\ 2x^2+3x-1 & 3x & 3x-3 \\ x^2+2x+3 & 2x-1 & 2x-1 \end{vmatrix}$$

$R_2 \to R_2 - R_1 - R_3$

$$\Delta = \begin{vmatrix} x^3+x & x+1 & x-2 \\ -4 & 0 & 0 \\ x^3+2x+3 & 2x-1 & 2x-1 \end{vmatrix}$$

$$\Delta = \begin{vmatrix} x^2+1 & x+1 & x-2 \\ -4 & 0 & 0 \\ x+2 & x-1 & x+1 \end{vmatrix}$$

$C_3 \to C_3 - C_2$

$$\Delta = 3\begin{vmatrix} x^2+x & x+1 & -1 \\ -4 & 0 & 0 \\ x+3 & x+1 & 1 \end{vmatrix}$$

$R_1 \to R_1 - R_3$

$$\Delta = \begin{vmatrix} x^3-3 & 3 & -2 \\ -4 & 0 & 0 \\ x+3 & x-2 & 1 \end{vmatrix}$$

$$\Delta = \begin{vmatrix} x^3-3 & 3 & -2 \\ -4 & 0 & 0 \\ x & x & 0 \end{vmatrix} + \begin{vmatrix} x^3-3 & 3 & -2 \\ -4 & 0 & 0 \\ 3 & -2 & 1 \end{vmatrix}$$

$$= x\begin{vmatrix} x^2 & 0 & 0 \\ -4 & 0 & 0 \\ 1 & 1 & 0 \end{vmatrix} + x\begin{vmatrix} -3 & 3 & 0 \\ -4 & 0 & 0 \\ 1 & 1 & 0 \end{vmatrix} + \begin{vmatrix} x^3 & 0 & 0 \\ -4 & 0 & 0 \\ 3 & -2 & 1 \end{vmatrix} + \begin{vmatrix} -3 & 3 & -2 \\ -4 & 0 & 0 \\ 3 & -2 & 1 \end{vmatrix}$$

$$= x\begin{vmatrix} -3 & 3 & 2 \\ -4 & 0 & 0 \\ 1 & 1 & 0 \end{vmatrix} + \begin{vmatrix} 3 & 3 & 0 \\ -4 & 0 & 0 \\ 3 & -2 & 1 \end{vmatrix}.$$

$= xA + B$

Example 33. *Let* $f(x) = \begin{vmatrix} 2\cos^2 x & \cos 2x & -\sin x \\ \sin 2x & 2\sin^2 x & \cos x \\ \sin x & -\cos x & 0 \end{vmatrix}$ *then, find the value of*

$\int_0^{\pi/2} \{f(x) + f'(x)\}\, dx = dx = 2\int_0^{\pi/2} dx = \pi.$

Sol. $C_1 \to C_1 + 2 \sin x C_2$
$C_2 \to C_3 + 2 \cos x C_3$

$$\begin{vmatrix} 2 & 0 & -\sin x \\ 0 & 2 & \cos x \\ \sin x & -\cos x & 0 \end{vmatrix} = \sin^2 x + 2\cos^2 x = 1$$

$$= \int_0^{\pi/2} (f(x) + f'(x) = dx = 2\int_0^{\pi/2} dx = \pi.$$

Example 34. *If a, b, c ≠ 0 and* $\begin{vmatrix} 1+a & 1 & 1 \\ 1 & 1+b & 1 \\ 1 & 1 & 1+x \end{vmatrix}$ *= 0, then find the value of 1/a + 1/b + 1/c.*

Sol. Let $\begin{vmatrix} 1+a & 1 & 1 \\ 1 & 1+b & 1 \\ 1 & 1 & 1+c \end{vmatrix}$

$$\Delta = abc \begin{vmatrix} \frac{1}{a}+1 & \frac{1}{b} & \frac{1}{c} \\ \frac{1}{a} & \frac{1}{b}+1 & \frac{1}{c} \\ \frac{1}{a} & \frac{1}{b} & \frac{1}{c}+1 \end{vmatrix} \quad \text{(taking } a, b, c \text{ common from } C_1)$$

$$= abc \begin{vmatrix} 1+\frac{1}{a}+\frac{1}{b}+\frac{1}{c} & \frac{1}{a} & 1+\frac{1}{b} \\ 1+\frac{1}{a}+\frac{1}{b}+\frac{1}{c} & \frac{1}{b}+1 & \frac{1}{c} \\ 1+\frac{1}{a}+\frac{1}{b}+\frac{1}{c} & \frac{1}{b} & \frac{1}{c}+1 \end{vmatrix} \quad (C_1 \to C_1 + C_2 + C_3)$$

$$= (abc)\,(1 + 1/a + 1/b + 1/c) \begin{vmatrix} 1 & \frac{1}{a} & 1+\frac{1}{b} \\ 1 & 1+\frac{1}{c} & \frac{1}{a} \\ 1 & \frac{1}{b} & 1+\frac{1}{c} \end{vmatrix}$$

$$= (abc)\,(1 + 1/a + 1/b + 1/c).1$$

But given that $\Delta = 0$, abc ∴ $(1 + 1/a + 1/b + 1/c)$

But $a, b, c \neq 0$ therefore, $1/a + 1/b + 1/c = -1$.

Example 35. *Find values of k, if area of triangle is 4 square units and vertices are*
(i) (k, 0), (4, 0), (0, 2) *(ii) (–2, 0), (0, 4), (0, k).*

Sol. We know that the area of a triangle whose vertices are (x_1, y_1), (x_2, y_2) and (x_3, y_3) is the absolute value of the determinant (Δ), where

$$\Delta = \frac{1}{2}\begin{vmatrix} x_1 & y_1 & 1 \\ x_2 & y_2 & 1 \\ x_3 & y_3 & 1 \end{vmatrix}$$

It is given that the area of triangle is 4 square units.

$\therefore \qquad \Delta = \pm 4$

(*i*) The area of the triangle with vertices $(k, 0)$, $(4, 0)$, $(0, 2)$ is given by the relation,

$$\Delta = \frac{1}{2}\begin{vmatrix} k & 0 & 1 \\ 4 & 0 & 1 \\ 0 & 2 & 1 \end{vmatrix}$$

$$= \frac{1}{2}[k(0-2) - 0(4-0) + 1(8-0)]$$

$$= \frac{1}{2}[-2k + 8] = -k + 4$$

$\therefore \qquad -k + 4 = \pm 4$

When $\qquad -k + 4 = -4, k = 8.$

When $\qquad -k + 4 = 4, k = 0.$

Thus, $\qquad k = 0, 8.$

(*ii*) The area of the triangle with vertices $(-2, 0)$, $(0, 4)$, $(0, k)$ is given by the relation,

$$\Delta = \frac{1}{2}\begin{vmatrix} -2 & 0 & 1 \\ 0 & 4 & 1 \\ 0 & k & 1 \end{vmatrix}$$

$$= \frac{1}{2}[-2(4-k)]$$

$$= k - 4$$

$\therefore \qquad k - 4 = \pm 4$

When $\qquad k - 4 = -4, k = 0.$

When $\qquad k - 4 = 4, k = 8.$

Hence, $\qquad k = 0, 8.$

Example 36. *Find equation of line joining (1, 2) and (3, 6) using determinants.*

Sol. Let $P(x, y)$ be any point on the line joining points A (1, 2) and B (3, 6). Then, the points *A*, *B*, and *P* are collinear. Therefore, the area of triangle ABP will be zero.

$$\therefore \qquad \frac{1}{2}\begin{vmatrix} 1 & 2 & 1 \\ 3 & 6 & 1 \\ x & y & 1 \end{vmatrix} = 0$$

$$\Rightarrow \qquad \frac{1}{2}[1(6-y) - 2(3-x) + 1(3y-6x)] = 0$$

$$\Rightarrow \qquad 6 - y - 6 + 2x + 3y - 6x = 0$$

$$\Rightarrow \qquad 2y - 4x = 0$$

$$\Rightarrow \qquad y = 2x$$

Thus, the equation of the line joining the given points is $y = 2x$.

Example 37. *Examine the consistency of the system of equations.*

$$x + 2y = 2$$
$$2x + 3y = 3.$$

Sol. The given system of equations are

$$x + 2y = 2$$
$$2x + 3y = 3$$

The given system of equations can be written in the form of $AX = B$, where

$$A = \begin{bmatrix} 1 & 2 \\ 2 & 3 \end{bmatrix}, X = \begin{bmatrix} x \\ y \end{bmatrix} \text{ and } B = \begin{bmatrix} 2 \\ 3 \end{bmatrix}.$$

Now, $A = 1(3) - 2(2) = 3 - 4 = -1 \neq 0$

$\therefore$ A is non-singular.

Therefore, A^{-1} exists.

Thus, the given system of equation is consistent.

Example 38. *Examine the consistency of the system of equations*

$$2x - y = 5$$
$$x + y = 4.$$

Sol. The given system of equations are

$$2x - y = 5$$
$$x + y = 4$$

The given system of equations can be written in the form of $AX = B$, where

$$A = \begin{bmatrix} 2 & -1 \\ 1 & 1 \end{bmatrix}, \text{X} = \begin{bmatrix} x \\ z \end{bmatrix} \text{ and B} = \begin{bmatrix} 5 \\ 4 \end{bmatrix}.$$

Now, $A = 2\,(1) - (-1)\,(1) = 2 + 1 = 3 \neq 0$

$\therefore$ A is non-singular.

Therefore, A^{-1} exists.

Thus, the given system of equation is consistent.

Example 39. *Examine the consistency of the system of equations.*

$$x + 3y = 5$$
$$2x + 6y = 8.$$

Sol. The given system of equations are

$$x + 3y = 5$$
$$2x + 6y = 8$$

The given system of equations can be written in the form of AX = B, where

$$\text{A} = \begin{bmatrix} 1 & 3 \\ 2 & 6 \end{bmatrix}, \text{X} = \begin{bmatrix} x \\ y \end{bmatrix} \text{ and B} = \begin{bmatrix} 5 \\ 8 \end{bmatrix}.$$

Now, $\text{A} = 1(6) - 3\,(2) = 6 - 6 = 0$

$\therefore$ A is a singular matrix.

Now, $(\text{adj A}) = \begin{bmatrix} 6 & -3 \\ -2 & 1 \end{bmatrix}$

$$(\text{adj A})\ B = \begin{bmatrix} 6 & -3 \\ -2 & 1 \end{bmatrix}\begin{bmatrix} 5 \\ 8 \end{bmatrix} = \begin{bmatrix} 30-24 \\ -10+8 \end{bmatrix}\begin{bmatrix} 6 \\ -2 \end{bmatrix} \neq 0$$

Thus, the solution of the given system of equations does not exist. Hence, the system of equations is consistent.

Example 40. *Examine the consistency of the system of equations.*

$$x + y + z = 1$$
$$2x + 3y + 2z = 2$$
$$ax + ay + 2az = 4.$$

Sol. The given system of equations are

$$x + y + z = 1$$
$$2x + 3y + 2z = 2$$
$$ax + ay + 2az = 4$$

This system of equations can be written in the form AX = B, where

$$A = \begin{bmatrix} 1 & 1 & 1 \\ 2 & 3 & 2 \\ a & a & 2a \end{bmatrix}, X = \begin{bmatrix} x \\ y \\ z \end{bmatrix} \text{ and } B = \begin{bmatrix} 1 \\ 2 \\ 4 \end{bmatrix}$$

Now, $A = (6a + 2a) + (4a - 2a) + 1\,(2a + 3a)$

$$= 4a - 2a - a = 4a - 3a = a \neq 0$$

$\therefore$ A is non-singular.

Therefore, A^{-1} exists.

Hence, the given system of equations is consistent.

Example 41. *Examine the consistency of the system of equations.*

$$3x - y - 2z = 2$$
$$2y - z = -1$$
$$3x - 5y = 3.$$

Sol. The given system of equations are

$$3x - y - 2z = 2$$
$$2y - z = -1$$
$$3x - 5y = 3$$

This system of equations can be written in the form of AX = B, where

$$A = \begin{bmatrix} 3 & -1 & -2 \\ 0 & 2 & -1 \\ 3 & -5 & 0 \end{bmatrix}, X = \begin{bmatrix} x \\ y \\ z \end{bmatrix} \text{ and } B - \begin{bmatrix} 2 \\ -1 \\ 3 \end{bmatrix}$$

Now,

$$|A| = 3(0 - 5) - 0 + 3\,(1 + 4) = -15 + 15 = 0$$

$\therefore$ A is a singular matrix.

Now, $$(\text{adj A}) = \begin{bmatrix} -5 & 10 & 5 \\ 3 & 6 & 3 \\ -6 & 12 & 6 \end{bmatrix}\begin{bmatrix} 5 & 10 & 5 \\ 3 & 6 & 3 \\ -6 & 12 & 6 \end{bmatrix}$$

$$\therefore \qquad (\text{adj A})\,B = \begin{bmatrix} -5 & 10 & 5 \\ -3 & 6 & 3 \\ -6 & 12 & 6 \end{bmatrix}\begin{bmatrix} 2 \\ -1 \\ 3 \end{bmatrix}$$

$$= \begin{bmatrix} -10-10+15 \\ -6-6+9 \\ -12-12+18 \end{bmatrix} = \begin{bmatrix} -5 \\ -3 \\ -6 \end{bmatrix} \neq 0$$

Thus, the solution of the given system of equations does not exist. Thus, the system of equations is inconsistent.

Example 42. *Examine the consistency of the system of equations*

$$5x - y + 4z = 5$$
$$2x + 3y + 5z = 2$$
$$5x - 2y + 6z = -1.$$

Sol. The given system of equations are

$$5x - y + 4z = 5$$
$$2x + 3y + 5z = 2$$
$$5x - 2y + 6z = -1$$

This system of equations can be written in the form of AX = B, where

$$A = \begin{bmatrix} 5 & -1 & 4 \\ 2 & 3 & 5 \\ 5 & -2 & 6 \end{bmatrix}, X = \begin{bmatrix} x \\ y \\ z \end{bmatrix} \text{ and } B = \begin{bmatrix} 5 \\ 2 \\ -1 \end{bmatrix}.$$

Now,
$$A = 5\,(18 + 10) + 1\,(12 - 25) + 4\,(-4 - 15)$$
$$= 5(28) + 1\,(-13) + 4\,(-19)$$
$$= 140 - 13 - 76$$
$$= 51 \neq 0$$

$\therefore$ A is non-singular.

Therefore, A^{-1} exists.

Thus, the given system of equations is consistent.

EXERCISE 1.1.4.

A. Evaluate the determinants.

1. $\begin{vmatrix} 2 & 4 \\ -5 & -1 \end{vmatrix}$ **[Ans.** 18]

2. (*i*) $\begin{vmatrix} \cos\theta & -\sin\theta \\ \sin\theta & \cos\theta \end{vmatrix}$ **[Ans.** 1] (*ii*) $\begin{vmatrix} x^2 - x + 1 & x - 1 \\ x + 1 & x + 1 \end{vmatrix}$ **[Ans.** $x^3 - x + 2$]

3. If $A = \begin{vmatrix} 1 & 2 \\ 4 & 2 \end{vmatrix}$, then show that $|2A| = 4|A|$.

B. Prove that by the using of property of determinants and without expanding.

1. $\begin{vmatrix} x & a & x+a \\ y & b & y+b \\ z & c & z+c \end{vmatrix} = 0$

2. $\begin{vmatrix} a-b & b-c & c-a \\ b-c & c-a & a-b \\ c-a & a-b & b-c \end{vmatrix} = 0$

3. $\begin{vmatrix} 2 & 7 & 65 \\ 3 & 8 & 75 \\ 5 & 9 & 86 \end{vmatrix} = 0$

4. $\begin{vmatrix} 1 & bc & a(b+c) \\ 1 & ca & b(c+a) \\ 1 & ab & c(a+b) \end{vmatrix} = 0$

5. $\begin{vmatrix} b+c & q+r & y+z \\ c+a & r+p & z+x \\ a+b & p+q & x+y \end{vmatrix} = 2\begin{vmatrix} a & p & x \\ b & q & y \\ c & r & z \end{vmatrix}$

6. Find area of the triangle with vertices at the point given in each of the following.
   (*i*) (1, 0), (6, 0), (4, 3) (*ii*) (2, 7), (1, 1), (10, 8)
   (*iii*) (–2, –3), (3, 2), (–1, –8).

7. Find values of k if area of triangle is 4 sq. units and vertices are
   (*i*) (k, 0), (4, 0), (0, 2) (*ii*) (–2, 0), (0, 4), (0, k).

8. (*i*) Find equation of line joining (1, 2) and (3, 6) using determinats.
   (*ii*) Find equation of line joining (3, 1) and (9, 3) using determinants.

9. Using Cofactors of elements of third column, evaluate $\Delta = \begin{vmatrix} 1 & x & yz \\ 1 & y & zx \end{vmatrix}$.

10. Find the value of x

    (*i*) If $\begin{vmatrix} x & 2 \\ 18 & x \end{vmatrix} = \begin{vmatrix} 6 & 2 \\ 18 & 6 \end{vmatrix}$ **[Ans.** ± 6]

    (*ii*) $\begin{vmatrix} 2 & 3 \\ 4 & 5 \end{vmatrix} = \begin{vmatrix} x & 3 \\ 2x & 5 \end{vmatrix}$ **[Ans.** $x = 2$]

    (*iii*) $\begin{vmatrix} 2 & 4 \\ 5 & 1 \end{vmatrix} = \begin{vmatrix} 2x & 4 \\ 6 & x \end{vmatrix}$ **[Ans.** $\pm\sqrt{3}$]

11. If $A = \begin{vmatrix} 1 & 0 & 1 \\ 0 & 1 & 2 \\ 0 & 0 & 4 \end{vmatrix}$, then show that $|3A| = 27|A|$.

12. By using properties of determinants, then show that

    (*i*) $\begin{vmatrix} a^2+1 & ab & ac \\ ab & b^2+1 & bc \\ ca & cb & c^2+1 \end{vmatrix} = 1 + a^2 + b^2 + c^2$

    (*ii*) $\begin{vmatrix} 1+a^2-b^2 & 2ab & -2b \\ 2ab & 1-a^2+b^2 & 2a \\ 2b & -2a & 1-a^2-b^2 \end{vmatrix} = (1 + a^2 + b^2)^3$

    (*iii*) $\begin{vmatrix} 1 & x & x^2 \\ x^2 & 1 & x \\ x & x^2 & 1 \end{vmatrix} = (1 - x^3)^2$

(*iv*) $\begin{vmatrix} x+y+2z & x & y \\ z & y+z+2x & y \\ z & x & z+x+2y \end{vmatrix} = 2(x+y+z)^3$

(*v*) $\begin{vmatrix} y+k & y & y \\ y & y+k & y \\ y & y & y+k \end{vmatrix} = k^2\,(3y+k)$

(*vi*) $\begin{vmatrix} a-b-c & 2a & 2a \\ 2b & b-c-a & 2b \\ 2c & 2c & c-a-b \end{vmatrix} = (a+b+c)^3$

(*vii*) $\begin{vmatrix} x+4 & 2x & 2x \\ 2x & x+4 & 2x \\ 2x & 2x & x+4 \end{vmatrix} = (5x+4)\,(4-x)^2$

(*viii*) $\begin{vmatrix} 0 & a & -b \\ -a & 0 & -c \\ b & c & 0 \end{vmatrix} = 0$

(*ix*) $\begin{vmatrix} -a^2 & ab & ac \\ ba & -b^2 & bc \\ ca & cb & -c^2 \end{vmatrix} = 4a^2\,b^2\,c^2$

(*x*) $\begin{vmatrix} 1 & a & a^2 \\ 1 & b & b^2 \\ 1 & c & c^2 \end{vmatrix} = (a-b)\,(b-c)\,(c-a).$

13. (*i*) Find equation of line joining (3, 1) and (9, 3) using determinants. [**Ans.** $x - 3y = 0$]

(*ii*) Find equation of line joining (1, 2) and (3, 6) using determinants. [**Ans.** $y = 2x$]

14. If area of triangle is 35 sq. units with vertices (2, –6), (5, 4) and (k, 4). Then, what is the value of k ? [**Ans.** (12, –2)]

15. Show that points (c, $a + b$), (b, $c + a$), (a, $b + c$) are collinear.

16. Find area of triangle:

(*i*) (–2, –3), (3, 2), (–1, – 8) [**Ans.** 15/2]

(*ii*) (2, 7), (1, 1), (10, 8) [**Ans.** 47/2]

(*iii*) (1, 0). (6, 0), (4, 3). [**Ans.** 15]

17. Prove that $\Delta = \begin{vmatrix} a+bx & c+dx & p+qx \\ ax+b & cx+d & px+q \\ u & v & w \end{vmatrix} = (1-x^2)\begin{vmatrix} a & c & p \\ b & d & q \\ u & v & w \end{vmatrix}$.

18. Evaluate: $\begin{vmatrix} {}^mc_1 & {}^mc_2 & {}^mc_3 \\ {}^nc_1 & {}^nc_2 & {}^nc_3 \\ {}^pc_1 & {}^pc_2 & {}^pc_3 \end{vmatrix}$. $\left[\textbf{Ans. } \frac{7}{12}(mnp\,(m-n)\,(n-p)\,(p-m)\right]$

19. $\begin{vmatrix} (m+n)^2 & l^2 & mn \\ (n+l)^2 & m^2 & ln \\ (l+m)^2 & n^2 & lm \end{vmatrix} = (l-m)(m-n)(n-l)(l+m+n)(l^2+m^2+n^2).$

20. $\begin{vmatrix} -a(a^2+b^2+c^2) & 2b^3 & 2c^3 \\ 2a^3 & -b(c^2+a^2-b^2) & 2c^3 \\ 2a^3 & 2b^3 & -c(a^2+b^2-c^2) \end{vmatrix} = abc\,(a^2+b^2+c^2)^3.$

21. Show that $\begin{vmatrix} (a+1)(a+2) & a+2 & 1 \\ (a+2)(a+3) & a+3 & 1 \\ (a+3)(a+4) & a+4 & 1 \end{vmatrix} = -2.$

22. Show that $\begin{vmatrix} 1 & a & a^2 \\ 1 & b & b^2 \\ 1 & c & c^2 \end{vmatrix} = \begin{vmatrix} 1 & bc & b+c \\ 1 & ca & c+a \\ 1 & ab & a+b \end{vmatrix}.$

23. Prove that $\begin{vmatrix} a^2 & b^2 & c^2 \\ (a+1)^2 & (b+1)^2 & (c+1)^2 \\ (a-1)^2 & (b-1)^2 & (c-1)^2 \end{vmatrix} = -4(a-b)(b-c)(c-a).$

24. Prove that $\begin{vmatrix} 1 & a^2+bc & a^3 \\ 1 & b^2+ca & b^3 \\ 1 & c^2+ab & c^3 \end{vmatrix} = -(a-b)(b-c)(c-a)(a^2+b^2+c^2).$

25. Prove that $\begin{vmatrix} b^2-ab & b-c & bc-ac \\ ab-a^2 & a-b & b^2-ab \\ bc-ac & c-a & ab-a^2 \end{vmatrix} = 0.$

26. Prove that $\begin{vmatrix} 1 & x & x^2 \\ x^2 & 1 & x \\ x & x^2 & 1 \end{vmatrix} = (1-x^3)^2.$

27. Let P (1, 3), Q (0, 0) and R(k, 0), be three joints such that area (ΔPQR) = 2 sq. units. Find the value of k.

28. Evaluate $\begin{vmatrix} 2 & 3 & 4 \\ 5 & 6 & 8 \\ 6x & 9x & 12x \end{vmatrix}$ [**Ans.** 0]

29. Let P (–2, 0), Q (0, 4) and R (0, k) are three points of a triangle than show that area of ΔPQR is 4 sq. units.

30. Use Cramer's Rule to solve:

$$2x + 3y - z = 1$$
$$4x + y - 3z = 11$$
$$3x - 2y + 5z = 21$$

[**Ans.** $x = -4, y = -2, z = 1$]

31. Use Cramer's Rule to solve:

$$-x + 3y - 2z = 5$$

$$4x - y - 3z = -8$$

$$2x + 2y - 5z = 7$$

$\left[\textbf{Ans.} -10, \frac{-79}{11}, \frac{-91}{11}\right]$

32. Solve by Cramer's rule the system of equations

$$2p + 3q + r - s = 1$$
$$p + q - s = -1$$
$$3q + r + s = 3$$
$$p + r - s = 0$$

Verifying $p = 1, q = 0, r = 1, s = 2$.

33. Solve the following using Cramer's rule

(*i*) $2x - 3y = 1$

$4x + 4y = 2$ [**Ans.** $x = 1/2, y = 0$]

(*ii*) $2x - 5y = 2$

$-4x + 10y = 1$ [**Ans.** $\Delta = 0$, no solution]

(*iii*) $6x - y = 0$

$2x - 1y = 1$ [**Ans.** $x = -1/22, y = -3/11$]

34. Using Cramer's rule. Solve the following questions.

(*i*) $2x_1 + x_2 - x_3 = 0$

$x_1 + x_3 = k$

$x_1 + x_2 + x_3 = 0$ $\left[\textbf{Ans.}\ x_1 = 8/3, x_2 = -4, x_3 = \frac{4}{3}\right]$

(*ii*) $x - y + z = 1$

$-x + z = 1$

$x + y - z = 0$ $\left[\textbf{Ans.}\ x = \frac{1}{2}, y = \frac{1}{2}, z = 1\right]$

35. Solving an independent system with Cramer's rule

$$x + y + z = 4$$
$$x - y = -3$$
$$x + 2y - z = 0$$

[**Ans.** –1, 2, 3]

36. Using Cramer's rule to solve the system

$-4x + 2y - 9z = 2$

$3x + 4y + z = 5$

$x - 3y + 2z = 8$ [**Ans.** 7, –3, –4]

37. Using Cramer's find x, y, z.

(*i*) $x - y = 2$

$x + 4y = 5$ [**Ans.** 13/5, 3/5]

(*ii*) $3x - 4y = -2$

$x + y = 6$ [**Ans.** 22/7, 20/7]

(*iii*) $x - 2y + z = -1$

$3x + y - 2z = 4$

$y - z = 1$ [**Ans.** 1, 1, 0]

(*iv*) $x - 2y - 2z = 3$

$2x - 4y + 4z = 1$

$3x - 3y - 3z = 4$ [**Ans.** –1/3, –25/24, –5/8]

(*v*) $2x + 2y + z = 1$

$x - y + 6z = 21$

$3x + 2y - z = -4$ [**Ans.** 1, –2, 3]

(*vi*) $x + y + z = 0$

$2x - y - 4z = 15$

$x - 2y - z = 7$ [**Ans.** 3, –1, –2]

1.4. MATRICES

To understand the meaning of a matrix, consider the stock position of different bulbs of red, blue and green colours of 25, 40, 60 and 100 watts, available in an eletrical store and displayed in a tabular form as given below :

$$\begin{array}{c} \\ 25W \\ 40W \\ 60W \\ 100W \end{array}\begin{array}{c} \begin{array}{ccc} \text{Red} & \text{Blue} & \text{Green} \end{array} \\ \begin{bmatrix} 21 & 49 & 37 \\ 80 & 95 & 106 \\ 16 & 30 & 25 \\ 11 & 23 & 31 \end{bmatrix} \end{array}$$

Thus, table consists of four rows and three columns. So there are $4 \times 3 = 12$ entries or elements in the table. The first row informs that there are 21 red, 49 blue and 37 green bulbs of 25 watts each available in the store and the second row informs that there are 80 red, 95 blue and 106 green bulbs of 40 watts each available in the store and so on. Each entry has a definite place and meaning in the table. For example, according to the last entry in the table, there are 31 green bulbs of 100 watts available in the store. In nut-shell the above table stores information. Such an arrangement of numbers, real or complex or of functions, in an array of rows and columns is called a matrix. The above table gives a 4×3 matrix, since it has 4 rows and 3 columns.

1.4.1. Definition of Matrix

A set of $m \times n$ numbers of functions, arranged in a rectangular array of m rows and n columns is called a matrix of order $m \times n$, read as m by n. To indicate the position of an element in a matrix, we sometimes denote each element by a letter followed by two suffixes. A matrix of order $m \times n$ with double-suffix *notation* is given below :

$$\begin{bmatrix} a_{11} & a_{12} & a_{13} \cdots\cdots & a_{1n} \\ a_{21} & a_{22} & a_{23} \cdots\cdots & a_{2n} \\ \vdots & \vdots & \vdots & \vdots \\ \vdots & \vdots & \vdots & \vdots \\ a_{m1} & a_{m2} & a_{m3} & a_{mn} \end{bmatrix}$$

The suffixes i and j in the element a_{ij} indicate that the element a_{ij} belongs to the ith row and the jth column. Thus a_{23} belongs to second row and third column.

Matrices are conveniently denoted by capital bold-faced letter A, B, C etc. They are also denoted by $[a_{ij}]$, $[b_{ij}]$, $[c_{ij}]$ etc. Thus we can denote the $m \times n$ matrix defined above either by A or $[a_{ij}]$.

A matrix is a single entity with a number of components. It is not merely a collection of numbers. The *stress* at a point inside an elastic solid has nine components and it can be expressed as a 3×3 matrix. Unlike a determinant, a matrix does not reduce to a single number.

1.4.2. Types of Matrices

***(i)* Row Matrix.** A matrix having only one row is called a *row matrix* or *row vector*. Thus the matrix $(a_{11}\ a_{12}\ a_{13}\ \ldots.\ a_{1n})$ is a row matrix of order $1 \times n$

***(ii)* Column Matrix.** A matrix having only one column is called a *column matrix* or *column vector.*

***(iii)* Square Matrix.** If the number of rows (m) and columns (n) are equal, the matrix is called a square matrix of order $m \times n$ or briefly n.

The diagonal containing the elements.

$$a_{11}, a_{22}, a_{33}, \ldots.. a_{nn}$$

is called **principal** or **leading diagonal**.

***(iv)* Diagonal Matrix.** It is a square matrix all of whose elements except those in the leading diagonal are zero, *i.e.,* $a_{ij} = 0$, when $i \neq j$.

A diagonal matrix whose diagonal elements are all equal is a called a **scalar matrix**.

A diagonal matrix whose diagonal elements are all equal to unity is called a **unit matrix** denoted by I. Examples of diagonal, scalar and unit matrices are respectively.

$$\begin{bmatrix} a & 0 & 0 \\ 0 & b & 0 \\ 0 & 0 & c \end{bmatrix}, \begin{bmatrix} a & 0 & 0 \\ 0 & a & 0 \\ 0 & 0 & a \end{bmatrix} \text{ and } \begin{bmatrix} 1 & 0 & 0 \\ 0 & 1 & 0 \\ 0 & 0 & 1 \end{bmatrix}$$

***(v)* Null matrix.** A matrix whose elements are all zero is called a **null matrix** or a **zero matrix**. It is denoted by **0** (zero). In this case $a_{ij} = 0$ for all i, j.

***(vi)* Triangular matrix.** A square matrix whose elements either above or below the leading diagonal, are all zero, is called a **triangular matrix.** Thus

$$\begin{bmatrix} a_{11} & a_{12} & a_{13} \\ 0 & a_{23} & a_{23} \\ 0 & 0 & a_{23} \end{bmatrix} \text{ and } \begin{bmatrix} a_{11} & 0 & 0 \\ a_{21} & a_{22} & 0 \\ a_{31} & a_{32} & a_{33} \end{bmatrix}$$

are examples of **upper** and **lower triangular matrices** respectively.

***(vii)* Transposed Matrix.** The matrix obtained from any given matrix. **A** by changing rows into columns and columns into rows is called *transposed matrix* or transpose of **A** and it is denoted by A' or A^t. Thus if,

$$A' = \begin{bmatrix} 1 & 2 & 3 \\ 4 & 5 & 6 \\ 7 & 8 & 9 \\ 10 & 11 & 12 \end{bmatrix}, \text{ then } A^t = \begin{bmatrix} 1 & 4 & 7 & 10 \\ 2 & 5 & 8 & 11 \\ 3 & 6 & 9 & 12 \end{bmatrix}$$

If follows from the definition that $(A^t)t = A$.

***(viii)* Symmetric Matrix.** A matrix is said to be symmetric if it is equal to its transpose. *i.e.,* $A = A^t$. Thus it is a square matrix with $a_{ij} = a_{ji}$ for all $i, j,$ *i.e.,* $a_{12} = a_{21}, a_{13}, a_{23} = a_{32}$ etc.

A skew-symmetric matrix is a matrix that is equal to the negative of its transpose *i.e.,* $\mathbf{A = - A^t}$.

Thus it is square matrix having $a_{ij} = - a_{ji}$ for all i, j. Clearly in this case $a_{11} = a_{22}, a_{33} = \ldots = 0$ *i.e.,* the main diagonal elements of a skew-symmetric matrix are all zero. The examples of symmetric and skew-symmetric matrices are respectively.

$$\begin{bmatrix} a & d & e \\ d & b & f \\ e & f & c \end{bmatrix} \text{ and } \begin{bmatrix} 0 & d & -e \\ -d & 0 & -f \\ e & f & 0 \end{bmatrix}$$

(*ix*) **Singular Matrix.** If the determinant $|A|$ of a square matrix **A** vanishes *i.e.*, when $|A| = 0$, **A** is called a singular matrix.

(*x*) **Non-singular Matrix.** If the determinant $|A|$ of a square matrix **A**. Vanishes *i.e.*,

If $|A| \neq 0$, A is said to be a **non-singular** matrix.

For example, consider the matrix.

$$A = \begin{bmatrix} 1 & 2 & 3 \\ 4 & 1 & 5 \\ 3 & 6 & 9 \end{bmatrix}$$

It is determinant $|A|$ is given by

$$|A| = \begin{bmatrix} 1 & 2 & 3 \\ 4 & 1 & 5 \\ 3 & 6 & 9 \end{bmatrix} = \begin{bmatrix} 1 & 0 & 0 \\ 4 & -7 & -7 \\ 3 & 0 & 0 \end{bmatrix} = 0$$

(Applying $C_2 - 2C_1$ and $C_3 - 3C_1$).

Hence A is a singular matrix.

(*xi*) **Adjoint of a Matrix.** Let A be the square matrix.

$$\begin{bmatrix} a_{11} & a_{12} & a_{13} \\ a_{21} & a_{22} & a_{23} \\ a_{31} & a_{32} & a_{33} \end{bmatrix}.$$

Now consider the determinant $|A|$. Let $A_{11}, A_{12}, A_{13}, A_{21}, \ldots$ be the co-factors of the elements $a_{11}, a_{12}, a_{13}, a_{21}, \ldots$ Then the matrix formed by these co-factors is given by

$$\begin{bmatrix} A_{11} & A_{12} & A_{13} \\ A_{21} & A_{22} & A_{23} \\ A_{31} & A_{32} & A_{33} \end{bmatrix}$$

The transpose of this co-factor matrix is called ajoint of the matrix A.

Thus, $$adj.\ A = \begin{bmatrix} A_{11} & A_{12} & A_{13} \\ A_{21} & A_{22} & A_{23} \\ A_{31} & A_{32} & A_{33} \end{bmatrix}$$

(*xii*) **Equal Matrices.** Two matrices A and B are said to be equal when they are of the same order and the elements in the corresponding positions are equal. In other words the two matrices are identical. We write then $A = B$.

Thus if $\begin{bmatrix} a & b & c \\ d & e & f \end{bmatrix} = \begin{bmatrix} 4 & 0 & 7 \\ 3 & 1 & 6 \end{bmatrix}$

then $a = 4, b = 0, c = 7, d = 3, e = 1$ and $f = 6$.

Also note that the matrices are of the same order *i.e.,* 2×3.

1.4.3. Addition, Subtraction, Multiplication of Matrices (up to 3 order)

(*i*) **Addition of Matrices.** The sum of two matrices A and B each of order $m \times n$ is defined as the matrix C of the same order $m \times n$, each elements of which is the sum of the corresponding elements of A and B. We write $\mathbf{C = A + B}$.

Let $$\mathbf{A} = \begin{bmatrix} 3 & 4 \\ 2 & -1 \end{bmatrix} \text{ and } \mathbf{B} = \begin{bmatrix} 5 & 0 \\ 6 & 8 \end{bmatrix}$$

then $$\mathbf{C = A + B} = \begin{bmatrix} 3 & 4 \\ 2 & -1 \end{bmatrix} + \begin{bmatrix} 5 & 0 \\ 6 & 8 \end{bmatrix}$$

$$= \begin{bmatrix} 3+5 & 4+0 \\ 2+6 & -1+8 \end{bmatrix} = \begin{bmatrix} 8 & 4 \\ 8 & 7 \end{bmatrix}$$

Properties of Matrix Addition. Let all the matrices **A, B, C** involved have the same order than (*a*) **A + B** is also a matrix of the same order. This is *closure property* of matrices under addition.

(*b*) $$\mathbf{A + B = B + A}$$

i.e., matrix addition is *commutative.*

(*c*) $$\mathbf{(A + B) + C = A + (B + C)}$$

i.e., matrix addition is *associative.*

(*d*) **B** is called *additive inverse* of **A** if **A + B = 0**, where 0 is the null matrix of the same order as **A** or **B**.

(*e*) **B** is called additive identity if **A + B = A**. Additive identity is a null or zero matrix **O**.

(*f*) $x(\mathbf{A + B}) = x\mathbf{A} + x\mathbf{B}$ and $(x + y)\,\mathbf{A} = x\mathbf{A} + y\mathbf{A}$, where x, y are scalars.

(*ii*) **Subtraction of Matrices.** If **A** and **B** are matrices of the same order, then the difference, $A — B$ may be regarded as the sum of the matrices **A** and (– **B**) *i.e.,*

$$\mathbf{A - B = A + (- B)}$$

Two matrices of different orders cannot be added or subtracted.

(*iii*) **Multiplication of a Matrices up to .... orders by a scalar.** If k is a scalar and A is a matrix, the product kA is a matrix, each element of which is k times the corresponding element of A. Thus

$$7\begin{bmatrix} 3 & 0 \\ 2 & -4 \end{bmatrix} = \begin{bmatrix} 21 & 0 \\ 14 & -28 \end{bmatrix}$$

(*iv*) **Matrix Multiplication.** The product $A \times B$, of two matrices is defined only if the number of columns in A is equal to the numbers of rows in B. A and B are then said to be *conformable* for the product AB. Let A be a 3×3 matrix and B a 3×2 matrix as given below. Then the product.

$$\begin{bmatrix} a_{11} & a_{12} & a_{13} \\ a_{21} & a_{22} & a_{23} \\ a_{31} & a_{32} & a_{33} \end{bmatrix} \times \begin{bmatrix} b_{11} & b_{12} \\ b_{21} & b_{22} \\ b_{31} & b_{33} \end{bmatrix}$$

is defined as the matrix

$$\begin{bmatrix} a_{11}b_{11}+a_{12}b_{21}+a_{13}b_{31} & a_{11}b_{12}+a_{12}b_{22}+a_{13}b_{32} \\ a_{21}b_{11}+a_{22}b_{21}+a_{23}b_{31} & a_{21}b_{12}+a_{22}b_{22}+a_{23}b_{32} \\ a_{31}b_{11}+a_{32}b_{21}+a_{33}b_{31} & a_{31}b_{12}+a_{32}b_{22}+a_{33}b_{32} \end{bmatrix}$$

The first element $a_{11}b_{11} + a_{12}b_{21} + a_{13}b_{31} = \sum_{k=1}^{3} a_{1k}b_{k1}$ is obtained by taking the first row of A and the first column of B and multiplying each element in the column and then summing them up. Similarly, the second element of the third row of the product $\sum_{k=1}^{3} a_{3k}b_{k2}$ is obtained by taking the third row of A and second column of B and then performing the indicated operations. Thus in the matrix $A \times B$, the (i, j)th element, say c_{ij}, is the sum of the products of the elements of the ith row of A with the corresponding element of the jth column of B.

Thus $$c_{ij} = \sum_{k=1}^{3} a_{ik}b_{kj} = a_{ij}b_{1j} + a_{i2}b_{2j} + a_{i3}b_{3j}$$

In general, let $A = [a_{ij}]$ and $B = [b_{ij}]$ be the matrices of orders $m \times n$ and $n \times p$ respectively. Let $C = [c_{ij}]$ be the matrix of their product of order $m \times p$. Then

$$c_{ij} = \sum_{k=1}^{n} a_{ik}b_{kj} \quad i = 1, 2, ...m; \quad j = 1, 2, ...p$$

which is called *inner product* of the ith row of A with jth column of B.

As an illustration, consider the product.

$$\begin{bmatrix} 0 & 1 & 0 \\ 0 & 2 & 1 \\ 2 & 3 & 1 \end{bmatrix} \times \begin{bmatrix} 1 & 2 \\ 3 & 0 \\ 4 & 1 \end{bmatrix}$$

$$\begin{bmatrix} 0\times1+1\times3+0\times4 & 0\times2+1\times0+0\times1 \\ 0\times1+2\times3+1\times4 & 0\times2+1\times0+1\times1 \\ 2\times1+3\times3+1\times4 & 2\times2+3\times0+1\times1 \end{bmatrix} = \begin{bmatrix} 3 & 0 \\ 10 & 1 \\ 15 & 5 \end{bmatrix}$$

- **Properties of Matrix Multiplication**

(*a*) **Matrix multiplication is not commutative.** In general **AB** $\neq$ **BA**. In fact, following cases arise.

(*i*) The product **AB** is defined but **BA** is not defined

(*ii*) The products **AB** and **BA** both are defined but they have different orders. So the questions of their equality does not arise.

(*iii*) The product **AB** and **BA** both are defined and also of the same order yet **AB** $\neq$ **BA**.

(*iv*) The product **AB** and **BA** both are defined, have the same order, and also **AB = BA.**

(*b*) **Matrix multiplication is associative.** The associative property of matrix multiplication.

(AB) C = A (BC)

holds good if the matrices **ABC** are conformable for the products.

(*c*) **Matrix multiplication is distributive w.r.t. matrix addition.**

The distributive property.

$$\mathbf{A\,(B + C) = AB + AC}$$

holds good for the matrices **A, B, C**, provided that they are conformable for the product and the sum.

Remarks. (*i*) Matrix **A** is said to be *post-multiplied* by the matrix **B** when **AB = A × B** is computed, while matrix **A** is said to be pre-multiplied by the matrix **B** when **BA = B × A** is computed.

If, **A, I** and **O** be square matrices of the same order, it can be seen that

$$\mathbf{AI = IA = A}$$

and $$\mathbf{AO = OA = O},$$

where **I** is a unit matrix and **O** is a null matrix.

(*ii*) If A is a square matrix, we write

$$AA = A^2,\ AAA = (A^2)\,A = A\,(A^2) = A^3,$$

AA ... p times $= A^p$, where A^2, A^3, A^p are also square matrices of the same order

Also $$A^pA^q = A^qA^p \text{ and } (A^p)^q = A^{pq} = (A^q)^p$$

where p, q are positive integers.

(*iii*) The sum of the elements of the main diagonal of a square matrix is called the **Trace** or **spur** of the matrix. Thus if $A = (a_{ij})$ is a square matrix of order n, then

$$\text{Trace } A = a_{11} + a_{22} + \ldots a_{nn} = \sum_{i=1}^{n} a_{ij}$$

In particular, Trace $I_n = 1 + 1 + \ldots + 1 = n$

(*iv*) The product of two non-zero matrices can be a zero matrix. For example.

$$\begin{bmatrix} 1 & -1 & 1 \\ -3 & 2 & -1 \\ -2 & 1 & 0 \end{bmatrix} \times \begin{bmatrix} 1 & 2 & 3 \\ 2 & 4 & 6 \\ 1 & 2 & 3 \end{bmatrix} = \begin{bmatrix} 0 & 0 & 0 \\ 0 & 0 & 0 \\ 0 & 0 & 0 \end{bmatrix}$$

Example 1. *Evaluate the continued product.*

$$\begin{bmatrix} 2 & 1 & -1 \\ 4 & -5 & 6 \\ -3 & 7 & 3 \end{bmatrix} \times \begin{bmatrix} 1 & 3 \\ 4 & -6 \\ -2 & 5 \end{bmatrix} \times \begin{bmatrix} 5 & 3 \\ -2 & 1 \end{bmatrix}$$

Sol. Since the matrices are 3 × 3, 3 × 2, 2 × 2, they are comformable, for the multiplication. Now the product of the first two matrices is given by:

$$\begin{bmatrix} 2+4+2 & 6-6-5 \\ 4-20-12 & 12+30+30 \\ -3+28-6 & -9-42+15 \end{bmatrix} \begin{bmatrix} 8 & -5 \\ -28 & 72 \\ 19 & -36 \end{bmatrix}$$

Hence the final product.

$$\begin{bmatrix} 8 & -5 \\ -28 & 72 \\ 19 & -36 \end{bmatrix} \times \begin{bmatrix} 5 & 3 \\ -2 & 1 \end{bmatrix}$$

$$= \begin{bmatrix} 40+10 & 24-5 \\ -140-144 & -84+72 \\ 95+72 & 57-36 \end{bmatrix} \times \begin{bmatrix} 50 & 19 \\ -284 & -12 \\ 167 & 21 \end{bmatrix}$$

1.4.4. Inverse or Reciprocal of a Matrix by Adjoint Method

If A be a square matrix and B be another square matrix of the same order such that $AB = BA = I$, then B is called the inverse matrix of A and is denoted by A^{-1} *i.e.*, $B = A^{-1}$. From above, it follows that if B is inverse of A, then A is inverse of B *i.e.*,

$$A = B^{-1} = (A^{-1})^{-1}$$

On multiplying the square matrix A and its adjoint, we have

$$A \times adj.\, A = \begin{bmatrix} a_{11} & a_{12} & a_{13} \\ a_{21} & a_{22} & a_{23} \\ a_{31} & a_{32} & a_{33} \end{bmatrix} \times \begin{bmatrix} A_{11} & A_{21} & A_{31} \\ A_{12} & A_{22} & A_{32} \\ A_{13} & A_{23} & A_{33} \end{bmatrix}$$

$$= \begin{bmatrix} \Delta & 0 & 0 \\ 0 & \Delta & 0 \\ 0 & 0 & \Delta \end{bmatrix} = \Delta \begin{bmatrix} 1 & 0 & 0 \\ 0 & 1 & 0 \\ 0 & 0 & 1 \end{bmatrix} = \Delta I.$$

It can be easily seen that each diagonal element of the product is equal to Δ while all other elements are zero. For example, the first element of the first row of the product is

$$a_{11}A_{11} + a_{12}A_{12} + a_{13}A_{13}$$

which is equal to Δ, and the second element of the first row is

$$= a_{11}A_{21} + a_{12}A_{22} + a_{13}A_{23}$$

$$= a_{11}\begin{vmatrix} a_{12} & a_{13} \\ a_{32} & a_{33} \end{vmatrix} - a_{12}\begin{vmatrix} a_{11} & a_{13} \\ a_{31} & a_{33} \end{vmatrix} + a_{13}\begin{vmatrix} a_{11} & a_{12} \\ a_{31} & a_{32} \end{vmatrix}$$

$$= \begin{vmatrix} a_{11} & a_{12} & a_{13} \\ a_{21} & a_{22} & a_{23} \\ a_{31} & a_{32} & a_{33} \end{vmatrix} = 0$$

since first two rows are identical.

Hence, $\qquad A \times adj\, A = \Delta I$

or $\qquad A \times \left(\dfrac{adj.\, A}{\Delta}\right) = I \qquad$ ...(1)

provided $\Delta \neq 0$ *i.e.*, A is a non-singular matrix.

From (*i*), we conclude that $A^{-1} = \dfrac{adj.\, A}{\Delta}$, since $AA^{-1} = I$

Note. (*i*). The inverse of a matrix is unique.

Let A be the given matrix and let us assume that B and C are its two inverses. Then

$$AB = BA = I \quad \text{and} \quad AC = CA = I$$

Now $\qquad B = IB = (CA)\,B = C\,(AB) = CI = C.$

This implies that inverse of a matrix is unique.

(*ii*) If A and B are non-singular matrices of the same order then

$$(AB)^{-1} = B^{-1}A^{-1}$$

Consider $\quad AB\,(B^{-1}A^{-1}) = A\,(BB^{-1})A^{-1} = AIA^{-1}$

$\therefore$ $(B^{-1}A^{-1})$ is inverse of AB *i.e.*, $(AB)^{-1} = B^{-1}A^{-1}$

(*iii*) The inverse of the transpose is the transpose of the inverse *i.e.*, $(A')^{-1} = (A^{-1})'$, where A is a non-singular matrix.

Example 2. *Obtain* A', *adj* A *and* A^{-1} *when*

$$A = \begin{bmatrix} 1 & 0 & -1 \\ 3 & 4 & 5 \\ 0 & -6 & -7 \end{bmatrix}$$

Sol. By changing rows into columns and columns in rows, we obtain

$$A' = \begin{bmatrix} 1 & 3 & 0 \\ 0 & 4 & -6 \\ -1 & 5 & -7 \end{bmatrix}$$

To obtain *adj* A, consider determinant of the matrix A

$$|A| = \begin{vmatrix} 1 & 0 & -1 \\ 3 & 4 & 5 \\ 0 & -6 & -7 \end{vmatrix}$$

Cofactors of the first row are

$$\begin{vmatrix} 4 & 5 \\ -6 & -7 \end{vmatrix} = -28 + 30 = 2; -\begin{vmatrix} 3 & 5 \\ 0 & -7 \end{vmatrix} = 21; \begin{vmatrix} 3 & 4 \\ 0 & -6 \end{vmatrix} = -18$$

Similarly, cofactor of the second row are 6, –7, 6 and of the third row area 4, –8, 4
Hence *adjoint* A can be given

$$\begin{bmatrix} 2 & 6 & 4 \\ 21 & -7 & -8 \\ -18 & 6 & 4 \end{bmatrix}$$

Further, $|A| = 1.(2) - 0.(21) + (-1) . (-18) = 20$

$$\therefore \qquad A^{-1} = \frac{1}{\Delta} \text{ adjoint } A = \frac{1}{20}\begin{bmatrix} 2 & 6 & 4 \\ 21 & -7 & -8 \\ -18 & 6 & 4 \end{bmatrix}$$

$$= \begin{bmatrix} \frac{1}{10} & \frac{3}{10} & \frac{1}{5} \\ \frac{21}{20} & -\frac{7}{20} & -\frac{2}{5} \\ -\frac{9}{10} & \frac{3}{10} & \frac{1}{5} \end{bmatrix}$$

Example 3. If $A = \begin{bmatrix} 1 & 1 & -1 \\ 2 & 0 & 3 \\ 3 & -1 & 2 \end{bmatrix}$, $B = \begin{bmatrix} 1 & 3 \\ 0 & 2 \\ -1 & 4 \end{bmatrix}$ and $C = \begin{bmatrix} 1 & 2 & 3 & -4 \\ 2 & 0 & -2 & 1 \end{bmatrix}$

Find $A(BC)$, $(AB)C$ and show that $(AB)C = A(BC)$

Sol. We have $AB = \begin{bmatrix} 1 & 1 & -1 \\ 2 & 0 & 3 \\ 3 & -1 & 2 \end{bmatrix} \begin{bmatrix} 1 & 3 \\ 0 & 2 \\ -1 & 4 \end{bmatrix} = \begin{bmatrix} 1+0+1 & 3+2-4 \\ 2+0-3 & 6+0+12 \\ 3+0-2 & 9-2+8 \end{bmatrix} = \begin{bmatrix} 2 & 1 \\ -1 & 18 \\ 1 & 15 \end{bmatrix}$

$$(AB)(C) = \begin{bmatrix} 2 & 1 \\ -1 & 18 \\ 1 & 15 \end{bmatrix} \begin{bmatrix} 1 & 2 & 3 & 4 \\ 2 & 0 & -2 & 1 \end{bmatrix} = \begin{bmatrix} 2+2 & 4+0 & 6-2 & -8+1 \\ -1+36 & -2+0 & -3-36 & 4+18 \\ 1+30 & 2+0 & 3-30 & -4+15 \end{bmatrix}$$

$$\begin{bmatrix} 4 & 4 & 4 & -7 \\ 35 & -2 & -39 & 22 \\ 31 & 2 & -27 & 11 \end{bmatrix}$$

$$BC = \begin{bmatrix} 1 & 3 \\ 0 & 2 \\ -1 & 4 \end{bmatrix} \begin{bmatrix} 1 & 2 & 3 & -4 \\ 2 & 0 & -2 & 1 \end{bmatrix} = \begin{bmatrix} 1+6 & 2+0 & 3-6 & -4+3 \\ 0+4 & 0+0 & 0-4 & 0+2 \\ -1+8 & -2+0 & -3-8 & 4+4 \end{bmatrix}$$

$$= \begin{bmatrix} 7 & 2 & -3 & -1 \\ 4 & 0 & -4 & 2 \\ 7 & -2 & -11 & 8 \end{bmatrix}$$

$$\therefore A(BC) = \begin{bmatrix} 1 & 1 & -1 \\ 2 & 0 & 3 \\ 3 & -1 & 2 \end{bmatrix} \begin{bmatrix} 7 & 2 & -3 & -1 \\ 4 & 0 & -4 & 2 \\ 7 & -2 & -11 & 8 \end{bmatrix}$$

$$= \begin{bmatrix} 7+4-7 & 2+0+2 & -3-4+11 & -1+2-8 \\ 14+0+21 & 4+0-6 & -6+0-33 & -2-0+24 \\ 21-4+14 & 6+0-4 & -9+4-22 & -3-2+16 \end{bmatrix} = \begin{bmatrix} 4 & 4 & 4 & -7 \\ 35 & -2 & -39 & 22 \\ 31 & 2 & -27 & 11 \end{bmatrix}$$

Clearly, $(AB)\,C = A(BC)$

Example 4. If $A = \begin{bmatrix} 0 & 6 & 7 \\ -6 & 0 & 8 \\ 7 & -8 & 0 \end{bmatrix}$, $B = \begin{bmatrix} 0 & 1 & 1 \\ 1 & 0 & 2 \\ 1 & 2 & 0 \end{bmatrix}$, $C = \begin{bmatrix} 2 \\ -2 \\ 3 \end{bmatrix}$

Calculate AC, BC and $(A + B)\,C$. Also, verify that $(A + B)\,C = AC + BC$

Sol. Now, $A + B = \begin{bmatrix} 0 & 7 & 8 \\ -5 & 0 & 10 \\ 8 & -6 & 0 \end{bmatrix}$

So, $(A + B)\,C = \begin{bmatrix} 0 & 7 & 8 \\ -5 & 0 & 10 \\ 8 & -6 & 0 \end{bmatrix}\begin{bmatrix} 2 \\ -2 \\ 2 \end{bmatrix} = \begin{bmatrix} 0-14+24 \\ -10+0+30 \\ 16+12+0 \end{bmatrix} = \begin{bmatrix} 10 \\ 20 \\ 28 \end{bmatrix}$

Further $AC = \begin{bmatrix} 0 & 6 & 7 \\ -6 & 0 & 8 \\ 7 & -8 & 0 \end{bmatrix}\begin{bmatrix} 2 \\ -2 \\ 2 \end{bmatrix} = \begin{bmatrix} 0-12+21 \\ -12+0+24 \\ 14+16+0 \end{bmatrix} = \begin{bmatrix} 9 \\ 12 \\ 30 \end{bmatrix}$

and $BC = \begin{bmatrix} 0 & 1 & 1 \\ 1 & 0 & 2 \\ 1 & 2 & 0 \end{bmatrix}\begin{bmatrix} 2 \\ -2 \\ 3 \end{bmatrix} = \begin{bmatrix} 0-2+3 \\ 2+0+6 \\ 2-4+0 \end{bmatrix} = \begin{bmatrix} 1 \\ 8 \\ -2 \end{bmatrix}$

So $AC + BC = \begin{bmatrix} 9 \\ 12 \\ 30 \end{bmatrix} + \begin{bmatrix} 1 \\ 8 \\ -2 \end{bmatrix} = \begin{bmatrix} 10 \\ 20 \\ 28 \end{bmatrix}$

Clearly, $(A + B)\,C = AC + BC$

Example 5. If $A = \begin{bmatrix} 1 & 2 & 3 \\ 3 & -2 & 1 \\ 4 & 2 & 1 \end{bmatrix}$, then show that $A^3 - 23A - 40\,I = 0$

Sol. We have, $A^2 = A.A = \begin{bmatrix} 1 & 2 & 3 \\ 3 & -2 & 1 \\ 4 & 2 & 1 \end{bmatrix}\begin{bmatrix} 1 & 2 & 3 \\ 3 & -2 & 1 \\ 4 & 2 & 1 \end{bmatrix} = \begin{bmatrix} 19 & 4 & 8 \\ 1 & 12 & 8 \\ 14 & 6 & 15 \end{bmatrix}$

$$A^3 = A.A^2 = \begin{bmatrix} 1 & 2 & 3 \\ 3 & -2 & 1 \\ 4 & 2 & 1 \end{bmatrix}\begin{bmatrix} 19 & 4 & 8 \\ 1 & 12 & 8 \\ 14 & 6 & 15 \end{bmatrix} = \begin{bmatrix} 63 & 46 & 69 \\ 69 & -6 & 23 \\ 92 & 46 & 63 \end{bmatrix}$$

$$-23\,A - 40I = \begin{bmatrix} 63 & 46 & 69 \\ 69 & -6 & 23 \\ 92 & 46 & 63 \end{bmatrix} - 23\begin{bmatrix} 1 & 2 & 3 \\ 3 & -2 & 1 \\ 4 & 2 & 1 \end{bmatrix} - 40\begin{bmatrix} 1 & 0 & 0 \\ 0 & 1 & 0 \\ 0 & 0 & 1 \end{bmatrix}$$

$$= \begin{bmatrix} 63 & 46 & 69 \\ 69 & -6 & 23 \\ 92 & 46 & 63 \end{bmatrix} - 23\begin{bmatrix} -23 & -46 & -69 \\ -69 & 46 & -23 \\ -92 & -46 & -23 \end{bmatrix} + \begin{bmatrix} -40 & 0 & 0 \\ 0 & -40 & 0 \\ 0 & 0 & -40 \end{bmatrix}$$

$$= \begin{bmatrix} 63-23-40 & 46-46+0 & 69-69+0 \\ 69-69+0 & -6+46-40 & 23-23+0 \\ 92-92+0 & 46-46+0 & 63-23-40 \end{bmatrix} = \begin{bmatrix} 0 & 0 & 0 \\ 0 & 0 & 0 \\ 0 & 0 & 0 \end{bmatrix} = 0$$

Example 6. If $A = \begin{bmatrix} 0 & 0 \\ 1 & -1 \end{bmatrix}$ and $B\begin{bmatrix} 0 & 1 \\ 1 & 0 \end{bmatrix}$, then S.T. $AB \neq BA$

Sol. Here, $AB = \begin{bmatrix} 0 & 1 \\ -1 & 0 \end{bmatrix}$, $BA = \begin{bmatrix} 0 & -1 \\ 1 & 0 \end{bmatrix}$

So, $AB \neq BA$

Example 7. Obtain the inverse of the following matrix using $A = \begin{bmatrix} 0 & 1 & 2 \\ 1 & 2 & 3 \\ 3 & 1 & 1 \end{bmatrix}$

Sol. $A = IA$, *i.e.*, $\begin{bmatrix} 0 & 1 & 2 \\ 1 & 2 & 3 \\ 3 & 1 & 1 \end{bmatrix} = \begin{bmatrix} 1 & 0 & 0 \\ 0 & 1 & 0 \\ 0 & 0 & 1 \end{bmatrix} A$

$R_1 \to R_1 - R_2$

$$\begin{bmatrix} 1 & 2 & 3 \\ 0 & 1 & 2 \\ 3 & 1 & 1 \end{bmatrix} = \begin{bmatrix} 0 & 1 & 0 \\ 1 & 0 & 0 \\ 0 & 0 & 1 \end{bmatrix} A$$

$R_3 \to R_3 - 3R_1$

$$\begin{bmatrix} 1 & 1 & 3 \\ 0 & 1 & 2 \\ 0 & -5 & -8 \end{bmatrix} = \begin{bmatrix} 0 & 1 & 0 \\ 1 & 0 & 0 \\ 0 & -3 & 1 \end{bmatrix} A$$

$R_1 \to R_1 - 2R_2$

$$\begin{bmatrix} 1 & 0 & -1 \\ 0 & 1 & 2 \\ 0 & -5 & -8 \end{bmatrix} = \begin{bmatrix} -2 & 1 & 0 \\ 1 & 0 & 0 \\ 0 & -3 & 1 \end{bmatrix} A$$

$R_3 \to R_3 + 5R_2$

$$\begin{bmatrix} 1 & 0 & -1 \\ 0 & 1 & 2 \\ 0 & 0 & 2 \end{bmatrix} = \begin{bmatrix} -2 & 1 & 0 \\ 1 & 0 & 0 \\ 5 & -3 & 1 \end{bmatrix}$$

$R_3 \to ½R_3$

$$\begin{bmatrix} 1 & 0 & -1 \\ 0 & 1 & 2 \\ 0 & 0 & 1 \end{bmatrix} = \begin{bmatrix} -2 & 1 & 0 \\ 1 & 0 & 0 \\ 5/2 & -3/2 & 1/2 \end{bmatrix}$$

$R_1 \to R_1 + R_3$

$$\begin{bmatrix} 1 & 0 & 0 \\ 0 & 1 & 2 \\ 0 & 0 & 1 \end{bmatrix} = \begin{bmatrix} 1/2 & -1/2 & 1/2 \\ 1 & 0 & 0 \\ 5/2 & -3/2 & 1/2 \end{bmatrix} A$$

$R_2 \to R_2 + 2R_3$

$$\begin{bmatrix} 1 & 0 & 0 \\ 0 & 1 & 0 \\ 0 & 0 & 1 \end{bmatrix} = \begin{bmatrix} 1/2 & -1/2 & 1/2 \\ -4 & -3 & -1 \\ 5/2 & -3/2 & 1/2 \end{bmatrix} A$$

$$A^{-1} = \begin{bmatrix} 1/2 & -1/2 & 1/2 \\ -4 & 3 & -1 \\ 5/2 & -3/2 & 1/2 \end{bmatrix}$$

Example 8: *For the following matrices:* $A = \begin{pmatrix} 2 & -4 & 5 \\ 3 & 2 & -7 \\ -7 & 0 & 6 \end{pmatrix}$ and $B = \begin{pmatrix} 0 & -5 & 6 \\ 2 & -3 & 4 \\ -7 & 2 & 1 \end{pmatrix}$, *show that matrix multiplication of a scalar holds good under addition operation.*

Sol. Given the matrices

$$A = \begin{pmatrix} 2 & -4 & 5 \\ 3 & 2 & -7 \\ -7 & 0 & 6 \end{pmatrix}_{3\times 3} \quad \text{and} \quad B = \begin{pmatrix} 0 & -5 & 6 \\ 2 & -3 & 4 \\ -7 & 2 & 1 \end{pmatrix}_{3\times 3}$$

Suppose 'q' be any scalar then we need to prove that:

$$q(A + B) = qA + qB$$

Now,
$$A + B = \begin{pmatrix} 2 & -4 & 5 \\ 3 & 2 & -7 \\ -7 & 0 & 6 \end{pmatrix} + \begin{pmatrix} 0 & -5 & 6 \\ 2 & -3 & 4 \\ -7 & 2 & 1 \end{pmatrix}$$

$$\Rightarrow \quad A + B = \begin{pmatrix} 2+0 & -4-5 & 5+6 \\ 3+2 & 2-3 & -7+4 \\ -7-7 & 0+2 & 6+1 \end{pmatrix}$$

$$\Rightarrow \quad A + B = \begin{pmatrix} 2 & -9 & 11 \\ 5 & -1 & -3 \\ -14 & 2 & 7 \end{pmatrix}$$

$$\Rightarrow \quad q(A + B) = \begin{pmatrix} 2 & -9 & 11 \\ 5 & -1 & -3 \\ -14 & 2 & 7 \end{pmatrix} = \begin{pmatrix} 2q & -9q & 11q \\ 5q & -1q & -3q \\ -14q & 2q & 7q \end{pmatrix} \quad \text{...}(I)$$

Now,
$$qA = q\begin{pmatrix} 2 & -4 & 5 \\ 3 & 2 & -7 \\ -7 & 0 & 6 \end{pmatrix} = \begin{pmatrix} 2q & -4q & 5q \\ 3q & 2q & -7q \\ -7q & 0 & 6q \end{pmatrix}$$

and
$$qB = q\begin{pmatrix} 0 & -5 & 6 \\ 2 & -3 & 4 \\ -7 & 2 & 1 \end{pmatrix} = \begin{pmatrix} 0 & -5q & 6q \\ 2q & -3q & 4q \\ -7q & 2q & q \end{pmatrix}$$

$$\Rightarrow \quad qA + qB = \begin{pmatrix} 2q & -4q & 5q \\ 3q & 2q & -7q \\ -7q & 0 & 6q \end{pmatrix} + \begin{pmatrix} 0 & -5q & 6q \\ 2q & -3q & 4q \\ -7q & 2q & q \end{pmatrix}$$

$$= \begin{pmatrix} 2q+0 & -4q-5q & 5q+6q \\ 3q+2q & 2q-3q & -7q+4q \\ -7q-7q & 0+2q & 6q+q \end{pmatrix} = \begin{pmatrix} 2q & -9q & 11q \\ 5q & -q & -3q \\ -14q & 2q & 7q \end{pmatrix} \quad ...(II)$$

From result (I) and (II), it is clear that

$$q(A + B) = qA + qB. \textbf{ Proved.}$$

Example 9. $A = \begin{pmatrix} 3 & 5 & -11 \\ 2 & -7 & 9 \\ 13 & 2 & 6 \end{pmatrix}, B = \begin{pmatrix} 3 & 1 & 2 \\ 5 & 0 & -9 \\ 7 & 11 & -14 \end{pmatrix}$ *and* $C = \begin{pmatrix} 5 & 0 & 2 \\ -13 & 0 & -21 \\ 17 & 9 & -4 \end{pmatrix}$ *then show that associativity under addition operation of matrices holds true.*

Sol. $A = \begin{pmatrix} 3 & 5 & -11 \\ 2 & -7 & 9 \\ 13 & 2 & 6 \end{pmatrix}, B = \begin{pmatrix} 3 & 1 & 2 \\ 5 & 0 & -9 \\ 7 & 11 & -14 \end{pmatrix}$ *and* $C = \begin{pmatrix} 5 & 0 & 2 \\ -13 & 0 & -21 \\ 17 & 9 & -4 \end{pmatrix}$ *then*

Given the matrices

$$A = \begin{pmatrix} 3 & 5 & -11 \\ 2 & -7 & 9 \\ 13 & 2 & 6 \end{pmatrix}, B = \begin{pmatrix} 3 & 1 & 2 \\ 5 & 0 & -9 \\ 7 & 11 & -14 \end{pmatrix} \text{ and } C = \begin{pmatrix} 5 & 0 & 2 \\ -13 & 0 & -21 \\ 17 & 9 & -4 \end{pmatrix} \text{ then}$$

Then we need to prove that

$$A + (B + C) = (A + B) + C$$

For this, we observe

$$B + C = \begin{pmatrix} 3 & 1 & 2 \\ 5 & 0 & -9 \\ 7 & 11 & -14 \end{pmatrix} + \begin{pmatrix} 5 & 0 & 2 \\ -13 & 0 & -21 \\ 17 & 9 & -4 \end{pmatrix}$$

$$= \begin{pmatrix} 3+5 & 1+0 & 2+2 \\ 5+13 & 0+0 & -9-21 \\ 7+17 & 11+9 & -14-4 \end{pmatrix} = \begin{pmatrix} 8 & 1 & 4 \\ -8 & 0 & -30 \\ 24 & 20 & -18 \end{pmatrix}$$

Then, $$A + (B + C) = \begin{pmatrix} 3 & 5 & -11 \\ 2 & -7 & 9 \\ 13 & 2 & 6 \end{pmatrix} = \begin{pmatrix} 8 & 1 & 4 \\ -8 & 0 & -30 \\ 24 & 20 & -18 \end{pmatrix}$$

$$= \begin{pmatrix} 3+8 & 5+1 & -11+4 \\ 2-8 & -7+0 & 9-30 \\ 13+24 & 2+20 & 6-18 \end{pmatrix} = \begin{pmatrix} 11 & 6 & -7 \\ -6 & -7 & -21 \\ 37 & 22 & -12 \end{pmatrix} \quad ...(I)$$

Similarly, we have

$$A + B = \begin{pmatrix} 3 & 5 & -11 \\ 2 & -7 & 9 \\ 13 & 2 & 6 \end{pmatrix} = \begin{pmatrix} 3 & 1 & 2 \\ 5 & 0 & -9 \\ 7 & 11 & -14 \end{pmatrix}$$

$$= \begin{pmatrix} 3+3 & 5+1 & -11+2 \\ 2+5 & -7+0 & 9-9 \\ 13+7 & 2+11 & 6-14 \end{pmatrix} = \begin{pmatrix} 6 & 6 & -9 \\ 7 & -7 & 0 \\ 20 & 13 & -8 \end{pmatrix}$$

$$\Rightarrow \quad (A+B)+C = \begin{pmatrix} 6 & 6 & -9 \\ 7 & -7 & 0 \\ 20 & 13 & -8 \end{pmatrix} + \begin{pmatrix} 5 & 0 & 2 \\ -13 & 0 & -21 \\ 17 & 9 & -4 \end{pmatrix} = \begin{pmatrix} 6+5 & 6+0 & -9+2 \\ 7-13 & -7+0 & 0-21 \\ 20+17 & 13+9 & -8-4 \end{pmatrix}$$

$$= \begin{pmatrix} 11 & 6 & -7 \\ -6 & -7 & -21 \\ 37 & 22 & -12 \end{pmatrix} \qquad ...(II)$$

From the results (*I*) and (*II*), it is clear

$A + (B + C) = (A + B) + C.$ **Proved.**

Example 10. *Evaluate (i) – 3A + 4B (ii) 4B – 3A and check whether*

– 3A + 4B = 4B – 3A

where, $$A = \begin{pmatrix} 3 & 4 & -2 \\ -11 & 5 & -7 \\ 7 & 2 & -14 \end{pmatrix} \text{ and } B = \begin{pmatrix} 2 & 8 & 0 \\ -5 & -9 & 1 \\ 7 & -13 & 10 \end{pmatrix}$$

Sol.

Given the matrices: $$A = \begin{pmatrix} 3 & 4 & -2 \\ -11 & 5 & -7 \\ 7 & 2 & -14 \end{pmatrix} \text{ and } B = \begin{pmatrix} 2 & 8 & 0 \\ -5 & -9 & 1 \\ 7 & -13 & 10 \end{pmatrix}$$

So, by scalar multiplication of matrices

$$(-3)A = -3A = (-3)\begin{pmatrix} 3 & 4 & -2 \\ -11 & 5 & -7 \\ 7 & 2 & -14 \end{pmatrix} = \begin{pmatrix} -9 & -12 & 6 \\ 33 & -15 & 21 \\ -21 & -6 & 42 \end{pmatrix}$$

and $$4B = 4\begin{pmatrix} 2 & 8 & 0 \\ -5 & -9 & 1 \\ 7 & -13 & 10 \end{pmatrix} = \begin{pmatrix} 8 & 32 & 0 \\ -20 & -36 & 4 \\ 28 & -52 & 40 \end{pmatrix}$$

$$\Rightarrow \quad -3A + 4B = \begin{pmatrix} -9 & -12 & 6 \\ 33 & -15 & 21 \\ -21 & -6 & 42 \end{pmatrix} + \begin{pmatrix} 8 & 32 & 0 \\ -20 & -36 & 4 \\ 28 & -52 & 40 \end{pmatrix}$$

$$= \begin{pmatrix} -9+8 & -12+32 & 6+0 \\ 33-20 & -15-36 & 21+4 \\ -21+28 & -6-52 & 42+40 \end{pmatrix} = \begin{pmatrix} -1 & 20 & 6 \\ 13 & -51 & 25 \\ 7 & -58 & 82 \end{pmatrix} \qquad ...(I)$$

and
$$4B - 3A = \begin{pmatrix} 8 & 32 & 0 \\ -20 & -36 & 4 \\ 28 & -52 & 40 \end{pmatrix} - 3\begin{pmatrix} 3 & 4 & -2 \\ -11 & 5 & -7 \\ 7 & 2 & -14 \end{pmatrix}$$

$$\Rightarrow \quad (4B - 3A) = \begin{pmatrix} 8 & 32 & 0 \\ -20 & -36 & 4 \\ 28 & -52 & 40 \end{pmatrix} - \begin{pmatrix} 9 & 12 & -6 \\ -33 & 15 & -21 \\ 21 & 6 & -42 \end{pmatrix}$$

$$= \begin{pmatrix} 8-9 & 32-12 & 0+6 \\ -20+33 & -36-15 & 4+21 \\ 28-21 & -52-6 & 40+42 \end{pmatrix} = \begin{pmatrix} -1 & 20 & 6 \\ 13 & -51 & 25 \\ 7 & -58 & 82 \end{pmatrix} \quad ...(II)$$

From the results (*I*) and (*II*) it is clear that

$$-3A + 4B = 4B - 3A.$$

1.4.5. Product of Matrices (Matrix Multiplication) Theorem

Matrix multiplication is a binary operation of a pair of matrices which when operated with each other subjected to certain conditions, produces another matrix. The main characteristics of any matrix multiplication include the order of matrices and specifying how the entries of matrices produces a new one.

Definition: Let $A = (a_j)_{m\times n}$ and $B = (b_{jk})_{n\times p}$ be two matrices such that the number of columns of matrix A equals the number of rows of matrix B *i.e.*, n. Then, a new matrix.

$C = (c_{ik})_{m\times p}$ can be generated such that

$$c_{ik} = \sum_{j=1}^{n} a_{ij}.b_{jk} \quad ...(I)$$

This '*c*' is called the product matrix of A and B in that order and is expressed as

$$C = AB.$$

Here

(*i*) Summation as reflected in *I* is w.r.t. the repeated suffix.

(*ii*) In the expression $C = AB$

A is pre-factor and

B is post-factor

(*iii*) AB is said to exist or conformable iff

= Number of columns of matrix A

= Number of rows of matrix B.

(*iv*) In the above summation:

$$c_{ik} = \sum_{j=1}^{n} a_{ij}.b_{jk} = a_{i1}b_{1k} + a_{i2}b_{2k} + ... + a_{in}b_{nk}$$

$\Rightarrow$ $(i, k)^{th}$ element of matrix $C = AB$ can be obtained by multiplying the corresponding elements of the ith row of A and the kth column of B and then adding the so obtained products.

1. Associativity under Multiplication of Matrices: Let A, B and C are any three matrices of the order $m \times n$, $n \times p$ and $p \times q$ respectively such that conformability is assured, then:

$$A(BC) = (AB)C$$

Proof: Consider the following matrices:

$$A = (a_{ij})_{m\times n};\ B = (b_{jk})_{n\times p} \quad \text{and} \quad C = (c_{kl})_{p\times q}$$

Then from the definition of matrix multiplication we may suppose.

$$AB = (u_{ik})_{m\times p} \text{ where } u_{ik} = \sum_{j=1}^{n} a_{ij}.b_{jk}$$

and

$$BC = (v_{jl})_{n\times q} \quad \text{where } v_{jl} = \sum_{k=1}^{p} a_{jk}.c_{kl}$$

Clearly, order of $A = (m \times n)$ and

order of $BC = (n \times q)$

So, $A(BC)$ is conformable.

Similarly, order of $AB = (m \times p)$ and

order of $C = (p \times q)$

So, $(AB)C$ is conformable.

Now, we may suppose

$$A(BC) = (w_{il})_{m\times q}$$

where $w_{il} = (i, l)^{\text{th}}$ element of $A\,(BC)$

where,

$$w_{il} = \sum_{j=1}^{n} a_{ij}.b_{jl}$$

$$= \sum_{j=1}^{n}\left[a_{ij}\left\{\sum_{k=1}^{p} b_{jk}c_{kl}\right\}\right] \qquad \text{(Putting the value of } v_{jl})$$

$$= \sum_{k=1}^{p}\left[\left\{\sum_{j=1}^{n}(a_{ij}\ b_{jk})\ c_{kl}\right\}\right]$$

(Finite summation interchangeability property)

$$= \sum_{k=1}^{p}\left[\left\{\sum_{j=1}^{n}(a_{ij}b_{jk})\right\}c_{kl}\right]$$

$$= \sum_{k=1}^{p} u_{ik}c_{kl} \qquad \left(\because u_{ik} = \sum_{j=1}^{n} c_{ij}b_{jk}\right)$$

$= (i, l)^{\text{th}}$ element of $(AB)\,C$

Thus, applying equality of matrices.

$A(BC) = (AB)C$. Proved.

2. Distributivity under addition of matrices: Let A, B and C are any three matrices of the order $(m \times n)$, $(n \times p)$ and $(n \times p)$ respectively then

$$A(B + C) = AB + AC$$

Note: For having $(B + C)$ feasible, the orders of B and C must be the same.

Obviously, order of $A = (m \times n)$ and

order of $(B + C) = (n \times p)$

$\Rightarrow A(B + C)$ is conformable and is of the order $(m \times p)$.

Again order AB and $AC = (m \times p)$.

Now, let $A = (a_{ij})_{m\times n}$

$B = (b_{jk})_{n\times p}$ and $C = (c_{jk})_{n\times p}$

$\Rightarrow \quad B + C = (b_{jk} + c_{jk})_{n\times p}$

and $(i, k)^{\text{th}}$ element of

$$A(B + C) = \sum_{j=1}^{n} a_{ij}(b_{jk} + c_{jk})$$

$$\therefore (i, k)\text{th element of} A(B + C) = \sum_{j=1}^{n}(a_{ij}b_{jk} + a_{jk}c_{jk})$$

$$= \sum_{j=1}^{n} a_{ij}b_{jk} + \sum_{j=1}^{n} a_{ij}c_{jk}$$

$= (i, k)^{\text{th}}$ element $AB + (i, k)^{\text{th}}$ element of A

$= (i, k)^{\text{th}}$ element of $(AB + AC)$

So, by equality of matrices

$A(B + C) = AB + AC$. Proved.

This distributivity is shown for left distributivity under addition of matrices.

In the same fashion, right distributivity can be established.

• **Commutativity (Not necessary):** The product of matrices does not hold good always. The following cases can be considered.

Consider, A and B are two matrices such that:

(I) AB exists and BA does not exist.

Let the order of $A = (m \times n)$ and

the order of $B = (n \times p)$

Obviously if $p \neq m$ then BA does not exist and AB exists.

In such case, commutativity does not hold good.

(II) AB and BA both are conformable.

Let order of $A = (m \times n)$ and

order of $B = (n \times m)$

Then obviously AB and BA both exist such that

order of $AB = (m \times n)$ and

order of $BA = (n \times n)$

So, if $m \neq n$

$\Rightarrow$ Order of AB and BA are different.

$\therefore AB \neq BA$.

3. AB and BA both exist and A and B are having same orders. Even then it is not necessary that A and B commute.

It can be shown by the following illustration.

Let, $A = \begin{pmatrix} 1 & 0 \\ 0 & 2 \end{pmatrix}_{2\times 2}$ and $B = \begin{pmatrix} 0 & 1 \\ 1 & 0 \end{pmatrix}_{2\times 2}$

Then, $$AB = \begin{pmatrix} 1 & 0 \\ 0 & 2 \end{pmatrix}\begin{pmatrix} 0 & 1 \\ 1 & 0 \end{pmatrix} = \begin{pmatrix} 1.0+0.1 & 1.1+0.0 \\ 0.0+2.1 & 0.1+2.0 \end{pmatrix} = \begin{pmatrix} 0 & 1 \\ 2 & 0 \end{pmatrix}$$

and $$BA = \begin{pmatrix} 0 & 1 \\ 1 & 0 \end{pmatrix}\begin{pmatrix} 1 & 0 \\ 0 & 2 \end{pmatrix} = \begin{pmatrix} 0.1+1.0 & 0.0+1.2 \\ 1.1+0.0 & 1.0+0.2 \end{pmatrix} = \begin{pmatrix} 0 & 2 \\ 1 & 0 \end{pmatrix}$$

Thus $AB \neq BA$.

4. Might be possible in many cases A and B commute; e.g.,

Let $$A = \begin{pmatrix} -2 & 3 & -1 \\ -1 & 2 & -1 \\ -6 & 9 & -4 \end{pmatrix} \text{ and } B = \begin{pmatrix} 1 & 3 & -1 \\ 2 & 2 & -1 \\ 3 & 0 & -1 \end{pmatrix}$$

We can have $$AB = \begin{pmatrix} -2 & 3 & -1 \\ -1 & 2 & -1 \\ -6 & 9 & -4 \end{pmatrix}\begin{pmatrix} 1 & 3 & -1 \\ 2 & 2 & -1 \\ 3 & 0 & -1 \end{pmatrix}$$

$\Rightarrow$ $$AB = \begin{pmatrix} -2+6-3 & -6+6+0 & 2-3+1 \\ -1+4+3 & -3+4+0 & 1-2+1 \\ -6+18-12 & -18+18+0 & 6-9+4 \end{pmatrix} = \begin{pmatrix} 1 & 0 & 0 \\ 0 & 1 & 0 \\ 0 & 0 & 1 \end{pmatrix}$$

and $$BA = \begin{pmatrix} 1 & 3 & -1 \\ 2 & 2 & -1 \\ 3 & 0 & -1 \end{pmatrix}\begin{pmatrix} -2 & 3 & -1 \\ -1 & 2 & -1 \\ -6 & 9 & -4 \end{pmatrix} = \begin{pmatrix} -2-3+6 & 3+6-9 & -1-3+1 \\ -4-2+6 & 6+4-9 & -2-2+2 \\ -6+0+6 & 9+0-9 & -3+0+1 \end{pmatrix}$$

$\Rightarrow$ $$BA = \begin{pmatrix} 1 & 0 & 0 \\ 0 & 1 & 0 \\ 0 & 0 & 1 \end{pmatrix} = AB. \textbf{ Ans.}$$

Representation of matrix products: Consider the two matrices A and B of the order $(m \times n)$ and $(n \times p)$ respectively.

Let, we consider:

$$A = (a_{ij})_{m\times n} \quad \text{and} \quad B = (b_{jk})_{n\times p}$$

where, $i = 1, 2, 3, ..., m$

$j = 1, 2, 3, ..., n$ and

$k = 1, 2, 3, ..., p.$

Also, say

Rows of matrix $A = R_1, R_2, ..., R_m$

and Columns of matrix $B = C_1, C_2, ..., C_p$

Obviously, order of $R_i = (1 \times n);\ i = 1, 2, ..., m$

and order of $C_k = (n \times 1);\ i = 1, 2, ..., p$

Hence, product R_iC_k is defined giving (1×1) matrix. Thus from product of matrix.

$$\begin{pmatrix} R_1 \\ R_2 \\ \vdots \\ R_m \end{pmatrix}(C_1C_2...C_p) = \begin{pmatrix} R_1C_1 & R_1C_2 & ... & R_1C_p \\ R_2C_1 & R_2C_2 & ... & R_2C_p \\ ... & ... & ... & ... \\ R_mC_1 & R_mC_2 & ... & R_mC_p \end{pmatrix}_m$$

Now, $(i, k)^{\text{th}}$ element of $AB = \sum_{j=1}^{n} a_{ij}b_{jk} = a_{i1}b_{1k} + a_{i2}b_{2k} + ... + a_{in}b_{nk}$

$$= [a_{i1}\; a_{i2}\; ...a_{in}]\begin{bmatrix} b_{1k} \\ b_{2k} \\ \vdots \\ b_{nk} \end{bmatrix} = R_iC_k = (i, k)\text{th element of matrix } [R_iC_k]$$

$$\Rightarrow \qquad AB = \begin{bmatrix} R_1 \\ R_2 \\ \vdots \\ R_m \end{bmatrix}[C_1\; C_2\; ...\; C_p] = \begin{pmatrix} R_1C_1 & R_1C_2 & ... & R_1C_p \\ R_2C_1 & R_2C_2 & ... & R_2C_p \\ ... & ... & ... & ... \\ R_mC_1 & R_mC_2 & ... & R_mC_p \end{pmatrix}_{m\times p}$$

1.4.6. Triangular Diagonal Matrices

Depending upon the triangular structure of the matrix, there are two types of triangular matrices:

(*a*) Upper Triangular Matrix (*b*) Lower Triangular Matrix

Definition

1. Upper Triangular Matrix: Let $A = (a_{ij})_{n\times n}$ be a square matrix of order n. Then A is said to be an upper triangular matrix if

$$a_{ij} = 0 \text{ if } i > j$$

i.e.,

$$A = \begin{pmatrix} a_{11} & a_{12} & ... & a_{1n} \\ 0 & a_{22} & ... & a_{2n} \\ 0 & 0 & ... & a_{3n} \\ 0 & 0 & ... & a_{nm} \end{pmatrix} \quad \text{Principal diagonal}$$

In this way in the upper triangular matrix. The elements below the principal are zero i.e., first row will have max. number of non-zero element and last row will have minimum number of non-zero element.

2. Lower Triangular Matrix: Let $A = (a_{ij})_{n\times n}$ is a square matrix of order n given A is said to be a lower triangular matrix if:

$$a_{ij} = 0 \;\forall\; i < j$$

i.e.

$$A = \begin{pmatrix} a_{11} & 0 & 0 & ... & 0 \\ a_{21} & a_{22} & 0 & ... & 0 \\ a_{31} & a_{32} & a_{33} & ... & 0 \\ ... & ... & ... & ... & ... \\ 0 & 0 & 0 & ... & \end{pmatrix} \text{Principal diagonal.}$$

In this way in lower triangular matrix, all the elements above the principal diagonal are zero.

First row *i.e.*, will have the minimum number of non-zero elements and the last row will have maximum number of non-zero elements.

A triangular matrix $A = (a_{ij})_{n \times n}$ is said to be strictly triangular if:

$$a_{ij} = 0 \ \forall \ i = 1, 2, ..., n$$

3. Diagonal Matrix: Let $A = (a_{ij})_{n \times n}$ be an n-rowed matrix such that the elements below and above the principal diagonals are all zero. Then 'A' is said to be a diagonal matrix.

It should be noted that the some or all elements along the principal diagonal of the diagonal matrix may be zero.

In such a case, if all the principal diagonal elements are zero then specifically it becomes null matrix.

e.g. $\begin{pmatrix} 2 & 0 & 0 \\ 0 & 5 & 0 \\ 0 & 0 & 9 \end{pmatrix}, \begin{pmatrix} 0 & 0 & 0 \\ 0 & 2 & 0 \\ 0 & 0 & -3 \end{pmatrix}, \begin{pmatrix} 0 & 0 & 0 \\ 0 & 0 & 0 \\ 0 & 0 & 0 \end{pmatrix}$

all are diagonal matrices.

As a representation, if the elements along the principal diagonal of the diagonal matrix D be $d_1, d_2, ..., d_n$ then we usually represent D by diagonal $(d_1, d_2, ..., d_n)$.

4. Scalar Matrix: It is a special type of diagonal matrix with the condition that:

$$a_{11} = a_{22} = ... = a_{nn} \text{ i.e.}$$

all the principal diagonal elements should be equal.

Example 11. *Suppose* $A = \begin{pmatrix} 2 & 1 \\ -1 & 2 \end{pmatrix}$ *and* $B = \begin{pmatrix} 2 & 3 & 1 \\ 1 & -3 & 2 \end{pmatrix}$. *Then express the feasibility of matrix multiplication from both ends* i.e., *AB and BA.*

Sol. Given the matrix:

$$A = \begin{pmatrix} 2 & 1 \\ -1 & 2 \end{pmatrix}_{2\times 2} \text{ and } B = \begin{pmatrix} 2 & 3 & 1 \\ 1 & -3 & 2 \end{pmatrix}_{2\times 3}$$

where the subscripts show the order of the matrices. Here, it can be observed that,

The number of columns of matrix $A = 2$ = number of rows of matrix B.

So the matrix multiplication of A with B is feasible i.e., AB exists as follows:

$$AB = \begin{pmatrix} 2 & 1 \\ -1 & 2 \end{pmatrix}\begin{pmatrix} 2 & 3 & 1 \\ 1 & -3 & 2 \end{pmatrix}.$$

$$\Rightarrow \quad AB = \begin{pmatrix} 2\times 2+1\times 1 & 2\times 3-1\times 3 & 2\times 1+1\times 2 \\ -1\times 2+2\times 1 & -1\times 3-2\times 3 & -1\times 1+2\times 2 \end{pmatrix}$$

$$\Rightarrow \quad AB = \begin{pmatrix} 5 & 3 & 4 \\ 0 & -9 & 3 \end{pmatrix}$$

which is the product matrix of the order 2×3.

However, for the product of matrix B with A, we observe,

The number of columns of matrix $B = 3 \neq$ No. of rows of matrix A.

So the product matrix BA does not exist.
This confirms that the matrix multiplication is not always commutative.

Example 12. *Find the product of the following two matrices*

$$A = \begin{pmatrix} 0 & c & -b \\ -c & 0 & a \\ b & -a & 0 \end{pmatrix} \text{ and } B = \begin{pmatrix} a^2 & ab & ac \\ ab & b^2 & bc \\ ac & bc & c^2 \end{pmatrix}$$ *in both ways* i.e., *AB and BA.*

Sol. Given the matrices:

$$\begin{pmatrix} 0 & c & -b \\ -c & 0 & a \\ b & -a & 0 \end{pmatrix}_{3\times 3} \quad \text{and } B = \begin{pmatrix} a^2 & ab & ac \\ ab & b^2 & bc \\ ac & bc & c^2 \end{pmatrix}_{3\times 3}$$

Evidently both are square matrices of the same order so AB and BA both exist.

Now $$AB = \begin{pmatrix} 0 & c & -b \\ -c & 0 & a \\ b & -a & 0 \end{pmatrix}\begin{pmatrix} a^2 & ab & ac \\ ab & b^2 & bc \\ ac & bc & c^2 \end{pmatrix}$$

$$\Rightarrow \quad AB = \begin{pmatrix} 0.a^2 + c.ab + (-b).ac & 0.ab + c.b^2 + (-b).bc & 0.ac + c.bc - bc^2 \\ -c.a^2 + 0.ab + a.ac & -c.ab + 0.b^2 + a.bc & -c.ac + 0.bc + ac^2 \\ b.a^2 + (-a).ab + 0.ac & b.ab + (-a).b^2 + 0.bc & b.ac - a.bc + ac^2 \end{pmatrix}$$

$$\Rightarrow \quad AB = \begin{pmatrix} abc - abc & b^2c - b^2c & bc^2 - bc^2 \\ -a^2c + a^2c & -abc + abc & -ac^2 + ac^2 \\ a^2b - a^2b & ab^2 - ab^2 & abc - abc \end{pmatrix}$$

$$\Rightarrow \quad AB = \begin{pmatrix} 0 & 0 & 0 \\ 0 & 0 & 0 \\ 0 & 0 & 0 \end{pmatrix} = \text{Null matrix.}$$

Now,

$$BA = \begin{pmatrix} a^2 & ab & ac \\ ab & b^2 & bc \\ ac & bc & c^2 \end{pmatrix}\begin{pmatrix} 0 & c & -b \\ -c & 0 & a \\ b & -a & 0 \end{pmatrix}$$

$$\Rightarrow \quad BA = \begin{pmatrix} a^2.0 + ab.(-c) + ac.b & a^2.c + ab.0 + ac.(-a) & a^2.(-b) + aba + ac.0 \\ ab.0 + b^2.(-c) + bc.b & ab.c + b^2.0 + bc(-a) & ab.(-b) + b^2.a + bc.0 \\ ac.0 + bc.(-c) + c^2.b & ac.c + bc.0 + c^2.(-a) & ac.(-b) + bc.a + c^2.0 \end{pmatrix}$$

$$\Rightarrow \quad BA = \begin{pmatrix} -abc + abc & a^2c + a^2c & -a^2b + a^2b \\ -b^2c + b^2c & abc - abc & -ab^2 + ab^2 \\ -bc^2 + bc^2 & ac^2 - ac^2 & -abc + abc \end{pmatrix}$$

$$\Rightarrow \qquad BA = \begin{pmatrix} 0 & 0 & 0 \\ 0 & 0 & 0 \\ 0 & 0 & 0 \end{pmatrix} = \text{Null matrix.}$$

Thus, we observe AB and BA both exist and

$$\boldsymbol{AB = BA = \textbf{Null matrix.}}$$

Example 13. If $A = \begin{pmatrix} 2 & 3 & 4 \\ 1 & 2 & 3 \\ -1 & 1 & 2 \end{pmatrix}$ *and* $B = \begin{pmatrix} 1 & 3 & 0 \\ -1 & 2 & 1 \\ 0 & 0 & 2 \end{pmatrix}$ *check whether* $AB = BA$?

Sol. Given, $A = \begin{pmatrix} 2 & 3 & 4 \\ 1 & 2 & 3 \\ -1 & 1 & 2 \end{pmatrix}$ *and* $B = \begin{pmatrix} 1 & 3 & 0 \\ -1 & 2 & 1 \\ 0 & 0 & 2 \end{pmatrix}$

Clearly, both are square matrices of order 3.
So, AB and BA both exist as follows:

$$AB = \begin{pmatrix} 2 & 3 & 4 \\ 1 & 2 & 3 \\ -1 & 1 & 2 \end{pmatrix}\begin{pmatrix} 1 & 3 & 0 \\ -1 & 2 & 1 \\ 0 & 0 & 2 \end{pmatrix}$$

$$\Rightarrow \qquad AB = \begin{pmatrix} 2.1 - 3.1 + 4.0 & 2.3 + 3.2 + 4.0 & 2.0 + 3.1 + 4.2 \\ 1.1 - 2.1 + 3.0 & 1.3 + +2.2 + 3.0 & 1.0 + 2.1 + 3.2 \\ -1.1 - 1.1 + 2.0 & -1.3 + 1.2 + 2.0 & -1.0 + 1.1 + 2.2 \end{pmatrix}$$

$$\Rightarrow \qquad AB = \begin{vmatrix} -1 & 12 & 11 \\ -1 & 7 & 8 \\ -2 & -1 & 5 \end{vmatrix} \qquad \ldots(i)$$

$$AB = \begin{pmatrix} 1 & 3 & 0 \\ -1 & 2 & 1 \\ 0 & 0 & 2 \end{pmatrix}\begin{pmatrix} 2 & 3 & 4 \\ 1 & 2 & 3 \\ -1 & 1 & 2 \end{pmatrix}$$

$$\Rightarrow \qquad BA = \begin{pmatrix} 1.2 - 3.1 + 0.(-1) & 1.3 + 3.2 + 0.1 & 1.4 + 3.3 + 0.2 \\ -1.2 + 2.1 + 1.(-1) & -1.3 + +2.2 + 1.1 & -1.4 + 2.3 + 1.2 \\ 0.2 + 0.1 + 2.(-1) & 0.3 + 0.2 + 2.1 & 0.4 + 0.3 + 2.2 \end{pmatrix}$$

$$\Rightarrow \qquad BA = \begin{pmatrix} 5 & 9 & 13 \\ -1 & 2 & 4 \\ -2 & 2 & 4 \end{pmatrix} \qquad \ldots(ii)$$

Thus, it is observed from (*i*) and (*ii*) that

$$\boldsymbol{AB \neq BA.}$$

Example 14. *Suppose A and B are two matrices such that AB and (A + B) are both defined, then prove that A and B are square matrices of the same order.*

Sol. To show that any matrix is a square matrix, it is pre-requisite to show:

Number of rows = Number of columns.

Now, suppose A be any matrix of the order $p \times q$ i.e.

$$\text{Number of rows} = p \quad \text{and}$$

$$\text{Number of columns} = q.$$

Also, $(A + B)$ is defined where B is another matrix.

From the addition of matrices we know that both matrices should be of the same order.

So, 'B' is also of the order $p \times q$.

Moreover, AB (product of matrices) is also defined. So, from the condition of product of matrices.

Number of columns of A = Number of rows of B

$$\Rightarrow \qquad q = p$$

In this way the order of matrices A and B becomes $p \times q$ *i.e.*, $p \times p$.

It shows that the matrices A and B are square matrices of the same order.

Example 15: *If* $A = \begin{pmatrix} \cos\alpha & -\sin\alpha \\ \sin\alpha & \cos\alpha \end{pmatrix}$ *and* $B = \begin{pmatrix} \cos\beta & -\sin\beta \\ \sin\beta & \cos\beta \end{pmatrix}$ *show that product of matrices is commutative.*

Sol. Given the matrices:

$$A = \begin{pmatrix} \cos\alpha & -\sin\alpha \\ \sin\alpha & \cos\alpha \end{pmatrix}_{2\times 2} \quad \textit{and } B = \begin{pmatrix} \cos\beta & -\sin\beta \\ \sin\beta & \cos\beta \end{pmatrix}$$

Then to show product of matrices is commutative, we have to establish

$$AB = BA$$

Now,
$$AB = \begin{pmatrix} \cos\alpha & -\sin\alpha \\ \sin\alpha & \cos\alpha \end{pmatrix}\begin{pmatrix} \cos\beta & -\sin\beta \\ \sin\beta & \cos\beta \end{pmatrix}$$

$$\Rightarrow \qquad AB = \begin{pmatrix} \cos\alpha\,.\,\cos\beta - \sin\alpha\,.\,\sin\beta & -\cos\alpha\,.\,\sin\beta - \sin\alpha\,.\,\cos\beta \\ \sin\alpha\,.\,\cos\beta + \cos\alpha\,.\,\sin\beta & -\sin\alpha\,.\,\sin\beta + \cos\alpha\,.\,\cos\beta \end{pmatrix}$$

$$\Rightarrow \qquad AB = \begin{pmatrix} \cos(\alpha+\beta) & -\sin(\alpha+\beta) \\ \sin(\alpha+\beta) & \cos(\alpha+\beta) \end{pmatrix} \qquad ...(i)$$

and
$$BA = \begin{pmatrix} \cos\beta & -\sin\beta \\ \sin\beta & \cos\beta \end{pmatrix}\begin{pmatrix} \cos\alpha & -\sin\alpha \\ \sin\alpha & \cos\alpha \end{pmatrix}$$

$$\Rightarrow \qquad BA = \begin{pmatrix} \cos\beta\,.\,\cos\alpha - \sin\beta\,.\,\sin\alpha & -\sin\alpha\,.\,\cos\beta - \sin\beta\,.\,\cos\alpha \\ \sin\beta\,.\,\cos\alpha + \cos\beta\,.\,\sin\alpha & -\sin\alpha\,.\,\sin\beta + \cos\alpha\,.\,\cos\beta \end{pmatrix}$$

$$\Rightarrow \qquad BA = \begin{pmatrix} \cos(\alpha+\beta) & -\sin(\alpha+\beta) \\ \sin(\alpha+\beta) & \cos(\alpha+\beta) \end{pmatrix} \qquad ...(ii)$$

Thus, from (*i*) and (*ii*)

$$\boldsymbol{AB = BA.}$$

Example 16: *If the matrix A is defined as:* $A = \begin{pmatrix} \cos hx & \sin hx \\ \sin hx & \cos hx \end{pmatrix}$ *then prove that:*

$$A^n = \begin{pmatrix} \cos h\, nx & \sin h\, nx \\ \sin h\, nx & \cos h\, nx \end{pmatrix} \text{ for some positive integer } n,$$

Sol. Given the matrix: $A = \begin{pmatrix} \cos hx & \sin hx \\ \sin hx & \cos hx \end{pmatrix}$

To show for some positive integer n,

$$A^n = \begin{pmatrix} \cos h\, nx & \sin h\, nx \\ \sin h\, nx & \cos h\, nx \end{pmatrix} \quad ...(I)$$

For this, we shall proceed through mathematical induction

For this $$A^2 = A \times A = \begin{pmatrix} \cos hx & \sin hx \\ \sin hx & \cos hx \end{pmatrix} \begin{pmatrix} \cos hx & \sin hx \\ \sin hx & \cos hx \end{pmatrix}$$

$$\Rightarrow \quad A^2 = \begin{pmatrix} \cos hx \,.\, \cos hx + \sin hx \,.\, \sin hx & \cos hx \,.\, \sin hx + \sin hx \,.\, \cos hx \\ \sin hx \,.\cos hx + \cos hx \,.\, \sin hx & \sin hx \,.\, \sin hx + \cos hx \,.\, \cos hx \end{pmatrix}$$

$$\Rightarrow \quad A^2 = \begin{pmatrix} \cos h^2x + \sin h^2x & 2\sin hx \,.\, \cos hx \\ 2\sin hx \,.\cos hx & \sin h^2x + \cos h^2x \end{pmatrix}$$

$$\Rightarrow \quad A^2 = \begin{pmatrix} \cos h\, 2x & \sin h\, 2x \\ \sin h\, 2x & \cos h\, 2x \end{pmatrix}$$

Thus I holds good for $n = 2$.

$$A^k \begin{pmatrix} \cos h\, kx & \sin h\, kx \\ \sin h\, kx & \cos h\, kx \end{pmatrix}$$

Then, our purpose would be served if we show that I holds good for $(k + 1)$ also *i.e.*

$$A^{k+1} = \begin{pmatrix} \cos h\, (k+1)x & \sin h\, (k+1)x \\ \sin h\, (k+1)x & \cos h\, (k+1)x \end{pmatrix}$$

For this L.H.S. $$A^{k+1} = \begin{pmatrix} \cos h\, kx & \sin h\, kx \\ \sin h\, kx & \cos h\, kx \end{pmatrix} \begin{pmatrix} \cos hx & \sin hx \\ \sin hx & \cos hx \end{pmatrix}$$

$$\Rightarrow \quad A^{k+1} = \begin{pmatrix} \cos h\, kx \,.\, \cos hx + \sin h\, kx \,.\, \sin hx & \cos h\, kx \,.\, \sin hx + \sin hx \,.\, \cos hx \\ \sin h\, kx \,.\cos hx + \cos h\, kx \,.\, \sin hx & \sin h\, kx \,.\, \sin hx + \cos h\, kx \,.\, \cos hx \end{pmatrix}$$

$$\Rightarrow \quad A^{k+1} = \begin{pmatrix} \cos h(k+1)x & \sin h\, (k+1)x \\ \sin h\, (k+1)x & \cos h\, (k+1)x \end{pmatrix} \text{ (Using trigonometrical identity)}$$

Thus, we observe that, I holds good for $(k + 1)$ also. So, inducing mathematically

$$A^n = \begin{pmatrix} \cos h\, nx & \sin h\, nx \\ \sin h\, nx & \cos h\, nx \end{pmatrix} \text{ for some positive integer } n. \quad \textbf{Proved.}$$

Example 17: *Suppose* $A = \begin{pmatrix} 0 & 1 \\ 1 & 0 \end{pmatrix}$; $B = \begin{pmatrix} 0 & -i \\ i & 0 \end{pmatrix}$ *and* $C = \begin{pmatrix} 1 & 0 \\ 0 & -1 \end{pmatrix}$. *Then, show that* $A^2 = B^2 = C^2$ $= I$ *(Identity matrix) and also*

$$AB = -BA$$

$$BC = -CB \text{ and}$$
$$AC = -CA.$$

Sol. Given the matrices:

$$A = \begin{pmatrix} 0 & 1 \\ 1 & 0 \end{pmatrix}; B = \begin{pmatrix} 0 & -i \\ i & 0 \end{pmatrix} \text{ and } C = \begin{pmatrix} 1 & 0 \\ 0 & -1 \end{pmatrix}$$

Then, $$A^2 = A\,.\,A\begin{pmatrix} 0 & 1 \\ 1 & 0 \end{pmatrix}\begin{pmatrix} 0 & 1 \\ 1 & 0 \end{pmatrix} = \begin{pmatrix} 0.0+1.1 & 0.1+1.0 \\ 1.0+0.1 & 1.1+0.0 \end{pmatrix}$$

$\Rightarrow$ $$A^2 = \begin{pmatrix} 1 & 0 \\ 0 & 1 \end{pmatrix} = I_2 \qquad (I_2 \text{ refers to identity matrix of the order 2})$$

Similarly, $$B^2 = B \times B = \begin{pmatrix} 0 & -i \\ i & 0 \end{pmatrix}\begin{pmatrix} 0 & -i \\ i & 0 \end{pmatrix}$$

$\Rightarrow$ $$B^2 = \begin{pmatrix} 0.0-i\,.\,i & 0(-i)-i\,.\,0 \\ i\,.\,0+0\,.\,i & i.(-i)+0.0 \end{pmatrix} = \begin{pmatrix} -i^2 & 0 \\ 0 & -i^2 \end{pmatrix}$$

$\Rightarrow$ $$B^2 = \begin{pmatrix} 1 & 0 \\ 0 & 1 \end{pmatrix} = I_2 \qquad (\because i^2 = -1)$$

and $$C^2 = C \times C = \begin{pmatrix} 1 & 0 \\ 0 & -1 \end{pmatrix}\begin{pmatrix} 1 & 0 \\ 0 & -1 \end{pmatrix} = \begin{pmatrix} 1.1+0.0 & 1.0-0.1 \\ 0.1-1.0 & 0.0+1.1 \end{pmatrix}$$

$\Rightarrow$ $$C^2 = \begin{pmatrix} 1 & 0 \\ 0 & 1 \end{pmatrix} = I_2$$

Thus, we find: $A^2 = B^2 = C^2 = I$, (Identity matrix). **Proved.**

Again, $$AB = \begin{pmatrix} 0 & 1 \\ 1 & 0 \end{pmatrix}\begin{pmatrix} 0 & -i \\ i & 0 \end{pmatrix} = \begin{pmatrix} 0.0+1.i & -0.i+1.0 \\ 0.1+0.i & -1.i+0.0 \end{pmatrix}$$

$\Rightarrow$ $$AB = \begin{pmatrix} i & 0 \\ 0 & -i \end{pmatrix}$$

and $$BA = \begin{pmatrix} 0 & -i \\ i & 0 \end{pmatrix}\begin{pmatrix} 0 & 1 \\ 1 & 0 \end{pmatrix} = \begin{pmatrix} 0.0-i.1 & 0.1-i.0 \\ i.0+0.1 & i.1+0.0 \end{pmatrix}$$

$\Rightarrow$ $$BA = \begin{pmatrix} -i & 0 \\ 0 & i \end{pmatrix} = -\begin{pmatrix} i & 0 \\ 0 & -i \end{pmatrix} = -AB$$

$\Rightarrow$ $AB = -BA.$ **Proved.**

Similarly, $$BC = \begin{pmatrix} 0 & -i \\ i & 0 \end{pmatrix}\begin{pmatrix} 1 & 0 \\ 0 & -1 \end{pmatrix} = \begin{pmatrix} 0.1-i.0 & 0.0+i.1 \\ i.1+0.0 & i.0-0.1 \end{pmatrix}$$

$\Rightarrow$ $$BC = \begin{pmatrix} 0 & i \\ i & 0 \end{pmatrix} \text{ and}$$

$$CB = \begin{pmatrix} 1 & 0 \\ 0 & -1 \end{pmatrix}\begin{pmatrix} 0 & -i \\ i & 0 \end{pmatrix} = \begin{pmatrix} 1.0+0.i & -i.1+0.0 \\ 0-1.i & -0.i-1.0 \end{pmatrix}$$

$\Rightarrow$ $$CB = \begin{pmatrix} 0 & -i \\ -i & 0 \end{pmatrix} = -\begin{pmatrix} 0 & i \\ i & 0 \end{pmatrix} = -BC$$

$\Rightarrow$ $\quad$ **$BC = -CB$. Proved.**

And finally,
$$AC = \begin{pmatrix} 0 & 1 \\ 1 & 0 \end{pmatrix}\begin{pmatrix} 1 & 0 \\ 0 & -1 \end{pmatrix} = \begin{pmatrix} 1.0+1.0 & 0.0-1.1 \\ 1.1+0.0 & 1.0-1.0 \end{pmatrix}$$

$$\Rightarrow \quad AC = \begin{pmatrix} 0 & -1 \\ 1 & 0 \end{pmatrix} \text{ and}$$

$$CA = \begin{pmatrix} 1 & 0 \\ 0 & -1 \end{pmatrix}\begin{pmatrix} 0 & 1 \\ 1 & 0 \end{pmatrix} = \begin{pmatrix} 1.0+0.1 & 1.1+0.0 \\ 0.0-1.1 & 0.1-1.0 \end{pmatrix} = \begin{pmatrix} 0 & 1 \\ -1 & 0 \end{pmatrix}$$

$$= -\begin{pmatrix} 0 & -1 \\ 1 & 0 \end{pmatrix} = -AC$$

$\Rightarrow$ $\quad$ **$AC = -CA$. Proved.**

Example 18. *Show that the product of two triangular matrices is itself a triangular.*

Sol. Suppose A and B are two triangular matrices of the order n. (As triangular matrix is a square matrix).

i.e. We may take

$$A = (a_{ij})_{n \times n} \text{ and } B = (b_{jk})_{n \times n}$$

Obviously, being triangular,

$$a_{ij} = 0;\ i > j$$

and
$$b_{jk} = 0;\ j > k$$

Now, suppose the product AB is represented by

$$AB = (c_{ik})_{n \times n}$$

Then obviously, by product of matrices:

$$c_{ik} = \sum_{j=1}^{n} a_{ij} b_{jk}$$

Now, let $i > k$.

If $j < i$; $a_{ij} = 0$ and hence $c_{ik} = 0$

However $j > i$,

$$\Rightarrow \quad j > k \qquad (\because i > k)$$

Then, $b_{jk} = 0$ and hence $c_{ik} = 0$.

So, whenever $i > k$.

$$\Rightarrow \quad c_{ik} = 0$$

Hence, the product matrix $AB = (c_{ik})_{n\times n}$ is also a triangular matrix. **Proved.**

Example 19. *If the matrix* $A = \begin{pmatrix} 0 & 1 \\ 0 & 0 \end{pmatrix}$, *show that* $(aI + bA)^n = a^n I + na^{n-1} bA$ *where* I = *Identity matrix of order 2 and n is a positive integer.*

Sol. To show $\quad (aI + bA)^n = a^n I + na^{n-1} bA \qquad$ ...(*i*)

for some scalars a and b and positive integer n.

$$A = \begin{pmatrix} 0 & 1 \\ 0 & 0 \end{pmatrix}, I = \begin{pmatrix} 1 & 0 \\ 0 & 1 \end{pmatrix}$$

Then, for some scalars a and b:

$$aI + bA = a\begin{pmatrix}1 & 0\\0 & 1\end{pmatrix} + b\begin{pmatrix}0 & 1\\0 & 0\end{pmatrix} = \begin{pmatrix}a & 0\\0 & a\end{pmatrix} + \begin{pmatrix}0 & b\\0 & 0\end{pmatrix} = \begin{pmatrix}a & b\\0 & a\end{pmatrix}$$

Then, for $n = 1$.

$$(aI + bA)^n = aI + bA = \begin{pmatrix}a & b\\0 & a\end{pmatrix}$$

and $$a^nI + na^{n-1}b\times A = a^n.\begin{pmatrix}1 & 0\\0 & 1\end{pmatrix} + na^{n-1}b\begin{pmatrix}0 & 1\\0 & 0\end{pmatrix}$$

$$= \begin{pmatrix}a^n & 0\\0 & a^n\end{pmatrix} + na^{n-1}\begin{pmatrix}0 & na^{n-1}b\\0 & 0\end{pmatrix} = \begin{pmatrix}a^n & na^{n-1}b\\0 & a^n\end{pmatrix}$$

For $n = 1$:

$$a^nI + na^{n-1}b\ .\ A = \begin{pmatrix}a & b\\0 & a\end{pmatrix}$$

i.e., for $n = 1$.

$(aI + bA)^n = a^nI + na^{n-1}bA$ holds good.

So, for $n = k$ (some positive integer). We may suppose that I holds good giving,

$$(aI + bA)^k = a^kI + ka^{k-1}b \times A \qquad ...(ii)$$

Our purpose would be served if we show that 'I' holds good for $n = (k + 1)$ also, i.e., we have to establish:

$$(aI + bA)^{k+1} = a^{k+1}I + (k + 1)\, a^k\, bA$$

Now, L.H.S. $(aI + bA)^{k+1} = (aI + bA)^k \times (aI + bA)$

$= (a^kI + ka^{k-1}b\ .\ A)\ (aI + bA)$

$\Rightarrow$ $(aI + bA)^{k+1} = (a^kI)\ (aI) + (a^kI)\ (bA) + (ka^{k-1}b\ .\ A)\ .\ aI + (ka^{k-1}b\ .\ A)\ .\ (bA)$

$\Rightarrow$ $(aI + bA)^{k+1} = a^{k+1}\ .\ I^2 + a^k\ .\ b\ (IA) + (ka^kb)\ (AI) + (ka^{k-1}\ .\ b^2)A^2$

$\Rightarrow$ $(aI + bA)^{k+1} = a^{k+1}\ .\ I + a^kb\ .\ A + ka^kbA + (ka^{k-1}\ .\ b^2)\ A^2 \qquad ...(iii)$

$(\therefore\ IA = A = AI : I^2 = I)$

Now, $$A^2 = A\ .\ A = \begin{pmatrix}0 & 1\\0 & 0\end{pmatrix}\begin{pmatrix}0 & 1\\0 & 0\end{pmatrix} = \begin{pmatrix}0.0+1.0 & 0.1+1.0\\0.0+0.0 & 0.1+0.0\end{pmatrix}$$

$\Rightarrow$ $$A^2 = \begin{pmatrix}0 & 0\\0 & 0\end{pmatrix} = 0 \qquad \text{(Null matrix).}$$

Putting the value of $A^2 = 0$ in (III) we get:

$$(aI + bA)^{k+1} = a^{k+1}I + a^kb\ .\ A + ka^kbA + ka^{k-1}\ .\ b^2\ .\ 1$$

$\Rightarrow$ $(aI + bA)^{k+1} = a^{k+1}I + (k + 1)\ a^kb\ .\ A$

Thus, I holds good for $n = k + 1$. **Proved.**

Example 20: *If for matrices A and B*

$AB = A$ and $BA = B$, then show that:

$A^2 = A$ and $B^2 = B$.

Sol. Given, for the matrices A and B

$AB = A$ and $BA = B$

Then, $B(AB) = BA$ (Pre-multiplying by B)

$\Rightarrow \quad (BA)B = BA$ ($\therefore$ Product of matrices is associative)

$\Rightarrow \quad B \,.\, B = B$ ($\because BA = B$)

$\Rightarrow \quad B^2 = B.$ **Proved.**

Similarly, $\quad BA = B$

$\Rightarrow \quad A(BA) = AB$ (Premultiplying by A)

$\Rightarrow \quad (AB)A = AB$ (Θ Product of matrices is associative)

$\Rightarrow \quad AA = A$ ($\because AB = A$)

$\Rightarrow \quad A^2 = A.$ **Proved.**

Example 21. *Let's have some function defined at $f(A) = n^2 - aA + b$.*

Then find $f(n)$ where $A = \begin{pmatrix} 2 & 1 & 2 \\ 1 & 2 & -2 \\ 5 & 3 & 1 \end{pmatrix}$*; a, b are scalars.*

Sol. Given function $\quad f(A) = A^2 - A_n + b$ for scalars a and b.

Then, for the matrix: $\quad A = \begin{pmatrix} 2 & 1 & 2 \\ 1 & 2 & -2 \\ 5 & 3 & 1 \end{pmatrix}$

We have to find out:

$f(A)$ *i.e.*, $A^2 - aA + bI$.

For this:

$$A^2 = \begin{pmatrix} 2 & 1 & 2 \\ 1 & 2 & -2 \\ 5 & 3 & 1 \end{pmatrix}\begin{pmatrix} 2 & 1 & 2 \\ 1 & 2 & -2 \\ 5 & 3 & 1 \end{pmatrix}$$

$$\Rightarrow \quad A^2 = \begin{pmatrix} 2.2+1.1+2.5 & 2.1+1.2+2.3 & 2.2-1.2+2.1 \\ 1.2+2.1-2.5 & 1.1+2.2-2.3 & 1.2-2.2-2.1 \\ 5.2+3.1+1.5 & 5.1+3.2+1.3 & 5.2-3.2+1.1 \end{pmatrix}$$

$$\Rightarrow \quad A^2 = \begin{pmatrix} 15 & 10 & 4 \\ -6 & -1 & -4 \\ 18 & 14 & 5 \end{pmatrix}$$

and $$aA = a\begin{pmatrix} 2 & 1 & 2 \\ 1 & 2 & -2 \\ 5 & 3 & 1 \end{pmatrix} = \begin{pmatrix} 2a & a & 2a \\ a & 2a & -2a \\ 5a & 3a & a \end{pmatrix}$$

and $$bI = b\begin{pmatrix} 1 & 0 & 0 \\ 0 & 1 & 0 \\ 0 & 0 & 1 \end{pmatrix} = \begin{pmatrix} b & 0 & 0 \\ 0 & b & 0 \\ 0 & 0 & b \end{pmatrix}$$

$\therefore \quad f(A) = A^2 - aA + bI$

$$\Rightarrow \quad f(A) = \begin{pmatrix} 15 & 10 & 4 \\ -6 & -1 & -4 \\ 18 & 14 & 5 \end{pmatrix} - \begin{pmatrix} 2a & a & 2a \\ a & 2a & -2a \\ 5a & 3a & a \end{pmatrix} + \begin{pmatrix} b & 0 & 0 \\ 0 & b & 0 \\ 0 & 0 & b \end{pmatrix}$$

$$\Rightarrow \qquad f(A) = \begin{pmatrix} 15-2a+b & 10-a & 4-2a \\ -6-a & -1-2a-b & -4+2a \\ 18-5a & 14-3a & 5-a+b \end{pmatrix} \textbf{ Ans.}$$

1.4.7. Order of a Matrix

A matrix having m rows and n columns is called a matrix of *order* $m \times n$ or simply $m \times n$ matrix (read as an m by n matrix).

In general, an $m \times n$ matrix has the following rectangular array:

$$\begin{bmatrix} a_{11} & a_{11} & a_{11} & \cdots & a_{11} & \cdots & a_{11} \\ a_{11} & a_{11} & a_{11} & \cdots & a_{11} & \cdots & a_{11} \\ \vdots & \vdots & \vdots & & \vdots & & \vdots \\ a_{11} & a_{11} & a_{11} & \cdots & a_{11} & \cdots & a_{11} \\ \vdots & \vdots & \vdots & & \vdots & & \vdots \\ a_{11} & a_{11} & a_{11} & \cdots & a_{11} & \cdots & a_{11} \end{bmatrix}$$

or $A = [a_{ij}]_{m \times n}$, $1 \le i \le m$, $1 \le j \le n \quad i, j \in \mathrm{N}$

Thus the ith row consists of the elements $a_{i1}, a_{i2}, a_{i3}, \ldots, a_{in}$, while the jth column consists of the elements $a_{1j}, a_{2j}, a_{3j}, \ldots, a_{mj}$

In general a_{ij}, is an element lying in the ith row and j^{th} column. We can also call it as the $(i, j)^{th}$ element of A. The number of elements in an $m \times n$ matrix will be equal to mn notation is given below.

1. We shall follow the notation, namely $A = [a_{ij}]_{m \times n}$ to indicate that A is a matrix or order $m \times n$.
2. We shall consider only those matrices whose elements are real numbers or functions taking real values.

We can also represent any point (x, y) in a plane by a matrix (column or row) as $\begin{bmatrix} x \\ y \end{bmatrix}$ (or $[x, y]$). For example point $P(0, 1)$ as a matrix representation may be given as

$$P = \begin{bmatrix} 0 \\ 1 \end{bmatrix} \text{ or } [0 \;\; 1].$$

Observe that in this way we can also express the vertices of a closed rectilinear figure in the form of a matrix. For example, consider a quadrilateral $ABCD$ with vertices A (1, 0), B (3, 2), C (1, 3), D (–1, 2).

Now, quadrilateral $ABCD$ in the matrix form, can be represented as

$$\begin{matrix} & A & B & C & D \\ \mathrm{X} = & \end{matrix} \begin{bmatrix} 1 & 3 & 1 & -1 \\ 0 & 2 & 3 & 2 \end{bmatrix}_{2 \times 4}$$

Example 22. *Consider the following information regarding the number of men and women workers in three factories I, II and III*

Factories	*Men workers*	*Women workers*
I	30	25
II	25	31
III	27	26

Represent the above information in the form of a 3 × 2 matrix. What does the entry in the third row and second column represent?

Sol. The information is represented in the form of a 3 × 2 matrix as follows:

$$A = \begin{bmatrix} 30 & 25 \\ 25 & 31 \\ 27 & 26 \end{bmatrix}$$

The entry in the third row and second column represents the number of women workers in factory III.

Example 23. *If a matrix has 8 elements, what are the possible orders it can have?*

Sol. We know that if a matrix is of order $m \times n$, it has mn elements. Thus, to find all possible orders of a matrix with 8 elements, we will find all ordered pairs of natural numbers, whose product is 8. Thus, all possible ordered pairs are (1, 8), (8, 1), (4, 2), (2, 4)

Hence, possible orders are 1 × 8, 8 ×1, 4 × 2, 2 × 4

Example 24. *Construct a 3 × 2 matrix whose elements are given by* $a_{ij} = \frac{1}{2}|i - 3j|$.

Sol. In general a 3 × 2 matrix is given by $A = \begin{bmatrix} a_{11} & a_{12} \\ a_{21} & a_{22} \\ a_{31} & a_{32} \end{bmatrix}$.

Now $\quad a_{ij} = \frac{1}{2} | i - 3j |, i = 1, 2, 3 \text{ and } j = 1, 2.$

Therefore $\quad a_{11} = \frac{1}{2} |1 - 3 \times 1| = 1 \qquad a_{12} = \frac{1}{2} |1 - 3 \times 2| = \frac{5}{2}$

$$a_{21} = \frac{1}{2} | 2 - 3 \times 1| = \frac{1}{2} \qquad a_{22} = \frac{1}{2} |2 - 3 \times 2| = 2$$

$$a_{31} = \frac{1}{2} | 3 - 3 \times 1 | = 0 \qquad a_{32} = \frac{1}{2} |3 - 3 \times 2| = \frac{3}{2}$$

Hence the required matrix is given by $A = \begin{bmatrix} 1 & \frac{5}{2} \\ \frac{1}{2} & 2 \\ 0 & \frac{3}{2} \end{bmatrix}$.

EXERCISE 1.1.5.

1. Find the values of x, y, z and a which satisfy the matrix equation.

$$\begin{bmatrix} x+3 & 2y+x \\ z-1 & 4a-6 \end{bmatrix} = \begin{bmatrix} 0 & -7 \\ 3 & 2a \end{bmatrix}$$

Ans. $[x = -3, y = -2, z = 4, a = 3]$

2. Find x, y, z and w given that

$$3\begin{bmatrix} x & y \\ z & w \end{bmatrix} = \begin{bmatrix} x & 6 \\ -1 & 2w \end{bmatrix} + \begin{bmatrix} 4 & x+y \\ z+w & 3 \end{bmatrix}$$

Ans. $[x = 2, y = 4, z = 1, w = 3]$

3. (*a*) Show that every square matrix A can be expressed as the sum of a symmetric matrix $\frac{1}{2}(A + A')$ and skw-symmetric matrix $\frac{1}{2}(A - A')$,

(*b*) Express the matrices given below as the sum of symmetric and a skew symmetric matrix.

(*a*) $\begin{bmatrix} -1 & 7 & 1 \\ 2 & 3 & 4 \\ 5 & 0 & 5 \end{bmatrix}$;

Ans. $\begin{bmatrix} -1 & 9/2 & 3 \\ 9/2 & 3 & 2 \\ 3 & 2 & 5 \end{bmatrix} + \begin{bmatrix} 0 & 5/2 & -2 \\ -5/2 & 0 & 2 \\ 2 & -2 & 0 \end{bmatrix}$

(*b*) $\begin{bmatrix} 8 & 6 & 12 \\ -6 & 7 & 44 \\ 2 & -4 & 3 \end{bmatrix}$;

Ans. $\begin{bmatrix} 8 & 0 & 7 \\ 0 & 7 & 20 \\ 7 & 20 & 3 \end{bmatrix} + \begin{bmatrix} 0 & 6 & 5 \\ -6 & 0 & 24 \\ -5 & -24 & 0 \end{bmatrix}$

4. Find for value of x, in the given matrix.

$$\begin{bmatrix} 3-x & 2 & 2 \\ 2 & 4-x & 1 \\ -2 & -4 & -1-x \end{bmatrix} \text{ is singular?}$$

Ans. $[x = 0, 3]$

5. From the product AB and BA, if possible, given that

(*a*) $A = \begin{bmatrix} 0 & 1 & 2 \\ 1 & 2 & 3 \\ 2 & 3 & 4 \end{bmatrix}$, $B = \begin{bmatrix} 1 & -2 \\ -1 & 0 \\ 2 & -1 \end{bmatrix}$

Ans. $\begin{bmatrix} 3 & -2 \\ 5 & -5 \\ 7 & -8 \end{bmatrix}$; BA is not defined (*b*) $A = \begin{bmatrix} 1 & -2 & 3 \\ 2 & 3 & -1 \\ -3 & 1 & 2 \end{bmatrix}, B = \begin{bmatrix} 1 & 0 & 2 \\ 0 & 1 & 2 \\ 1 & 2 & 0 \end{bmatrix}$ **Ans.** $\begin{bmatrix} 4 & 4 & -2 \\ 1 & 1 & 10 \\ -1 & 5 & -4 \end{bmatrix}, \begin{bmatrix} -5 & 0 & 7 \\ -4 & 5 & 5 \\ 5 & 4 & 1 \end{bmatrix}$

(*c*) $A = \begin{bmatrix} 0 & 1 \\ 1 & 0 \\ 0 & 1 \end{bmatrix}, B = \begin{bmatrix} 0 & 1 & 2 \\ 2 & 1 & 0 \end{bmatrix}$ **Ans.** $\begin{bmatrix} 2 & 1 & 0 \\ 0 & 1 & 2 \\ 2 & 1 & 0 \end{bmatrix}; \begin{bmatrix} 1 & 2 \\ 1 & 2 \end{bmatrix}$

(*d*) $A = \begin{bmatrix} 1 & 3 & 0 \\ -1 & 2 & 1 \\ 0 & 0 & 2 \end{bmatrix}; B = \begin{bmatrix} 2 & 3 & 4 \\ 1 & 2 & 3 \\ -1 & 1 & 2 \end{bmatrix}$ **Ans.** $\begin{bmatrix} 5 & 9 & 13 \\ -1 & 2 & 4 \\ -2 & 2 & 4 \end{bmatrix}; \begin{bmatrix} -1 & 12 & 11 \\ -1 & 7 & 8 \\ -2 & -1 & 5 \end{bmatrix}$

Show that $AB \neq BA$ in each case.

6. If $A = \begin{bmatrix} 2 & 3 \\ 3 & 10 \end{bmatrix}$, show that $(2I - A)(10I - A) = 9I$

where I is a unit matrix of order 2.

7. Prove that $(AB)C = A(BC)$ where A, B, C are matrices conformable for the products.

8. If $A = \begin{bmatrix} 1 & 2 \\ -2 & 3 \end{bmatrix}, B = \begin{bmatrix} 2 & 1 \\ 2 & 3 \end{bmatrix}, C = \begin{bmatrix} -3 & 1 \\ 2 & 0 \end{bmatrix}$

verify that (*a*) $(AB)C = A(BC)$ (*b*) $A(B + C) = AB + AC$

9. Evaluate

(*a*) $(x \;\; y \;\; z) \begin{bmatrix} a & h & g \\ h & b & f \\ g & f & c \end{bmatrix} \begin{bmatrix} x \\ y \\ z \end{bmatrix}$ **Ans.** $ax^2 + by^2 + cz^2 + 2fyz + 2gzx + 2hxy$

(*b*) $\begin{bmatrix} 1 \\ -2 \\ 3 \end{bmatrix} \times (4 \;\; 5 \;\; 2) \times \begin{bmatrix} 2 \\ -3 \\ 5 \end{bmatrix} \times (3 \;\; 2)$ **Ans.** $\begin{bmatrix} 9 & 6 \\ -18 & -12 \\ 27 & 18 \end{bmatrix}$

10. If $E = \begin{bmatrix} 0 & 1 & 0 \\ 0 & 0 & 1 \\ 0 & 0 & 0 \end{bmatrix}$ and $F \begin{bmatrix} 0 & 0 & 0 \\ 1 & 0 & 0 \\ 0 & 1 & 0 \end{bmatrix}$

Find EF and FE and show that $E^2F + F^2E = E$

Ans. $EF = \begin{bmatrix} 1 & 0 & 0 \\ 0 & 1 & 0 \\ 0 & 0 & 0 \end{bmatrix}$ and $FE \begin{bmatrix} 0 & 0 & 0 \\ 0 & 0 & 1 \\ 0 & 0 & 1 \end{bmatrix}$

11. If A and B are two matrices such that $AB = A$ and $BA = B$, then show that $A^2 = A$ and $B^2 = B$.

12. If $A = \begin{bmatrix} \cos\theta & \sin\theta \\ -\sin\theta & \cos\theta \end{bmatrix}$, show that

$$A^n = \begin{bmatrix} \cos n\theta & \sin n\theta \\ -\sin n\theta & \cos n\theta \end{bmatrix}, \text{ if } n \text{ is a +ve integer.}$$

13. If $A = \begin{bmatrix} 4 & 3 \\ 2 & 5 \end{bmatrix}$, find the matrix.

$A^2 - 9A + 14I$, where I is a unit matrix of second order. **Ans.** $\begin{bmatrix} 0 & 0 \\ 0 & 0 \end{bmatrix}$

14. If $A = \begin{bmatrix} 1 & -2 & 3 \\ 2 & 3 & -1 \\ -3 & 1 & 2 \end{bmatrix}$, find the value of matrix $A^2 - 3A + 9I$. **Ans.** $\begin{bmatrix} -6 & 1 & 2 \\ 5 & 4 & 4 \\ 2 & 8 & -3 \end{bmatrix}$

15. Show that $\begin{bmatrix} \cos^2\theta & \cos\theta\sin\theta \\ \cos\theta\sin\theta & \sin^2\theta \end{bmatrix} \times \begin{bmatrix} \cos^2\phi & \cos\phi\sin\phi \\ \cos\phi\sin\phi & \sin^2\phi \end{bmatrix} = \begin{bmatrix} 0 & 0 \\ 0 & 0 \end{bmatrix}$

when θ and ϕ differe by an odd multiple of $\frac{\pi}{2}$

16. If $A = \begin{bmatrix} 0 & 1 & 0 \\ 0 & 0 & 1 \\ p & q & r \end{bmatrix}$, show that $A^3 = pI + qA + rA^2$

where I is the unit matrix of order 3.

17. Obtain *adjoint* A and A^{-1} when A is

(*a*) $\begin{bmatrix} 2 & 5 & 3 \\ 3 & 1 & 2 \\ 1 & 2 & 1 \end{bmatrix}$ **Ans.** $Adj\ A = \begin{bmatrix} -3 & 1 & 7 \\ -1 & -1 & 5 \\ 5 & 1 & -13 \end{bmatrix}$; $A^{-1} = 1/4\ Adj\ A$.

(*b*) $\begin{bmatrix} 1 & 0 & 0 \\ \alpha & 1 & 0 \\ \beta & \gamma & 1 \end{bmatrix}$ **Ans.** $A^{-1} = \begin{bmatrix} 1 & 0 & 0 \\ -\alpha & 1 & 0 \\ -\gamma & -\beta & -\gamma \end{bmatrix}$

(*c*) $\begin{bmatrix} 3 & -1 & 1 \\ -15 & 6 & -5 \\ 5 & -2 & 2 \end{bmatrix}$ **Ans.** $A^{-1} = \begin{bmatrix} 2 & 0 & -1 \\ 5 & 1 & 0 \\ 0 & 1 & 3 \end{bmatrix}$

(*d*) $\begin{bmatrix} 1 & 1 & 3 \\ 1 & 3 & -3 \\ -2 & -4 & -4 \end{bmatrix}$ **Ans.** $A^{-1} = \begin{bmatrix} 3 & 1 & 3/2 \\ -5/4 & -1/4 & -3/4 \\ -1/4 & -1/4 & -1/4 \end{bmatrix}$

(*e*) $\begin{bmatrix} 1 & -3 & 2 \\ 2 & 0 & 0 \\ 1 & 4 & 1 \end{bmatrix}$ **Ans.** $A^{-1} = \begin{bmatrix} 0 & 1/2 & 0 \\ -1/11 & -1/22 & 2/11 \\ 4/11 & -7/22 & 3/11 \end{bmatrix}$

(*f*) $\begin{bmatrix} 3 & -3 & 4 \\ 2 & -3 & 4 \\ 0 & -1 & 1 \end{bmatrix}$ **Ans.** $A^{-1} = \begin{bmatrix} 1 & -1 & 0 \\ -2 & 3 & -4 \\ -2 & 3 & -3 \end{bmatrix}$

18. Show that the matrix $\begin{bmatrix} -4 & -3 & -3 \\ 1 & 0 & 1 \\ 4 & 4 & 3 \end{bmatrix}$ is its own adjoint.

19. If $A = \begin{bmatrix} 3 & -3 & 4 \\ 2 & -3 & 4 \\ 0 & -1 & 1 \end{bmatrix}$, show that $A^3 = A^{-1}$ and Compute *adj. A.*

20. Show that $\begin{bmatrix} \cos\theta & -\sin\theta \\ \sin\theta & \cos\theta \end{bmatrix} = \begin{bmatrix} 1 & -\tan\frac{\theta}{2} \\ \tan\frac{\theta}{2} & 1 \end{bmatrix}\begin{bmatrix} 1 & -\tan\frac{\theta}{2} \\ -\tan\frac{\theta}{2} & 1 \end{bmatrix}^{-1}$

21. If $D = \text{diag.}\ [d_1, d_2, \ldots.\ d_n]$. $d_1, d_2, \ldots.\ d_n \neq 0$, prove that

$D^{-1} = \text{diag.}\ [[d_1^{-1}, d_2^{-1}, \ldots.\ d_n^{-1}]$

22. If A is a non-singular matrix of order n, show that $|\text{adj. A}| = |A|^{n-1}$

23. Prove that

(*i*) $(AB)^{-1} = B^{-1}A^{-1}$ (*ii*) $(ABC)^{-1} = C^{-1}B^{-1}A^{-1}$

24. If A and B are square matrices of the same order and A is symmetrical, show that $B'AB$ is also symmetrical.

1.5. SETS

Set serves as a fundamental part of the mathematics. Now, this concept is being used in almost every branch of mathematics. Sets are used to define the concepts of relations and functions. The study of math *i.e.,* geometry, sequences, probability, etc. requires the knowledge of sets.

In our life, we often speak of collections of objects of a particular kind, such as, a pack of cards, a crowd of people, a cricket team, etc. In mathematics also, we come across collections, for example, prime number, even number, odd number, natural numbers, points, etc. More specially, we examine the following collections:

Do you know?

The theory of sets was developed by German mathematician Georg Cantor (1845-1918). He first encountered sets while working on "problems on trigonometric series".

(*i*) The rivers of India

(*ii*) The solution of the equation: $x^2 - 9 = 0$, viz, +3 and –3.

(*iii*) Odd natural numbers less than 20, *i.e.*, 1, 3, 5, 7, 9, 11, 13, 15, 17, 19

(*iv*) Prime factors of 190, namely, 2, 3, 5 and 7

(*v*) Various kinds of triangles

(*vi*) The vowels in the English alphabet, namely, *a, e, i, o, u*

We note that each of the above example is a well-defined collection of objects in the sense that definitely decide whether a given particular object belongs to a given collection or not. For example, we can say that the table does not belong to the collection of playing cards. On the other hand, the table does belong to colleciton of furniture.

We give below a few more examples of sets used particularly in mathematics, *viz.*

N : the set of all natural numbers,

Z : the set of all integers,

Q : the set of all rational numbers,

R : the set of real numbers,

$\mathbf{Z}^+$: the set of positive integers,

$\mathbf{Q}^+$: the set of positive rational numbers, and

$\mathbf{R}^+$: the set of positive real numbers.

These are also represent following types :

$\mathbb{Z}$ The set of integers

$\mathbb{N}$ The set of natural numbrs (positive integers)

$\mathbb{Q}$ The set of rational numbers

$\mathbb{R}$ The set of real numbers

For example : $x \in \mathbb{Z}$ means that x in the set of integers, *i.e.*, x is an integer.

$\in$ means 'is a member of'

Now, the criterion for determining a mathematician as most renowned may vary from person to person. Thus, it is not a well-defined collection now conclude that the ***a set is a well-defined collection of objects.***

The following points may be noted :

(*i*) Objects, elements and members of a set are synonymous terms.

(*ii*) Sets are usually denoted by capital letters A, B, C, X, Y, Z, etc.

(*iii*) The elements of a set are represented by small letters *a, b, c, x, y, z,* etc.

If a is an element of a set A, we say that "a belongs to A" the Greek symbol ε (epsilon) is used to denote the phrase '*belongs to*'. Thus, we write $a \in A$. If 'b' is not an element of a set A, we write $b \notin$ A and read "b does not belong to A".

Thus, in the set V of vowels in the English alphabet, $a \in V$ but $b \notin V$. In the set P of prime factors of 30, 3 ∈ P but 15 ∉ P.

1.5.1. Sets and their Representation

There are two methods of representing a set :

(*i*) Roster or tabular form (*ii*) Set-builder form.

(*i*) **Roster or tabular form.** In roster form, all the elements of a set are listed, the elements are being separated by commas (,) and are enclosed within curly braces { }. For example, the set of all even positive integers less than 7 is described in roster form as {2, 4, 6}. Some more examples of representing a set in roster form are given below :

(*a*) The set of all natural numbers which divide 56 is {1, 2, 4, 14, 56}.

(*b*) The set of all vowels in the English alphabet is {*a, e, i, o, u*}.

(*c*) The set of odd natural numbers is represented by {1, 3, 5, . . .}. The dots tell us that the list of odd numbers continue indefinitely.

(*ii*) **Set-builder form.** In set-builder form, all the elements of a set possess a single common property which is not possessed by any element outside the set. For example, in the set {*a, e, i, o, u*}, all the elements possess a common property, namely, each of them is a vowel in the English alphabet, and no other letter possess this property. Denoting this set by V, we write

$V = \{x : x$ is a group of fruits$\}$

• Basic concepts and notation

Set theory begins with a fundamental binary relation between an object o and a set A. If o is a **member** (or **element**) of A, the notation $o \in A$ is used. Since sets are objects, the membership relation can relate sets as well.

A derived binary relation between two sets is the subset relation, also called **set inclusion.** If all the members of set A are also members of set B, then A is a **subset** of B, denoted $A \subseteq B$. For example, {1, 2} is a subset of {1, 2, 3}, and so is {2} but {1, 4} is not. From this definition, a set is a subset of itself. For cases where this possibility is unsuitable or would make sense to be rejected, the term **proper subset** is defined. A is called a **proper subset** of B if and only if A is a subset of B, but A is not equal to B. Also that 1, 2, and 3 are members (elements) of the set {1, 2, 3} but are not subsets of it; and in turn, the subsets, such as {1}, are not members of the set {1, 2, 3}.

Just as arithmetic features binary operations on numbers, set theory features binary operations on sets.

Union of the sets A and B, denoted $A \cup B$, is the set of all objects that are a member of A, or B, or both. The union of {1, 2, 3} and {2, 3, 4} is the set {1, 2, 3, 4} .

Intersection of the sets **A** and **B**, denoted $A \cap B$, is the set of all objects that are members of both A and B. The intersection of {1, 2, 3} and {2, 3, 4} is the set {2, 3}.

Set difference of U and A, denoted $U \setminus A$, is the set of all members of U that are not members of A. The set difference {1, 2, 3}\{2, 3, 4} is {1} , while, conversely, the set difference {2, 3, 4}\{1, 2, 3} is {4}. When A is a subset of U, the set difference $U \setminus A$ is also called the *complement* of A in U. In this case, if the choice of U is clear from the context, the notation A^c is sometimes used instead of $U \setminus A$, particularly if U is a universal set as in the study of Venn diagrams.

Symmetric difference of sets A and B, denoted $A \Delta B$ or $A \varnothing B$, is the set of all objects that are a member of exactly one of A and B (elements which are in one of the sets, but not in both). For instance, for the sets {1, 2, 3} and {2, 3, 4}, the symmetric difference set is {1, 4} . It is the set difference of the union and the intersection, $(A \cup B) \setminus (A \cap B)$ or $(A \setminus B) \cup (B \setminus A)$.

Cartesian product of A and B, denoted $A \times B$, is the set whose members are all possible ordered pairs (a, b) where a is a member of A and b is a member of B. The cartesian product of {1, 2} and {red, white} is {(1, red), (1, white), (2, red), (2, white)}.

Power set of a set A is the set whose members are all of the possible subsets of A. For example, the power set of {1, 2} is { {}, {1}, {2}, {1, 2} }.

Some basic sets of central importance are the empty set (the unique set containing no elements; occasionally called the null set though this name is ambiguous), the set of natural numbers, and the set of real numbers.

It may be observed that we describe the element of the set by using a symbol x (any other symbol like the letters y, z, etc. could be used) which is followed by a colon " : ". After the sign of colon, we write the characteristic property possessed by the elements of the set and then enclose the whole description within braces. The above description of the set V is read as "the set of all x such that x is a vowel of the English alphabet". In this description the braces stand for "the set of all", the colon stands for "such that".

Example 1. A = $\{x : x$ is a natural number and $3 < x < 10\}$ is read as "the set of all x such that x is a natural number and x lies between 3 and 10". Hence, the numbers 4, 5, 6, 7, 8 and 9 are the elements of the set A.

Example 2. $A = \{x : x$ is a cricket team members whose age is more than 34 years}. Hence the player is "M.S. Dhoni" so $\{x :$ M.S. Dhoni}

If we denote the sets described in (a), (b) and (c) above in roster form by A, B, C, respectively, then A, B, C can also be represented in set-builder form as follows:

$A = \{x : x$ is a natural number which divides 42}

$B = \{y : y$ is a vowel in the English alphabet}

$C = \{z : z$ is an odd natural number}.

Example 3. Write the solution set of the equation $x^2 + x - 2 = 0$ in roster form.

Sol. The given equation can be written as $(x - 1)(x + 2) = 0$, *i.e.*, $x = 1, -2$

Therefore, the solution set of the given equation can be written in roster form as $\{1, -2\}$.

Example 4. Write the set $\{x : x$ is a positive integer and $x^2 < 40\}$ in the roster form.

Sol. The required numbers are 1, 2, 3, 4, 5, 6. So, the given set in the roster form is {1, 2, 3, 4, 5, 6}.

Example 5. Match each of the set on the left described in the roster form with the same set on the right described in the set-builder form :

(i) {P, R, I, N, C, A, L}	*(a)* $\{x : x$ is a positive integer and is a divisor of 18}
(ii) { 0 }	*(b)* $\{x : x$ is an integer and $x^2 - 9 = 0\}$
(iii) {1, 2, 3, 6, 9, 18}	*(c)* $\{x : x$ is an integer and $x + 1 = 1\}$
(iv) {3, –3}	*(d)* $\{x : x$ is a letter of the word PRINCIPAL}

Sol. Since in (d), there are 9 letters in the word PRINCIPAL and two letters P and I are repeated, so (i) matches (d). Similarly, (ii) matches (c) as $x + 1 = 1$ implies $x = 0$. Also, 1, 2, 3, 6, 9, 18 are all divisors of 18 and so (iii) matches (a). Finally, $x^2 - 9 = 0$ implies $x = 3, -3$ and so (iv) matches (b).

Example 6. Write the set $A = \{1, 4, 9, 16, 25, \ldots\}$ in set-builder form.

Sol. We may write the set A as

$$A = \{x : x \text{ is the square of a natural number}\}$$

Alternatively, we can write

$$A = \{x : x = n^2, \text{ where } n \in \mathbf{N}\}.$$

Example 7. Write the set $\left\{\frac{1}{2}, \frac{2}{3}, \frac{3}{4}, \frac{4}{5}, \frac{5}{6}, \frac{6}{7}\right\}$ in the set-builder form.

Sol. We see that each member in the given set has the numerator one less than the denominator. Also, the numerator begin from 1 and do not exceed 6. Hence, in the set-builder form the given set is

$$\left\{x : x = \frac{n}{n+1}, \text{ where } n \text{ is a natural number and } 1 \le n \le 6\right\}$$

EXERCISE 1.1.6.

1. Write the following sets in roster form:

(*i*) $D = \{x : x$ is a prime number which is divisor of $60\}$ **Ans.** $D = \{2, 3, 5\}$

(*ii*) $C = \{x : x$ is a two-digit natural number such that the sum of its digits is $8\}$ **Ans.** $C = \{17, 26, 35, 44, 53, 62, 71, 80\}$

(*iii*) $B = \{x : x$ is a natural number less than $6\}$ **Ans.** $\{1, 2, 3, 4, 5\}$

(*iv*) $A = \{x : x$ is an integer and $-3 < x < 7\}$ **Ans.** $A = \{-3, -2, -1, 0, 1, 2, 3, 4, 5, 6\}$

(*v*) E = The set of all letters in the word TRIGONOMETRY **Ans.** {T, R, I, G, O, N, M, E, Y}

(*vi*) F = The set of all letters in the word BETTER **Ans.** F = {B, E, T, R}

2. Which of the following are sets ? Justify your answer.

(*i*) The collection of all the months of a year beginning with the letter J.

(*ii*) The collection of ten most talented writers of India.

(*iii*) A team of eleven best-cricket batsmen of the world.

(*iv*) The collection of all boys in your class.

(*v*) The collection of all natural numbers less than 100.

(*vi*) A collection of novels written by the writer Munshi Prem Chand.

(*vii*) The collection of all even integers.

(*viii*) The collection of questions in this chapter

(*ix*) A collection of most dangerous animals of the world.

Ans. {(*i*), (*iv*), (*v*), (*vi*), (*vii*) and (*viii*) are sets}

3. Match each of the set on the left in the roster form with the same set on the right described in set-builder form:

(*i*) {1, 2, 3, 6} (*a*) $\{x : x$ is a prime number and a divisor of $6\}$ **Ans.** (*i*) (*c*)

(*ii*) {2, 3}	(*b*) $\{x : x$ is an odd natural number less than 10}	**Ans.** (*ii*) (*a*)
(*iii*) {M,A,T,H,E,I,C,S}	(*c*) $\{x : x$ is natural number and divisor of 6}	**Ans.** (*iii*) (*d*)
(*iv*) {1, 3, 5, 7, 9}	(*d*) $\{x : x$ is a letter of the word MATHEMATICS}.	**Ans.** (*iv*) (*b*)

4. List all the elements of the following sets :

(*i*) $A = \{x : x$ is an odd natural number} **Ans.** (*i*) $A = \{1, 3, 5\}$

(*ii*) $B = \{x : x$ is an integer, $-\frac{1}{2} < x < \frac{9}{2}\}$ **Ans.** (*ii*) $B = \{0, 1, 2, 3, 4\}$

(*iii*) $C = \{x : x$ is an integer, $x^2 \leq 4\}$ **Ans.** (*iii*) $C = \{-2, -1, 0, 1, 2\}$

(*iv*) $D = \{x : x$ is a letter in the word "LOYAL"} **Ans.** (*iv*) $D = \{$ L, O, Y, A}

(*v*) $E = \{x : x$ is a month of a year not having 31 days}

Ans. (*v*) E = {February, April, June, September, November}

(*vi*) $F = \{x : x$ is a consonant in the English alphabet which precedes $k\}$.

Ans. (*vi*) $F = \{b, c, d, f, g, h, i\}$

5. Let A = {1, 2, 3, 4, 5, 6}. Insert the appropriate symbol $\in$ or $\notin$ in the blank spaces:

(*i*) 10. . .A	**Ans.** (*i*) $\notin$
(*ii*) 2. . .A	**Ans.** (*ii*) $\in$
(*iii*) 4. . . A	**Ans.** (*iii*) $\in$
(*iv*) 0. . .A	**Ans.** (*iv*) $\notin$
(*v*) 8 . . . A	**Ans.** (*v*) $\in$
(*vi*) 5. . .A	**Ans.** (*vi*) $\in$

6. Write the following sets in the set-builder form :

(*i*) {2, 4, 8, 16, 32}	**Ans.** (*i*) $\{x : x = 2^n, n \in N$, and $1 \leq n \leq 5\}$
(*ii*) {5, 25, 125, 625}	**Ans.** (*ii*) $\{x : x = 5^n, n \in N$, and $1 \leq n \leq 4\}$
(*iii*) {3, 6, 9, 12}	**Ans.** (*iii*) $\{x : x = 3^n, n \in N$, and $1 \leq n \leq 4\}$
(*iv*) {1, 4, 9, . . ., 100}	**Ans.** (*iv*) $\{x : x = n^2, n \in N$, and $1 \leq n \leq 10\}$
(*v*) {2, 4, 6, . . .}	**Ans.** (*v*) $\{x : x$ is an even natural number}

1.5.2. The Empty Set

A set which does not contain any element is called the *empty set* or the *null set* or the *void set*.

Consider the set

$A = \{x : x$ is a student of Class II presently studying in a school}

We can go to the school and count the number of students presently studying in Class II in the school. Thus, the set A contains a finite number of elements.

We now write another set B as follows:

$B = \{x : x$ is a student presently studying in both Classes I and II}

We observe that a student cannot study simultaneously in both Classes I and II.

Thus, the set B contains no element at all.

According to this definition, B is an empty set while A is not an empty set. The empty set is denoted by the symbol ϕ or { }.

We give below a few examples of empty sets.

(*i*) Let $A = \{x : 1 < x < 2, x \text{ is a natural number}\}$. Then A is the empty set, because there is no natural number between 1 and 2.

(*ii*) $B = \{x : x^2 - 2 = 0 \text{ and } x \text{ is rational number}\}$. Then B is the empty set because the equation $x^2 - 2 = 0$ is not satisfied by any rational value of x.

(*iii*) $C = \{x : x \text{ is an even prime number greater than } 2\}$. Then C is the empty set, because 2 is the only even prime number.

(*iv*) $D = \{ x : x^2 = 4, x \text{ is odd} \}$. Then D is the empty set, because the equation $x^2 = 4$ is not satisfied by any odd value of x.

1.5.3. Finite and Infinite Sets

A set which is empty or consists of a definite number of elements is called *finite* otherwise, the set is called *infinite*.

Consider $A = \{1, 2, 3, 4, 5, 6, 7\}$, $B = \{a, b, c, d, e, f, g, h\}$ and C = {men living presently in different parts of the world}

Let A contains 7 elements and B contains 8 elements. How many elements does C contain? As it is, we do not know the number of elements in C, but it is some natural number which may be quite a big number. By number of elements of a set S, we mean the number of distinct elements of the set and we denote it by $n(S)$. If $n(S)$ is a natural number, then S is *non-empty finite* set.

Let the set of natural numbers. Now, the number of elements of this set is not finite since there are infinite number of natural numbers. Now, the set of natural numbers is an infinite set. The sets A, B and C given above are finite sets and $n(A) = 7$, $n(B) = 8$ and $n(C)$ = some finite number.

Do you know?

Infinites sets can not be described in the roster form. For example, the set of real numbers cannot be described in this form, because the elements of this set donot follow any particular pattern.

Consider some examples :

(*i*) Let G be the set of points on a line. Then G is infinite.

(*ii*) Let S be the set of solutions of the equation $x^2 - 16 = 0$. Then S is finite.

(*iii*) Let W be the set of the days of the week. Then W is finite.

When we represent a set in the roster form, we write all the elements of the set within braces { }. It is not possible to write all the elements of an infinite set within braces { } because the numbers of elements of such a set is not finite. So, we represent some infinite set in the roster form by writing a few elements which clearly indicate the structure of the set followed (or preceded) by three dots.

For example, $\{1, 2, 3 \ldots\}$ is the set of natural numbers, $\{1, 3, 5, 7, \ldots\}$ is the set of odd natural numbers, $\{\ldots, -3, -2, -1, 0, 1, 2, 3, \ldots\}$ is the set of integers. All these sets are infinite.

Example 8. Check the following sets are finite or infinite:

(*i*) $\{x : x \in \text{N and } (x - 1)(x - 2) = 0\}$

(*ii*) $\{x : x \in \text{N and } x^2 = 4\}$

(*iii*) $\{x : x \in \text{N and } 2x - 1 = 0\}$

(*iv*) $\{x : x \in \text{N and } x \text{ is prime}\}$

(*v*) $\{x : x \in \text{N and } x \text{ is odd}\}$

Sol.

(*i*) Given set = {1, 2}. Hence, it is finite.

(*ii*) Given set = {2}. Hence, it is finite.

(*iii*) Given set = ϕ. Hence, it is finite.

(*iv*) The given set is the set of all prime numbers and since set of prime numbers is infinite. Hence the given set is infinite

(*v*) Since there are infinite number of odd numbers, hence, the given set is infinite.

1.5.4. Equal Sets

Let two sets P and Q are said to be *equal* if they have exactly the same elements and we write $P = Q$. Otherwise, the sets are said to be *unequal* and we write $P \neq Q$.

Let two sets P and Q, if every element of P is also an element of Q and if every element of Q is also an element of P, then the sets P and Q are said to be equal.

Clearly, the two sets have exactly the same elements.

We consider the following examples :

(*i*) Let $A = \{1, 2, 3, 4\}$ and $B = \{3, 1, 4, 2\}$. Then $A = B$.

(*ii*) Let A be the set of prime numbers less than 6 and P the set of prime factors of 30. Then A and P are equal, since 2, 3 and 5 are the only prime factors of 30 and also these are less than 6.

A set does not change if one or more elements of the set are repeated.

For example, the sets A = $\{p, q, r\}$ and B = $\{q, q, p, r, r\}$ are equal, since each element of A is in B and vice-versa. That is generally do not repeat any element in describing a set.

Example 9. Find the pairs of equal sets, if any, give reasons:

$A = \{0\}$, $B = \{x : x > 15 \text{ and } x < 5\}$,

$C = \{x : x - 5 = 0\}$, $D = \{x : x^2 = 25\}$,

$E = \{x : x$ is an integral positive root of the equation $x^2 - 2x - 15 = 0\}$.

Sol. Since 0 $\in$ A and 0 does not belong to any of the sets B, C, D and E, it follows that, A $\neq$ B, A $\neq$ C, A $\neq$ D, A $\neq$ E.

Since $B = \phi$, but none of the other sets are empty. Therefore B $\neq$ C, B $\neq$ D and B $\neq$ E. Also C = {5} but –5 $\in$ D, hence C $\neq$ D.

Since E = {5}, C = E. Further, D = {–5, 5} and E = {5}, we find that, D $\neq$ E.

Thus, the only pair of equal sets is C and E.

Example 10. Which of the following pairs of sets are equal? Justify your answer.

(*i*) X, the set of letters in "ALLOY" and B, the set of letters in "LOYAL".

(*ii*) A = $\{n : n \in Z \text{ and } n^2 \leq 4\}$ and B = $\{x : x \in R \text{ and } x^2 - 3x + 2 = 0\}$.

Sol. (*i*) We have, X = {A, L, L, O, Y}, B = {L, O, Y, A, L}. Then X and B are equal sets as repetition of elements in a set do not change a set. Thus,

X = {A, L, O, Y} = B

(*ii*) A = {–2, –1, 0, 1, 2}, B = {1, 2}. Since 0 $\in$ A and 0 $\in$ B, A and B are not equal sets.

EXERCISE 1.1.7.

1. Which of the following are examples of the null set

(*i*) {y : y is a point common to any two parallel lines}

(*ii*) {x : x is a natural numbers, $x < 5$ and $x > 7$}

(*iii*) Set of even prime numbers

(*iv*) Set of odd natural numbers divisible by 2 **Ans.** (*i*), (*ii*), (*iv*)

2. State whether each of the following set is finite or infinite:

(*i*) The set of lines which are parallel to the x-axis **Ans.** Infinite

(*ii*) The set of letters in the English alphabet **Ans.** Finite

(*iii*) The set of numbers which are multiple of 5 **Ans.** Infinite

(*iv*) The set of animals living on the earth **Ans.** Finite

(*v*) The set of circles passing through the origin (0, 0) **Ans.** Infinite

3. Which of the following sets are finite or infinite

(*i*) The set of months of a year **Ans.** Finite

(*ii*) {1, 2, 3, . . .} **Ans.** Infinite

(*iii*) {1, 2, 3, . . .99, 100} **Ans.** Finite

(*iv*) The set of positive integers greater than 100 **Ans.** Infinite

(*v*) The set of prime numbers less than 99 **Ans.** Finite

4. In the following, state whether A = B or not:

(*i*) A = {a, b, c, d} B = {d, c, b, a} **Ans.** (*i*) Yes

(*ii*) A = {4, 8, 12, 16} B = {8, 4, 16, 18} **Ans.** (*ii*) No

(*iii*) A = {2, 4, 6, 8, 10} B = {x : x is positive even integer and $x \leq 10$} **Ans.** (*iii*) Yes

(*iv*) A = {x : x is a multiple of 10}, B = { 10, 15, 20, 25, 30, . . . } **Ans.** (*iv*) No

5. Are the following pair of sets equal ? Give reasons.

(*i*) A = {2, 3}, B = {x : x is solution of $x^2 + 5x + 6 = 0$} **Ans.** (*i*) No

(*ii*) A = { x : x is a letter in the word FOLLOW} **Ans.** (*i*) Yes

B = { y : y is a letter in the word WOLF}

6. From the sets given below, select equal sets :

A = { 2, 4, 8, 12}, B = { 1, 2, 3, 4}, C = { 4, 8, 12, 14}, D = { 3, 1, 4, 2}

E = {–1, 1}, F = { 0, a}, G = {1, –1}, H = { 0, 1} **Ans.** [$B = D$, $E = G$]

1.5.5. Subsets

A set A is said to be a subset of a set B if every element of A is also an element of B.

Let the sets : X = set of all students in your school, Y = set of all students in your class.

Now, every element of Y is also an element of X; we say that Y is a subset of X. The fact that Y is subset of X is expressed in symbols as Y ⊂ X. The symbol ⊂ stands for 'is a subset of' or 'is contained in'.

In other words, A ⊂ B if whenever $a \subset A$, then $a \subset B$. It is often convenient to use the symbol "⇒" which means *implies*. Using this symbol, we can write the definiton of *subset* as follows:

$$A \subset B \text{ if } a \in A \Rightarrow a \in B$$

Now, the statement as "*A is a subset of B if a is an element of A implies that a is also an element of B*". If A is not a subset of B, we write A ⊄ B.

We may note that for A to be a subset of B, all that is needed is that every element of A is in B. It is possible that every element of B may or may not be in A. If it so happens that every element of B is also in A, then we shall also have $B \subset A$. In this case, A and B are the same sets so that we have $A \subset B$ and $B \subset A \Leftrightarrow A = B$, where " $\Leftrightarrow$ " is a symbol for two way implications, and is usually read as *if and only if* (briefly written as "iff").

It follows from the above definition that every set *A is a subset of itself, i.e.*, $A \subset A$. Since the empty set ϕ has no elements, we agree to say that ***ϕ is a subset of every set***. We now consider some examples :

(*i*) Let A = { *a, e, i, o, u*} and B = { *a, b, c, d*}. Then A is not a subset of B, also B is not a subset of A.

(*ii*) Let A = {1, 3, 5} and B = {*x* : *x* is an odd natural number less than 6}. Then $A \subset B$ and $B \subset A$ and hence A = B.

(*iii*) If A is the set of all divisors of 56 and B the set of all prime divisors of 56, then B is a subset of A and we write $B \subset A$.

(*iv*) The set **Q** of rational numbers is a subset of the set **R** of real numbes, and we write $\mathbf{Q} \subset$ R.

Suppose A and B be two sets. If $A \subset B$ and $A \neq B$, then A is called a *proper subset of* B and B is called *superset* of A. For example,

A = {1, 2, 3} is a proper subset of B = {1, 2, 3, 4}.

If a set A has only one element, we call it a *singleton set*. Thus, {*a*} is a singleton set.

Example 11. *Consider the sets*

$$\phi,\ A = \{1, 3\},\ B = \{1, 5, 9\},\ C = \{1, 3, 5, 7, 9\}.$$

Insert the symbol $\subset$ or $\not\subset$ between each of the following pair of sets:

(*i*) ϕ . . . B (*ii*) A . . . B (*iii*) A . . . C (*iv*) B . . . C

Sol. (*i*) $\phi \subset$ B as ϕ is a subset of every set.

(*ii*) $A \not\subset B$ as $3 \subset A$ and $3 \notin B$

(*iii*) $A \subset C$ as $1, 3 \in A$ also belongs to C

(*iv*) $B \subset C$ as each element of B is also an element of C.

Example 12. *Let A = { a, e, i, o, u} and B = { a, b, c, d}. Is A a subset of B ? No. (Why?). Is B a subset of A? No. (Why?)*

Sol. Given, A = {*a, e, i, o, u*}; B = {*a, b, c, d*}

Here, every element of set 'A' is not present in 'set B'. So 'A' is not subset of 'B'.

Example 13. *Let A, B and C be three sets. If $A \in B$ and $B \subset C$, is it true that $A \subset C$?. If not, give an example.*

Sol. No. Let A = {1}, B = {{1}, 2} and C = {{1}, 2, 3}. Here $A \in B$ as A = {1} and $B \subset C$. But $A \not\subset C$ as $1 \in A$ and $1 \notin C$.

So, an element of a set can never be a subset of itself.

Subsets of set of real numbers

There are many important subsets of **R**. We give below the names of some of these subsets.

The set of natural numbers **N** = {1, 2, 3, 4, 5, . . .}

The set of integers **Z** = {. . ., –3, –2, –1, 0, 1, 2, 3, . . .}

The set of rational numbers $\mathbf{Q} = \{ x : x = \frac{p}{q}, p, q, \in \mathbf{Z}$ and $q \neq 0\}$ which is read " **Q** is the set of all numbers x such that x equals the quotient $\frac{p}{q}$, where p and q are integers and q is not zero". Members of **Q** include –5 (which can be expressed as $-\frac{5}{1}$), $\frac{5}{7}$, $3\frac{1}{2}$ (which can be expressed as $\frac{7}{2}$), and $-\frac{11}{3}$.

The set of irrational numbers, denoted by **T**, is composed of all other real numbers. Thus **T** $= \{x : x \in \mathbf{R}$ and $x \notin \mathbf{Q}\}$, *i.e.*, all real numbers that are not rational. Members of **T** include $\sqrt{2}$, $\sqrt{5}$ and π.

Some of subsets are:

$\mathbf{N} \subset \mathbf{Z} \subset \mathbf{Q}$, $\mathbf{Q} \subset \mathbf{R}$, $\mathbf{T} \subset \mathbf{R}$, $\mathbf{N} \not\subset \mathbf{T}$.

Intervals as subsets of R Let $a, b \in \mathbf{R}$ and $a < b$. Then the set of real numbers $\{ y : a < y < b\}$ is called an *open interval* and is denoted by (a, b). All the points between a and b belong to the open interval (a, b) but a, b themselves do not belong to this interval.

The interval which contains the end points also is called *closed interval* and is denoted by $[a, b]$. Thus

$$[a, b] = \{x : a \leq x \leq b\}$$

We can also have intervals closed at one end and open at the other, *i.e.*,

$[a, b) = \{x : a \leq x < b\}$ is an *open interval* from a to b, including a but excluding b.

$(a, b] = \{ x : a < x \leq b \}$ is an *open interval* from a to b including b but excluding a.

These notations provide an alternative way of designating the subsets of set of real numbers. For example, if A = (–3, 5) and B = [–7, 9], then $A \subset B$. The set $[0, \infty)$ defines the set of non-negative real numbers, while set $(-\infty, 0)$ defines the set of negative real numbers. The set $(-\infty, \infty)$ describes the set of real numbers in relation to a line extending from $-\infty$ to ∞.

Here, we note that an interval contains infinitely many points.

For example, the set $\{x : x \in \mathbf{R}, -5 < x \leq 7\}$, written in set-builder form, can be written in the form of interval as (–5, 7] and the interval [–3, 5) can be written in set builder form as $\{x : -3 \leq x < 5\}$.

The number $(b - a)$ is called the *length of any of the intervals* (a, b), $[a, b]$, $[a, b)$ or $(a, b]$.

1.5.6. Power Set

The collection of all subsets of a set Q is called the *power set* of Q. It is denoted by P(Q). In P(Q), every element is a set.

Consider the set {1, 2}. Let us write down all the subsets of the set {1, 2}. We know that ϕ is a subset of every set. So, ϕ is a subset of {1, 2}. We see that {1} and {2} are also subsets of {1, 2}. Also, we know that every set is a subset of itself. So, {1, 2} is a subset of {1, 2}. Thus, the set {1, 2} has, in all, four subsets, viz. ϕ, {1}, {2} and {1, 2}. The set of all these subsets is called the *power set* of {1, 2}.

Thus, as in above, if Q = {1, 2}, then

$$P(Q) = \{\phi, \{ 1 \}, \{ 2 \}, \{ 1,2 \}\}$$

Also, note that $n\, [P(Q)] = 4 = 2^2$

In general, if Q is a set with $n(Q) = m$, then it can be shown that $n\ [P(Q)] = 2^m$.

1.5.7. Universal Set

Suppose, we have to deal with the elements and subsets of a basic set which is relevant to that particular context. For example, while studying the system of numbers, we are interested in the set of natural numbers and its subsets such as the set of all prime numbers, the set of all even numbers, and so forth. This basic set is called the **"*Universal Set*"**. The universal set is usually denoted by U, and all its subsets by the letters A, B, C, etc.

For example, for the set of all integers, the universal set can be the set of rational numbers or, for that matter, the set **R** of real numbers. For another example, in human population studies, the universal set consists of all the people in the world.

EXERCISE 1.1.8.

1. Examine whether the following statements are true or false:

(*i*) $\{x : x$ is an even natural number less than $6\} \subset \{x : x$ is a natural number which divides $36\}$ **Ans.** (*i*) T

(*ii*) $\{a\} \notin \{a, b, c\}$ **Ans.** (*ii*) F

(*iii*) $\{a\} \subset \{a, b, c\}$ **Ans.** (*iii*) T

(*iv*) $\{1, 2, 3\} \subset \{1, 3, 5\}$ **Ans.** (*iv*) F

(*v*) $\{a, e\} \subset \{x : x$ is a vowel in the English alphabet$\}$ **Ans.** (*v*) T

(*vi*) $\{a, b\} \not\subset \{b, c, a\}$ **Ans.** (*vi*) F

2. Make correct statements by filling in the symbols $\subset$ or $\not\subset$ in the blank spaces :

(*i*) $\{x : x$ is an even natural number$\}$ . . . $\{x : x$ is an integer$\}$ **Ans.** (*i*) $\subset$

(*ii*) $\{x : x$ is an equilateral triangle in a plane$\}$ . . . $\{x : x$ is a triangle in the same plane$\}$ **Ans.** (*ii*) $\subset$

(*iii*) $\{x : x$ is a triangle in a plane$\}$ . . . $\{x : x$ is a rectangle in the plane$\}$ **Ans.** (*iii*) $\not\subset$

(*iv*) $\{x : x$ is a circle in the plane$\}$ . . .$\{x : x$ is a circle in the same plane with radius 1 unit$\}$ **Ans.** (*iv*) $\not\subset$

(*v*) $\{x : x$ is a student of Class XI of your school$\}$. . .$\{x : x$ student of your school$\}$ **Ans.** (*v*) $\subset$

(*vi*) $\{2, 3, 4\}$ . . . $\{1, 2, 3, 4,5\}$ **Ans.** (*vi*) $\subset$

(*vii*) $\{a, b, c\}$ . . . $\{b, c, d\}$ **Ans.** (*ii*) $\not\subset$

3. Let A = { 1, 2, { 3, 4 }, 5 }. Which of the following statements are incorrect and why?

(*i*) $\{3, 4\} \subset A$ (*ii*) $\{3, 4\} \in A$

(*iii*) $\{\{3, 4\}\} \subset A$ (*iv*) $1 \in A$

(*v*) $1 \subset A$ (*vi*) $\{1, 2, 5\} \subset A$

(*vii*) $\{1, 2, 5\} \in A$ (*viii*) $\{1, 2, 3\} \subset A$

(*ix*) $\phi \in A$ (*x*) $\phi \subset A$

(*xi*) $\{\phi\} \subset A$ **Ans.** {(*i*), (*v*), (*vii*), (*viii*), (*ix*), (*xi*)}

4. Write all subsets of the following sets:

(*i*) $\{a\}$ **Ans.** (*i*) ϕ, $\{a\}$

(*ii*) $\{a, b\}$ **Ans.** (*ii*) ϕ, $\{a\}$, $\{b\}$, $\{a, b\}$

(*iii*) {1, 2, 3} **Ans.** (*iii*) ϕ, {1}, {2}, {3}, {1, 2}, {1, 3}, {2, 3}, {1, 2, 3}

(*iv*) ϕ **Ans.** (*iv*) ϕ

5. How many elements has P(A), if A = ϕ? **Ans.** 1

6. Given the sets A = {1, 3, 5}, B = {2, 4, 6} and C = {0, 2, 4, 6, 8}, which of the following may be considered as universal set (s) for all the three sets A, B and C

(*i*) {0, 1, 2, 3, 4, 5, 6} (*ii*) ϕ

(*iii*) {0, 1, 2, 3, 4, 5, 6, 7, 8, 9, 10} (*iv*) {1, 2, 3, 4, 5, 6, 7, 8} **Ans.** (*iii*)

7. What universal set(s) would you propose for each of the following :

(*i*) The set of right triangles. (*ii*) The set of isosceles triangles.

8. Write the following intervals in set-builder form :

(*i*) (– 3, 0) **Ans.** $\{x : x \in R, -3 < x < 0\}$

(*ii*) [6 , 12] **Ans.** $\{x : x \in R, 6 \le x \le 12\}$

(*iii*) (6, 12] **Ans.** $\{x : x \in R, 6 < x \le 12\}$

(*iv*) [–23, 5) **Ans.** $\{x : x \in R, -23 \le x < 5\}$

9. Write the following as intervals :

(*i*) $\{x : x \in R, -4 < x \le 6\}$ (*ii*) $\{x : x \in R, -12 < x < -10\}$

(*iii*) $\{x : x \in R, 0 \le x < 7\}$ (*iv*) $\{x : x \in R, 3 \le x \le 4\}$

Ans. (*i*) (– 4, 6), (*ii*) (–12, 10), (*iii*) [0, 7), (*iv*) [3, 4]

• Other Notation

If A is a subset of B, then all of the elements of A are also in B. For example, if A = {1, 2, 3} and B = {1, 2, 3, 4, 5} then A $\in$ B ($\in$ means is a subset of).

Number of Members

If A = {1, 2, 4, 8}, then n(A) = 4. This is because *n*(A) means the number of members in set A.

The Universal Set

The universal set is the set of all sets. All sets are therefore subsets of the universal set.

• Venn Diagrams

Venn diagrams are used to represent sets. Here, the set A{1, 2, 4, 8} is shown using a circle. In Venn diagrams, sets are usually represented using circles. The universal set is the rectangle. The set A is a subset of the universal set and so it is within the rectangle.

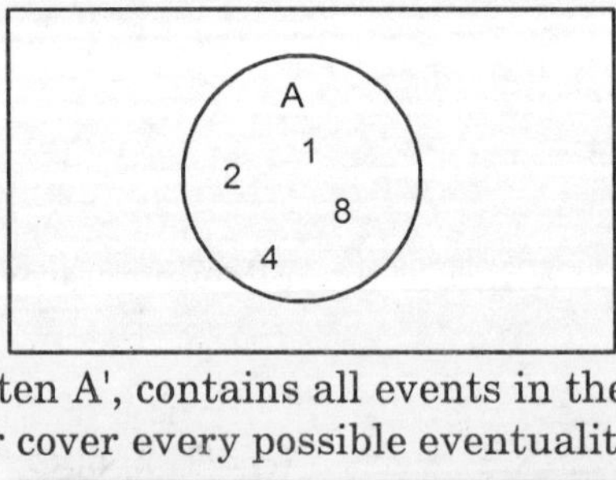

The **complement** of A, written A', contains all events in the sample space which are not members of A. A and A' together cover every possible eventuality.

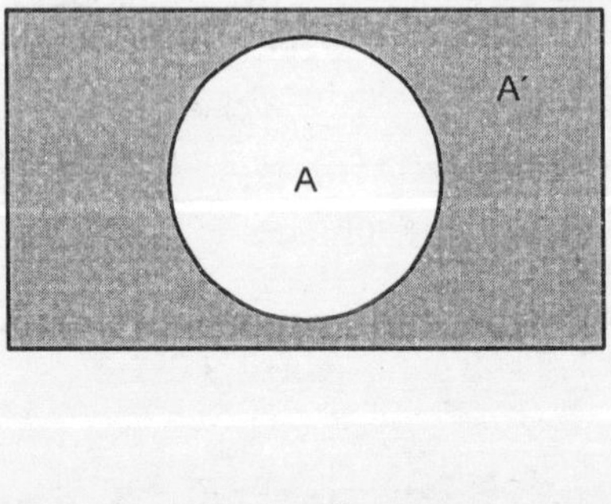

A∈B means the union of sets A and B and contains all of the elements of both A and B. This can be represented on a Venn Diagram as follows:

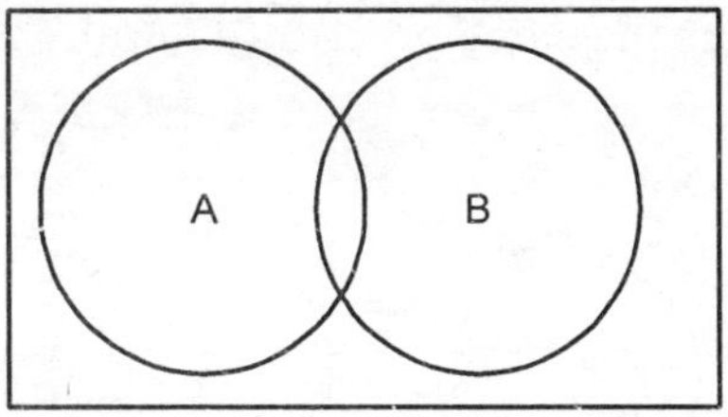

A∩B means the intersection of sets A and B. This contains all of the elements which are in both A and B. A∩B is shown on the Venn Diagram below:

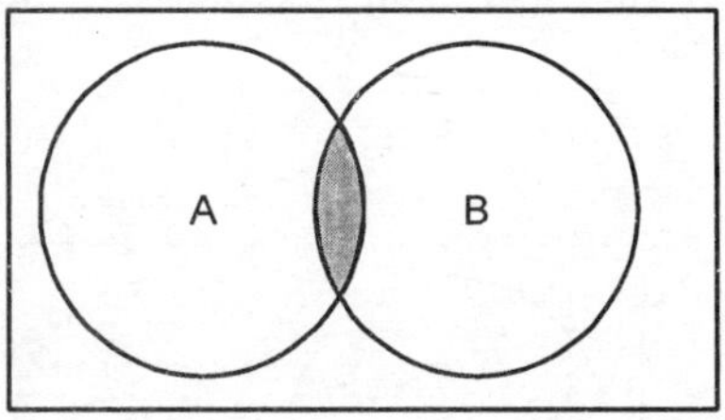

An important result connecting the number of members in sets and their unions and intersections is:

$n(A) + n(B) - n(A \cap B) = n(A \cup B)$

The relationships between sets can be represented by means of different type of figure like rectangle, square, circle etc. and also, set is represented by rectangle and subsets by circle diagrams which are known as *Venn diagrams*. These diagrams are named after the English logician, John Venn (1834-1883). These diagrams consist of rectangles and closed curves usually circles. The universal set is represented usually by a rectangle and its subsets by circles. In Venn diagrams, the elements of the sets are written in their respective circles.

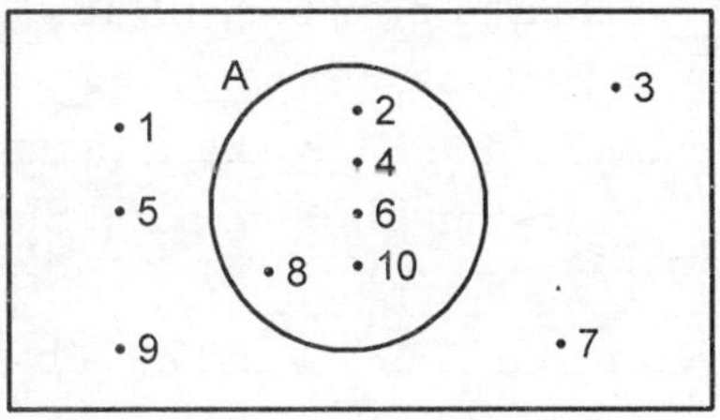

Example 14. In above Figure, U = {1, 2, 3, ..., 10} is the universal set of which A = {2, 4, 6, 8, 10} is a subset.

Example 15. In the below Figure, U = {1, 2, 3, ..., 10} is the universal set of which

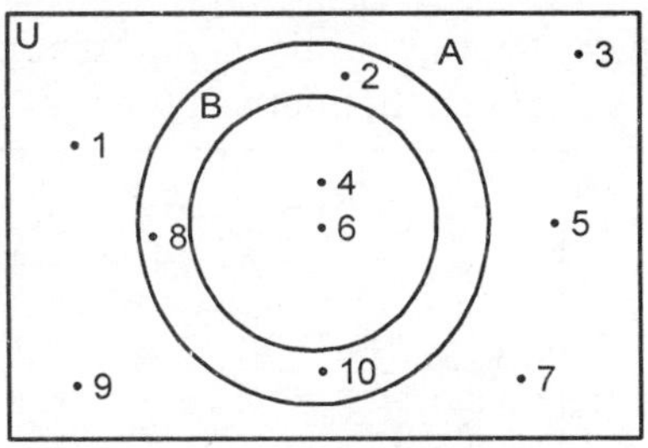

A = {2, 4, 6, 8, 10} and B = {4, 6} are subsets, and also B∈A.

1.5.8. Operations on Sets

Let two sets we can combine to form new sets. There is nothing particular about this number. Now, we can carry out the same procedure for any number of sets, finite or infinite. Consider the following figure

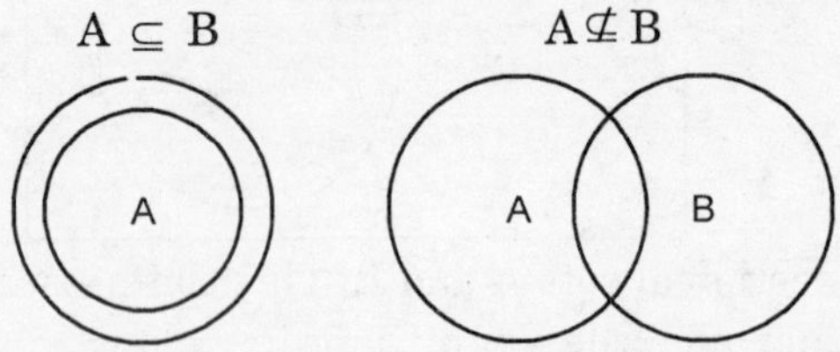

the diagram as shown above, which are used to show relationship between sets, called *Venn diagram.*

In Venn diagram the universal set U is represneted by the interior of rectangle and sets by circle inside the rectangle.

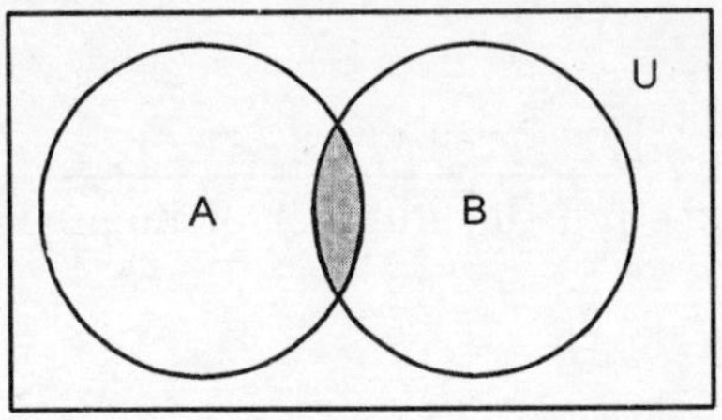

Union of sets.

Suppose P and Q be any two sets. The union of P and Q is the set which consists of all the elements of P and all the elements of Q, the common elements being taken only once. The symbol '∪' is used to denote the *union. Symbolically, we write* P ∪ Q *and usually read as 'P union Q'.*

Example 16. Let A = { 2, 4, 6, 8} and B = { 6, 8, 10, 12, 14}. Find A ∪ B.

Sol. We have A ∪ B = { 2, 4, 6, 8, 10, 12, 14}

Note that the common elements 6 and 8 have been taken only once while writing A ∪ B.

Example 17. Let A = { *a, e, i, o, u* } and B = { *a, i, u* }. Show that A ∪ B = A

Sol. We have, A ∪ B = { *a, e, i, ɔ, u* } = A.

This example illustrates that union of sets A and its subset B is the set A itself, i.e., if B ⊂ A, then A ∪ B = A.

Example 18. Let X = {Aman, Geeta, Shalu} be the set of students of Class XI, who are in school hockey team. Let Y = {Geeta, David, Ashok} be the set of students from Class XI who are in the school football team. Find X ∪ Y and interpret the set.

Sol. We have, X ∪ Y = {Aman, Geeta, Shalu, David, Ashok}. This is the set of students from Class XI who are in the hockey team or the football team or both.

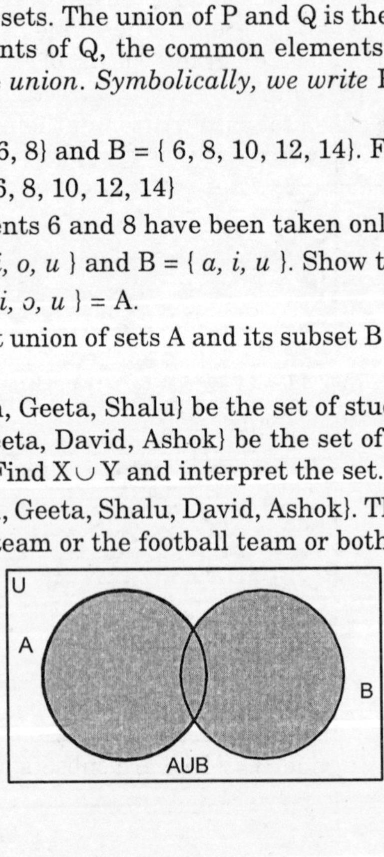

Thus, we can define the union of two sets as follows:

The union of two sets A and B is the set C which consists of all those elements which are either in A or in B (including those which are in both). In symbols, we write.

$$A \cup B = \{ x : x \in A \text{ or } x \in B \}$$

The union of two sets can be represented by a Venn diagram as shown in Figure given in previous page.

The shaded portion in Figure represents $A \cup B$.

• Properties of Union

(*i*) $A \cup B = B \cup A$	(Commutative law)
(*ii*) $(A \cup B) \cup C = A \cup (B \cup C)$	(Associative law)
(*iii*) $A \cup \phi = A$	(Law of identity element, ϕ is the identity of $\cup$)
(*iv*) $A \cup A = A$	(Idempotent law)
(*v*) $U \cup A = U$	(Law of U)

Intersection of sets. The intersection of sets A and B is the set of all elements which are common to both A and B. The symbol '$\cap$' is used to denote the *intersection*. The intersection of two sets A and B is the set of all those elements which belong to both A and B. Symbolically, we write $A \cap B = \{x : x \in A \text{ and } x \in B\}$.

The intersection of two sets A and B is the set of all those elements which belong to both A and B. Symbolically, we write $A \cap B = \{x : x \in A \text{ and } x \in B\}$. The shaded portion in below Figure indicates the intersection of A and B.

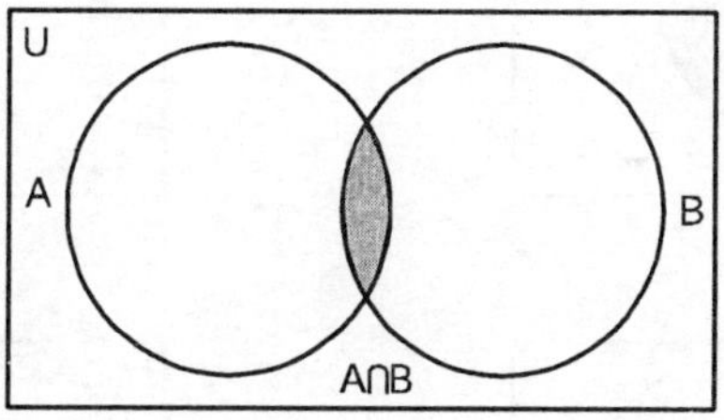

Example 19. Consider the sets P and Q of $P = \{1, 3, 5, 7, 8\}$ and $Q = \{2, 3, 4, 7, 8\}$. Find $P \cap Q$.

Sol. We see that 3, 7, 8 are the only elements which are common to both P and Q.

Hence $P \cap Q = \{ 3, 7, 8\}$.

Example 20. Consider the sets X and Y of $X = \{20, 25, 22, 28, 30\}$ and $Y = \{25, 22, 27, 29\}$. Find $X \cap Y$.

Sol. We see that element '25 and 22' is the only element common to both. Hence, $X \cap Y = \{25, 22\}$.

Example 21. Let $A = \{1, 2, 3, 4, 5, 6, 7, 8, 9, 10\}$ and $B = \{ 1, 2, 3, 4 \}$. Find $A \cap B$ and hence show that $A \cap B = B$.

Sol. We have $A \cap B = \{ 1, 2, 3, 4 \} = B$. We note that $B \subset A$ and that $A \cap B = B$.

If A and B are two sets such that $A \cap B = \phi$, then A and B are called *disjoint sets*.

For example, let $A = \{2, 4, 6, 8\}$ and $B = \{1, 3, 5, 7\}$. Then A and B are disjoint sets, because there are no elements which are common to A and B. The disjoint sets can be represented by means of Venn diagram as shown in the below. In the diagram, A and B are disjoint sets.

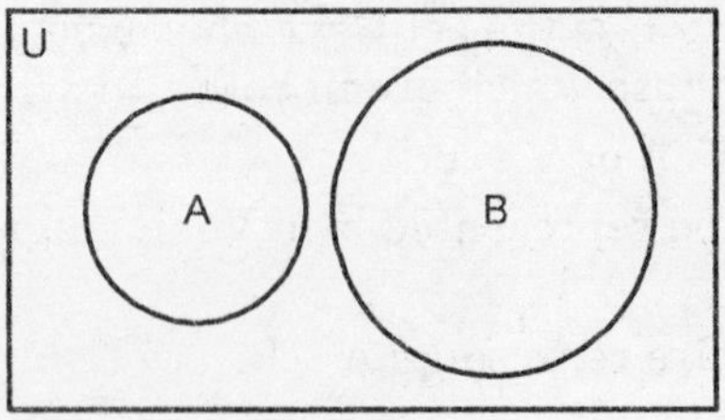

Some Properties of Operation of Intersection

(*i*) $A \cap B = B \cap A$ (Commutative law).

(*ii*) $(A \cap B) \cap C = A \cap (B \cap C)$ (Associative law).

(*iii*) $\phi \cap A = \phi$, $U \cap A = A$ (Law of ϕ and U).

(*iv*) $A \cap A = A$ (Idempotent law)

(*v*) $A \cap (B \cup C) = (A \cap B) \cup (A \cap C)$ (Distributive law) i.e., $\cap$ distributes over $\cup$.

This can be seen easily from the following Venn diagrams.

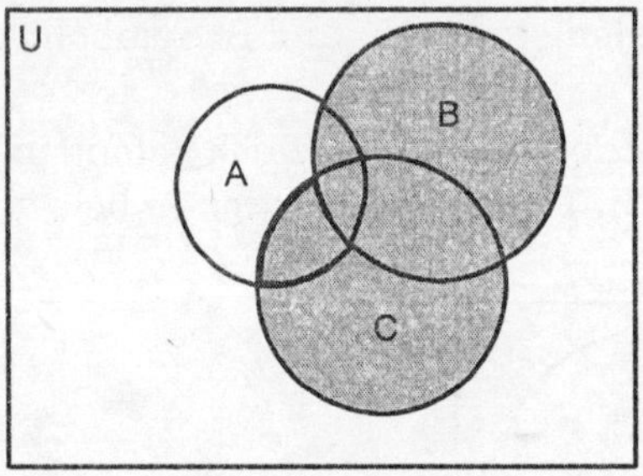

(*i*) $(B \cup C)$

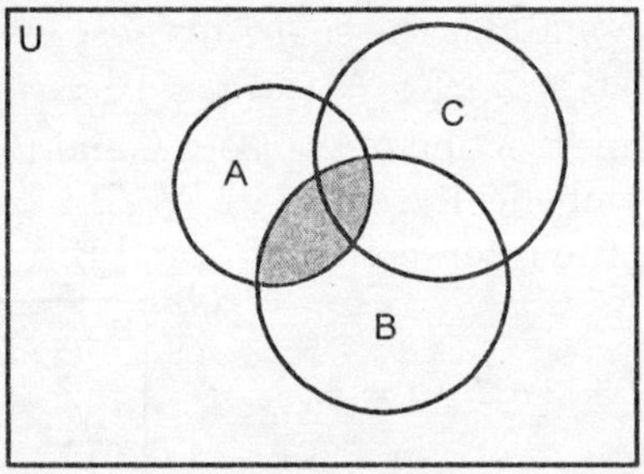

(*ii*) $(A \cap B)$

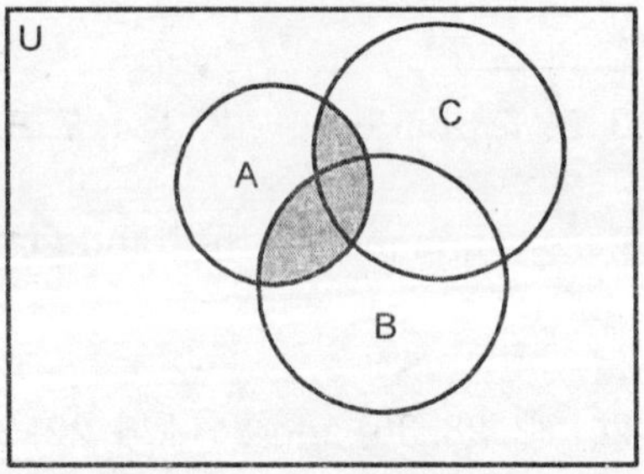

(*iii*) $A \cap (B \cup C)$

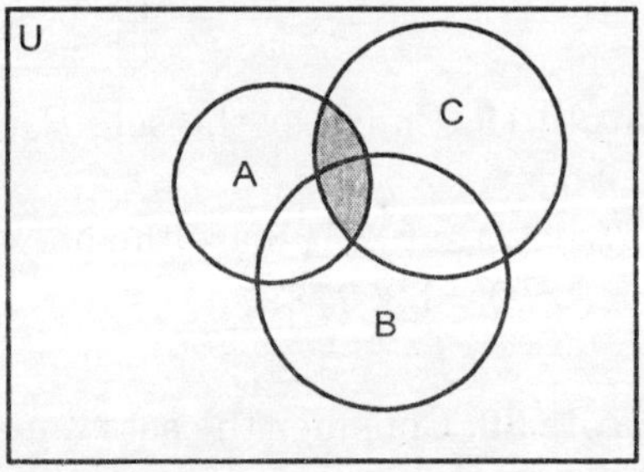

(*iv*) $(A \cap C)$

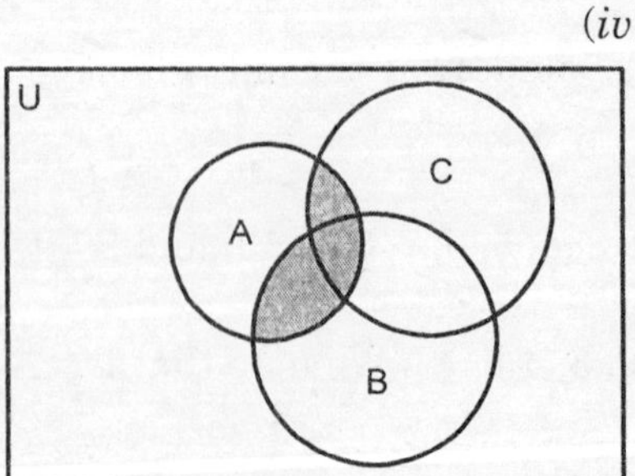

(*iv*) $(A \cap B) \cup (A \cap C)$

• Difference of sets

The difference of the sets P and Q in this order is the set of elements which belong to P but not to Q. Symbolically, we write P – Q and read as "P minus Q".

Example 22. Let P = { 1, 2, 3, 4, 5, 6}, Q = { 2, 4, 6, 8}. Find P – Q and Q – P.

Sol. We have, P – Q = { 1, 3, 5 }, since the elements 1, 3, 5 belong to P but not to Q and Q – P = { 8 }, since the element 8 belongs to Q and not to P.

We note that P – Q $\neq$ Q – P.

Example 23. Let A = { a, e, i, o, u } and B = { a, i, k, u}. Find A – B and B – A

Sol. We have, A – B = { e, o }, since the elements e, o belong to A but not to B and B – A = { k }, since the element k belongs to B but not to A.

Do you know?

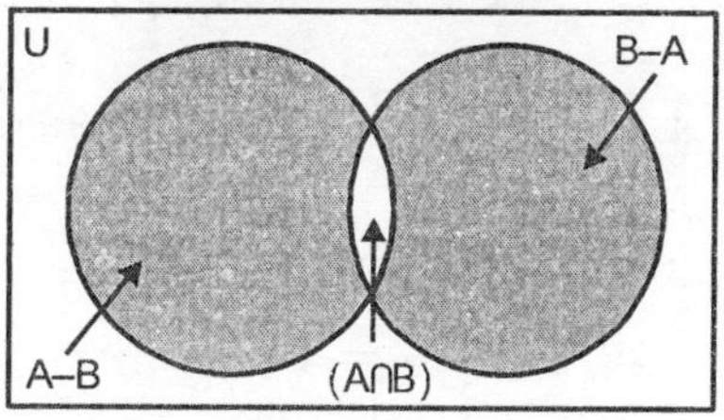

The sets A=B, $A \cap B$ and B–A are mutually disjoint sets, i.e., the intersection of any of these two sets is the null set as shown in Figure.

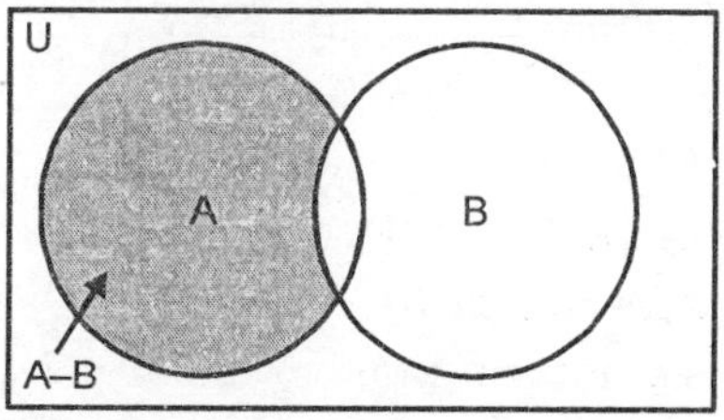

We note that A – B $\neq$ B – A. Using the setbuilder notation, we can rewrite the definition of difference as A – B = { x : $x \in$ A and $x \notin$ B }

The difference of two sets A and B can be represented by Venn diagram as shown in above figure. The shaded portion represents the difference of the two sets A and B.

EXERCISE 1.1.9.

1. If A and B are two sets such that A $\subset$ B, then what is A $\cup$ B ? **Ans.** B

2. Let A = { a, b }, B = {a, b, c}. Is A $\subset$ B ? What is A $\cup$ B ? **Ans.** Yes, A $\cup$ B = {a, b, c, ϕ}

3. Find the union of each of the following pairs of sets :

(*i*) A = {a, e, i, o, u} B = {a, b, q} **Ans.** (*i*) {a, e, i, o, u, abq}

(*ii*) X = {1, 3, 4, 5} Y = {1, 2, 3} **Ans.** (*ii*) {1, 2, 3,4, 5}

(*iii*) A = {x : x is a natural number and $1 < x \leq 6$ } **Ans.** (*iii*) {$x : 1 < x < 10, x \in$ N}
B = {x : x is a natural number and $6 < x < 10$ }

(*iv*) A = {x : x is a natural number and multiple of 5} **Ans.** (*iv*) {*x:x 1, 5, or multiply of 5*}
B = {x : x is a natural number less than 6}

(*v*) A = {1, 2, 3}, B = ϕ **Ans.** A $\cup$ B = {1, 2, 3}

4. If A = {1, 2, 3, 4}, B = {3, 4, 5, 6}, C = {5, 6, 7, 8} and D = {7, 8, 9, 10}; find

(*i*) A $\cup$ B (*ii*) A $\cup$ C

(*iii*) B $\cup$ C (*iv*) B $\cup$ D

(*v*) A $\cup$ B $\cup$ C (*vi*) A $\cup$ B $\cup$ D

(*vii*) B $\cup$ C $\cup$ D

Ans. (*i*) { 1, 2, 3, 4, 5, 6}, (*ii*) { 1, 2, 3, 4, 5, 6, 7, 8}, (*iii*) {3, 4, 5, 6, 7, 8}
(*iv*) {3, 4, 5, 6, 7, 8, 9,10}, (*v*) {1, 2, 3, 4, 5, 6, 7, 8}, (*vi*) {1, 2, 3, 4, 5, 6, 7, 8, 9, 10}
(*vii*) { 3, 4, 5, 6, 7, 8, 9, 10}

5. Find the intersection of each pair of sets of question 3 above.

Ans. (*i*) $X \cap Y = \{1, 3\}$, (*ii*) $A \cap B = \{9\}$ (*iii*) {3}, (*iv*) ϕ, (*v*) ϕ

6. If A = { 3, 5, 7, 9, 11 }, B = {7, 9, 11, 13}, C = {11, 13, 15}and D = {15, 17}; find

(*i*) $A \cap B$ (*ii*) $B \cap C$
(*iii*) $A \cap C \cap D$ (*iv*) $A \cap C$
(*v*) $B \cap D$ (*vi*) $A \cap (B \cup C)$
(*vii*) $A \cap D$ (*viii*) $A \cap (B \cup D)$
(*ix*) $(A \cap B) \cap (B \cup C)$ (*x*) $(A \cap D) \cap (B \cup C)$

Ans. (*i*) {7, 9, 11}, (*ii*) {11, 13}, (*iii*) ϕ, (*iv*) {11}, (*v*) ϕ,
(*vi*) {7, 9, 11}, (*vii*) ϕ, (*viii*) {7, 9, 11}, (*ix*) { 7, 9, 11}, (*x*) (7, 9, 11, 15}

7. If A = {*x* : *x* is a natural number }, B = {*x* : *x* is an even natural number}
C = {*x* : *x* is an odd natural number} and D = {*x* : *x* is a prime number }, find

(*i*) $A \cap B$ (*ii*) $A \cap C$
(*iii*) $A \cap D$ (*iv*) $B \cap C$
(*v*) $B \cap D$ (*vi*) $C \cap D$

Ans. (*i*) B. (*ii*) C, (*iii*) D, (*iv*) ϕ, (*v*) {2}, (*vi*) {*x*:*x* is an odd prime number}

8. Which of the following pairs of sets are disjoint

(*i*) {1, 2, 3, 4} and {*x* : *x* is a natural number and $4 \le x \le 6$ }
(*ii*) { *a*, *e*, *i*, *o*, *u* } and { *c*, *d*, *e*, *f* }
(*iii*) {*x* : *x* is an even integer } and {*x* : *x* is an odd integer} **Ans.** (*iii*)

9. If A = {3, 6, 9, 12, 15, 18, 21}, B = { 4, 8, 12, 16, 20 },
C = { 2, 4, 6, 8, 10, 12, 14, 16 }, D = {5, 10, 15, 20 }; find

(*i*) A – B (*ii*) A – C
(*iii*) A – D (*iv*) B – A
(*v*) C – A (*vi*) D – A
(*vii*) B – C (*viii*) B – D
(*ix*) C – B (*x*) D – B
(*xi*) C – D (*xii*) D – C

Ans. (*i*) {3, 6, 9, 15, 18, 21}, (*ii*) {3, 9, 15, 18, 21}, (*iii*) {3, 6, 9, 12, 18}, (*iv*) {4, 8, 16, 20},
(*v*) {2, 4, 8, 10, 14, 16}, (*vi*) {5, 10, 20}, (*vii*) {20}, (*viii*) {4, 8, 12, 16},
(*ix*) { 2, 6, 10, 14}, (*x*) (5, 10, 15}, (*xi*) {2, 4, 6, 8, 12, 14, 16}, (*xii*) {5, 15, 20}

10. If X= {*a*, *b*, *c*, *d*} and Y = { *f*, *b*, *d*, *g*}, find

(*i*) X – Y (*ii*) Y – X
(*iii*) $X \cap Y$

Ans. (*i*) {*a*, *c*}, (*ii*) {*f*, *g*}, (*iii*) {*b*, *d*}

11. If **R** is the set of real numbers and **Q** is the set of rational numbers, then what is **R** – **Q**?

Ans. Set of irrational numbers

12. State whether each of the following statement is true or false. Justify your answer.

(*i*) { 2, 3, 4, 5 } and { 3, 6} are disjoint sets. **Ans.** (*i*) F

(*ii*) { *a*, *e*, *i*, *o*, *u* } and { *a*, *b*, *c*. *d* } are disjoint sets. **Ans.** (*ii*) F

(*iii*) { 2, 6, 10, 14 } and { 3, 7, 11, 15} are disjoint sets. **Ans.** (*iii*) T

(*iv*) { 2, 6, 10 } and { 3, 7, 11} are disjoint sets. **Ans.** (*iv*) T

1.5.9. Complement of a Set

Suppose U be the universal set and A a subset of U. Then the complement of A is the set of all elements of U which are not the elements of A. Symbolically, we write A′ to denote the complement of A with respect to U. Thus,

$A' = \{x : x \in U \text{ and } x \notin A\}$. Obviously $A' = U - A$.

We note that the complement of a set A can be looked upon, alternatively, as the difference between a universal set U and the set A.

Suppose U be the universal set which consists of all prime numbers and A be the subset of U which consists of all those prime numbers that are not divisors of 42. So, $A = \{x : x \in U$ and x is not a divisor of 42}. Now, that $2 \in U$ but $2 \notin A$, because 2 is divisor of 42. Similarly, $3 \in U$ but $3 \notin A$, and $7 \notin U$ but $7 \notin A$. Now 2, 3 and 7 are the only elements of U which do not belong to A. The set of these three prime numbers, *i.e.*, the set {2, 3, 7} is called the *Complement* of A with respect to U, and is denoted by A′. So we have A′ = {2, 3, 7}. Thus, we see that

$A' = \{x : x \in U \text{ and } x \notin A\}$.

Example 24. Let U = {p, 1, 2, q, 3, 4, 5, r, 6, 7, 8, 9, 10} and A = {1, 3, 5, 7, 9}. Find A′.

Sol. We note that p, q, r, 2, 4, 6, 8, 10 are the only elements of U which do not belong to A. Hence A′ = { 2, 4, 6, 8,10 }.

Example 25. Let U be universal set of all the students of Class 10th of a coeducational school and A be the set of all girls in Class 10th. Find *A*′.

Sol. Since A is the set of all girls, A′ is clearly the set of all boys in the class.

Example 26. Let U = {1, 2, 3, 4, 5, 6}, A = {2, 3} and B = {3, 4, 5}.

Find A′, B′, $A' \cap B'$, $A \cap B$ and hence show that $(A \cap B)' = A' \cup B'$.

> **Do you know?**
>
> If A is a subset of the universal set U, then its complement A′ is also a subset of U.
>
> Again in Example 24, we have A′ = {p, q, r, 2, 4, 6, 8, 10}
>
> Hence $(A')' = \{x : x \in U \text{ and } x \notin A'\}$ = {1, 3, 5, 7, 9} = A
>
> It is clear from the definition of the complement that for any subset of the universal set U, we have (A′)′ = A

Sol. Now we calculate A′ = {1, 4, 5, 6}, B′ = { 1, 2, 6 }. Hence $A' \cap B'$ = {1, 6}

Also $A \cap B$ = { 2, 3, 4, 5 }, so that $(A \cup B)'$ = { 1, 6 } $(A \cup B)'$ = {1, 6} = $A' \cap B'$.

It can be shown that the above result is true in general. If A and B are any two subsets of the universal set U, then $(A \cup B)' = A' \cap B'$. Similarly, $(A \cap B)' = A' \cap B'$. These two results are stated in words as follows :

The complement of the union of two sets is the intersection of their complements and the complement of the intersection of two sets is the union of their complements. These are called De Morgan's laws. These are named after the mathematician De Morgan.

The complement A′ of a set A can be represented by a Venn diagram as shown in Fig. The shaded portion represents the complement of the set A.

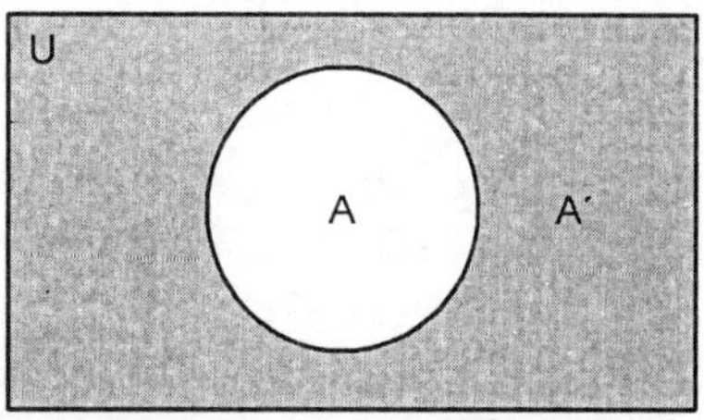

Properties of Complement Sets

1. Complement laws: (*i*) $A \cup A' = U$ (*ii*) $A \cap A' = \phi$

2. De Morgan's law: (*i*) $(A \cup B)' = A' \cap B'$ (*ii*) $(A \cap B)' = A' \cup B'$

3. Law of double complementation : $(A')' = A$

4. Laws of empty set and universal set $\phi' = U$ and $U' = \phi$.

These laws can be verified by using Venn diagrams.

EXERCISE 1.1.10.

1. If U = {*a, b, c, d, e, f, g, h*}, find the complements of the following sets :

(*i*) A = {*a, b, c*} (*ii*) B = {*d, e, f, g*}

(*iii*) C = {*a, c, e, g*} (*iv*) D = { *f, g, h, a*}

Ans. (*i*) {*d, e, f, g, h*} (*ii*) {*a, b, c, h*}, (*iii*) {*b, d, f, h*}, (*iv*) {*b, c, d, e*}

2. Let U = { 1, 2, 3, 4, 5, 6, 7, 8, 9 }, A = { 1, 2, 3, 4}, B = { 2, 4, 6, 8 } and C = { 3, 4, 5, 6 }. Find (*i*) A′ (*ii*) B′ (*iii*) $(A \cup C)'$ (*iv*) $(A \cup B)'$ (*v*) $(A')'$ (*vi*) $(B - C)'$

Ans. (*i*) {5, 6, 7, 8, 9}, (*ii*) (1, 3, 5, 7, 9}, (*iii*) {7, 8, 9}
(*iv*) {5, 7, 9}, (*v*) {1, 2, 3, 4}, (*vi*) {1, 3, 4, 5, 6, 7, 9}

3. If U = {1, 2, 3, 4, 5, 6, 7, 8, 9 }, A = {2, 4, 6, 8} and B = { 2, 3, 5, 7}. Verify that

(*i*) $(A \cup B)' = A' \cap B'$ (*ii*) $(A \cap B)' = A' \cup B'$

4. Taking the set of natural numbers as the universal set, write down the complements of the following sets:

(*i*) {x : x is an even natural number} **Ans.** (*i*) x : x is an odd natural number

(*ii*) { x : x is an odd natural number } **Ans.** (*ii*) x : x is an even natural number

(*iii*) {x : x is a positive multiple of 3} **Ans.** (*iii*) x : $x \in N$ and x is not a multiple of 3

(*iv*) { x : x is a prime number } **Ans.** (*iv*) x : x is a positive composite number and $x = 1$

(*v*) {x : x is a natural number divisible by 3 and 5}

Ans. (*v*) {x : x is a positive integer which is not divisible by 3 or not divisible by 5}

(*vi*) { x : x is a perfect square } **Ans.** (*vi*) {x : $x \in N$ and n is not a perfect square}

(*vii*) { x : x is a perfect cube} **Ans.** (*vii*) { x : $x \in N$ and n is not a perfect cube}

(*viii*) { x : $x + 5 = 8$ } **Ans.** (*viii*) { x : $x \in N$ and $x \neq 3$}

(*ix*) { x : $2x + 5 = 9$} **Ans.** (*ix*) { x : $x \in N$ and $n \neq 2$}

(*x*) { x : $x \geq 7$} **Ans.** (*x*) { x : $x \in N$ and $n < 7$}

(*xi*) { x : $x \in N$ and $2x + 1 > 10$} **Ans.** (*xi*) { x : $x \in N$ and $n \leq 9/2$}

5. Draw appropriate Venn diagram for each of the following :

(*i*) $(A \cup B)'$, (*ii*) $A' \cap B'$,

(*iii*) $(A \cap B)'$, (*iv*) $A' \cup B'$

6. Let U be the set of all triangles in a plane. If *A* is the set of all triangles with at least one angle different from 60°, what is A′?

Ans. A′ is the set of all equilateral triangles.

7. Fill in the blanks to make each of the following true statement:

(*i*) $A \cup A' = \ldots$ (*ii*) $\phi' \cap A = \ldots$

(*iii*) $A \cap A' = \ldots$ (*iv*) $U' \cap A = \ldots$

Ans. (*i*) U (*ii*) *A*, (*iii*) ϕ, (iv) ϕ

• Union and Intersection of Two Sets

Now, we can use of union, intersection and difference of two sets. We derived this form in general form.

Let A and B be finite sets. If $A \cap B = \phi$, then

(*i*) $n\ (A \cap B) = n\ (A) + n\ (B)$ ... (*a*)

The elements in $A \cup B$ are either in A or in B but not in both as $A \cap B = \phi$. So, (*a*) follows immediately.

In general, if A and B are finite sets, then

(*ii*) $n\ (A \cup B) = n\ (A) + n\ (B) - n\ (A \cap B)$ ... (*b*)

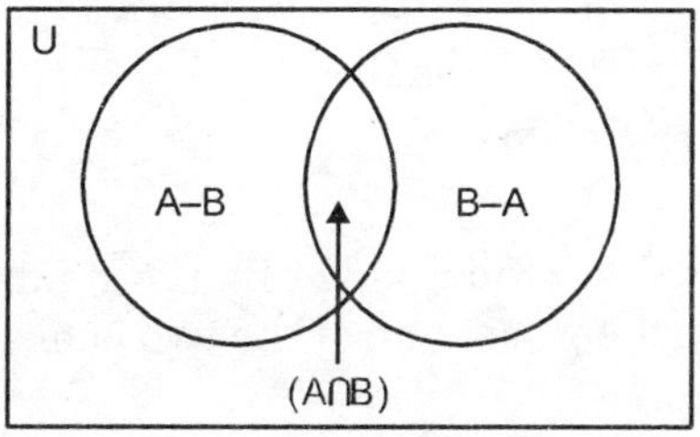

Now the sets A – B, $A \cap B$ and B – A are disjoint and their union is $A \cup B$ in above figure. Therefore,

$$n(A \cup B) = n\ (A - B) + n\ (A \cap B) + n\ (B - A)$$
$$= n\ (A - B) + n\ (A \cap B) + n\ (B - A) + n\ (A \cap B) - n\ (A \cap B)$$
$$= n\ (A) + n\ (B) - n\ (A \cap B), \text{ which verifies } (b)$$

(*iii*) If A, B and C are finite sets, then

$$n\ (A \cup B \cup C) = n\ (A) + n\ (B) + n\ (C) - n\ (A \cap B) - n\ (B \cap C) - n\ (A \cap C) + n\ (A \cap B \cap C) \quad ...(c)$$

In fact, we have

$$n\ (A \cup B \cup C) = n\ (A) + n\ (B \cup C) - n\ [A \cap (B \cup C)]\ [\text{from } (b)]$$
$$= n\ (A) + n\ (B) + n\ (C) - n\ (B \cap C) - n\ [A \cap (B \cup C)]\ [\text{from } (b)]$$

Since $A \cap (B \cup C) = (A \cap B) \cup (A \cap C)$, we get

$$n[A \cap (B \cup C)] = n\ (A \cap B) + n\ (A \cap C) - n\ [(A \cap B) \cap (A \cap C)]$$
$$= n\ (A \cap B) + n\ (A \cap C) - n\ (A \cap B \cap C)$$

$$\Rightarrow\ n(A \cup B \cup C) = n\ (A) + n\ (B) + n\ (C) - n\ (A \cap B) - n\ (B \cap C) - n\ (A \cap C) + n\ (A \cap B \cap C)$$

This proves (*c*).

Example 27. If X and Y are two sets such that $X \cup Y$ has 60 elements, X has 32 elements and Y has 26 elements, how many elements does $X \cap Y$ have ?

Sol. Given that

$$n(X \cup Y) = 60,\ n(X) = 32,\ n\ (Y) = 36,$$
$$n\ (X \cap Y) = ?$$

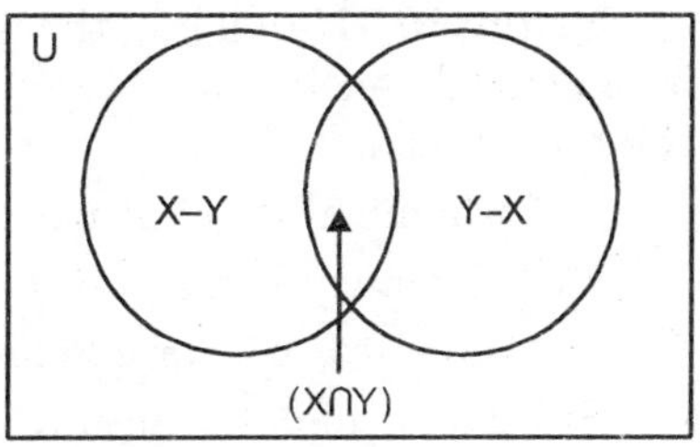

By using the formula

$$n\ (X \cup Y) = n\ (X) + n\ (Y) - n\ (X \cap Y),$$

we find that

$$n\ (X \cap Y) = n\ (X) + n\ (Y) - n\ (X \cup Y)$$
$$= 32 + 36 - 60 = 8$$

Example 28. In a school there are 25 teachers who teaches Art or Economics. Of these, 11 teaches art and 3 teaches both Economics and Art. How many teaches economics ?

Sol. Let A denote the set of teachers who teach arts and E denote the set of teachers who teach economics. In the statement of the problem, the word 'or' gives us a clue of union and the word 'and' gives us a clue of intersection. We, therefore, have

$$n(A \cup E) = 25,\ n\ (A) = 11 \text{ and } n\ (A \cap E) = 3$$

We wish to determine $n(E)$.

Using the result

$$n(A \cup E) = n(A) + n\ (E) - n\ (A \cap E),$$

we obtain $25 = 11 + n(E) - 3$

Thus $n(E) = 17$

Hence 17 teachers teaches Economics.

Example 29. In a class of 35 students, 24 like to read math and 16 like to read novels. Also, each student likes to read at least one of the two books. How many students like to read both math and novel ?

Sol. Let X be the set of students who like to read math and Y be the set of students who like to read novels. Then $X \cup Y$ is the set of students who like to read at least one book, and $X \cap Y$ is the set of students who like to read both books.

Given $n\ (X) = 24,\ n\ (Y) = 16,\ n\ (X \cup Y) = 35,\ n\ (X \cap Y) = ?$

Using the formula $n\ (X \cup Y) = n\ (X) + n\ (Y) - n\ (X \cap Y)$, we get

$$35 = 24 + 16 - n\ (X \cap Y)$$

Thus, $n\ (X \cap Y) = 5$

i.e., 5 students like to read both books.

Example 30. In a survey of 500 students in a school, 200 were listed as playing cricket, 150 as playing football and 75 were listed as playing both cricket and football. Find how many students were playing neither cricket nor football.

Sol. Let U denote the set of surveyed students and A denote the set of students playing cricket and B denote the set of students playing football. Then

$$n\ (U) = 500,\ n\ (A) = 200,\ n\ (B) = 150 \text{ and } n\ (A \cap B) = 75.$$

Now
$$\begin{aligned} n\ (A' \cap B') &= n\ (A \cup B)' \\ &= n\ (U) - n\ (A \cup B) \\ &= n\ (U) - n\ (A) - n\ (B) + n\ (A \cap B) \\ &= 500 - 200 - 150 + 75 = 225 \end{aligned}$$

Hence 225 students were playing neither cricket nor football.

Example 31. In a lab, there are 200 type of chemicals, 120 had been reacts with HCl and 50 chemical reacts with H_2SO_4, and 30 chemicals reacts both HCl and H_2SO_4. Find which chemical reacts

(*i*) Chemical HCl but not chemical H_2SO_4

(*ii*) Chemical H_2SO_4 but not chemical HCl

(*iii*) Chemical HCl or chemical H_2SO_4.

Sol. Let U denote the universal set consisting of chemicals in a lab, A denote the set of individuals reacts to the chemical HCl and B denote the set of individuals reacts to the chemical H_2SO_4.

Here $n(U) = 200$, $n(A) = 120$, $n(B) = 50$ and $n(A \cap B) = 30$

(*i*) From the Venn diagram given in figure, we have

$$A = (A - B) \cup (A \cap B).$$

$$n(A) = n(A - B) + n(A \cap B) \text{ (Since A – B) and } A \cap B \text{ are disjoint.)}$$

or $$n(A - B) = n(A) - n(A \cap B) = 120 - 30 = 90$$

Hence, the number of individuals reacts to chemical HCl but not to chemical H_2SO_4 is 90.

(*ii*) From the figure, we have

$$B = (B - A) \cup (A \cap B).$$

and so, $$n(B) = n(B - A) + n(A \cap B)$$

(Since B – A and $A \cap B$ are disjoint.)

or $$n(B - A) = n(B) - n(A \cap B)$$

$$= 50 - 30 = 20$$

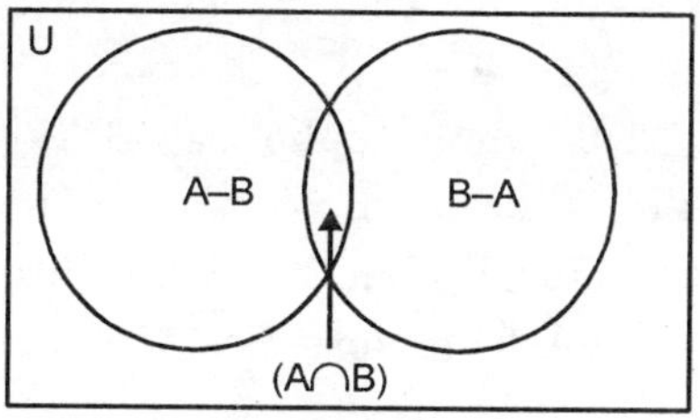

Thus, the number of chemicals reacts with H_2SO_4 and not to HCl is 20.

(*iii*) The number of chemical reacts with either to chemical HCl or to chemical H_2SO_4, *i.e.*, $n(A \cup B) = n(A) + n(B) - n(A \cap B) = 120 + 50 - 30 = 140$.

EXERCISE 1.1.11.

1. In a group, 50 people play cricket, 20 play hockey and 10 play both cricket and hockey. How many play at least one of these two games?

2. In a committee of 65 people, 40 speak english, 10 speak both hindi and urdu. How many speak urdu only and not english? How many speak Urdu?

3. If P and Q are two sets such that $n(P) = 17$, $n(Q) = 23$ and $n(P \cup Q) = 38$, find $n(P \cap Q)$.

4. If X and Y are two sets such that $X \cup Y$ has 18 elements, X has 8 elements and Y has 15 elements ; how many elements does $X \cap Y$ have?

5. In a group of 400 people, 250 can speak Hindi and 200 can speak English. How many people can speak both Hindi and English?

6. In a group of 70 people, 37 like coffee, 52 like tea and each person likes at least one of the two drinks. How many people like both coffee and tea?

7. If X and Y are two sets such that X has 40 elements, $X \cup Y$ has 60 elements and $X \cap Y$ has 10 elements, how many elements does Y have?

8. If S and T are two sets such that S has 21 elements, T has 32 elements, and $S \cap T$ has 11 elements, how many elements does $S \cup T$ have?

ANSWERS

1. 60,	2. 25, 35	3. 2	4. 5
5. 50	6. 19	7. 30	8. 42

Example 32. List all the subsets of the set { –1, 0, 1 }.

Sol. Let A = { –1, 0, 1 }. The subset of A having no element is the empty set ϕ. The subsets of A having one element are { –1 }, { 0 }, { 1 }. The subsets of A having two elements are {–1, 0}, {–1, 1} ,{0, 1}. The subset of A having three elements of A is A itself. So, all the subsets of A are ϕ, {–1}, {0}, {1}, {–1, 0}, {–1, 1}, {0, 1} and {–1, 0, 1}.

Example 33. Show that the set of letters needed to spell " CATARACT " and the set of letters needed to spell " TRACT" are equal.

Sol. Let X be the set of letters in "CATARACT". Then

$$X = \{C, A, T, R\}$$

Let Y be the set of letters in " TRACT". Then

$$Y = \{T, R, A, C, T\} = \{T, R, A, C\}$$

Since every element in X is in Y and every element in Y is in X. It follows that X = Y.

Example 34. Show that $A \cup B = A \cap B$ implies A = B

Sol. Let $a \in A$. Then $a \in A \cup B$. Since $A \cup B = A \cap B$, $a \in A \cap B$. So $a \in B$.

Therefore, $A \subset B$. Similarly, if $b \in B$, then $b \in A \cup B$. Since

$A \cup B = A \cap B$, $b \in A \cap B$. So, $b \subset A$. Therefore, $B \subset A$. Thus, A = B

Example 35. For any sets A and B, show that $P(A \cap B) = P(A) \cap P(B)$.

Sol. Let $X \in P(A \cap B)$. Then $X \subset A \cap B$. So, $X \subset A$ and $X \subset B$. Therefore, $X \in P(A)$ and $X \in P(B)$ which implies $X \in P(A) \cap P(B)$. This gives $P(A \cap B) \subset P(A) \cap P(B)$. Let $Y \in P(A) \cap P(B)$. Then $Y \in P(A)$ and $Y \in P(B)$. So, $Y \subset A$ and $Y \subset B$. Therefore, $Y \subset A \cap B$, which implies $Y \in P(A \cap B)$. This gives $P(A) \cap P(B) \subset P(A \cap B)$.

Hence $P(A \cap B) = P(A) \cap P(B)$.

Example 36. A market research group conducted a survey of 1000 consumers and reported that 720 consumers like yellow colour and 450 consumers like red coloure, what is the least number that must have liked both colours?

Sol. Let U be the set of consumers questioned, S be the set of consumers who liked the colour yellow and T be the set of consumers who like the coloure red. Given that

$$n(U) = 1000,\ n(S) = 720,\ n(T) = 450$$

So
$$n(S \cup T) = n(S) + n(T) - n(S \cap T)$$
$$= 720 + 450 - n(S \cap T) = 1170 - n(S \cap T)$$

Therefore, $n(S \cup T)$ is maximum when $n(S \cap T)$ is least. But $S \cup T \subset U$ implies $n(S \cup T) \le n(U) = 1000$. So, maximum values of $n(S \cup T)$ is 1000. Thus, the least value of $n(S \cap T)$ is 170. Hence, the least number of consumers who liked both colours are 170.

Example 37. Out of 500 car owners investigated, 400 owned car A and 200 owned car B, 50 owned both A and B cars. Is it possible?

Sol. Let U be the set of car owners investigated, M be the set of persons who owned car A and S be the set of persons who owned car B.

Given that $n(U) = 500$, $n(M) = 400$, $n(S) = 200$ and $n(S \cap M) = 50$.

Then $n(S \cup M) = n(S) + n(M) - n(S \cap M) = 200 + 400 - 50 = 550$

But $S \cup M \subset U$ implies $n(S \cup M) \le n(U)$.

This is a contradiction. So, it is not possible.

Example 38. A college awarded 38 medals in football, 15 in basketball and 20 in cricket. If these medals went to a total of 58 men and only three men got medals in all the three sports, how many received medals in exactly two of the three sports ?

Sol. Let F, B and C denote the set of men who received medals in football, basketball and cricket, respectively.

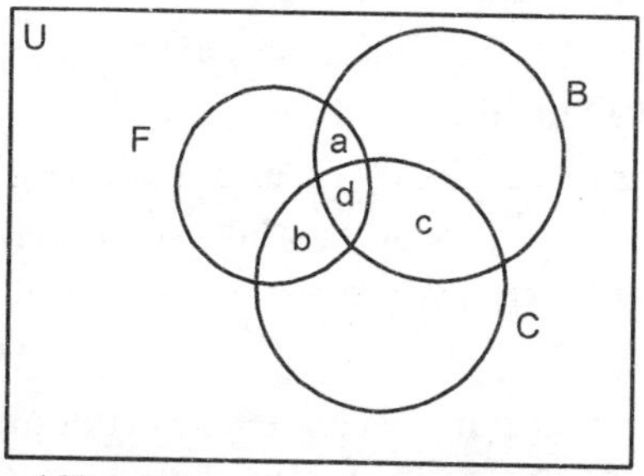

Then, n (F) = 38, n (B) = 15, n (C) = 20

n (F∪B∪C) = 58 and n (F∩B∩C) = 3

Therefore, $n(F \cup B \cup C) = n(F) + n(B) + n(C) - n(F \cap B) - n(F \cap C) - n(B \cap C) + n(F \cap B \cap C)$, gives, $n(F \cap B) + n(F \cap C) + n(B \cap C) = 18$

Consider the Venn diagram as given in above Figure.

Here, a denotes the number of men who got medals in football and basketball only, b denotes the number of men who got medals in football and cricket only, c denotes the number of men who got medals in basket ball and cricket only and d denotes the number of men who got medal in all the three. Thus, $d = n(F \cap B \cap C) = 3$ and $a + d + b + d + c + d = 18$

Therefore $a + b + c = 9$,

which is the number of people who got medals in exactly two of the three sports.

EXERCISE 1.1.12.

1. In each of the following, determine whether the statement is true or false. If it is true, prove it. If it is false, give an example.

(*i*) If $x \in$ A and A $\in$ B, then $x \in$ B — **Ans.** F

(*ii*) If A $\subset$ B and B $\in$ C, then A $\in$ C — **Ans.** T

(*iii*) If A $\subset$ B and B $\subset$ C, then A $\in$ C — **Ans.** T

(*iv*) If A $\not\subset$ B and B $\not\subset$ C, then A $\not\subset$ C — **Ans.** F

(*v*) If $x \in$ A and A $\not\subset$ B, then $x \in$ B — **Ans.** F

(*vi*) If A $\subset$ B and $x \notin$ B, then $x \notin$ A — **Ans.** T

2. Let A, B, and C be the sets such that A∪B = A∪C and A∩B = A∩C. Show that B = C.

3. Decide, among the following sets, which sets are subsets of one and another:

A = { x : $x \in$ **R** and x satisfy $x^2 - 8x + 12 = 0$ },

B = { 2, 4, 6 }, C = { 2, 4, 6, 8, . . . }, D = { 6 }.

Ans. A⊂B, A⊂C, B⊂C, D⊂A, D⊂B, D⊂C

4. Show that the following four conditions are equivalent :

(*i*) A⊂B (*ii*) A – B = φ

(*iii*) A ∪ B = B (*iv*) A ∩ B = A

5. Show that if A⊂B, then C – B⊂C – A.

6. Assume that P (A) = P (B). Show that A = B

7. Is it true that for any sets A and B, P (A)∪P (B) = P (A∪B)? Justify your answer.

8. Show that for any sets A and B,

A = (A∩B)∪(A – B) and A∪(B – A) = (A∪B)

9. Using properties of sets, show that

(*i*) A∪(A∩B) = A (*ii*) A∩(A∪B) = A.

10. Show that A∩B = A∩C need not imply B = C.

11. Let A and B be sets. If A∩X = B∩X = ϕ and A∪X = B∪X for some set X, show that A = B.

(**Hints** A = A∩(A∪X) , B = B∩(B∪X) and use Distributive law)

12. Find sets A, B and C such that A∩B, B∩C and A∩C are non-empty sets and A∩B∩C = ϕ.

Ans. A = {1, 2}, B = {1, 3}, C = {2, 3}

13. In a survey of 600 students in a school, 150 students were found to be taking tea and 225 taking coffee, 100 were taking both tea and coffee. Find how many students were taking neither tea nor coffee?

Ans. 325

14. In a group of students, 100 students know Hindi, 50 know English and 25 know both. Each of the students knows either Hindi or English. How many students are there in the group? **Ans.** 125

15. In a survey of 60 people, it was found that 25 people read newspaper H, 26 read newspaper T, 26 read newspaper I, 9 read both H and I, 11 read both H and T, 8 read both T and I, 3 read all three newspapers. Find:

(*i*) the number of people who read at least one of the newspapers. **Ans.** 52,

(*ii*) the number of people who read exactly one newspaper. **Ans.** 30

16. In a survey it was found that 21 people liked product A, 26 liked product B and 29 liked product C. If 14 people liked products A and B, 12 people liked products C and A, 14 people Liked products B and C and 8 liked all the three products. Find how many liked product C only. **Ans.** 11.

17. Is each of the following a well-defined set? Give brief reasons for each of your answers.

(*a*) The collection of all alphanumeric characters.

Ans. Yes; alphanumeric characters are A...Z, a...z, and 0...9

(*b*) The collection of all tall people. **Ans.** No; 'tall' is not well defined

(*c*) The collection of all real numbers x for which: $2x - 9 = 16$. **Ans.** Yes; the st is {12.5}

(*d*) The collection of all integers x for which: $2x - 9 = 16$. **Ans.** the empty set

(*e*) The collection of all good tennis players. **Ans.** No; 'good' is not well-defined

18. U = {natural numbers}; A = {2, 4, 6, 8, 10}; B = {1, 3, 6, 7, 8}

State whether each of the following is true or false:

(*a*) 2 ∈ A (*b*) 11 ∈ B

(*c*) 4 ∉ B (*d*) A ∈ U

(*e*) A = {even numbers}

Ans. (*a*) T, (*b*) F, (*c*) T, (*d*), F; A is a subset of U

(*e*) F; {even numbers} means the set of all the even numbers, not just those between 2 and 10.

19. U = R; A = {4, $\sqrt{2}$, 2/3, –2.5, –5, 33, $\sqrt{9}$, π}

Using the {...} set notation, write the sets of:

(*a*) natural numbers in A **Ans.** (*a*) {4, 33, $\sqrt{9}$ }

(*b*) integers in A **Ans.** (*a*) {4, –5, 33, $\sqrt{9}$ }

(*c*) rational numbers in A **Ans.** (*a*) {4, 2/3, –2.5, –5, 33, $\sqrt{9}$ }

(*d*) irrational numbers in A **Ans.** (*a*) { $\sqrt{2}$, π }

20. True or false?

(*a*) $\phi = \{0\}$ (*b*) $x \in \{x\}$

(*c*) $\phi = \{\phi\}$ (*d*) $\phi \in \{\phi\}$ **Ans.** (*a*) F, (*b*) T, (*c*) F, (*d*) T

21. The following sets have been defined using the | notation. Re-write them by listing some of the elements.

(*a*) {*p* | *p* is a capital city, *p* is in Europe} **Ans.** {London, Paris, Rome, ...}

(*b*) {*x* | *x* = 2*n* – 5, *x* and *n* are natural numbers} **Ans.** {1, 3, 5, 7, ...}, but not –3 or –1}

(*c*) {*y* | $2y^2 = 50$, y is an integer} **Ans.** (5, –5}

(*d*) {*z* | $3z = n^2$, *z* and *n* are natural numbers} **Ans.** {3, 27, 243, ...}

22. If *U* = {letters of the alphabet}, A = {*a, a, a, b, b, a, c*}, B = {*c, b, a, c*} and C = {*a, b, c*}, what can be said about A, B and C? **Ans.** (They are equal)

23. U = {natural numbers}; A = {2, 4, 6, 8, 10}; B = {1, 3, 6, 7, 8}. State whether each of the following is true or false:

(*a*) A ⊂ U (*b*) B ⊆ A

(*c*) ø ⊂ U **Ans.** (*a*) T, (*b*) F, (*c*) T

24. U = {*a, b, c, d, e, f, g, h*}; P = {*c, f*}; Q = {*a, c, d, e, f, h*}; R = {*c, d, h*}

(*a*) Draw a Venn diagram, showing these sets with all the elements entered into the appropriate regions. If necessary, redraw the diagram to eliminate any empty regions.

(*b*) Which of sets *P, Q* and *R* are proper subsets of others? Write your answer(s) using the ⊂ symbol.

(*c*) P and R are disjoint sets. True or False?

Ans. (*a*)

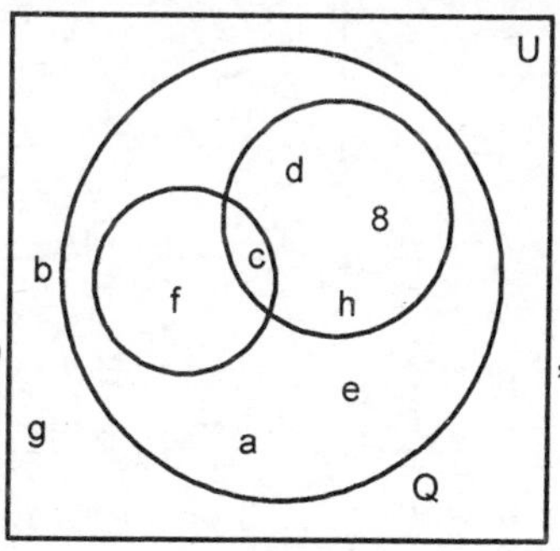

, (*b*) P⊂ Q, R⊂ Q; (*c*) False

25. Sketch Venn diagrams that show the universal set, U, the sets A and B, and a single element *x* in each of the following cases: **Ans.** (*a*)

(*a*) $x \in$ A; A ⊂ B

(*b*) $x \in$ A; A and B are disjoint

(*c*) $x \in$ A; $x \notin$ B; B ⊂ A

(*d*) $x \in A$; $x \in B$; A is not a subset of B; B is not a subset of A

Ans. (*a*)

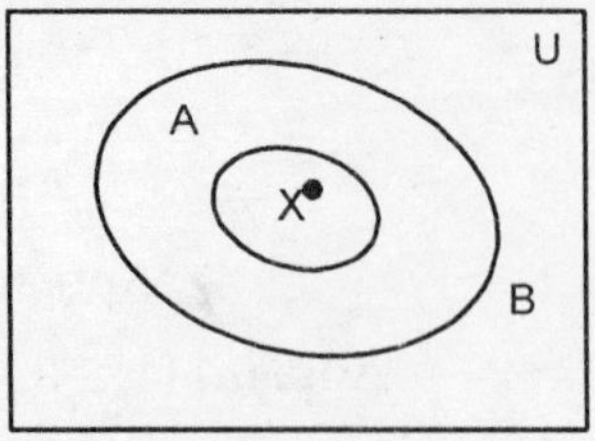

, (*b*)

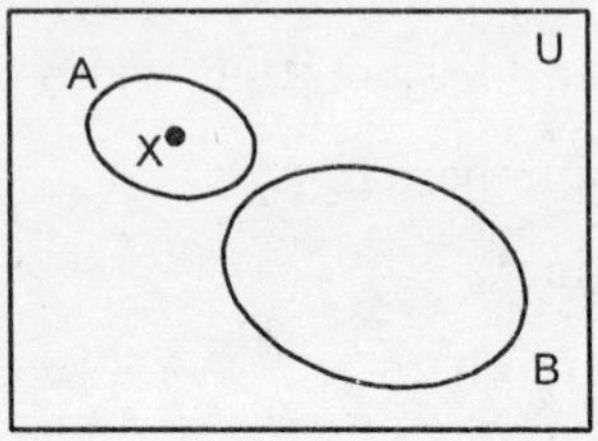

(*c*)

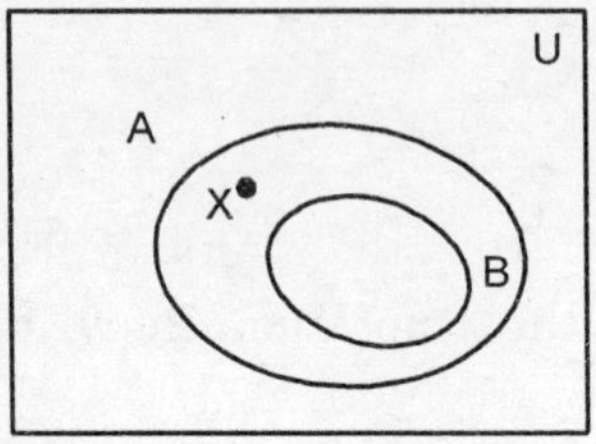

, (*d*)

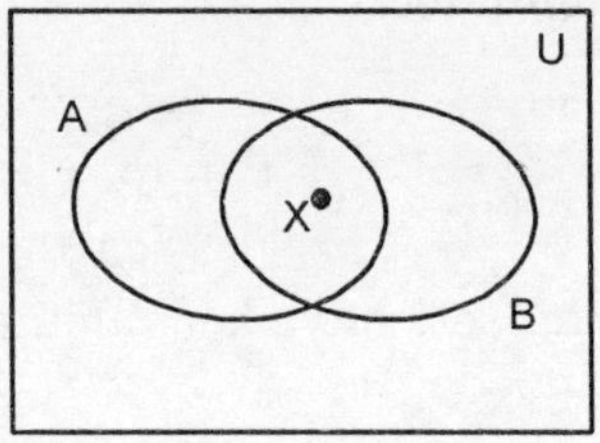

26. U = {1, 2, 3, 4, 5, 6, 7, 8, 9, 10}

A = {2, 4, 6, 8, 10}

B = {1, 3, 6, 7, 8}

C = {3, 7}

(*a*) Illustrate the sets U, A, B and C in a Venn diagram, marking all the elements in the appropriate places.

(**Note:** If any region in your diagram does not contain any elements, re-draw the set loops to correct this.)

(*b*) Complete the statement using a single symbol: C – B = ...

Ans. (*a*)

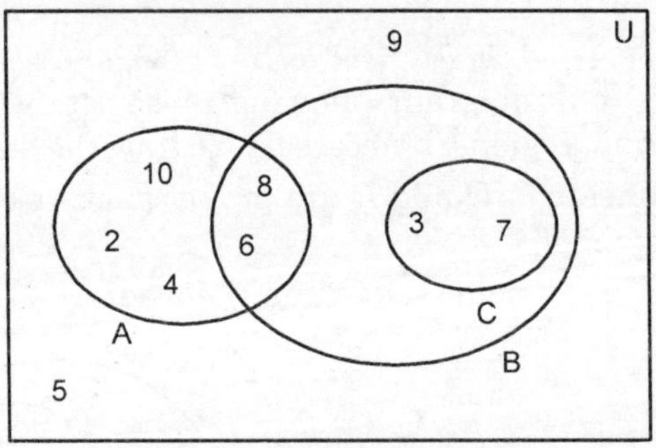

, (*b*) C – B = ϕ

27. True or false?

(*a*) $|\phi| = 1$ (*b*) $|\{x, x\}| = 2$

(*c*) $|U \cap \phi| = 0$ **Ans.** (*a*) F, (*b*) F, (*c*) T

28. What can you say about two sets P and Q if:

(*a*) $P \cap Q' = \emptyset$ (*b*) $P \cup Q = P$? **Ans.** (*a*) $P \subseteq Q$ (*b*) $Q \subseteq P$

29. Identify the sets represented by each of the shaded areas below, using the set notation symbols $\cap$, $\cup$ and ′ only:

(a)

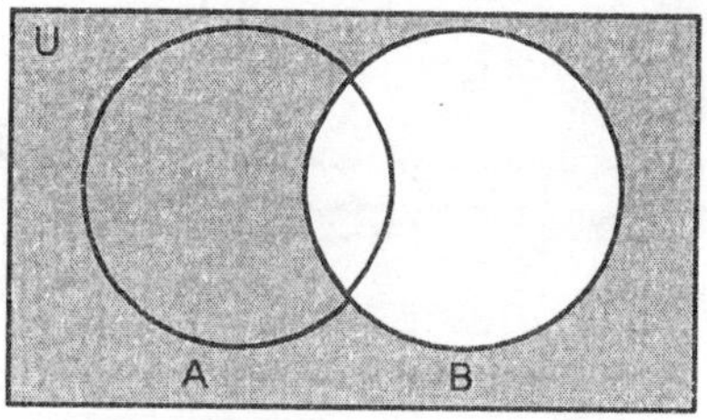

(b)

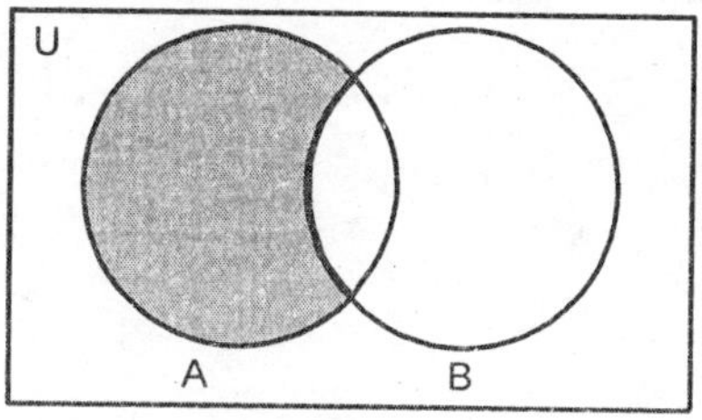

(c)

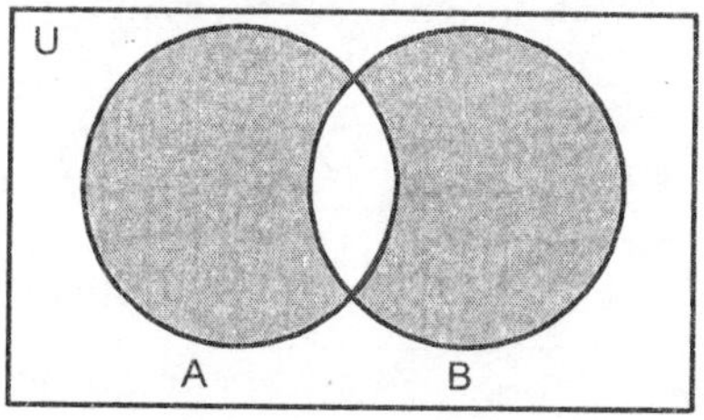

(d)

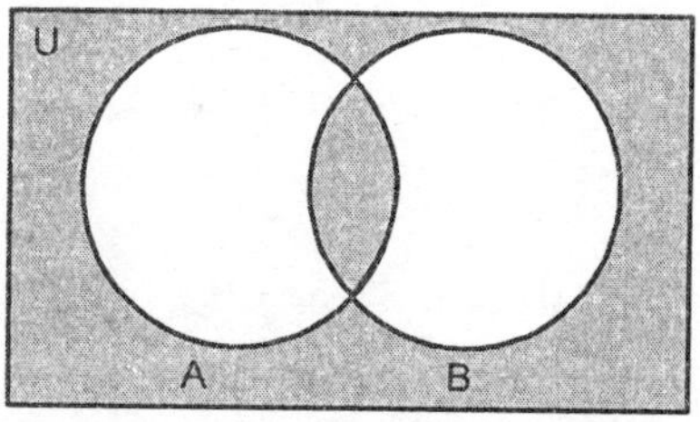

Ans. (*a*) B', (*b*) $A \cap B'$, (*c*) $(A \cup B) \cap (A \cap B)'$ or $(A \cap B') \cup (A' \cap B)$,

(*d*) $(A \cap B) \cup (A' \cap B')$ or $(A \cup B)' \cup (A \cap B)$ or ...?

30. (*a*) One of the shaded regions in question 29 represents the set A – B. Identify which one it is, and hence write a definition of A – B using only symbols from the list $\cap$, $\cup$ and $'$.

(*b*) Again using one of your answers to question 29, write a definition of $A \Delta B$ using only symbols from the list $\cap$, $\cup$ and $'$. (There are two possibilities here – see if you can find them both)

Ans. (*a*) Region (*b*) represents A – B. So $A - B = A \cap B'$

(*b*) Region (*c*) represents $A \Delta B$.

So $A \Delta B = (A \cap B) \cup (A' \cap B')$ or $(A \cup B)' \cup (A \cap B)$

31. (*a*) If A = {1, 2, 3, 4}, write down *P*(*A*) by listing its elements. What is the value of $|P(A)|$?

(*b*) If $|A| = 5$, what is the value of $|P(A)|$?

(*c*) If $|A| = 10$, what is the value of $|P(A)|$?

Ans. (*a*) $P(A) = \{\phi, \{1\}, \{2\}, \{3\}, \{4\}, \{1, 2\}, \{1, 3\}, \{1, 4\}, \{2, 3\}, \{2,4\},$

$\{3, 4\}, \{2, 3, 4\}, \{1, 3, 4\}, \{1, 2, 4\}, \{1, 2, 3\}, \{1, 2, 3, 4\}\}$ $|P(A)| = 16$

(*b*) 32, (*c*) $2^{10} = 1024$

32. Prove the following identities, stating carefully which of the set laws you are using at each stage of the proof.

(*a*) $B \cup (\phi \cap A) = B$

(*b*) $(A' \cap U)' = A$

(*c*) $(C \cup A) \cap (B \cup A) = A \cup (B \cap C)$

(*d*) $(A \cap B) \cup (A \cap B') = A$

(*e*) $(A \cap B) \cup (A \cup B')' = B$

(*f*) $A \cap (A \cup B) = A$

Ans.

(*a*) $B \cup (\phi \cap A)$	$= B \cup (A \cap \phi)$	Commutative
	$= B \cup \phi$	Identity
	$= B$	Identity
(*b*) $(A' \cap U)'$	$= (A')' \cup U'$	De Morgan
	$= A \cup U'$	Involution
	$= A \cup \phi$	Complement
	$= A$	Identity
(c) $(C \cup A) \cap (B \cup A)$	$= (A \cup C) \cap (B \cup A)$	Commutative
	$= (A \cup C) \cap (A \cup B)$	Commutative
	$= ((A \cup C) \cap B) \cup ((A \cup C) \cap A)$	Distributive
	$= ((A \cap B) \cup (C \cap B)) \cup (A \cup (C \cap A))$	Distributive
	$= ((B \cap C) \cup (A \cap B)) \cup (A \cup (A \cap C))$	Commutative (2*x*)
	$= ((B \cap C) \cup (A \cap B)) \cup ((A \cap U) \cup (A \cap C))$	Identity
	$= ((B \cap C) \cup (A \cap B)) \cup (A \cap (U \cup C))$	Distributive
	$= ((B \cap C) \cup (A \cap B)) \cup (A \cap U)$	Identity
	$= ((B \cap C) \cup (A \cap B)) \cup A$	Identity
	$= (B \cap C) \cup ((A \cap B) \cup A))$	Associative
	$= (B \cap C) \cup ((A \cap B) \cup (A \cap U))$	Identity
	$= (B \cap C) \cup (A \cap (B \cup U))$	Distributive
	$= (B \cap C) \cup (A \cap U)$	Identity
	$= (B \cap C) \cup A$	Identity
	$= A \cup (B \cap C)$	Commutative
(*d*) $(A \cap B) \cup (A \cap B')$	$= A \cap (B \cup B')$	Distributive
	$= A \cap U$	Complement
	$= A$	Identity
(*e*) $(A \cap B) \cup (A \cup B')'$	$= (A \cap B) \cup (A' \cap (B')')$	De Morgan
	$= (A \cap B) \cup (A' \cap B)$	Involution
	$= (B \cap A) \cup (B \cap A')$	Commutative (× 2)
	$= B \cap (A \cup A')$	Distributive
	$= B \cap U$	Complement
	$= B$	Identity

(*f*) $A \cap (A \cup B)$	$= (A \cup \phi) \cap (A \cup B)$	Identity
	$= A \cup (\phi \cap B)$	Distributive
	$= A \cap (B \cap \phi)$	Commutative
	$= A \cup \phi$	Identity
	$= A$	Identity

33. $X = \{a, c\}$ and $Y = \{a, b, e, f\}$. Write down the elements of:

(*a*) $X \times Y$

(*b*) $Y \times X$

(*c*) X^2 $(= X \times X)$

(*d*) What could you say about two sets *A* and *B* if $A \times B = B \times A$?

Ans. (*a*) $X \times Y = \{(a, a), (a, b), (a, e), (a, f), (c, a), (c, b), (c, e), (c, f)\}$

(*b*) $Y \times X = \{(a, a), (a, c), (b, a), (b, c), (e, a), (e, c), (f, a), (f, c)\}$

(*c*) $X \times X = \{(a, a), (a, c), (c, a), (c, c)\}$,

(*d*) They are equal: A = B

34. A chess board's 8 rows are labelled 1 to 8, and its 8 columns a to h. Each square of the board is described by the ordered pair (column letter, row number).

(*a*) A knight is positioned at (*d*, 3). Write down its possible positions after a single move of the knight.

(*b*) If $R = \{1, 2, ..., 8\}$, $C = \{a, b, ..., h\}$, and *P* = {coordinates of all squares on the chess board}, use set notation to express *P* in terms of *R* and *C*.

(*c*) A rook is positioned at (*g*, 2). If $T = \{2\}$ and $G = \{g\}$, express its possible positions after one move of the rook in terms of *R, C, T* and *G*.

Ans. (*a*) (b, 2), (b, 4), (c, 1), (c, 5), (e, 1), (e, 5), (f, 2), (f, 4)

(*b*) $P = C \times R$,

(*c*) $((G \times R) \cup (C \times T)) - (G \times T)$

35. In a certain programming language, all variable names have to be 3 characters long. The first character must be a letter from 'a' to 'z'; the others can be letters or digits from 0 to 9.

If $L = \{a, b, c, ... , z\}$, $D = \{0, 1, 2, ..., 9\}$, and *V* = {permissible variable names}, use a Cartesian product to complete:

$V = \{pqr \mid (p, q, r) \in ... \}$

Ans. $V = \{pqr \mid (p, q, r) \in L \times (L \cup D) \times (L \cup D)\}$

36. It is believed that, for any sets A, B and C, $A \times (B \cap C) = (A \times B) \cap (A \times C)$. (Note that, if this is true, it says that $\times$ is distributive over $\cap$.)

Copy and complete the two Cartesian diagrams shown below – one for the expression on each side of the equation – to investigate this.

Do you think that the proposition is true?

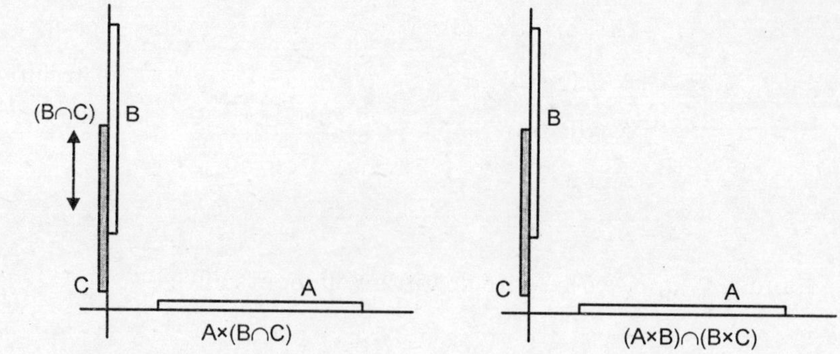

Ans.

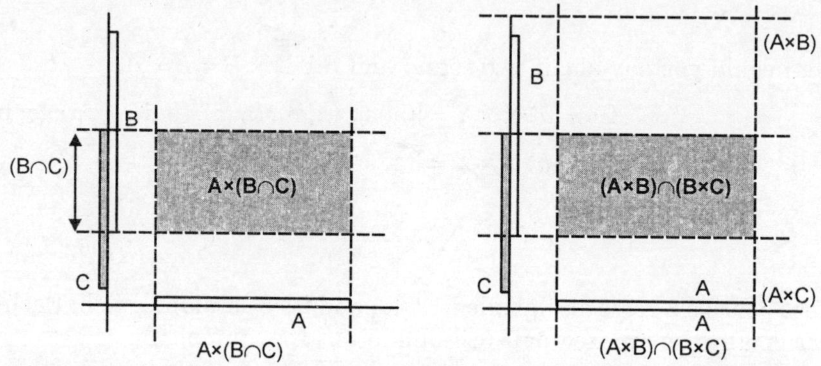

The shaded area is the same in each case, so it looks as though the proposition is true.

Unit—II

Trigonometry

1.1. REVIEW OF RATIOS OF SOME STANDARD ANGLES (0°, 30°, 45°, 60°, 90°)

Trigonometry is derived from the Greek words 'trigon and metron, and it means measuring the sides of triangle. The subject was originally developed to solve geometric problems involving triangles.

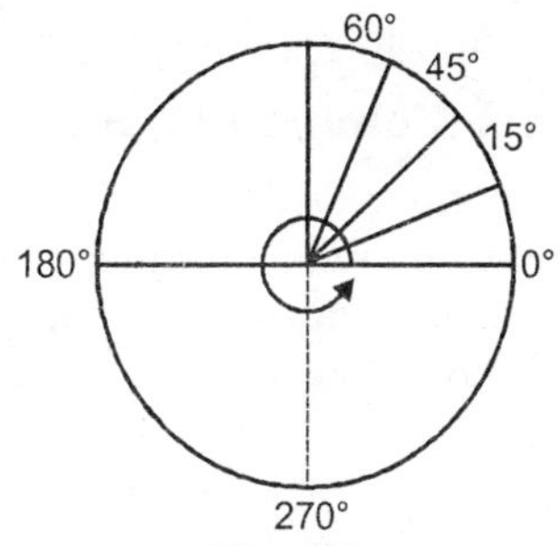

Fig. 1.1.

Trigonometric is used in many areas such as the science of seismology, designing electric circuits, describing, the state of an atom, predicting the heights.

An angle is a measure of rotation of given ray about its initial point, the original ray is called the initial side and the final position of the ray after rotation is called the terminal side of the angle. The point of rotation is called the vertix.

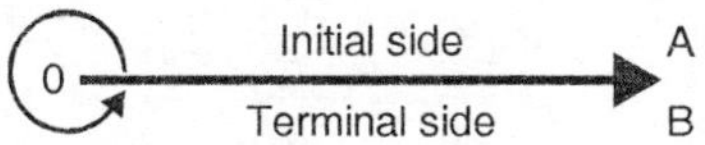

Fig. 1.2.

If the direction of rotation is anticlockwise, the angle is said to be positive and if the direction of rotation is clockwise, then the angle is negative.

This is often convenient for large angles. For example, we can say that a rapidly spinning wheel is making an angle of say 15 revolution per second. We shall describe two other units of measurement of an angle which are most commonly uses, *viz.* degree measure and radian measure.

Hence, the concept of angle is one of the most important concepts in geometry. The concepts of equality, sums, and differences of angles are important and used throughout geometry, but the subject of trigonometry is based on the measurement of angles.

There are two commonly used units of measurement for angles. The more familiar unit of measurement is that of degrees. A circle is divided into 360 equal degrees, so that a right angle is 90°.

For the time being, we will only consider angles between 0° and 360°, but later, in the section on trigonometric functions, we will consider angles greater than 360° and negative angles.

Degrees may be further divided into minutes and seconds, but that division is not as universal as it used to be. Each degree is divided into 60 equal parts called minutes. So seven and a half degree can be called 7 degrees and 30 minutes, written 7° 30′. Each minute is further divided into 60 equal parts called seconds, and, for instance, 2 degree 5 minutes 30 seconds is written 2°5′30″. The division of degrees into minutes and seconds of angle is analogous to the division of hours into minutes and seconds of time.

Parts of a degree are now usually referred to decimally. For instance seven and a half degrees is now usually written 7.5°.

When a single angle is drawn on a xy-plane for analysis, we'll draw it in standard position with the vertex at the origin (0, 0), one side of the angle along the x-axis, and the other side above the x-axis.

Radians

The other common measurement for angles is radians. For this measurement, consider the unit circle (a circle of radius 1) whose center is the vertex of the angle in question. Then the angle cuts off an arc of the circle, and the length of that arc is the radian measure of the angle. It is easy to convert between degree measurement and radian measurement. The circumference of the entire circle is 2π, so it follows that 360° equals to 2π radians.

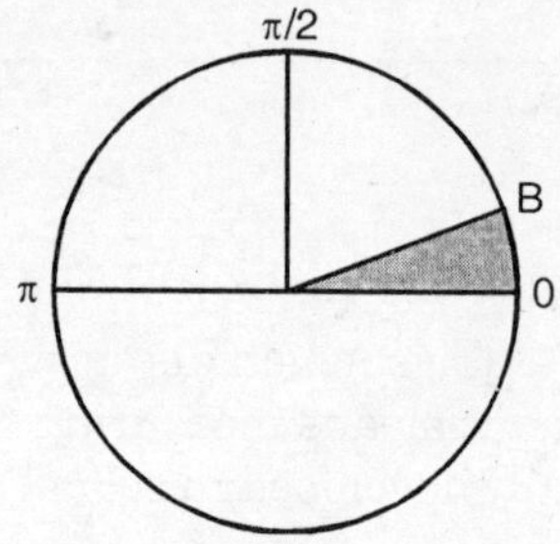

Fig. 1.3.

Hence, 1° equal $\pi/180$ radians and 1 radian equals $180/\pi$.

1.1.1. Common Angles

Below is a table of common angles in both degree measurement and radian measurement. Note that the radian measurement is given in terms of π. It could, of course, be given decimally, but radian measurement often appears with a factor of π.

Angle	Degree	Radians
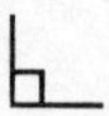	90°	$\pi/2$
	60°	$\pi/3$
	45°	$\pi/4$
	30°	$\pi/6$

Fig. 1.4.

1.1.2. T-(Trigonometrical) Ratios of Allied Angle

We will learn how to find the trigonometrical ratios of any angle using the following step-by-step procedure.

Step I : To find the trigonometrical ratios of angles $(n \cdot 90° \pm \theta)$; where n is an integer and θ is a positive acute angle, we will follow the below procedure.

First, we need to determine the sign of the given trigonometrical ratio. Now to determine the sign of the given trigonometrical ratio, we need to find the quadrant in which the angle $(n \cdot 90° + \theta)$ or $(n \cdot 90° - \theta)$ lies.

Now, using the rule "*All, sin, tan, cos*" we will find the sign of the given trigonometrical ratio.

Therefore,

(*i*) All trigonometrical ratios are positive, if the given angle $(n \cdot 90° + \theta)$ or $(n . 90° + \theta)$ lies in the I quadrant (first quadrant);

(*ii*) Only sin and csc ratios is positive, if the given angle $(n \cdot 90° + \theta)$ or $(n \cdot 90° - \theta)$ lies in the II quadrant (second quadrant);

(*iii*) Only tan and cot ratios is positive, if the given angle $(n \cdot 90° + \theta)$ or $(n \cdot 90° - \theta)$ lies in the III quadrant (third quadrant);

(*iv*) Only cos and sec ratios is positive, if the given angle $(n \cdot 90° + \theta)$ or $(n \cdot 90° - \theta)$ lies in the IV quadrant (fourth quadrant).

Step II: Now determine whether n is an even or odd integer.

(*i*) If n is an even integer, the form of the given trigonometrical ratio will remain the same *i.e.*,

$\sin (n \cdot 90° + \theta) = \sin \theta$ $\quad$ $\csc (n \cdot 90° + \theta) = \csc \theta$;

$\sin (n \cdot 90° - \theta) = - \sin \theta$; $\quad$ $\csc (n \cdot 90° - \theta) = - \csc \theta$;

$\cos (n \cdot 90° + \theta) = \cos \theta$; $\quad$ $\sec (n \cdot 90° + \theta) = \sec \theta$;

$\cos (n \cdot 90° - \theta) = - \cos \theta$; $\quad$ $\sec (n \cdot 90° - \theta) = - \sec \theta$.

$\tan (n \cdot 90° + \theta) = \tan \theta$; $\quad$ $\cot (n \cdot 90° + \theta) = \cot \theta$

$\tan (n \cdot 90° - \theta) = - \tan \theta$. $\quad$ $\cot (n \cdot 90° - \theta) = - \cot \theta$;

(*ii*) If n is an odd integer, then the form of the given trigonometrical ratio is altered *i.e.*,

sin changes to cos;
i.e., $\sin (n \cdot 90° + \theta) = \cos \theta$
or, $\sin (n \cdot 90° - \theta) = - \cos \theta$

csc changes to sec;
i.e., $\csc (n \cdot 90° + \theta) = \sec \theta$
or, $\csc (n \cdot 90° - \theta) = - \sec \theta$

cos changes to sin;
i.e., $\cos (n \cdot 90° + \theta) = \sin \theta$
or, $\cos (n \cdot 90° - \theta) = - \sin \theta$

sec changes to csc;
i.e., $\sec (n \cdot 90° + \theta) = \csc \theta$
or, $\sec (n \cdot 90° - \theta) = - \csc \theta$

tan changes to cot;
i.e., $\tan (n \cdot 90° + \theta) = \cot \theta$
or, $\tan (n \cdot 90° - \theta) = - \cot \theta$

cot changes to tan;
i.e., $\cot (n \cdot 90° + \theta) = \tan \theta$
or, $\cot (n \cdot 90° - \theta) = - \tan \theta$

The trigonometrical ratios table will help us to find the values of trigonometric standard angles. The standard angles of trigonometrical ratio are 0°, 30°, 45°, 60°, and 90°.

The values of trigonometrical ratio of standard angles are very important to solve the trigonometrical problems. Therefore, it is necessary to remember the value of the trigonometircal ratios of these standard angles. The sine, cosine and tangent of the standard angles are given in the table.

Trigonometric Table

Angles of degrees	*0°*	*30° or $\pi/6$*	*45° or $\pi/4$*	*60° or $\pi/3$*	*90° or $\pi/2$*
sin	0	$\frac{1}{2}$	$\frac{\sqrt{2}}{2}$	$\frac{\sqrt{3}}{2}$	1
cos	1	$\frac{\sqrt{3}}{2}$	$\frac{\sqrt{2}}{2}$	$\frac{1}{2}$	0
tan	0	$\frac{\sqrt{3}}{3}$	1	$\sqrt{3}$	Not defined
cosec	Not defined	2	$\sqrt{2}$	$\frac{2\sqrt{3}}{3}$	1
sec	1	$\frac{2\sqrt{3}}{3}$	$\sqrt{2}$	2	Not defined
cot	Not defined	$\sqrt{3}$	1	$\frac{\sqrt{3}}{3}$	0

Relation between Degree and Radian

$$\pi \text{ radian} = 180°$$

$$1 \text{ radian} = \frac{180°}{\pi} = 57°16'$$

$$\Rightarrow \quad 1° = \frac{\pi}{180} \text{ radian} = 0.01746 \text{ radian}$$

Example 1. *Change 40°20′ into radian.*
Solution. We know that 180° = π radian

$$40°20' = 40\ 1/3° = \frac{\pi}{180} \times \frac{121}{3} \text{ radian} = \frac{121\pi}{540} \text{ radian.}$$

Example 2. *6 radians into degree.*

Solution. 6 radian $= \frac{180°}{\pi} \times 6 \text{ degree} = \frac{1080 \times 7}{22} \text{ degree}$

$$= 343\ \frac{7}{11} \text{ degree} = 340° + \frac{7 \times 60}{11} \text{ min.} \qquad [\because 1°60']$$

$$= 343° + 38' + \frac{2}{11} \text{ min.} = 343° + 38' + 10.9''$$

• Trigonometric Ratios of Positive and Negative angles

On a coordinate grid a general angle is measured from the positive x-axis and is represented by the angle through which a line OM rotates about the origin.

When we rotate anti-clockwise, the angle is positive while a clockwise rotation gives a negative angle.

Cases :

Case I. When $\theta = 120°$, $a = 180 - 120 = 60°$

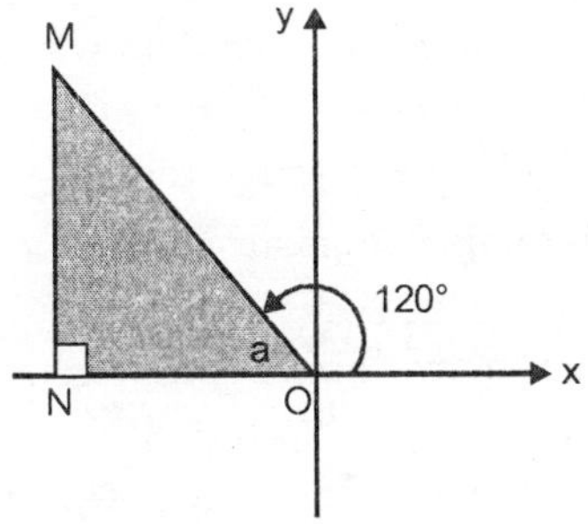

Fig. 1.5.

Case II. When $\theta = 300°$, $a = 360 - 300 = 60°$

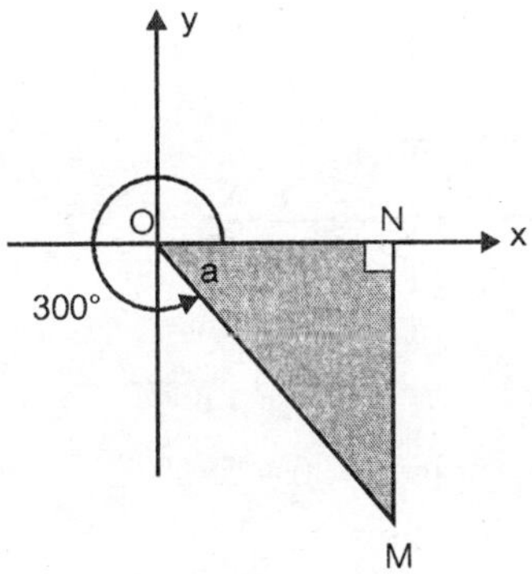

Fig. 1.6.

Case III. When $\theta = 520°$, $a = 540° - 520° = 20°$

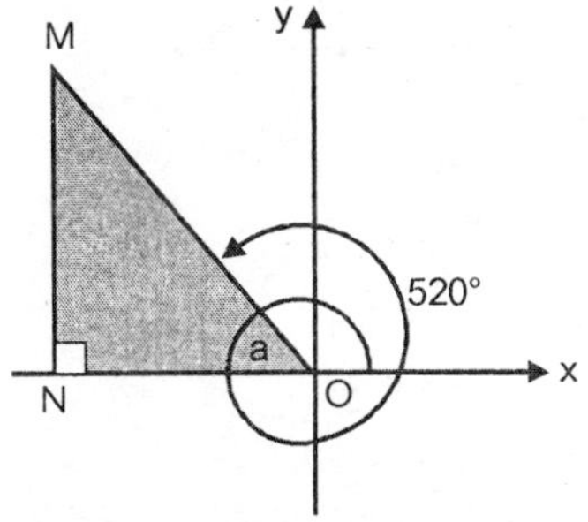

Fig. 1.7.

Trigonometric functions for any angle in radians and degrees

The four quadrants of the Cartesian axes are as follows:

(*i*) *Trigonometry with any angle*

As the line OM rotates, the point M moves to the first quadrant where its coordinates are both positive, and into the second quadrant, where the x-coordinate becomes negative.

In the third quadrant, both coordinates are negative and finally, in the fourth quadrant, the point has a positive x- and negative y-coordinate.

(*ii*) *Measure of central tendency*

You can see that the angle MON, called a, is always acute, and measured from the x-axis.

The signs of the trigonometric functions depend on which quadrant the point M lies in and represent the signs of the x and y-coordinates of M.

Learn the information in the following figures to help you understand the signs.

First quadrant : All the functions are positive.

The acute angle $a = \theta$

(Note. If OM is rotated through more than a complete revolution, then $a = \theta - 360°$)

Second quadrant : The acute angle $a = 180° - \theta$

By looking at the signs of the coordinates of M, we see that the trigonometric functions of are:

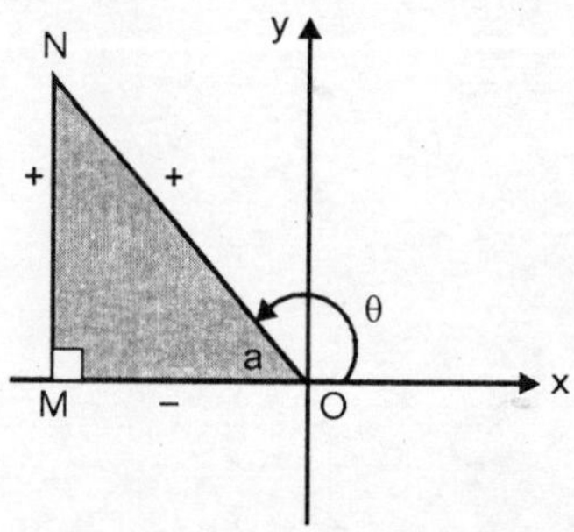

cos θ = + sin a
cos θ = – cos a
tan θ = – sin a

Fig. 1.8.

Third quadrant: The acute angle, $a = \theta - 180°$

The signs of the coordinates of M show us that the trigonometric functions are:

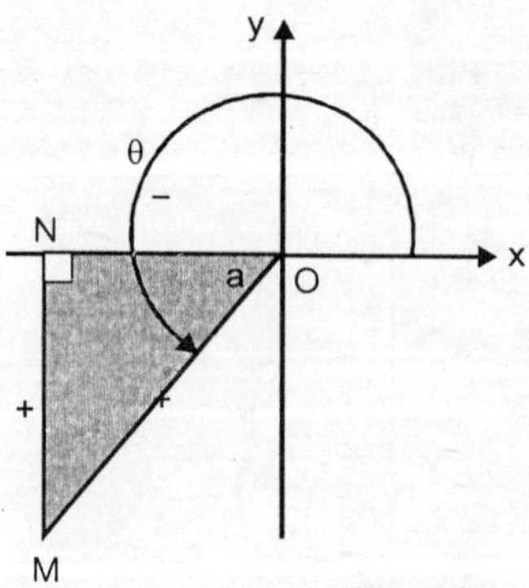

sin θ = – sin a
cos θ = – cos a
tan θ = + sin a

Fig. 1.9.

Fourth quadrant : The acute angle $a = 360° - \theta$

The signs of the coordinates of M show us that the trigonometric functions of arc:

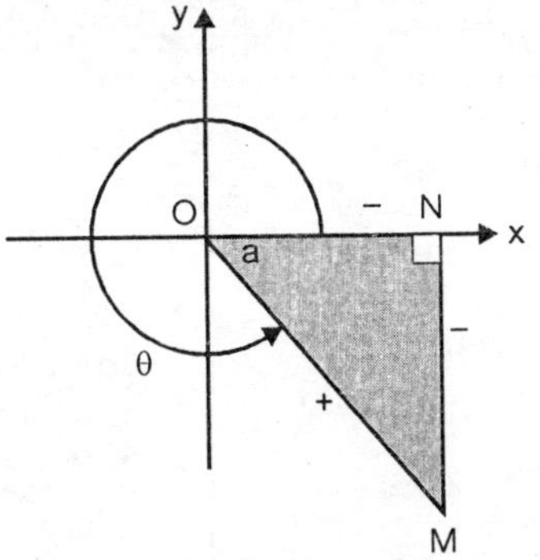

sin θ = − sin a
cos θ = + cos a
tan θ = − sin a

Fig.1.10.

This can be summarised as:

These sign rules and the value of the acute angle a allow you to find the value of any trigonometric function of any angle.

Example:

Find the values of sin 150°, sin 210° and sin 690°, if sin 30° = 0.5.

$\sin 150° = \sin 30° = 0.5,\quad \sin 210° = -\sin 30° = -0.5$

$\sin 690° = \sin 330° = -\sin 30° = -0.5$

You also need to be aware of negative angles created from the rotation of M in a clockwise direction.

Here, θ = − 35° or 325°

i.e., each position of line OM gives us two different values of θ (theta), one is positive and other is negative.

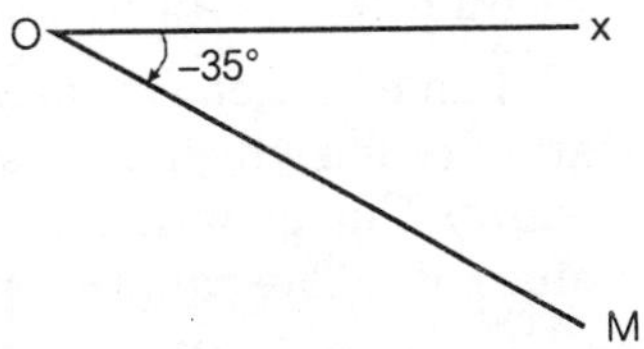

Fig. 1.11.

Signs and variations of t-ratios

Quadrant	*sin θ*	*cos θ*	*tan θ*
I	+ (0 to 1)	+ (1 to 0)	+ (0 to ∞)
II	+ (1 to 0)	− (0 to − 1)	− (− ∞ to 0)
III	− (0 to −1)	− (− 1 to 0)	+ (0 to ∞)
IV	− (−1 to 0)	+ (0 to 1)	− (− ∞ to 0)

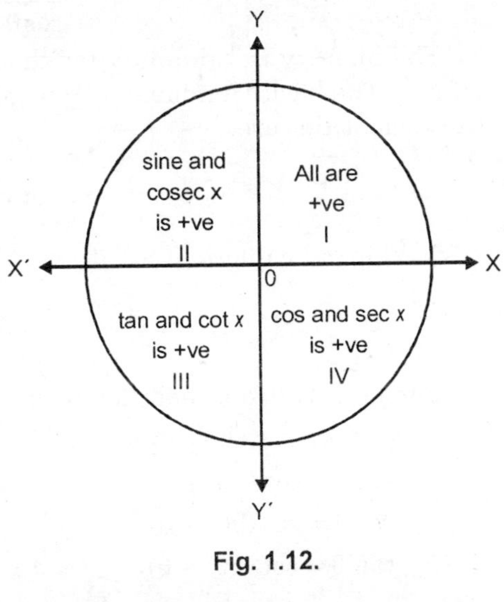

Fig. 1.12.

Any t-ratio of $(n \, . \, 90° \pm \theta) = \pm$ same ratio of θ, when n is even = ± co-ratio of θ, when n is odd

The sine + or – is to be decided from the quadrant in which $n \cdot 90° \pm \theta$ lies.

e.g.,

$$\sin 570° = \sin (6 \times 90° + 30°) = -\sin 30° = -1/2$$

$$\tan 315° = \tan (3 \times 90° + 45°) = -\cot 45° = -1$$

Signs of Trigonometric functions in various quadrants

Quadrant	*Signs of various T-functions*
I	All T-functions are positive
II	$\sin\theta$ and cosec θ are positive. All other are negative
III	$\tan\theta$ and $\cot\theta$ are positive. All other are negative
IV	$\cos\theta$ and $\sec\theta$ are positive All other are negative.

Do you know?

- In the sexagesimal system, we measure angles in degree, minute, and second.
- In the circular measure, we measure angle in radian π^c = 180°.
- If an arc of length l make an angle $\theta°$ at the centre of circle of radius r, $\theta = l/r$.

We understand that

I	II	III	IV
All are +ve	sin is +ve	tan is +ve	cos is +ve

• Fundamental Identities

If an equation contains one or more variables and is valid for all replacement value of the variables for which both sides of the equation are defined, then the equation is known as an *identity.* The equation $x^2 + 5x = x(x + 5)$, is an identity because it is valid for all replacement values of x. The equation is valid for only for certain replacement values of variable.

An equation that is said to be an identity without stating any restrictions is in reality an identity only for those replacement values for which both sides of the identity are defined. For example,

$$\frac{\sin\theta}{\cos\theta} = \tan\theta$$

The identity is valid only for those value of θ for which both sides of the equations are defined. The basic (fundamental) trigonometric identities can be devided into several groups. These identities are:

$$\cot\theta = \frac{1}{\tan\theta}, \sec\theta = \frac{1}{\cos\theta}, \cos\theta = \frac{1}{\sin\theta}$$

Next are the quotient identities. These include

$$\tan\theta = \frac{\sin\theta}{\cos\theta}, \cot\theta = \frac{\cos\theta}{\sin\theta}$$

Then it is called a conditional equation.

Then there are the cofunction identities. These include

$\sin\theta = \cos(90° - \theta)$ $\quad$ $\cos\theta = \tan(90° - \theta)$

$\cos\theta = \sin(90° - \theta)$ $\quad$ $\sec\theta = \csc(90° - \theta)$

$\tan\theta = \cot(90° - \theta)$ $\quad$ $\csc\theta = \sec(90° - \theta)$

Next there are the identities for negatives. These include

$\sin(-\theta) = -\sin\theta$ $\cos(-\theta) = -\cos\theta$

$\tan(-\theta) = -\tan\theta$

Finally there are the Pythagorean identities. These include

$\sin^2\theta + \cos^2\theta = 1$ $\cos^2\theta + 1 = \sec^2\theta$

$\tan^2\theta + 1 = \csc^2\theta$

The second identity is obtained by dividing the first by cos 2 θ, and the third identity is obtained by dividing the first by sin 2 θ. The process of showing the validity of one identity based on previously known facts is called proving the identity.

The validity of the foregoing identities follows directly from the definitions of the basic trigonometric functions and can be used to verify other identities.

• Procedure for Use of Fundamental Identities

Solving identities exits, but there are some general rules or strategies that can be followed to help guide the process:

1. Try to simplify the more complicated side of the identity until it is identical to the second side of the identity.
2. Try to transform both sides of an identity into an identical thrid expressoin.
3. Try to express both sides of the identity in terms of only sines and cosines; then try to make both sides identical.
4. Try to apply the Pythagorean identities as much as possible.
5. Try to use factoring and combining of terms, multiplying one side of the identity by an expression that is equal to 1, squaring both sides of the identity, and other algebraic techniques to manipulate equations.

• Examples Using Fundamental Identities and Table of Acute Angles

If θ be an acute angle, the value of sin θ and cos θ lies between 0 and 1.

sin $\sin 0^\circ = \sqrt{\frac{0}{4}} = 0 \quad \sin 30 = \sqrt{\frac{1}{4}} = \frac{1}{2}$

$$\sin 45^\circ = \sqrt{\frac{2}{4}} = \frac{1}{\sqrt{2}} = \frac{\sqrt{2}}{2}$$

$$\sin 60^\circ = \sqrt{\frac{3}{4}} = \frac{\sqrt{3}}{2}$$

cos $\cos 0^\circ = \sqrt{\frac{4}{4}} = 1$

$$\cos 30^\circ = \sqrt{\frac{3}{4}} = \frac{\sqrt{3}}{2}$$

$$\cos 45^\circ = 1$$

$$\cos 60^\circ = \sqrt{\frac{1}{4}} = \frac{1}{2}$$

tangent $\tan 0° = 0$

$$\tan 30° = \frac{\sqrt{3}}{3}$$

$$\tan 45° = \frac{\sqrt{2}}{4} = \frac{1}{\sqrt{2}} = \frac{\sqrt{2}}{2}$$

$$\tan 60° = \sqrt{3}$$

cosec $\csc 0°$ = not defined

$$\csc 30° = 2$$

$$\csc 45° = \sqrt{2}$$

$$\csc 60° = \frac{2\sqrt{3}}{3}$$

secant $\sec 0° = 1$

$$\sec 30° = \frac{2\sqrt{3}}{3}$$

$$\sec 45° = \sqrt{2}$$

$$\sec 60° = 2$$

contangent $\cot 0°$ = not defined

$$\cot 30° = \sqrt{3}$$

$$\cot 45° = 1$$

$$\cot 60° = \frac{\sqrt{3}}{3} = \frac{\sqrt{3}}{\sqrt{3} \times \sqrt{3}} = \frac{1}{\sqrt{3}}$$

Example 1. *Changing to sines and cosines to verify the identity : sec x cot x = csc x.*

Sol. The left hand side of the equation contains the more complicated expression. Thus, we work with the left side. Let us express this side of the identity in terms of sines and cosines. Perhaps this strategy will enable us to transform the left side into csc x, the expression on the right.

$$\therefore \qquad \sec x \cot x = \frac{1}{\cos x} \times \frac{\cos x}{\sin x} = \frac{1}{\sin x} = \csc x$$

Example 2. *Using factoring to verify an identity.* $\cos x - \cos x \sin^2 x = \cos^3 x$

Sol. We start with the more complicated side, the left hand side. Factor out the greatest common factor cos$x$, from each of the two terms.

$\cos x - \cos x \sin^2 x = \cos x\,(1 - \sin^2 x)$ Factor cos x from the two terms.

$= \cos x \cos^2 x$ Use a variation of $\sin^2 x + \cos^2 x = 1$.

Solving for $\cos^2 x$ $x = 1 - \sin^2 x$.

$= \cos^3 x$

Example 3. *Find the degree measure corresponding to each of the following radian measures.*

(*i*) $\left(\frac{3}{4}\right)^c$ (*ii*) $\left(\frac{7\pi}{12}\right)^c$

(*iii*) $(-2)^c$.

Sol. (*i*) $\pi^c = 180° \Rightarrow 1° = \left(\frac{180°}{\pi}\right)^° \Rightarrow \left(\frac{3}{4}\right) = \left(\frac{180°}{\pi} \times \frac{3}{4}\right)^° = \left(180° \times \frac{7}{22} \times \frac{3}{4}\right)^° = 42°57'16''$

$$\therefore \qquad \left(\frac{3}{4}\right)^c = 42°57'16''$$

(*ii*) $\pi^c = 180° \Rightarrow 1^c = \left(\frac{180}{\pi}\right)^° \Rightarrow \left(\frac{7\pi}{12}\right)^c = \left(\frac{180}{\pi} \times \frac{7\pi}{12}\right)^° = 105°$

(*iii*) $\pi^c = 180° \Rightarrow 1^c = \left(\frac{180}{\pi}\right)^°$

$$\Rightarrow \qquad (-2)^c = \left\{\frac{180}{\pi} \times (-2)\right\}^° = \left\{180 \times \frac{7}{22} \times (-2)^°\right\}$$

$$\Rightarrow \qquad = 114°32'44''$$

$$(-2)^c = -(114°32'44'')$$

Example 4. *Find the radian measure corresponding to each of the following degree measure (i) 30°, (ii) 240°, (iii) (– 37°30′).*

Sol.

(*i*) $180° = \pi^c \Rightarrow 1° = \left(\frac{\pi}{180}\right)^c$

$$\Rightarrow \qquad -37°30' = \left\{\frac{\pi}{180} \times \left(\frac{-75}{2}\right)\right\}^c \qquad \left[\because -37°30' = -(37\tfrac{1}{2})° = -\left(\frac{75}{2}\right)^°\right]$$

$$= \left(\frac{-5\pi}{24}\right)^c$$

Hence, $(-37°\ 30') = \left(\frac{-5\pi}{24}\right)^c$

(*ii*) $180° = \pi^c \Rightarrow 1° = \left(\frac{\pi}{180}\right)^c$

$$\Rightarrow \qquad 240° = \left(\frac{\pi}{180} \times 240\right)^c = \left(\frac{4\pi}{3}\right)^c$$

Hence, $240° = \left(\frac{4\pi}{3}\right)^c$

(*iii*) 180° $= \pi^c \Rightarrow 1° = \left(\dfrac{\pi}{180}\right)^c$

$$\Rightarrow 15° = \left(\frac{\pi}{180} \times 15\right)^c = \left(\frac{\pi}{12}\right)^c$$

Hence, $15° = \left(\dfrac{\pi}{12}\right)^c$

Example 5. *In a right angle triangle, the difference between the two acute angles is* $\left(\dfrac{\pi}{15}\right)^c$. *Find the angles in degree.*

Sol. Difference between the acute angle $= \left(\dfrac{\pi}{c}\right)^c = \dfrac{\pi}{15} \times \dfrac{180°}{\pi} = 12°$

$$x + y = 90 \qquad ...(i)$$

$$x - y = 12 \qquad ...(ii)$$

Solving (*i*) and (*ii*), we get $x = 51$ and $y = 39$

∴ angles of triangles are 51°, 39°, 90°.

Example 6. *A wheel makes 360° revolutions, in 1 minutes. How many radians does it turn in 1 seconds.*

Sol. Number of revolutions made in 60 seconds = 360°

Number of revolution made in 1 second = 360/60 = 6

Angle moved in 1 revolution = $(2\pi)^c = (2\pi \times 6)^c = (12\ \pi)^c$.

Example 7. *Verify the identify* $\cos\alpha + \sin\alpha \cdot \tan\alpha = \sec\alpha$.

Sol. Now, we can write as above question $\cos\alpha \cdot 1 + \sin\alpha \tan\alpha$

$$(\cos\alpha) \cdot \left(\frac{\cos\alpha}{\cos\alpha}\right) + (\sin\alpha)\left(\frac{\sin\alpha}{\cos\alpha}\right) = \sec\alpha$$

$$\frac{\cos^2\alpha + \sin^2\alpha}{\cos\alpha} = \frac{1}{\cos\alpha} = \sec\alpha$$

Example 8. *Verify the identity* $\dfrac{\tan\alpha}{\sec\alpha - 1} = \dfrac{\sec\alpha + 1}{\tan\alpha}$.

Sol.

$$\frac{\tan\alpha}{\sec\alpha - 1} = \frac{\sec\alpha + 1}{\tan\alpha}$$

$$\frac{(\tan\alpha)(\sec\alpha + 1)}{(\sec\alpha - 1)(\sec\alpha + 1)} = \frac{\sec\alpha + 1}{\tan\alpha}$$

$$\frac{(\tan\alpha)(\sec\alpha + 1)}{\sec^2\alpha - 1} = \frac{\sec\alpha + 1}{\tan\alpha}$$

$$\frac{(\tan\alpha)(\sec\alpha + 1)}{\tan^2\alpha} = \frac{\sec\alpha + 1}{\tan\alpha}$$

$$\frac{\sec\alpha + 1}{\tan\alpha} = \frac{\sec\alpha + 1}{\tan\alpha}.$$

Example 9. *Prove that* $\dfrac{\sin^2\alpha+2\cos\alpha-1}{\sin^2\alpha+3\cos\alpha-3}=\dfrac{1}{1-\sec\alpha}$.

Sol.
$$\frac{\sin^2\alpha+2\cos\alpha-1}{\sin^2\alpha+3\cos\alpha-3}=\frac{1}{1-\sec\alpha}$$
$$\frac{-(\cos^2\alpha)+2\cos\alpha-1}{(1-\cos^2\alpha)+3\cos\alpha-3}=\frac{1}{1-\sec\alpha}$$
$$\frac{-(\cos^2\alpha-2\cos\alpha)}{-(\cos^2\alpha-3\cos\alpha+2)}=\frac{1}{1-\sec\alpha}$$
$$\frac{\cos\alpha}{\cos\alpha-1}=\frac{1}{1-\sec\alpha}$$
$$\left(\frac{\dfrac{\cos\alpha}{\cos\alpha}}{\dfrac{\cos\alpha-1}{\cos\alpha}}\right)=\frac{1}{1-\sec\alpha}\Rightarrow\frac{1}{1-1/\cos\alpha}=\frac{1}{1-\sec\alpha}$$

Example 10. *If* $\cos x=-3/5$, *find the value of 5 trigonometric functions.*

Sol. $\cos x=-3/5$, $\sec x=-5/3$
$$\sin^2x+\cos^2x=1,\Rightarrow\sin^2x=1-9/25=16/25$$
$$\sin x=+4/5$$
$$\sin x=-4/5$$
$$\text{cosec } x=5/4.$$

Example 11. *Find the value of* $\sin\dfrac{31\pi}{3}$.

Sol. $\sin\dfrac{31\pi}{3}=\sin\left(10\pi+\dfrac{\pi}{3}\right)=\sin\dfrac{\pi}{3}=\dfrac{\sqrt{3}}{2}$.

Example 12. *Find the value of* $\cos(-1710°)$

Sol. $\cos(-1710)=\cos(-1710°+5\times360°)=\cos(-1710°+1800°)=\cos 90°=0$.

Example 13. *If* $\sin\theta=-4/5$ *and* $\pi<\theta<\dfrac{3\pi}{2}$, *find the values of all other five trigonometric functions.*

Sol. θ lies in the third quadrant in which $\tan\theta$ and $\cot\theta$ are positive and all the other function.
$$\sin\theta=-\frac{4}{5}\Rightarrow\text{cosec }\theta=\frac{1}{\sin\theta}=-\frac{5}{4}$$
$$\cot^2\theta=(\text{cosec}^2\theta-1)=\frac{25}{16}-1=\frac{9}{16}$$
$$\cot\theta=\sqrt{\frac{9}{16}}=\frac{3}{4};\ \tan\theta=\frac{4}{3}$$

$$\cos\theta = \cot\theta\sin\theta = \left(\frac{3}{4}\right)\times\left(-\frac{4}{5}\right) = -\frac{3}{5}$$

$\sec\theta = -5/3.$

$$\cos\theta = -\frac{3}{5};\ \tan\theta = \frac{4}{3};\ \cot\theta = \frac{3}{4}$$

Example 14. *Find minute hand and the hour hand of a clock when time is 7:20, then find the angle between them.*

Sol. Hour hand move in 12 hours = 360°

Angle in the 7 h 20 min in 22/3 hours = $\left(\frac{360}{12}\times\frac{22}{3}\right)^\circ$ = 220°

Angle by the minute hand in 60 min = 360°

Angle by the minute hand in 20 min = $\left(\frac{360}{60}\times 20\right)$ = 120°

Thus required angle between the hand = 220° – 120° = 100.

Example 15. (*i*) cosec $\left(-\frac{33\pi}{4}\right)$, (*ii*) sec $\left(-\frac{19\pi}{3}\right)$, (*iii*) cot $\left(\frac{29\pi}{4}\right)$, (*iv*) tan $\left(-\frac{16\pi}{3}\right)$, (*v*) cos $\left(\frac{41\pi}{4}\right)$, (*vi*) sin $\left(\frac{25\pi}{3}\right)$.

Sol.

(*i*)
$$\text{cosec}\left(\frac{-33\pi}{4}\right) = -\text{cosec}\,\frac{33\pi}{4} \qquad [\because \text{cosec}\,(-\theta) = -\text{cosec}\,\theta]$$

$$= -\text{cosec}\left(8\pi+\frac{\pi}{4}\right) = -\text{cosec}\,\frac{\pi}{4} \qquad [\because \text{cosec}\,(2n\pi+\theta) = \text{cosec}\,\theta]$$

$$= -\sqrt{2}$$

(*ii*)
$$\sec\left(-\frac{19\pi}{3}\right) = \sec\left(\frac{19\pi}{3}\right) \qquad [\therefore \sec(-\theta) = \sec\theta]$$

$$= \sec\left(6\pi+\frac{\pi}{3}\right) = \sec\left(\frac{\pi}{3}\right) \qquad [\because \sec(2n\pi+\theta) = \sec\theta]$$

$$= 2$$

(*iii*)
$$\cot\left(\frac{29\pi}{4}\right) = \cot\left(7\pi+\frac{\pi}{4}\right)$$

$$= \cot\frac{\pi}{4} \qquad [\because \cot(n\pi+\theta) = \cot\theta]$$

$$= 1$$

(*iv*) $\tan\left(-\frac{16\pi}{3}\right) = -\tan\frac{16\pi}{3}$ $[\because \tan(-\theta) = -\tan\theta]$

$= -\tan\left(5\pi + \frac{\pi}{3}\right) = -\tan\frac{\pi}{3}$ $[\because \tan(n\pi + \theta) = \tan\theta]$

$= -\sqrt{3}$

(*v*) $\cos\left(\frac{41\pi}{4}\right) = \cos\left(10\pi + \frac{\pi}{4}\right) = \cos\frac{\pi}{4} = \frac{1}{\sqrt{2}}$

$[\because \cos(2n\pi + \theta) = \cos\theta]$

(*vi*) $\sin\left(\frac{25\pi}{3}\right) = \sin\left(8\pi + \frac{\pi}{3}\right) = \sin\frac{\pi}{3} = \frac{\sqrt{3}}{2}.$ $[\because \sin(2n\pi + \theta) = \sin\theta]$

Example 16. *Find the value of (i) cosec (– 1110°), (ii) cot (– 600°), (iii) sin (765°)*

Sol. (*i*) $180° = \pi^c$

$\Rightarrow$ $(1110)° = \left(\frac{\pi}{180} \times 1110\right)^c = \left(\frac{37\pi}{6}\right)^c$

$\therefore$ $\text{cosec}(-1110°) = -\text{cosec}(1110°)$ $[\because \text{cosec}(-\theta) = -\text{cosec}\,\theta]$

$= -\text{cosec}\left(\frac{37\pi}{6}\right) = -\text{cosec}\left(6\pi + \frac{\pi}{6}\right)$

$= -\text{cosec}\frac{\pi}{6} = -2$ $[\because \text{cosec}(2n\pi + \theta) = \text{cosec}\,\theta]$

(*ii*) $180° = \pi^c$

$\Rightarrow$ $\cot(-600°) = \left(\frac{\pi}{180} \times 600\right)^c = \left(\frac{10\pi}{3}\right)^c$ $[\because \cot(-\theta) = -\cot\theta]$

$\therefore$ $\cot(-600°)$

$= -\cot\left(\frac{10\pi}{3}\right) = -\cot\left(3\pi + \frac{\pi}{3}\right) = -\cot\frac{\pi}{3}$

$[\because \cot(n\pi + \theta) = \cot\theta]$

$= -\frac{1}{\sqrt{3}}.$

(*iii*) $180° = \pi^c$

$\Rightarrow$ $765° = \left(\frac{\pi}{180} \times 765\right)^c = \left(\frac{17\pi}{4}\right)^c.$

$\therefore$ $\sin(765°) = \sin\left(\frac{17\pi}{4}\right) = \sin\left(\frac{\pi}{4}\right)$

$= \sin\frac{\pi}{4} = \frac{1}{\sqrt{2}}.$ $[\because \sin(2n\pi + \theta) = \sin\theta]$

Example 17. *Find the value of (i) sec 6 π, (ii) tan* $\left(\frac{5\pi}{4}\right)$, *(iii) sin 5π, (iv) cos (–π), (v) sin 16π, (vi) cos 15 π.*

Sol.

(*i*) $\sec 6\pi = \sec(6\pi + 0) = \sec 0° = 1.$ $\quad [\because \sec(2n\pi + \theta) = \sec\theta]$

(*ii*) $\tan\dfrac{5\pi}{4} = \tan\left(\pi + \dfrac{\pi}{4}\right) = \tan\dfrac{\pi}{4} = 1$ $\quad [\because \tan(n\pi + \theta) = \tan\theta]$

(*iii*) $\sin 5\pi = \sin(4\pi + \pi) = \sin\pi = 0$ $\quad [\because \sin(2n\pi + \theta) = \sin\theta]$

(*iv*) $\cos(-\pi) = \cos\pi = -1$ $\quad [\because \cos(-\theta) = \cos\theta]$

(*v*) $\sin 16\pi = \sin(16\pi + 0) = \sin 0° = 0$ $\quad [\because \sin(2n\pi + \theta) = \sin\theta]$

(*vi*) $\cos 15\pi = \cos(14\pi + \pi) = \cos\pi = -1$ $\quad [\because \cos(2n\pi + \theta) = \cos\theta]$

Example 18. *Prove that* $\sin^2 \pi/6 + \cos^2 \pi/3 - \tan^2 \pi/4 = -1/2$.

Sol. L.H.S. $= \sin^2 \pi/6 + \cos^2 \pi/3 - \tan^2 \pi/4$

$$= \left(\frac{1}{2}\right)^2 + \left(\frac{1}{2}\right)^2 - 1^2$$

$$\left[\because \sin\frac{\pi}{6} = \frac{1}{2}, \cos\frac{\pi}{3} = \frac{1}{2} \;\&\; \tan\frac{\pi}{4} = 1\right]$$

$$= \left(\frac{1}{4} + \frac{1}{4} - 1\right) = -\frac{1}{2} = \text{R.H.S.}$$

Example 19. *Calculate the angle subtended at the centre of a circle of diameter 50 cm by an arc of length 11 cm.*

Sol. Here, $r = 25$ cm and $l = 11$ cm.

Let the measure of the required angle by θ^c.

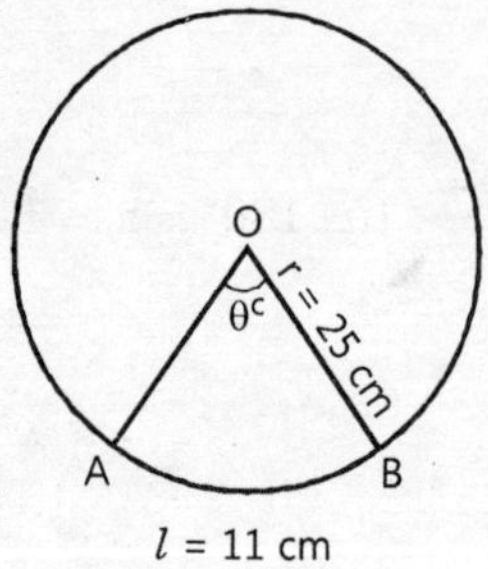

Fig. 1.13.

Then, $$\theta^c = \left(\frac{l}{r}\right)^c = \left(\frac{11}{25}\right)^c = \left(\frac{11}{25} \times \frac{180}{\pi}\right)^\circ$$

$$[\because \pi^c = 180°]$$

$$= \left(\frac{11}{25} \times \frac{7}{22} \times 180\right)^\circ$$

$$= \left(\frac{126}{5}\right)^\circ = 25°12'$$

Hence, the required angle is 25°12′.

Example 20. *If the arc of the same length in two circles subtend angles of 60° and 75° at their respective centres, find the ratio of their radii.*

Sol. Let r_1 and r_2 the radius of two circles.

$$\theta_1 = 60° = \left(60 \times \frac{\pi}{180}\right)^c = \left(\frac{\pi}{3}\right)^c$$

$$\theta_2 = 75° = \left(75 \times \frac{\pi}{180}\right)^c = \left(\frac{5\pi}{12}\right)^c$$

Let the length of each arc be l cm then

$$l = r_1\theta_1 = r_2\,\theta_2$$

$$\Rightarrow \quad \left(r_1 \times \frac{\pi}{3}\right) = \left(r_2 \times \frac{5\pi}{12}\right)$$

$$\Rightarrow \quad \frac{r_1}{r_2} = \frac{5}{4}$$

$$r_1 : r_2 = 5 : 4$$

Fig. 1.14.

Example 21. *The angles of triangle are in A.P. and the ratio of number of degrees in the least to the number of radians in the greatest in 60 : π. Find the angles in degrees and radius.*

Sol. Suppose the angle $(a - d)°$, and $a°$ and $(a + d°)$

$$(a - d) + a + (a + d) = 180 \Rightarrow 3a = 180 \Rightarrow a = 60°$$

The angles are $(60 - d°)$, $60°$, $(60 + d°)$

Number of degrees in the least angle $= (60 - d)$

$$\text{Number of greatest angle} = \left\{(60+d)\times\frac{\pi}{180}\right\} = \frac{60-d}{(60+d)\times\pi/180} = \frac{60}{\pi}\;(60-d)\times\pi$$

$$= 60 \times (60 + d) \times \frac{\pi}{180}$$

$$\Rightarrow \quad 3(60 - d) = 60 + d$$

$$\Rightarrow \quad 4d = 120 \Rightarrow d = 30$$

$$(60 - 30°),\ 60,\ (60 + 30°)$$

$$30,\ 60,\ 90.$$

EXERCISE 1.1.

1. Show each of the following angles in degrees:

(*i*) $(-4)^c$ (*ii*) $\left(\frac{5}{6}\right)^c$

(*iii*) $-\left(\frac{18\pi}{5}\right)^c$ (*iv*) $\left(\frac{5\pi}{12}\right)^c$.

[**Ans.** (*i*) – (229°5′27′′), (*ii*) 47°43′38′′, (*iii*) – 648°, (*iv*) 75°]

2. Express each of following angle in radians:

(*i*) – (22°30′) **Ans.** $-\left(\frac{\pi}{8}\right)^c$

(*ii*) – 270° **Ans.** $-\left(\frac{3\pi}{2}\right)^c$

(*iii*) 7°30′ **Ans.** $\left(\frac{\pi}{24}\right)^c$

(*iv*) + 400° **Ans.** $\left(\frac{20\pi}{9}\right)^c$

(*v*) 330° **Ans.** $\left(\frac{11\pi}{6}\right)^c$

(*vi*) 225° **Ans.** $\left(\frac{5\pi}{4}\right)^c$

(*vii*) 120° **Ans.** $\left(\frac{2\pi}{3}\right)^c$

(*viii*) 36° **Ans.** $\left(\frac{\pi}{5}\right)^c$

3. If the angles of triangle are in A.P. and greatest angle is double the least. Find all the angles in degree and radians. [**Ans.** (40°, 60°, 80°)], $\left\{\left(\frac{2\pi}{9}\right)^c, \left(\frac{\pi}{3}\right)^c, \left(\frac{4\pi}{9}\right)^c\right\}$]

4. The angle of a quadrilateral area in A.P., in the greatest angle is double the least. Express the least angle in radians. [**Ans.** $\left(\frac{\pi}{3}\right)^c$]

5. Length of wire 121 cm is bent so as to lie along the arc of a circle of radius 180 cm. Find in degrees, the angle subtended at the centre by the arc. [**Ans.** 38°30′]

6. A train is moving on a circular curve of radius 155 m at the rate of 66 km per hour. What angle has it turned in 10 seconds. [**Ans.** 7°]

7. A wheel makes 180 revolution in 1 minute. How many radians does it turn in second? [**Ans.** $(6\pi)^c$]

8. Find the degree measure of the angle subtended at centre of circle of diameter 60 cm by an arc length of 16.5 cm. [**Ans.** 8 : 5]

9. Diameter of a circle is 30 cm, length of chord is 15 cm. Find the length of the minor arc of the chord. [**Ans.** 31°30′]

10. Large hand of clock is 42 cm long. How many centimeters does its extremity moves in 20 minutes. [**Ans.** 88 centimeters]

11. Angles of quadrilateral are in A.P. and the greatest angle is double the least. Express the least angle in radians. [**Ans.** $\left(\frac{\pi}{3}\right)^c$]

12. Show that sin 5A + sin 3A = 2 sin 4A cos A

13. Prove that $\cos 8x - 5\cos x = \dfrac{-2\sin 13x}{2}\dfrac{\sin 3x}{2}$.

14. Prove that $\dfrac{\sin A + \sin 2A}{\cos A - \cos 2A} = \cot\dfrac{A}{2}$

15. Prove that $\dfrac{\sin A + \sin 3A + \sin 5A + \sin 7A}{\cos A + \cos 3A + \cos 5A + \cos A} = \tan 4A$

16. Prove that $\dfrac{\sin A + 2\sin 3A + \sin 5A}{\sin 3A + 2\sin 5A + \sin 7A} = \dfrac{\sin 3A}{\sin 5A}$

17. Prove that $\dfrac{\sin A - \sin 5A + \sin 9A - \sin 13A}{\cos A - \cos 5A - \cos 9A + \cos 13A} = \cot 4A$

18. Prove that $\dfrac{\sin 8\theta\cos\theta - \sin 6\theta\cos 3\theta}{\cos 2\theta\cos\theta - \sin 3\theta\sin 4\theta} = \tan 2\theta$

19. Prove that $\dfrac{\tan 5\theta + \tan 3\theta}{\tan 5\theta - \tan 3\theta} = \cos 2\theta$

20. Prove that $\sin\dfrac{\theta}{2}\sin\dfrac{7\theta}{2} + \sin\dfrac{3\theta}{2}\sin\dfrac{11\theta}{2} = \sin 2\theta\sin 5\theta$

21. Prove that $\cos A\sin(B-C) + \cos B\sin(C-A) + \cos C\sin(A-B) = 0$

22. Prove that $2\cos\dfrac{\pi}{13}\cos\dfrac{9\pi}{13} + \cos\dfrac{3\pi}{13} + \cos\dfrac{5\pi}{13} = 0$

23. Prove that $\dfrac{\sin\theta + \sin 2\theta}{1 + \cos\theta + \cos 2\theta} = \tan\theta$

24. Prove that $\sin 20° - \sin 40°\sin 60°\sin 80° = 3/16$

25. Prove that $\tan 3A\tan 2A\tan A = \tan 3A - \tan 2A - \tan A$

26. Prove that $\tan(45° + A/2) = \sec A + \tan A$

27. If A + B + C = 180°, Prove that

(i) $\sin(B+2C) + \sin(C+2A) + \sin(A+2B) = 4\sin\dfrac{(B-C)}{2}\sin\dfrac{C-A}{2}\sin\dfrac{A-B}{2}$

(ii) $\dfrac{\sin 2A + \sin 2B + \sin 2C}{\sin A + \sin B + \sin C} = 8\sin\dfrac{A}{2}\sin\dfrac{B}{2}\sin\dfrac{C}{2}$

Trigonometric ratios of allied angles

Any two angles whose sum or difference is either zero or is an integral multiple of $\left(\frac{\pi^c}{2}\right)$ are called *allied angles*.

Thus, if θ is the measures of a given angles, then the angle whose measures in terms of θ.

$\frac{\pi}{2} \pm \theta; \pi \pm \theta; \frac{3\pi}{2} \pm \theta; 2\pi \pm \theta$ ..... are its allied angles.

e.g., 1. $150° + 30° = 180° = 2 \times \frac{\pi^c}{2}$ = integral multiple of $\frac{\pi}{2}$

∴ 150° and 30° are allied angles.

2. $60° + 30° = 90° = 1 \times \frac{\pi}{2}$ = integral multiple of $\frac{\pi}{2}$

∴ 60° and 30° are allied angles.

Trigonometric Ratios of Allied Angles

Angle / Ratios	$-\theta$	$\frac{\pi}{2}-\theta$ 90–θ	$\frac{\pi}{2}+\theta$ 90+θ	$\pi-\theta$ 180–θ	$\pi+\theta$ 180+θ	$\frac{3\pi}{2}-\theta$ 270–θ	$\frac{3\pi}{2}+\theta$ 270+θ	$2\pi-\theta$ 360–θ	$2\pi+\theta$ 360+θ
sin	sin	cos θ	cosc θ	sinc θ	sinc θ	cosc θ	cosc θ	sinc θ	sinc θ
cos	cos	sin θ	sinc θ	cosc θ	cosc θ	sinc θ	sinc θ	cosc θ	cosc θ
tan	tan	cot θ	cotc θ	tanc θ	tanc θ	cotc θ	cotc θ	–tanc θ	tanc θ
cot	cot	tan θ	tanc θ	cotc θ	cotc θ	tanc θ	tanc θ	–cotc θ	cotc θ
cosec	cosec	sec θ	secc θ	cosecc θ	cosecc θ	secc θ	secc θ	–cosecc θ	cosec θ
sec	sec	cosec θ	cosecc θ	secc θ	secc θ	cosecc θ	cosecc θ	secc θ	secc θ

Terminal Angles

1. $\sin (n \times 360° + \theta) = \sin \theta$
2. $\cos (n \times 360° + \theta) = \cos \theta$
3. $\tan (n \times 360° + \theta) = \tan \theta$
4. $\sin \left(\frac{\pi}{2} \pm \theta\right) = \pm \sin \theta$, if n is odd
5. $\cos \left(\frac{\pi}{2} \pm \theta\right) = \pm \cos \theta$, if n is odd

Trigonometric Ratios of Compound Angles

These are also called addition formula.

If A and B are two angles the sum of $(A + B)$ or difference $(A - B)$ is called a **compound angle.**

e.g. 1. if $A = 150°$; $B = 210°$ then

$A + B = 150° + 210° = 360°$

and $A - B = 150° - 210°$

$= -60°$ are compound angles.

Formulae

1. $\sin(A + B) = \sin A \cos B + \cos A \sin B$
2. $\sin(A - B) = \sin A \cos B - \cos A \sin B$
3. $\cos(A + B) = \cos A \cos B - \sin A \sin B$
4. $\cos(A - B) = \cos A \cos B + \sin A \sin B$
5. $\tan(A + B) = \dfrac{\tan A + \tan B}{1 - \tan A \tan B}$
6. $\tan(A - B) = \dfrac{\tan A - \tan B}{1 + \tan A \tan B}$

Trigonometric ratios of $\left(\frac{\pi}{2} + \theta\right)$

$\sin\left(\frac{\pi}{2} + \theta\right)$

Using $\sin(A + B) = \sin A \cos B + \cos A \sin B$

Putting $A = \frac{\pi}{2}$ and $B = \theta$

$\therefore \sin\frac{\pi}{2} \cos\theta + \cos\frac{\pi}{2} \sin\theta$

But $\cos\frac{\pi}{2} = 0$; $\sin\frac{\pi}{2} = 1$

$\therefore \sin\left(\frac{\pi}{2} + \theta\right) = 1.\cos\theta + 0.\sin\theta = \cos\theta + 0$

$\therefore \sin\left(\frac{\pi}{2} + \theta\right) = \cos\theta$

1.1.3. Sum, Difference Formulae and their Applications (without proof)

The set of formulas which are useful in transforming sums and difference of trigonometric into their products and vice versa.

These sets of formulas are derived directly from trigonometric addition and sum formulas.

Here, we will derive the transformation formulas using following four formulas with trignometric addition and subtraction formulas:

If we add and subtract the first two and last two equations in the equations above theorem get the following equations:

$$\sin A \cos B + \cos A \sin B = \sin(A + B) \quad ...(i)$$

$$\sin A \cos B - \cos A \sin B = \sin(A - B) \quad ...(ii)$$

$$\cos A \cos B - \sin A \sin B = \cos(A + B) \quad ...(iii)$$

$$\cos A \cos B + \cos A \sin B = \cos(A - B) \quad ...(iv)$$

Now the above four equations if we replace "$A + B$" by "C" and "$A - B$" by "D" or $A + B$ and $A - B = D$ or, and then we can get the following equations:

$$2\sin A \cos B = \sin(A + B) + \sin(A - B)$$

$$\sin A \cos B = \frac{1}{2}[\sin(A+B)+\sin(A-B)]$$

$$2\cos A \sin B = \sin(A + B) - \sin(A - B)$$

$$\cos A \sin B = \frac{1}{2}[\sin(A+B)-\sin(A-B)]$$

$$2\cos A \cos B = \cos(A + B) + \cos(A - B)$$

$$\cos A \cos B = \frac{1}{2}[\cos(A+B)+\cos(A-B)]$$

$$2\sin A \sin B = \cos(A - B) - \cos(A + B)$$

$$\sin A \sin B = \frac{1}{2}[\cos(A-B)-\cos(A+B)]$$

The eight formulas derived here are called the trigonometric transformation form they can be used to transform the sums or difference of trigonometric functions to their and product of trigonometric functions to their sums or difference.

Theorem

(*i*) $\cos\left(\frac{\pi}{2}-x\right) = \sin x$ (*ii*) $\sin\left(\frac{\pi}{2}-x\right) = \cos x$

(*iii*) $\tan\left(\frac{\pi}{2}-x\right) = \cot x$ (*iv*) $\cot\left(\frac{\pi}{2}-x\right) = \tan x$

(*v*) $\sec\left(\frac{\pi}{2}-x\right) = \operatorname{cosec} x$ (*vi*) $\operatorname{cosec}\left(\frac{\pi}{2}-x\right) = \sec x$

Proof.

(*i*) $\cos\left(\frac{\pi}{2}-x\right) = \cos\frac{\pi}{2}\cos x + \sin x \frac{\pi}{2}\sin x$

$= (0 + \cos x) + (1 + \sin x) = \sin x$ $[\cos \pi/2 = 0, \sin \pi/2=1]$

(*ii*) $\sin\left(\frac{\pi}{2}-x\right) = \sin\theta,$ when $\theta = \left(\frac{\pi}{2}-x\right)$

$= \cos\left(\frac{\pi}{2}-\theta\right)$ $[\because \cos(\pi/2 - \theta) = \sin\theta]$

$= \cos\left\{\frac{\pi}{2}-\left(\frac{\pi}{2}-x\right)\right\} = \cos x$

(*iii*) $\tan\left(\frac{\pi}{2}-x\right) = \frac{\sin\left(\frac{\pi}{2}-x\right)}{\cos\left(\frac{\pi}{2}-x\right)} = \frac{\cos x}{\sin x} = \cot x$

(*iv*) $\cot\left(\frac{\pi}{2}-x\right) = \frac{\cos\left(\frac{\pi}{2}-x\right)}{\sin\left(\frac{\pi}{2}-x\right)} = \frac{\sin x}{\cos x} = \tan x$

(*v*) $\sec\left(\frac{\pi}{2}-x\right) = \frac{1}{\cos\left(\frac{\pi}{2}-x\right)} = \frac{1}{\sin x} = \operatorname{cosec} x$ $\quad\left[\because \cos\left(\frac{\pi}{2}-x\right) = \sin x\right]$

(*vi*) $\operatorname{cosec}\left(\frac{\pi}{2}-x\right) = \frac{1}{\sin\left(\frac{\pi}{2}-x\right)} = \frac{1}{\cos x} = \sec x$ $\quad\left[\because \sin\left(\frac{\pi}{2}-x\right) = \cos x\right]$

Theorem 2.

(*i*) $\cos\left(\frac{\pi}{2}+x\right) = -\sin x$ (*ii*) $\sin\left(\frac{\pi}{2}+x\right) = \cos x$

(*iii*) $\tan\left(\frac{\pi}{2}+x\right) = -\cot x$ (*iv*) $\cot\left(\frac{\pi}{2}+x\right) = -\tan x$

(*v*) $\sec\left(\frac{\pi}{2}+x\right) = -\operatorname{cosec} x$ (*vi*) $\operatorname{cosec}\left(\frac{\pi}{2}+x\right) = \sec x$

Proof.

(*i*) $\cos\left(\frac{\pi}{2}+x\right) = -\sin x$

$$= \cos\frac{\pi}{2}\cos x - \sin\frac{\pi}{2}\sin x$$

$$= (0 \times \cos x) - (1 \times \sin x) \quad \left[\because \cos\frac{\pi}{2} = 0 \text{ and } \sin\frac{\pi}{2} = 1\right]$$

$$= -\sin x$$

(*ii*) $\sin\left(\frac{\pi}{2}+x\right) = \sin\left[\frac{\pi}{2}-(-x)\right] = \cos(-x) = \cos x$ $\quad\left[\because \sin\left(\frac{\pi}{2}-\theta\right) = \cos\theta \text{ and } \cos(-x) = \cos\theta\right]$

(*iii*) $\tan\left(\frac{\pi}{2}+x\right) = \frac{\sin\left(\frac{\pi}{2}+x\right)}{\cos\left(\frac{\pi}{2}+x\right)} = \frac{\cos x}{-\sin x}$ $\quad\left[\because \sin\left(\frac{\pi}{2}+x\right) = \cos x\right.$

$= -\cot x$ $\quad \cos\left(\frac{\pi}{2}+x\right) = -\sin x$

(*iv*) $\cot\left(\frac{\pi}{2}+x\right) = \frac{1}{\tan\left(\frac{\pi}{2}+x\right)} = -\frac{1}{-\cot x} = -\tan x$ $\quad\left[\because \tan\left(\frac{\pi}{2}+x\right) = -\cot x\right.$

(*v*) $\sec\left(\frac{\pi}{2}+x\right) = \frac{1}{\cos\left(\frac{\pi}{2}+x\right)} = -\frac{1}{\sin x} = -\operatorname{cosec}x$ $\left[\because \cos\left(\frac{\pi}{2}+x\right) = -\sin x\right.$

(*vi*) $\operatorname{cosec}\left(\frac{\pi}{2}+x\right) = \frac{1}{\sin\left(\frac{\pi}{2}+x\right)} = \frac{1}{\cos x} = \sec x$ $\left[\because \sin\left(\frac{\pi}{2}+x\right) = \cos x\right.$

Theorem 3.

(*i*) $\cos(\pi - x) = -\cos x$

(*ii*) $\sin(\pi - x) = \sin x$

(*iii*) $\tan(\pi - x) = -\tan x$

(*iv*) $\cot(\pi - x) = -\cot x$

(*v*) $\sec(\pi - x) = -\sec x$

(*vi*) $\operatorname{cosec}(\pi - x) = \operatorname{cosec}x.$

Proof.

(*i*) $\cos(\pi - x) = \cos\pi\cos x + \sin\pi\sin x$

$= (-1 \times \cos x) + (0 \times \sin x) = -\cos x$ $[\because \cos\pi = -1 \text{ and } \sin\pi = 0]$

(*ii*) $\sin(\pi - x) = \sin\left[\frac{\pi}{2}+\left(\frac{x}{2}-x\right)\right] = \cos\left(\frac{x}{2}-x\right) = \sin x$ $\left[\because \sin\left(\frac{x}{2}-\theta\right) = \cos\theta\right]$

$[\because \cos(\pi/2 - x) = \sin x]$

(*iii*) $\tan(\pi - x) = \frac{\sin(\pi - x)}{\cos(\pi - x)} = \frac{\sin x}{-\cos x} = -\tan x$ $[\because \sin(\pi - x) = \sin x, \cos(\pi - x) = -\cos x]$

(*iv*) $\cot(\pi - x) = \frac{1}{\tan(\pi - x)} = \frac{1}{-\tan x}$ $[\because \tan(\pi - x) = -\tan x]$

$= -\cot x$

(*v*) $\sec(\pi - x) = \frac{1}{\cos(\pi - x)} = \frac{1}{-\cos x}$ $[\because \cos(\pi - x) = -\cos x]$

$= -\sec x$

(*vi*) $\operatorname{cosec}(\pi - x) = \frac{1}{\sin(\pi - x)} = \frac{1}{\sin x}$ $[\because \sin(\pi - x) = \sin x]$

$= \operatorname{cosec}x$

Theorem 4.

(*i*) $\cos(\pi + x) = -\cos x$

(*ii*) $\sin(\pi + x) = -\sin x$

(*iii*) $\tan(\pi + x) = \tan x$

(*iv*) $\cot(\pi + x) = \cot x$

(*v*) $\sec(\pi + x) = -\sec x$

(*vi*) $\operatorname{cosec}(\pi + x) = -\operatorname{cosec}x$

Proof.

(*i*) $\cos(\pi + x) = \cos\pi\cos x - \sin\pi\sin x$

$= (-1) \times \cos x - (0 \times \sin x) = -\cos x$ $[\because \cos\pi = -1 \text{ and } \sin\pi = 0]$

(*ii*) $\sin(\pi + x) = \sin\left[\frac{\pi}{2}+\left(\frac{x}{2}+x\right)\right] = \cos\left(\frac{x}{2}+x\right) = -\sin x$ $\left[\because \sin\left(\frac{x}{2}+\theta\right) = \cos\theta\right]$

(*iii*) $\tan(\pi + x) = \frac{\sin(\pi + x)}{\cos(\pi + x)} = \frac{-\sin x}{-\cos x} = \tan x$ $[\because \sin(\pi + x) = -\sin x, \cos(\pi + x) = -\cos x]$

(*iv*) $\cot(\pi + x) = \frac{1}{\tan(\pi + x)} = \frac{1}{\tan x} = \cot x$ $[\because \tan(\pi + x) = \tan x]$

(*v*) $\sec(\pi + x) = \frac{1}{\cos(\pi + x)} = \frac{1}{-\cos x} = -\sec x$ $[\because \cos(\pi + x) = -\cos x]$

(*vi*) $\text{cosec}(\pi + x) = \frac{1}{\sin(\pi + x)} = \frac{1}{-\sin x} = -\text{cosec}\, x$ $[\because \sin(\pi + x) = -\sin x]$

Theorem 5.

(*i*) $\cos(2\pi - x) = \cos x$ (*ii*) $\sin(2\pi - x) = -\sin x$

(*iii*) $\tan(2\pi - x) = -\tan x$ (*iv*) $\cot(2\pi - x) = -\cot x$

(*v*) $\sec(2\pi - x) = \sec x$ (*vi*) $\text{cosec}(2\pi - x) = -\text{cosec} x$

Proof.

(*i*) $\cos(2\pi - x) = \cos 2\pi \cos x + \sin 2\pi \sin x$

$= (1 \times \cos x) + (0 \times \sin x) = \cos x$ $[\because \cos 2\pi = 1 \text{ and } \sin 2\pi = 0]$

(*ii*) $\sin(2\pi - x) = \sin[\pi + (\pi + x)] = -\sin(\pi + x) = -[-\sin x] = -\sin x$

$$\left[\because \sin\left(\frac{x}{2} - \theta\right) = \cos\theta\right]$$

Similarly we can proof other indentities.

Summary

(*i*) $\cos(2\pi - x) = \cos x$ (*ii*) $\sin(2\pi - x) = -\sin x$

(*iii*) $\tan(2\pi - x) = -\tan x$ (*iv*) $\cot(2\pi - x) = -\cot x$

(*v*) $\sec(2\pi - x) = \sec x$ (*vi*) $\text{cosec}(2\pi - x) = -\text{cosec} x$

Similarly,

(*i*) $\cos(2\pi + x) = \cos x$ (*ii*) $\sin(2\pi + x) = \sin x$

(*iii*) $\tan(2\pi + x) = \tan x$ (*iv*) $\cot(2\pi + x) = \cot x$

(*v*) $\sec(2\pi + x) = \sec x$ (*vi*) $\text{cosec}(2\pi + x) = \text{cosec} x$

Also, we have

(*i*) $\cos\left(\frac{3\pi}{2} - x\right) = -\sin x$ (*ii*) $\sin\left(\frac{3\pi}{2} - x\right) = -\cos x$

(*iii*) $\cos\left(\frac{3\pi}{2} - x\right) = \sin x$ (*iv*) $\sin\left(\frac{3\pi}{2} - x\right) = -\cos x$

Summarised Table

	$(-x)$	$\left(\frac{\pi}{2} - x\right)$	$\left(\frac{\pi}{2} + x\right)$	$(\pi - x)$	$(\pi + x)$	$\left(\frac{3\pi}{2} - x\right)$	$\left(\frac{3\pi}{2} + x\right)$
sin	$-\sin x$	$\cos x$	$\cos x$	$\sin x$	$-\sin x$	$-\cos x$	$-\cos x$
cos	$\cos x$	$\sin x$	$-\sin x$	$-\cos x$	$-\cos x$	$-\sin x$	$\sin x$
tan	$-\tan x$	$\cot x$	$-\cot x$	$-\tan x$	$\tan x$	$\cot x$	$-\cot x$

Remarks : (*i*) The T-ratios for $(2\pi - x)$ and $(2n\pi - x)$ are the same as those for $(-x)$.
(*ii*) The T-ratios for $(2\pi + x)$ and $(2n\pi + x)$ are the same as those for x.

Theorem 6.

(*i*) $\sin(\theta + \phi) = \sin\theta \cos\phi + \cos\theta \sin\phi$

(*ii*) $\sin(\theta - \phi) = \sin\theta \cos\phi - \cos\theta \sin\phi$.

Proof. (*i*) $\sin(\theta + \phi) = \cos\left[\frac{\pi}{2} - (\theta + \phi)\right] = \cos\left[\left(\frac{\pi}{2} - \theta\right) - \phi\right]$

$$= \cos\left(\frac{\pi}{2} - \theta\right)\cos\phi + \sin\left(\frac{\pi}{2} - \theta\right)\sin\phi$$

$$= \sin\theta \cos\phi + \cos\theta \sin\phi.$$

Hence, $\sin(\theta + \phi) = \sin\theta \cos\phi + \cos\theta \sin\phi$

(*ii*) Replacing ϕ by $-\phi$ in (*i*), we get

$$\sin[\theta + (-\phi)] = \sin\theta \cos(-\phi) + \cos\theta \sin(-\phi)$$

$$\Rightarrow \quad \sin(\theta - \phi) = \sin\theta \cos\phi - \cos\theta \sin\phi$$

$[\because \cos(-\phi) = \cos\phi,\ \sin(-\phi) = -\sin\phi]$

Hence, $\sin(\theta - \phi) = \sin\theta \cos\phi - \cos\theta \sin\phi$.

Theorem 7. If name of the angles, θ, ϕ and $(\theta + \phi)$ is an odd multiple of $\pi/2$ then

(*i*) $\tan(\theta + \phi) = \dfrac{\tan\theta + \tan\phi}{1 - \tan\theta \tan\phi}$ (*ii*) $\tan(\theta - \phi) = \dfrac{\tan\theta - \tan\phi}{1 + \tan\theta \tan\phi}$.

Proof. $\tan(\theta + \phi) = \dfrac{\sin(\theta + \phi)}{\cos(\theta + \phi)} = \dfrac{(\sin\theta \cos\phi + \cos\theta \sin\phi)}{(\cos\theta \cos\phi - \sin\theta \sin\phi)} = \dfrac{\left(\dfrac{\sin\theta \cos\phi}{\cos\theta \cos\phi} + \dfrac{\cos\theta \sin\phi}{\cos\theta \cos\phi}\right)}{\left(\dfrac{\cos\theta \cos\phi}{\cos\theta \cos\phi} - \dfrac{\sin\theta \sin\phi}{\cos\theta \cos\phi}\right)}$

[dividing num. and denom. by $\cos\theta \cos\phi$]

$$= \frac{\tan\theta + \tan\phi}{1 - \tan\theta \tan\phi}$$

Since, $\tan(\theta + \phi) = \dfrac{\tan\theta + \tan\phi}{1 - \tan\theta \tan\phi}$

(*ii*) Replacing ϕ by $-\phi$'in (*i*), we get

$$\tan[\theta + (-\phi)] = \frac{\tan\theta + \tan(-\phi)}{1 - \tan\theta \tan(-\phi)}$$

$$\Rightarrow \quad \tan(\theta - \phi) = \frac{\tan\theta - \tan\phi}{1 + \tan\theta \tan\phi}$$

Hence, $\tan(\theta - \phi) = \dfrac{\tan\theta - \tan\phi}{1 + \tan\theta \tan\phi}$.

Theorem 8. If name of the angles θ, ϕ and $(\theta + \phi)$ is a multiple of π then

(*i*) $\cot(\theta + \phi) = \dfrac{\cot\theta + \cot\phi - 1}{\cot\phi + \cot\theta}$ (*ii*) $\cot(\theta - \phi) = \dfrac{\cot\theta + \cot\phi + 1}{\cot\phi - \cot\theta}$.

Proof. $\cot(\theta + \phi) = \dfrac{\cos(\theta + \phi)}{\sin(\theta + \phi)} = \dfrac{(\cos\theta + \cos\phi - \sin\theta + \sin\phi)}{(\sin\theta + \cos\phi + \cos\theta + \sin\phi} = \dfrac{\left(\dfrac{\cos\theta \cos\phi}{\sin\theta \cos\phi} - \dfrac{\sin\theta \sin\phi}{\sin\theta \cos\phi}\right)}{\left(\dfrac{\sin\theta \cos\phi}{\sin\theta \sin\phi} + \dfrac{\cos\theta \sin\phi}{\sin\theta \sin\phi}\right)}$

[dividing num. and denom. by $\sin\theta \sin\phi$]

$$= \frac{\cot\theta + \cot\phi - 1}{\cot\theta + \cot\phi}$$

Hence, $\cot(\theta + \phi) = \dfrac{\cot\theta + \cot\phi - 1}{\cot\theta + \cot\phi}$

(*ii*) Replacing ϕ by $-\phi$ in (*i*), we get

$$\cot[\theta + (-\phi)] = \frac{\cot\theta + \cot(-\phi) - 1}{\cot(-\phi) + \cot\theta}$$

$$\Rightarrow \quad \cot(\theta - \phi) = \frac{-\cot\theta\cot\phi - 1}{-\cot\phi + \cot\theta} \qquad [\because \cot(-\phi) = -\cot\phi]$$

$$\Rightarrow \quad \cot(\theta - \phi) = \frac{\cot\theta\cot\phi + 1}{\cot\phi - \cot\theta}$$

Hence, $\cot(\theta - \phi) = \dfrac{\cot\theta\cot\phi + 1}{\cot\phi - \cot\theta}$.

Summary

(*i*) $\tan(\theta + \phi) = \dfrac{\tan\theta + \tan\phi}{1 - \tan\theta\tan\phi}$

(*ii*) $\tan(\theta - \phi) = \dfrac{\tan\theta - \tan\phi}{1 + \tan\theta\tan\phi}$

(*iii*) $\cot(\theta + \phi) = \dfrac{\cot\theta + \cot\phi - 1}{\cot\phi + \cot\theta}$

(*ii*) $\cot(\theta - \phi) = \dfrac{\cot\theta\cot\phi + 1}{\cot\phi - \cot\theta}$

Theorem 9. For all $x, y \in R$, we have

(*i*) $\sin(x + y)\sin(x - y) = \sin^2 x - \sin^2 y$

(*ii*) $\cos(x + y)\cos(x - y) = \cos^2 x - \sin^2 y$

Proof. (*i*) $\sin(x + y)\sin(x - y)$

$$= (\sin x\cos y + \cos x\sin y)(\sin x\cos y - \cos x\sin y)$$
$$= \sin^2 x\cos^2 y - \cos^2 x\sin^2 y$$
$$= \sin^2 x(1 - \sin^2 y) - (1 - \sin^2 x)\sin^2 y$$
$$= \sin^2 x \quad \sin^2 x\sin^2 y - \sin^2 y + \sin^2 x\sin^2 y$$
$$= \sin^2 x - \sin^2 y$$

(*ii*) $\cos(x + y)\cos(x - y)$

$$= (\cos x\cos y - \sin x\sin y)(\cos x\cos y + \sin x\sin y)$$
$$= \cos^2 x\cos^2 y - \sin^2 x\sin^2 y$$
$$= \cos^2 x(1 - \sin^2 y) - (1 - \cos^2 x)\sin^2 y$$
$$= \cos^2 x - \cos^2 x\sin^2 y - \sin^2 y + \cos^2 x\sin^2 y = \cos^2 x - \sin^2 y$$

1.1.4. Formula to Transform the Sum or Difference into Product

$$\sin C + \sin D = 2\sin\left(\frac{C + D}{2}\right)\cos\left(\frac{C - D}{2}\right)$$

$$\sin C - \sin D = 2\sin\left(\frac{C - D}{2}\right)\cos\left(\frac{C + D}{2}\right)$$

$$\cos C + \cos D = 2\cos\left(\frac{C + D}{2}\right)\cos\left(\frac{C - D}{2}\right)$$

$$\cos C - \cos D = -2 \sin\left(\frac{C+D}{2}\right) \sin\left(\frac{C-D}{2}\right)$$

1.1.5. T (Trigonometric) Ratio of Multiple Angle and Sub-multiple Angles 2A, 3A, A/2)

Angle which is an multiplication of some angle is termed as a multiple angle *e.g.*, angle 2A, 3A, A/2 are the multiple angle, A/2, A/3 ... are the sub multiple angles.

- **Trigonometric Ratio of Angles 2A**

$$\sin 2A = \sin(A + A) [\text{Using } \sin(A + B) = \sin A \cos B + \cos A \sin B)$$
$$= \sin A \cos A + \cos A \sin A$$
$$\sin 2A = 2\sin A \cos A = \frac{2\tan A}{1+\tan^2 A}$$
$$\cos 2A = \cos(A + A) (\text{Using } \cos(A + B) = \cos A \cos B + \cos A \cos B)$$
$$= \cos A \cos A + \sin A \sin A$$
$$\cos 2A = \cos^2 A - \sin^2 A \quad ...(i)$$
$$= 1 - 2\sin A \cdot \sin A$$
$$\cos 2A = 1 - 2\sin^2 A \quad ...(ii)$$

from (i) $\cos^2 A = \cos^2 A - \sin^2 A$ $\quad = \cos^2 A - (1 - \cos^2 A) = \cos^2 A - 1 + \cos^2 A \quad ...(iii)$

$$= 2\cos^2 A - 1$$
$$\cos 2A = \cos^2 A - \sin^2 A = 1 - 2\sin^2 A = 2\cos^2 A - 1 = \frac{1-\tan^2\theta}{1+\tan^2\theta}$$

Now, $\tan 2A = \tan(A + A)$ $\quad \left[\text{Using } \tan(A + B) = \frac{\tan A + \tan B}{1 - \tan A \tan B}\right]$

$$= \frac{\tan A + \tan A}{1 - \tan A \times \tan A} = \frac{2\tan A}{1-\tan^2 A}$$

- **Trigonometric Ratio of Angle 3*A***

$$\sin 3A = 3\sin A - 4\sin^3 A$$
$$\cos 3A = 4\cos^3 A - 3\cos A$$
$$\tan 3A = \frac{3\tan A - \tan^3 A}{1 - 3\tan^2 A}$$

- **Trignometric Ratio of Angle *A*/2**

All the trigonometric ratio of angle *A*/2 are given below:

$$\sin A = 2\sin A/2 \cos A/2 = \frac{2\tan A/2}{1+\tan^2 A/2}$$

$$\cos A = \cos^2 A/2 - \sin^2 A/2 = 1 - 2\sin^2 A/2 = 2\cos^2 A/2 - 1 = \frac{1-\tan^2 A/2}{1+\tan^2 A/2}$$

Example 22. *Find the value of (i) tan 330°, (ii) cos (– 870°), (iii) cosec (– 1410), (iv) cot (–135°), (v) sin 1230°, (vi) cos 480°*

Sol.

(*i*) $\tan 330° = \tan (360° - 30°) = -\tan 30° = \frac{-1}{\sqrt{3}}$ $[\tan(360° - \theta) = -\tan\theta]$

(*ii*) $\cos (-870°) = \cos (-870) = -\cos (2 \times 360° + 150°) = \cos 150°$ $[\cos (-\theta) = \cos \theta]$

$= \cos (180° - 30°) = -\cos 30° = \frac{-\sqrt{3}}{2}$

(*iii*) $\text{cosec} (-1410°) = -\cos 1410° = -\text{cosec} (4 \times 360° - 30°)$

$= \text{cosec } 30 = 2$ $[\because \text{cosec} (2n\pi - \theta) = -\text{cosec } \theta)$

(*iv*) $\cot (-135°) = -\cot 135° = -\cot (90° + 45°) = \tan 45° = 1$ $[\because \cot (-\theta) = -\cot \theta)$

$[\cot (90 + \theta) = -\tan\theta]$

(*v*) $\sin 1230° = \sin (3 \times 360° + 150°) = \sin 150°$ $[\because \sin (2n\pi + \theta) = \sin \theta)$

$= \sin (180° - 30°) = \sin 30° = 1/2$

(*vi*) $\cos 480° = \cos (360° + 120°) = \cos 120°$ $[\because \cos (360° + \theta) = \cos \theta]$

$= -\cos 60° = \frac{-1}{2}$

Example 23. *Calculate tan* $\frac{3\pi}{12}$.

Sol. $\tan \frac{13\pi}{12} = \tan \left(\pi + \frac{\pi}{12}\right) = \tan \frac{\pi}{12}$ $[\because \tan (\pi + \theta) = \tan \theta]$

$$= \tan \left(\frac{\pi}{3} - \frac{\pi}{4}\right) = \frac{\tan \pi/3 - \tan \pi/4}{1 + \tan \pi/3 . \tan \pi/4} = \frac{\sqrt{3} - 1}{\sqrt{3} + 1}$$

$$= \frac{\sqrt{3} - 1 \times \sqrt{3} - 1}{\sqrt{3} + 1 \sqrt{3} - 1} = \frac{(\sqrt{3} - 1)^2}{(3 - 1)} = \frac{3 + 1 - 2\sqrt{3}}{2}$$

$$= \frac{4 - 2\sqrt{3}}{2} = (2 - \sqrt{3}).$$

Example. 24. *If cos θ = 4/5 and cos φ = 12/13, where θ and φ both lie in the fourth quadrant, find the values of (i) cos (θ + φ), (ii) sin (θ − φ), (iii) tan (θ + θ).*

Sol. Given θ = 4/5 and cosφ = 12/13.

Since θ lies in the fourth quadrant, we have

$\cos \theta > 0$, $\sin \theta < 0$ and $\tan \theta < 0$.

Again, since φ lies in the fourth quadrant, we have

$\cos \phi > 0$, $\sin \phi < 0$ and $\tan \phi < 0$.

Now, $\sin \theta = -\sqrt{1 - \cos^2 \theta} = -\sqrt{1 - \frac{16}{25}} = -\sqrt{\frac{9}{25}} = \frac{-3}{5}$

and $\tan \theta = \frac{\sin\theta}{\cos\theta} = \left(\frac{-3}{5}\right) \times \frac{5}{4} = \frac{-3}{4}$

Also, $\sin \phi = -\sqrt{1 - \cos^2 \phi} = -\sqrt{1 - \frac{144}{169}} = -\sqrt{\frac{25}{169}} = \frac{-5}{13}$

and $\tan \phi = \frac{\sin\phi}{\cos\phi} = \left(\frac{-5}{13}\right) \times \frac{13}{12} = \frac{-5}{12}$

$\therefore$ (*i*)
$$\cos(\theta + \phi) = \cos\theta\cos\phi - \sin\theta\sin\phi$$
$$= \left(\frac{4}{5}\times\frac{12}{13}\right) - \left\{\left(\frac{-3}{5}\right)\times\left(\frac{-5}{13}\right)\right\} = \left(\frac{48}{65} - \frac{15}{65}\right) = \frac{33}{65}$$

(*ii*)
$$\sin(\theta - \phi) = \sin\theta\cos\phi - \cos\theta\sin\phi$$
$$= \left\{\left(\frac{-3}{5}\right)\times\left(\frac{12}{13}\right)\right\} - \left\{\frac{4}{5}\times\left(\frac{-5}{13}\right)\right\} = \left(\frac{-36}{65} + \frac{20}{65}\right) = \frac{-16}{65}$$

(*iii*)
$$\tan(\theta + \phi) = \frac{\tan\theta + \tan\phi}{1 - \tan\theta.\tan\phi}$$
$$= \frac{\left(\frac{-3}{4}\right) + \left(\frac{-5}{12}\right)}{1 - \left\{\left(\frac{-3}{4}\right)\times\left(\frac{-5}{12}\right)\right\}} = \frac{\left(\frac{-7}{6}\right)}{\left(1 - \frac{5}{16}\right)} = \left(\frac{-7}{6}\right) = \frac{16}{11} = \frac{-56}{33}$$

Example 25. *Prove that*

(*i*) $\sin\frac{7\pi}{12}\cos\frac{\pi}{4} - \cos\frac{7\pi}{12}\sin\frac{\pi}{4} = \frac{\sqrt{3}}{2}$ (*ii*) $\sin\frac{\pi}{4}\cos\frac{\pi}{12} + \cos\frac{\pi}{4}\sin\frac{\pi}{12} = \frac{\sqrt{3}}{2}$

(*iii*) $\cos\frac{2\pi}{3}\cos\frac{\pi}{4} - \sin\frac{2\pi}{3}\sin\frac{\pi}{4} = \frac{-(\sqrt{3}+1)}{2\sqrt{2}}$.

Sol. (*i*) $\sin\frac{7\pi}{12}\cos\frac{\pi}{4} - \cos\frac{7\pi}{12}\sin\frac{\pi}{4}$

$$= \sin\left(\frac{7\pi}{12} - \frac{\pi}{4}\right) \quad [\because \sin x\cos y - \cos x\sin y = \sin(x - y)]$$
$$= \sin\frac{\pi}{3} = \frac{\sqrt{3}}{2}$$

(*ii*) $\sin\frac{\pi}{4}\cos\frac{\pi}{12} + \cos\frac{\pi}{4}\sin\frac{\pi}{12}$

$$= \sin\left(\frac{\pi}{4} + \frac{\pi}{12}\right) \quad [\because \sin x\cos y + \cos x\sin y = \sin(x + y)]$$
$$= \sin\frac{\pi}{3} = \frac{\sqrt{3}}{2}$$

(*iii*) $\cos\frac{2\pi}{3}\cos\frac{\pi}{4} - \sin\frac{2\pi}{3}\sin\frac{\pi}{4} = \cos\left(\pi - \frac{\pi}{3}\right)\cos\frac{\pi}{4} - \sin\left(\pi - \frac{\pi}{3}\right)\sin\frac{\pi}{4}$

$$= -\cos\frac{\pi}{3}\cos\frac{\pi}{4} - \sin\frac{\pi}{3}\cos\frac{\pi}{4}$$
$$[\because \sin(\pi - x)\sin x\cos(\pi - x) = -\cos x]$$
$$= \left(-\frac{1}{2}\times\frac{1}{\sqrt{2}}\right) - \left(\frac{\sqrt{3}}{2}\times\frac{1}{\sqrt{2}}\right) = -\left\{\frac{(\sqrt{3}+1)}{2\sqrt{2}}\right\}.$$

Example 26. *Evaluate (i) sin* $\frac{\pi}{12}$, *(ii)* $\sin\frac{5\pi}{12}$

Sol. (*i*) $\sin\frac{\pi}{12}$ $= \sin\left(\frac{\pi}{4}-\frac{\pi}{6}\right)$

$$= \sin\frac{\pi}{4}\cos\frac{\pi}{6} - \cos\frac{\pi}{4}\sin\frac{\pi}{6}$$

$$= \left(\frac{1}{\sqrt{2}}\times\frac{\sqrt{3}}{2}\right)-\left(\frac{1}{\sqrt{2}}\times\frac{1}{2}\right) = \frac{(\sqrt{3}-1)}{2\sqrt{2}}$$

(*ii*) $\sin\frac{5\pi}{12} = \sin\left(\frac{\pi}{4}+\frac{\pi}{6}\right)$ $= \sin\frac{\pi}{4}\cos\frac{\pi}{6} + \cos\frac{\pi}{4}\sin\frac{\pi}{6}$

$$= \left(\frac{1}{\sqrt{2}}\times\frac{\sqrt{3}}{2}\right)+\left(\frac{1}{\sqrt{2}}\times\frac{1}{2}\right) = \frac{(\sqrt{3}+1)}{2\sqrt{2}}.$$

Example 27. *Prove that* $\frac{\sin(x+y)}{\sin(x-y)} = \frac{\tan x+\tan y}{\tan x-\tan y}$

Sol. We have L.H.S. $= \frac{\sin(x+y)}{\sin(x-y)} = \frac{\sin x\cos y+\cos x\sin y}{\sin x\cos y-\cos x\sin y}$

$= \frac{\tan x+\tan y}{\tan x-\tan y}$ [dividing num. & denom. by cos$x$ cos$y$]

= R.H.S.

Example 28. *If sin* $\theta = 15/17$ *and cos* $\phi = 12/13$, *where* θ *and* ϕ *both lie in the first quadrant, find the values of (i) sin* $(\theta+\phi)$, *(ii) cos* $(\theta-\phi)$, *(iii) tan* $(\theta+\phi)$

Sol. Given θ = 15/17 and cosϕ = 12/13.

Since θ lies in the fourth quadrant, we have

$$\sin\theta > 0, \cos\theta > 0 \text{ and } \tan\theta > 0.$$

Again, since ϕ lies in the first quadrant, we have

$$\sin\phi > 0, \cos\phi > 0 \text{ and } \tan\phi > 0.$$

Now, $\cos^2\theta = (1-\sin^2\theta) = \left(1-\frac{225}{289}\right) = \frac{64}{289}$

$\Rightarrow$ $\cos\theta = +\sqrt{\frac{64}{289}} = \frac{8}{17}$

And, $\sin^2\phi = (1-\cos^2\phi) = \left(1-\frac{144}{169}\right) = \frac{25}{169}$

$\Rightarrow$ $\sin\phi = +\sqrt{\frac{25}{169}} = \frac{5}{13}$

$\therefore$ $\tan\theta = \frac{\sin\theta}{\cos\theta} = \left(\frac{15}{17}\times\frac{17}{8}\right) = \frac{15}{8}$

andtanϕ $= \dfrac{\sin\phi}{\cos\phi} = \left(\dfrac{5}{13} \times \dfrac{13}{12}\right) = \dfrac{5}{12}$

Now, (*i*) sin (θ + ϕ) = sinθ cosϕ + cosθ sinϕ

$$= \left(\frac{15}{17} \times \frac{12}{13}\right) + \left(\frac{8}{17} \times \frac{5}{13}\right) = \left(\frac{180}{221} \times \frac{40}{221}\right) = \frac{220}{221}$$

(*ii*) cos(θ – ϕ) = cosθ cosϕ + sinθ sinϕ

$$= \left(\frac{8}{17} \times \frac{12}{13}\right) + \left(\frac{15}{17} \times \frac{5}{13}\right) = \left(\frac{96}{221} \times \frac{75}{221}\right) = \frac{171}{221}$$

$$(iii)\ \tan(\theta + \phi) = \frac{\tan\theta + \tan\phi}{1 - \tan\theta\tan\phi} = \frac{\left(\frac{15}{8}\right) + \left(\frac{5}{12}\right)}{\left\{1 - \left(\frac{15}{8} \times \frac{5}{12}\right)\right\}} = \frac{\left(\frac{55}{24}\right)}{\left(1 - \frac{25}{32}\right)} = \left(\frac{55}{24} \times \frac{32}{7}\right) = \frac{220}{21}.$$

Example 29. *If sin θ = 3/5 and cos ϕ = –12/13, where θ and ϕ both lies in the second quadrant, find the values of (i) sin (θ – ϕ), (ii) cos (θ + ϕ), (iii) tan (θ – ϕ).*

Sol. Given : sinθ = 3/5 and cos ϕ = –12/13

Since θ lies in the fourth quadrant, we have

$$\sin\theta > 0,\ \cos\theta < 0 \text{ and } \tan\theta < 0.$$

Again, since ϕ lies in the second quadrant, we have

$$\sin\phi > 0,\ \cos\phi < 0 \text{ and } \tan\phi < 0.$$

Now, $\cos\theta = -\sqrt{1 - \sin^2\theta} = -\sqrt{1 - \frac{9}{25}} = -\sqrt{\frac{16}{25}} = \frac{-4}{5}$

and $\tan\theta = \dfrac{\sin\theta}{\cos\theta} = \dfrac{3}{5} \times \left(-\dfrac{5}{4}\right) = \dfrac{-3}{4}$

Also, $\sin\phi = +\sqrt{1 - \cos^2\phi} = +\sqrt{1 - \frac{144}{169}} = +\sqrt{\frac{25}{169}} = +\frac{5}{13}$

and $\tan\phi = \dfrac{\sin\phi}{\cos\phi} = \dfrac{5}{13} \times \left(\dfrac{-13}{12}\right) = \dfrac{-5}{12}$

∴ (*i*) sin (θ – ϕ) = sinθ cosϕ – cosθ sinϕ

$$= \left\{\frac{3}{5} \times \left(\frac{-12}{13}\right)\right\} - \left\{\left(\frac{-4}{5}\right) \times \frac{5}{13}\right\} = \left(\frac{-36}{65} + \frac{20}{65}\right) = \frac{-16}{65}$$

(*ii*) cos (θ + ϕ) = cosθ cosϕ – sinθ sinϕ

$$= \left\{\left(\frac{-4}{5}\right) \times \left(\frac{-2}{13}\right)\right\} - \left\{\left(\frac{3}{5} \times \frac{5}{13}\right)\right\} = \left(\frac{48}{65} + \frac{3}{13}\right) = \frac{33}{65}$$

$$(iii)\ \tan(\theta - \phi) = \frac{\tan\theta - \tan\phi}{1 + \tan\theta.\tan\phi} = \frac{\left(\frac{-3}{4}\right) - \left(\frac{-5}{12}\right)}{1 + \left\{\left(\frac{-3}{4}\right) \times \left(\frac{-5}{12}\right)\right\}} = \frac{\frac{-3}{4} + \frac{5}{12}}{1 + \frac{5}{16}} = \left(\frac{-1}{3}\right) \times \frac{16}{21} = -\frac{16}{63}.$$

Example 30. *Show that cot2x cotx – cot 3x cot 2x – cot 3x cotx = 1.*

Sol.

$$\cot 3x = \cot (2x + x)$$

$$\Rightarrow \quad \cot 3x = \frac{\cot 2x \cot x - 1}{\cot 2x + \cot x}$$

$$\Rightarrow \quad \cot 3x \cot 2x + \cot 3x \,.\, \cot x = \cot 2x \cot x - 1$$

$$\Rightarrow \quad \cot 2x \cot x - \cot 3x \cot 2x - \cot 3x \cot x = 1$$

Example 31. *Evaluate tan* $\frac{13\pi}{12}$

Sol. $\tan\left(\frac{\pi}{12}\right) = \tan\left(\frac{\pi}{3} - \frac{\pi}{4}\right)$ $\qquad [\because \tan(\pi + \theta) = \tan\theta]$

$$= \frac{\tan \pi/3 - \tan \pi/4}{1 + \tan \pi/3 . \tan \pi/4}$$

$$= \frac{(\sqrt{3}-1)\times(\sqrt{3}-1)}{\sqrt{3}+1\,\sqrt{3}-1} = \frac{(\sqrt{3}-1)^2}{(3-1)} = \frac{3+1-2\sqrt{3}}{2}$$

$$= \frac{4-2\sqrt{3}}{2} = (2-\sqrt{3})$$

Example 32. *Prove that sin 105° + cos 105°* = $\frac{1}{\sqrt{2}}$

Sol.

$$\sin 105° + \cos 105° = \sin (60° + 45°) + \cos (60° + 45°)$$

$$= \sin 60 \cos 45° + \cos 60° \sin 45° + (\cos 60° \cos 45° - \sin 60 \sin 45°)$$

$$= \left\{\left(\frac{\sqrt{3}}{3}\times\frac{1}{\sqrt{2}}\right)+\left(\frac{1}{2}\times\frac{1}{\sqrt{2}}\right)\right\}+\left\{\left(\frac{1}{2}\times\frac{1}{\sqrt{2}}\right)-\left(\frac{\sqrt{3}}{2}\times\frac{1}{2}\right)\right\}$$

$$= \left(\frac{\sqrt{3}}{2\sqrt{2}}+\frac{1}{2\sqrt{2}}+\frac{1}{2\sqrt{2}}-\frac{\sqrt{3}}{2\sqrt{2}}\right) = \frac{1}{2} \text{ R.H.S.}$$

Example 33. Prove that tan 56° $= \frac{\cos 11° + \sin 11°}{\cos 11° - \sin 11°}$

Sol.

$$\tan 56° = \tan (45° + 11°) = \frac{\tan 45° + \tan 11°}{1 - \tan 45° \tan 11°} = \frac{1 + \tan 11°}{1 - \tan 11°}$$

$$= \frac{1 + \frac{\sin 11}{\cos 11}}{1 - \frac{\sin 11}{\cos 11}} = \frac{\cos 11 + \sin 11}{\cos 11 - \sin 11} \text{ R.H.S.}$$

Example 34. *Prove that tan 70° = tan 20° + 2tan 50°*

Sol.

$$\tan (20 + 50°) = \frac{\tan 20° + \tan 50°}{1 - \tan 20° \tan 50°}$$

$$= \tan 70° - \tan 70° \tan 20° \tan 50° = \tan 20° + \tan 50°$$

$$= \tan 70° - \tan(90° - 20°) \tan 20° \tan 50° = \tan 20° + \tan 50°$$

$$= \tan 70° - \cot 20° \tan 20° \tan 50° = \tan 20° + \tan 50°$$

$$\Rightarrow \quad = \tan 70° - \tan 50° = \tan 20° + \tan 50°.$$

Example 35. *Prove that cot2x cotx – cot3x cot 2x – cot3x cotx = 1.*

Sol. $\cot 3x = \cot (2x \times x) = \cot 3x = \dfrac{\cot 2x \cot x - 1}{\cot 2x + \cot x}$

$\Rightarrow \quad \cot 3x \cot 2x + \cot 2x \cot 3x = \cot 2x \cot x - 1$

$\Rightarrow \cot 2x \cot x - \cot 3x \cot 2x - \cot 3x \cot x = 1.$

Example 36. *Prove that sin 15° = sin 45° – 30°*

Sol.
$$= \sin 45° \cos 30° - \cos 45° \sin 30°$$
$$= \frac{1}{\sqrt{2}} \times \frac{\sqrt{3}}{2} - \frac{1}{2} \times \frac{1}{2} = \frac{\sqrt{3}-1}{2\sqrt{2}}.$$

Example 37. *Prove that* $\dfrac{\cot 7x \times \cos 5x}{\sin 7x - \sin 5x} = \cot x.$

Sol. $\dfrac{2\cos\dfrac{7x+5x}{2}\cos\dfrac{7x-5x}{2}}{\dfrac{\cos 7x+5x}{2}\sin\dfrac{7x-5x}{2}} = \dfrac{\cos x}{\sin x} = \cot x = \text{R.H.S.}$

Example 38. Prove that

$$\frac{\sin 5x - 2\sin 3x + \sin x}{\cos 5x \,.\, \cos x} = \tan x$$

Sol.
$$\frac{\sin 5x - 2\sin 3x + \sin x}{\cos 5x \,.\, \cos x} = \frac{\sin 5x + \sin x - 2\sin 3x}{\cos 5x . \cos x}.$$
$$= \frac{2\sin 3x \cos 2x - 2\sin 3x}{-2\sin 3x \sin 2x} = \frac{\sin 3x\,(\cos 2x - 1)}{\sin 3x \sin 2x}$$
$$= \frac{1-\cos 2x}{\sin 2x} = \frac{2\sin 2x}{2\sin x \cos x} = \tan x = \text{R.H.S.}$$

Example 39. *Prove that* $\dfrac{\tan\alpha}{\sec\alpha - 1} = \dfrac{\sec\alpha + 1}{\tan\alpha}.$

Sol. $\dfrac{\tan\alpha}{\sec\alpha - 1} = \dfrac{\sec\alpha + 1}{\tan\alpha}$

$$\Rightarrow \quad \frac{(\tan\alpha)(\sec\alpha+1)}{(\sec\alpha-1)(\sec\alpha+1)} = \frac{\sec\alpha+1}{\tan\alpha}$$

$$\frac{(\tan\alpha)(\sec\alpha+1)}{\sec\alpha-1} = \frac{\sec\alpha+1}{\tan\alpha} \qquad \text{Factoring rule}$$

$$\frac{(\tan\alpha)(\sec\alpha+1)}{\tan^2\alpha} = \frac{\sec\alpha+1}{\tan\alpha} \qquad \text{Pythagorean identity}$$

$$\frac{\sec\alpha+1}{\tan\alpha} = \frac{\sec\alpha+1}{\tan\alpha} \qquad \text{Algebraic manipulation}$$

Example 40. *Verify the identity* $\dfrac{\sin^2\alpha + 2\cos\alpha - 1}{\sin^2\alpha + 3\cos\alpha - 3} = \dfrac{1}{1-\sec\alpha}.$

Sol.
$$\frac{\sin^2\alpha + 2\cos\alpha - 1}{\sin^2\alpha + 3\cos\alpha - 3} = \frac{1}{1-\sec\alpha} \qquad \text{Identity to be verfied}$$

$$\frac{(1-\cos^2\alpha) + 2\cos\alpha - 1}{(1-\cos^2\alpha) + 3\cos\alpha - 3} = \frac{1}{1-\sec\alpha} \qquad \text{Pytrhagorean identity}$$

$$\frac{-(\cos^2\alpha - 2\cos\alpha)}{-(\cos^2\alpha - 3\cos\alpha + 2)} = \frac{1}{1-\sec\alpha}$$ Combining term

$$\frac{-(\cos\alpha)(\cos\alpha - 2)}{(\cos\alpha - 1)(\cos\alpha - 2)} = \frac{1}{1-\sec\alpha}$$ Factoring

$$\frac{\cos\alpha}{\cos\alpha - 1} = \frac{1}{1-\sec\alpha}$$ Algebraic manipulation

$$\frac{\left(\dfrac{\cos\alpha}{\cos\alpha}\right)}{\left(\dfrac{\cos\alpha - 1}{\cos\alpha}\right)} = \frac{1}{1-\sec\alpha}$$ Algebraic manipulation

$$\frac{1}{1-\dfrac{1}{\cos\alpha}} = \frac{1}{1-\sec\alpha}$$ Algebraic manipulation

$$\frac{1}{1-\cos\alpha} = \frac{1}{1-\sec\alpha}$$ Reciprocal identity

Example 41. Prove that $\dfrac{\sin 3x - \sin x}{\cos 2x} = 2\sin x$.

Sol. We have L.H.S. $= \dfrac{\sin 3x - \sin x}{\cos 2x} = \dfrac{2\cos\left(\dfrac{3x+x}{2}\right)\sin\left(\dfrac{3x-x}{2}\right)}{\cos 2x}$

$\left[\because \sin C - \sin D = 2\cos\left(\dfrac{C+D}{2}\right)\sin\left(\dfrac{C-D}{2}\right)\right]$

$$= \frac{2\cos 2x \sin x}{\cos 2x} = 2\sin x = \text{R.H.S.}$$

Example 42. *Prove that* $\dfrac{\sin 5x - 2\sin 3x + \sin x}{\cos 5x - \cos x} = \text{cosec } 2x - \cot 2x$.

Sol. We have L.H.S. $= \dfrac{(\sin 5x + \sin x) - 2\sin 3x}{\cos 5x - \cos x}$

$$= \frac{2\sin\left(\dfrac{5x+x}{2}\right)\cos\left(\dfrac{5x-x}{2}\right) - 2\sin 3x}{-2\sin\left(\dfrac{5x+x}{2}\right)\sin\left(\dfrac{5x-x}{2}\right)}$$

$\left[\because \sin C + \sin D = 2\sin\left(\dfrac{C+D}{2}\right)\sin\left(\dfrac{C-D}{2}\right)\right.$ and

$\left.\cos C - \cos D = -2\sin\left(\dfrac{C+D}{2}\right)\sin\left(\dfrac{C-D}{2}\right)\right]$

$$= \frac{2\sin 3x\cos 2x - 2\sin 3x}{-2\sin 3x\sin 2x} = \frac{2\sin 3x(\cos 2x - 1)}{-2\sin 3x\sin 2x}$$

$$= \frac{(1-\cos 2x)}{\sin 2x} = \frac{1}{\sin 2x} - \frac{\cos 2x}{\sin 2x}$$

$= \text{cosec } 2x - \cot 2x = \text{R.H.S.}$

Example 43. If $\dfrac{\cos(A+B)}{\cos(A-B)} = \dfrac{\sin(C+D)}{\sin(C-D)}$, prove that $\tan A \tan B \tan C + \tan D = 0$.

Sol. We have $\dfrac{\cos(A+B)}{\cos(A-B)} = \dfrac{\sin(C+D)}{\sin(C-D)}$

$$\Rightarrow \quad \frac{\cos(A+B)+\cos(A-B)}{\cos(A+B)-\cos(A-B)} = \frac{\sin(C+D)+\sin(C-D)}{\sin(C+D)-\sin(C-D)}$$

$$\Rightarrow \quad \frac{2\cos A\cos B}{-2\sin A\sin B} = \frac{2\sin C\cos D}{2\cos C\sin D}$$

$$\Rightarrow \quad -\cot A\cot B = \tan C\cot D$$

$$\Rightarrow \quad \tan A\tan B\tan C = -\tan D$$

$$\Rightarrow \quad \tan A\tan B\tan C + \tan D = 0$$

Hence, $\tan A \tan B \tan C + \tan D = 0$.

Example 44. *Prove that* $\dfrac{\cos 4x + \cos 3x + \cos 2x}{\sin 4x + \sin 3x + \sin 2x} = \cot 3x$.

Sol. We have

$$\text{L.H.S.} = \frac{(\cos 4x + \cos 2x) + \cos 3x}{(\sin 4x + \sin 2x) + \sin 3x}$$

$$= \frac{2\cos\left(\frac{4x+2x}{2}\right)\cos\left(\frac{4x-2x}{2}\right) + \cos 3x}{2\sin\left(\frac{4x+2x}{2}\right)\cos\left(\frac{4x-2x}{2}\right) + \sin 3x}$$

[using formulae for ($\cos C + \cos D$) and ($\sin C + \sin D$)]

$$= \frac{2\cos 3x\cos x + \cos 3x}{2\sin 3x\cos x + \sin 3x} = \frac{\cos 3x(2\cos x + 1)}{\sin 3x(2\cos x + 1)}$$

$$= \frac{\cos 3x}{\sin 3x} = \cot 3x = \text{R.H.S.}$$

Example 45. *Prove that* $\sin x + \sin 3x + \sin 5x + \sin 7x = 4\sin 4x \cos 2x \cos x$.

Sol. We have

$$\text{L.H.S.} = (\sin 7x + \sin x) + (\sin 5x + \sin 3x)$$

$$= 2\sin\left(\frac{7x+x}{2}\right)\cos\left(\frac{7x-x}{2}\right) + 2\sin\left(\frac{5x+3x}{2}\right)\cos\left(\frac{5x-3x}{2}\right)$$

$$= 2\sin 4x\cos 3x + 2\sin 4x\cos x$$

$$= 2\sin 4x(\cos 3x + \cos x)$$

$$= (2\sin 4x)\times 2\cos\left(\frac{3x+x}{2}\right)\cos\left(\frac{3x-x}{2}\right)$$

$$= (2\sin 4x)\times 2\cos 2x\cos x$$

$$= 4\sin 4x\cos 2x\cos x = \text{R.H.S.}$$

Example 46. *Prove that* $\cos 2x \cos \dfrac{x}{2} - \cos 3x \cos \dfrac{9x}{2} = -\sin 5x \sin\left(\dfrac{5x}{2}\right)$

Sol. We have

$$\text{L.H.S.} = \frac{1}{2}\left[2\cos 2x \cos\frac{x}{2} - 2\cos 3x \cos\frac{9x}{2}\right]$$

$$= \frac{1}{2}\left[\cos\left(2x+\frac{x}{2}\right)+\cos\left(2x-\frac{x}{2}\right)\right]-\frac{1}{2}\left[\cos\left(\frac{9x}{2}+3x\right)+\cos\left(\frac{9x}{2}-3x\right)\right]$$

$$= \frac{1}{2}\left(\cos\frac{5x}{2}+\cos\frac{3x}{2}\right)-\frac{1}{2}\left(\cos\frac{15x}{2}+\cos\frac{3x}{2}\right)$$

$$= \frac{1}{2}\left(\cos\frac{5x}{2}-\cos\frac{15x}{2}\right)$$

$$= \frac{1}{2}\left[-2\sin\frac{\left(\frac{5x}{2}+\frac{15x}{2}\right)}{2}\sin\frac{\left(\frac{15x}{2}-\frac{5x}{2}\right)}{2}\right]$$

$$= -\sin 5x \sin\left(\frac{5x}{2}\right) = \text{R.H.S.}$$

Example 47. *Prove that* $\dfrac{\sin 8x\cos x - \sin 6x \cos 3x}{\cos 2x \cos x - \sin 4x \sin 3x} = \tan 2x.$

Sol. We have

$$\text{L.H.S.} = \frac{2\sin 8x\cos x - 2\sin 6x\cos 3x}{2\cos 2x\cos x - 2\sin 4x\sin 3x}$$

$$= \frac{[\sin(8x+x)+\sin(8x-x)]-[\sin(6x+3x)+\sin(6x-3x)]}{[\cos(2x+x)+\cos(2x-x)]-[\cos(4x-3x)-\cos(4x+3x)]}$$

$[\because 2\sin A \cos B = \sin(A + B) + \sin(A - B)$, etc.]

$$= \frac{(\sin 9x+\sin 7x)-(\sin 9x+\sin 3x)}{(\cos 3x+\cos x)-(\cos x-\cos 7x)}$$

$$= \frac{(\sin 7x-\sin 3x)}{(\cos 3x+\cos 7x)}$$

$$= \frac{2\cos\left(\frac{7x+3x}{2}\right)\sin\left(\frac{7x-3x}{2}\right)}{2\cos\left(\frac{7x+3x}{2}\right)\cos\left(\frac{7x-3x}{2}\right)}$$

$\left[\because \sin C - \sin D = 2\cos\dfrac{(C+D)}{2}\sin\dfrac{(C-D)}{2}\right]$

$$= \frac{\cos 5x\sin 2x}{\cos 5x\cos 2x} = \frac{\sin 2x}{\cos 2x} = \tan 2x = \text{R.H.S.}$$

Example 48. *Prove that* $\dfrac{(\sin x - \sin y)}{(\cos x + \cos y)} = \tan\left(\dfrac{x-y}{2}\right).$

Sol.

$$\text{L.H.S.} = \frac{\sin x - \sin y}{\cos x + \cos y}$$

$$= \frac{2\cos\left(\frac{x+y}{2}\right)\sin\left(\frac{x-y}{2}\right)}{2\cos\left(\frac{x+y}{2}\right)\cos\left(\frac{x-y}{2}\right)} = \tan\left(\frac{x-y}{2}\right) = \text{R.H.S.}$$

Example 49. *Prove that* $(\cos x + \cos y)^2 + (\sin x + \sin y)^2 = 4\cos^2\left(\frac{x-y}{2}\right)$

Sol. We have L.H.S. $= (\cos x + \cos y)^2 + (\sin x + \sin y)^2$

$$= \left\{2\cos\left(\frac{x+y}{2}\right)\cos\left(\frac{x-y}{2}\right)\right\}^2 + \left\{2\sin\left(\frac{x+y}{2}\right)\cos\left(\frac{x-y}{2}\right)\right\}^2$$

$$= 4\cos^2\left(\frac{x+y}{2}\right)\cos^2\left(\frac{x-y}{2}\right) + 4\sin^2\left(\frac{x+y}{2}\right)\cos^2\left(\frac{x-y}{2}\right)$$

$$= 4\cos^2\left(\frac{x-y}{2}\right).\left\{\cos^2\left(\frac{x+y}{2}\right) + \sin^2\left(\frac{x+y}{2}\right)\right\}$$

$$= 4\cos^2\left(\frac{x-y}{2}\right) = \text{R.H.S.}$$

Example 50. *Prove that* $\cos\alpha + \cos\beta + \cos\gamma + \cos(\alpha + \beta + \gamma)$

$$= 4\cos\left(\frac{\alpha+\beta}{2}\right)\cos\left(\frac{\beta+\gamma}{2}\right)\cos\left(\frac{\gamma+\alpha}{2}\right)$$

Sol. We have

$$\text{L.H.S.} = (\cos\alpha + \cos\beta) + \{\cos(\alpha + \beta + \gamma) + \cos\gamma\}$$

$$= 2\cos\left(\frac{\alpha+\beta}{2}\right)\cos\left(\frac{\alpha-\beta}{2}\right) + 2\cos\left(\frac{\alpha+\beta+2\gamma}{2}\right)\cos\left(\frac{\alpha+\beta}{2}\right)$$

$$= 2\cos\left(\frac{\alpha+\beta}{2}\right)\left\{\cos\left(\frac{\alpha-\beta}{2}\right) + \cos\left(\frac{\alpha+\beta+2\gamma}{2}\right)\right\}$$

$$= 2\cos\left(\frac{\alpha+\beta}{2}\right)\left\{2\cos\left(\frac{\gamma+\alpha}{2}\right)\cos\left(\frac{\beta+\gamma}{2}\right)\right\}$$

[Using $\cos C + \cos D = 2\cos\frac{(C+D)}{2}\cos\frac{(C-D)}{2}$]

$$= 4\cos\left(\frac{\alpha+\beta}{2}\right)\cos\left(\frac{\beta+\gamma}{2}\right)\cos\left(\frac{\gamma+\alpha}{2}\right) = \text{R.H.S.}$$

Example 51. *Prove that* $\sin^2 6x - \sin^2 4x = \sin 10x \sin 2x$.

Sol. L.H.S. $= \sin^2 6x - \sin^2 4x$

$= \sin(6x + 4x)\sin(6x - 4x)$

[$\because \sin^2 x - \sin^2 y = \sin(x + y)\sin(x - y)$]

$= \sin 10x \sin 2x = $ R.H.S.

Example 52. *Prove that cos 20° cos 40° cos 60° cos 80° = 1/16*

Sol. We have cos 20° cos 40° cos 60° cos 80°

$$= \frac{1}{2} \cos 60° \cos 40° (2\cos 80° \cos 20°)$$

$$= \frac{1}{2} \times \frac{1}{2} \cos 40° \{\cos (80° + 20°) + \cos (80° - 20°)\}$$

$$= \frac{1}{4} \cos 40° (\cos 100° + \cos 60°)$$

$$= \frac{1}{4} \cos 40° (\cos 100° + 1/2)$$

$$= \frac{1}{8} (2\cos (100° \cos 40°) + \frac{1}{8} \cos 40°$$

$$= \frac{1}{8} [\cos (100° + 40°) + \cos (100° - 40°)] + \frac{1}{8} \cos 40°$$

$$= \frac{1}{8} (\cos 140° + \cos 60°) + \frac{1}{8} \cos 40°$$

$$= \frac{1}{8} \cos 140° + \frac{1}{16} + \frac{1}{8} \cos 40°$$

$$= \frac{1}{8} \cos (180° - 40°) + \frac{1}{16} + \frac{1}{8} \cos 40°$$

$$= \frac{1}{8} \cos 40° + \frac{1}{16} + \frac{1}{8} \cos 40°$$

$$= \frac{1}{16} = \text{R.H.S.}$$

Example 53. *Prove that sin 10° sin 50° sin 60° sin 70°* $= \frac{\sqrt{3}}{16}$.

Sol. We have sin 10° sin 50° sin 60° sin 70°

$$= \frac{1}{2} \sin 60° \sin 10° (2\sin 70° \sin 50°)$$

$$= \frac{1}{2} \times \frac{\sqrt{3}}{2} \times \sin 10° \{\cos (70° - 50°) - \cos (70° + 50°)\}$$

$$= \frac{\sqrt{3}}{4} \sin 10° \{\cos 20° - \cos 120°\}$$

$$= \frac{\sqrt{3}}{4} \sin 10° \left(\cos 20° + \frac{1}{2}\right) \qquad [\because \cos 120° = -\frac{1}{2}]$$

$$= \frac{\sqrt{3}}{8}\,(2\cos 20° \sin 10°) + \frac{\sqrt{3}}{8}\sin 10°$$

$$= \frac{\sqrt{3}}{8}\,[\sin (20° + 10°) - \sin (20° - 10°)] + \frac{\sqrt{3}}{8}\sin 10°$$

$$= \frac{\sqrt{3}}{8}\sin 30° = \left(\frac{\sqrt{3}}{8} \times \frac{1}{2}\right) = \frac{\sqrt{3}}{16} = \text{R.H.S.}$$

Example 54. *Prove that (i)* $\frac{\sin 2x}{1-\cos 2x} = \cot x$, *(ii)* $\frac{1-\cos 2x}{1+\cos 2x} = \tan^2 x$.

Sol. *(i)* L.H.S. $= \frac{\sin 2x}{1-\cos 2x} = \frac{2\sin x\cos x}{2\sin^2 x}$ $[\because \sin 2x = 2\sin x\cos x, (1-\cos 2x) = 2\sin^2 x]$

$$= \frac{\cos x}{\sin x} = \cot x = \text{R.H.S.}$$

(ii) L.H.S. $= \frac{1-\cos 2x}{1+\cos 2x} = \frac{2\sin^2 x}{2\cos^2 x}$ $\left[\because (1-\cos 2x) = 2\sin^2 x. \quad (1+\cos 2x) = 2\cos^2 x\right]$

$= \tan^2 x = \text{R.H.S.}$

Hence, $\frac{1-\cos 2x}{1+\cos 2x} = \tan^2 x.$

Example 55. *Prove that* $\cos 4x = 1 - 8\sin^2 x\cos^2 x$.

Sol. We have L.H.S. $= \cos 4x$

$= \cos 2(2x) = \cos 2\theta$, where $2x = \theta$

$= \cos^2\theta - \sin^2\theta$ $[\because \cos 2\theta = \cos^2\theta - \sin^2\theta]$

$= (1 - 2\sin^2\theta) = \{1 - 2(\sin 2x)^2\}$

$= 1 - 2\,(2\sin x\cos x)^2$ $[\because \sin 2x = 2\sin x\cos x]$

$= 1 - 8\sin^2 x\cos^2 x = \text{R.H.S.}$

Hence, $\cos 4x = 1 - 8\sin^2 x\cos^2 x.$

Example 56. *Prove that* $\frac{1+\sin 2x-\cos 2x}{1+\sin 2x+\cos 2x} = \tan x$.

Sol. We have L.H.S. $= \frac{1+\sin 2x-\cos 2x}{1+\sin 2x+\cos 2x}$

$$= \frac{(1-\cos 2x)+\sin 2x}{(1+\cos 2x)+\sin 2x} = \frac{2\sin^2 x + 2\sin x\cos x}{2\cos^2 x + 2\sin x\cos x}$$

$\left[\because (1-\cos 2x) = 2\sin^2 x, (1+\cos 2x) = 2\cos^2 x \text{ and } \sin 2x = 2\sin x\cos x\right]$

$$= \frac{2\sin x(\sin x+\cos x)}{2\cos x(\sin x+\cos x)} = \tan x = \text{R.H.S.}$$

Example 57. *Prove that* $\dfrac{1-\sin 2x}{1+\sin 2x}=\tan^2\left(\dfrac{\pi}{4}-x\right)$.

Sol. We have $\quad$ L.H.S. $=\dfrac{1-\sin 2x}{1+\sin 2x}$

$$=\frac{1-\cos\left(\frac{\pi}{2}-2x\right)}{1+\cos\left(\frac{\pi}{2}-2x\right)} \qquad \left[\because \sin 2x=\cos\left(\frac{\pi}{2}-2x\right)\right]$$

$$=\frac{2\sin^2\left(\frac{\pi}{4}-x\right)}{2\cos^2\left(\frac{\pi}{4}-x\right)}=\tan^2\left(\frac{\pi}{4}-x\right)=\text{R.H.S.}$$

$[\because (1-\cos^2\theta)=2\sin^2\theta$ and $(1+\cos 2\theta)=2\cos^2\theta]$

$$\therefore \qquad \frac{1-\sin 2x}{1+\sin 2x}=\tan^2\left(\frac{\pi}{4}-x\right).$$

Example 58. *Prove that* $\dfrac{\cos x+\sin x}{\cos x-\sin x}-\dfrac{\cos x-\sin x}{\cos x+\sin x}=2\tan 2x$.

Sol. We have $\quad$ L.H.S. $=\dfrac{\cos x+\sin x}{\cos x-\sin x}-\dfrac{\cos x-\sin x}{\cos x+\sin x}$

$$=\frac{(\cos x+\sin x)^2-(\cos x-\sin x)^2}{\cos^2 x-\sin^2 x}$$

$$=\frac{4\sin x\cos x}{\cos^2 x-\sin^2 x}=\frac{2\sin 2x}{\cos 2x}=2\tan 2x=\text{R.H.S.}$$

Example 59. *Prove that* $\tan 4\theta=\dfrac{4\tan\theta(1-\tan^2\theta)}{1-6\tan^2\theta+\tan^4\theta}$.

Sol. We have, L.H.S. = $\tan 4\theta=\tan 2(2\theta)$

$$=\frac{2\tan\theta}{1-\tan^2\theta}=\frac{2\tan\theta}{1-\tan^2 2\theta}$$

$$=\frac{2\times\frac{2\tan\theta}{(1-\tan^2\theta)}}{1-\left(\frac{2\tan\theta}{1-\tan^2\theta}\right)^2}=\frac{4\tan\theta(1-\tan^2\theta)}{(1-\tan^2\theta)^2-4\tan^2\theta}$$

$$=\frac{4\tan\theta\,(1-\tan^2\theta)}{1-6\tan^2\theta+\tan^4\theta}=\text{R.H.S.}$$

Example 60. *Prove that* $\dfrac{\tan 5\theta+\tan 3\theta}{\tan 5\theta-\tan 3\theta}=4\cos 2\theta\cos 4\theta$.

Sol. We have,

$$\text{LHS} = \frac{\tan 5\theta + \tan 3\theta}{\tan 5\theta - \tan 3\theta}$$

$$= \frac{\left(\frac{\sin 5\theta}{\cos 5\theta} + \frac{\sin 3\theta}{\cos 3\theta}\right)}{\left(\frac{\sin 5\theta}{\cos 5\theta} - \frac{\sin 3\theta}{\cos 3\theta}\right)} = \frac{(\sin 5\theta \cos 3\theta + \cos 5\theta \sin 3\theta)}{(\sin 5\theta \cos 3\theta - \cos 5\theta \sin 3\theta)}$$

$$= \frac{\sin(5\theta + 3\theta)}{\sin(5\theta - 3\theta)} = \frac{\sin 8\theta}{\sin 2\theta} = \frac{2\sin 4\theta \cos 4\theta}{\sin 2\theta}$$

$$= \frac{4\sin 2\theta \cos 2\theta \cos 4\theta}{\sin 2\theta} = 4\cos 2\theta \cos 4\theta = \text{R.H.S.}$$

Example 61. *Prove that* $\frac{\sec 8\theta - 1}{\sec 4\theta - 1} = \frac{\tan 8\theta}{\tan 2\theta}$.

Sol. We have

$$\text{L.H.S.} = \frac{\sec 8\theta - 1}{\sec 4\theta - 1}$$

$$= \frac{\left(\frac{1}{\cos 8\theta} - 1\right)}{\left(\frac{1}{\cos 4\theta} - 1\right)} = \frac{(1 - \cos 8\theta)}{(1 - \cos 4\theta)} \cdot \frac{\cos 4\theta}{\cos 8\theta}$$

$$= \frac{(2\sin^2 4\theta)(\cos 4\theta)}{(2\sin^2 2\theta)(\cos 8\theta)} = \frac{(2\sin 4\theta \cos 4\theta)(\sin 4\theta)}{(2\sin^2 2\theta)(\cos 8\theta)}$$

$$= \frac{(\sin 8\theta)(2\sin 2\theta \cos 2\theta)}{(\cos 8\theta)(2\sin^2 2\theta)} = \tan 8\theta \cot 2\theta = \frac{\tan 8\theta}{\tan 2\theta}$$

$$= \text{R.H.S.}$$

Example 62. *Show that* $\sqrt{2 + \sqrt{2 + 2\cos 4\theta}} = 2\cos\theta$.

Sol.

$$\text{L.H.S.} = \sqrt{2 + \sqrt{2 + 2\cos 4\theta}}$$

$$= \sqrt{2 + \sqrt{2(1 + \cos 4\theta)}} = \sqrt{2 + \sqrt{4\cos^2 2\theta}}$$

$$= \sqrt{2 + 2\cos 2\theta} = \sqrt{2(1 + \cos 2\theta)}$$

$$= \sqrt{4\cos^2\theta} = 2\cos\theta = \text{R.H.S.}$$

Example 63. *Prove that* $\cos 5x = 16\cos^5 x - 20\cos^3 x + 5\cos x$.

Sol. We have

$$\cos 5x = \cos(3x + 2x)$$

$$= \cos 3x \cos 2x - \sin 3x \sin 2x$$

$$= (4\cos^3 x - 3\cos x)(2\cos^2 x - 1) - (3\sin x - 4\sin^3 x)(2\sin x \cos x)$$

$$= (8\cos^5 x - 10\cos^3 x + 3\cos x) - 6\sin^2 x \cos x + 8\sin^4 \cos x$$

$$= (8\cos^5 x - 10\cos^3 x + 3\cos x) - 6(1 - \cos^2 x)\cos x + 8(1 - \cos^2 x)^2 . \cos x$$

$$= 16\cos^5 x - 20\cos^3 x + 5\cos x.$$

Hence, $\cos 5x = 16\cos^5 x - 20\cos^3 x + 5\cos x$.

Example 64. *Prove that* $\cos 6x = 32\cos^6 x - 48\cos^4 x + 18\cos^2 x - 1$.

Sol. We have $\cos 6x = \cos 2(3x)$

$$= \cos 2\theta, \quad \text{where } 3x = \theta$$

$$= 2\cos^2\theta - 1 = 2\cos^2 3x - 1$$

$$= 2(4\cos^3 x - 3\cos x)^2 - 1 \quad [\because \cos 3x = (4\cos^3 x - 3\cos x)]$$

$$= 2(9\cos^2 x + 16\cos^6 x - 24\cos^4 x) - 1$$

$$= 32\cos^6 x - 48\cos^4 x + 18\cos^2 x - 1$$

Hence, $\cos 6x = 32\cos^6 x - 48\cos^4 x + 18\cos^2 x - 1$.

Example 65. *Prove that* $\cot 2x \cot x - \cot 3x \cot 2x - \cot 3x \cot x = 1$.

Sol. We have $\cot 3x = \cot(2x + x)$

$$\Leftrightarrow \quad \cot 3x = \frac{\cot 2x \cot x - 1}{\cot 2x + \cot x}$$

$$\Leftrightarrow \quad \cot 3x \cot 2x + \cot 3x \cot x = \cot 2x \cot x - 1$$

$$\Leftrightarrow \cot 2x \cot x - \cot 3x \cot 2x - \cot 3x \cot x = 1$$

Hence, $\cot 2x \cot x - \cot 3x \cot 2x - \cot 3x \cot x = 1$

Example 66. *Find the value of*

(*i*) *sin 18°* (*ii*) *cos 18°*

(*iii*) *cos 36°* (*iv*) *sin 36°*

(*v*) *sin 72°* (*vi*) *cos 72°*

(*vii*) *sin 54°* (*viii*) *cos 54°*

Sol. (*i*) Let $\theta = 18°$. Then

$$\theta = 18° \Rightarrow 5\theta = 90°$$

$$\Rightarrow 2\theta = (90° - 3\theta)$$

$$\Rightarrow \sin 2\theta = \sin(90° - 3\theta) = \cos 3\theta$$

$$\Rightarrow 2\sin\theta\cos\theta = 4\cos^3\theta + 3\cos\theta$$

$$\Rightarrow 2\sin\theta\cos\theta - 4\cos^3\theta - 3\cos\theta = 0$$

$$\Rightarrow \cos\theta(2\sin\theta - 4\cos^2\theta + 3) = 0$$

$$\Rightarrow 2\sin\theta - 4\cos^2\theta + 3 = 0 \quad [\because \cos\theta = \cos 18° \neq 0]$$

$$\Rightarrow 2\sin\theta - 4(1 - \sin^2\theta) + 3 = 0$$

$$\Rightarrow 4\sin^2\theta + 2\sin\theta - 1 = 0$$

$$\Rightarrow \sin 18° = \frac{-2 \pm \sqrt{4+16}}{8} = \frac{(-1 \pm \sqrt{5})}{4}$$

$$\Rightarrow \sin 18\,i = \frac{\sqrt{5}-1}{4} \qquad [\because \sin\theta = \sin 18° > 0]$$

$$\therefore \qquad \sin 18° = \frac{\sqrt{5}-1}{4}$$

(*ii*) $\cos^2 18° = (1 - \sin^2 18°) = \left\{1 - \frac{(\sqrt{5}-1)^2}{16}\right\} = \left\{1 - \frac{(6-2\sqrt{5})}{16}\right\} = \frac{10+2\sqrt{5}}{16}$

$$\Leftrightarrow \qquad \cos 18° = \frac{\sqrt{10+2\sqrt{5}}}{4} \qquad [\because \cos 18° > 0]$$

(*iii*) $\cos 36° = (1 - 2\sin^2 18°) \quad = \left\{1 - 2.\frac{(\sqrt{5}-1)^2}{16}\right\} = \left\{1 - \frac{(6-2\sqrt{5})}{8}\right\} = \frac{\sqrt{5}+1}{4}$

(*iv*) $\sin 36° = \sqrt{1-\cos^2 36°} \quad = \left\{1 - \frac{(\sqrt{5}+1)^2}{16}\right\}^{1/2} = \left\{\frac{10-2\sqrt{5})}{16}\right\}^{1/2} = \frac{\sqrt{10-2\sqrt{5}}}{4}$

(*v*) $\sin 72° = \sin(90° - 18°) = \cos 18° = \frac{\sqrt{10+2\sqrt{5}}}{4}$

(*vi*) $\cos 72° = \cos(90° - 18°) = \sin 18° = \frac{(\sqrt{5}-1)}{4}$

(*vii*) $\sin 54° = \sin(90° - 36°) = \cos 36° = \frac{(\sqrt{5}+1)}{4}$

(*viii*) $\cos 54° = \cos(90° - 36°) = \sin 36° = \frac{\sqrt{10-2\sqrt{5}}}{4}$.

EXERCISE 1.1.2.

1. Prove that $\frac{\cos(\pi+\theta)\cos(-\theta)}{\cos(\pi-\theta)\cos\left(\frac{\pi}{2}+\theta\right)} = -\cot\theta$

2. Prove that (*i*) $\sin 75° = \frac{(\sqrt{6}+\sqrt{2})}{4}$, (*ii*) $\frac{\cos 135° - \cos 120°}{\cos 135° + \cos 120°} = (3 - 2\sqrt{2})$,

(*iii*) $\tan 15° + \cot 15° = 4$.

3. Prove that $\frac{\tan\left(\frac{\pi}{4}+\pi\right)}{\tan\left(\frac{\pi}{4}-\pi\right)} = \left(\frac{1+\tan x}{1-\tan x}\right)^2$.

4. Prove that $\frac{\cos\theta}{\sin(90°-\theta)} + \frac{\sin(-\theta)}{\sin(180°+\theta)} - \frac{\tan(90°+\theta)}{\cot\theta} = 3$

5. Prove that (*i*) $\frac{\tan 69° + \tan 66°}{1 - \tan 69° \tan 66°} = -1$, (*ii*) $\cot 105° - \tan 105° = 2\sqrt{3}$, (*iii*) $\cos 15° - \sin 15° = \frac{1}{\sqrt{2}}$.

6. Prove that $\frac{\sin(180° + \theta)\cos(90° + \theta)\tan(270° - \theta)\cot(360° - \theta)}{\sin(360° - \theta)\cos(360° + \theta)\operatorname{cosec}(-\theta)\sin(270° + \theta)} = 1.$

7. Prove that $\frac{\cos 9° + \sin 9°}{\cos 9° - \sin 9°} = \tan 54°$

8. Prove that $\frac{\cos 8° - \sin 8°}{\cos 8° - \sin 8°} = \tan 37°$.

9. If θ and ϕ lie in the first quadrant such that $\sin\theta = \frac{8}{17}$ and $\cos\phi = \frac{12}{13}$, find the values of

(*i*) $\sin(\theta - \phi)$, (*ii*) $\cos(\theta + \phi)$, (*iii*) $\tan(\theta - \phi)$

$\left[\textbf{Ans.}(i)\frac{21}{221}, (ii)\frac{140}{221}, (iii)\frac{21}{220}\right]$

10. If x and y are acute angles such that $\cos x = \frac{13}{14}$ and $\cos y = \frac{1}{7}$, prove that $(x - y) = -\frac{\pi}{3}$.

11. If x and y are acute angles such that $\sin x = \frac{1}{\sqrt{5}}$ and $\sin y = \frac{1}{\sqrt{10}}$, prove that $(x + y) = \frac{\pi}{4}$.

12. Prove that (*i*) $2\sin\frac{5\pi}{12}\sin\frac{\pi}{12} = \frac{1}{2}$, (*ii*) $2\cos\frac{5\pi}{12}\cos\frac{\pi}{12} = \frac{1}{2}$, (*iii*) $2\sin\frac{5\pi}{12}\cos\frac{\pi}{12} = \frac{(2+\sqrt{3})}{2}$.

13. Prove that (*i*) $\cos\left(\frac{\pi}{3} + x\right) = \frac{1}{2}(\cos x - \sqrt{3}\sin x)$

(*ii*) $\sin\left(\frac{\pi}{4} + x\right) + \sin\left(\frac{\pi}{4} - x\right) = \sqrt{2}\cos x$ (*iii*) $\frac{1}{\sqrt{2}}\cos\left(\frac{\pi}{4} + x\right) = \frac{1}{2}(\cos x - \sin x)$

(*iv*) $\cos x + \cos\left(\frac{2\pi}{3} + x\right) + \cos\left(\frac{2\pi}{3} - x\right) = 0$

14. If $\cos x = \frac{3}{5}$ and $\cos y = \frac{-24}{25}$, where $\frac{3\pi}{2} < x < 2\pi$ and $x < y < \frac{3\pi}{2}$, find the values of

(*i*) $\sin(x + y)$, (*ii*) $\cos(x - y)$, (*iii*) $\tan(x + y)$. [**Ans.** (*i*) $\frac{3}{5}$ (*ii*) $\frac{-44}{125}$ (*iii*) $\frac{-3}{4}$]

15. If $\sin x = \frac{12}{13}$ and $\sin y \frac{4}{5}$, where $\frac{\pi}{2} < x < \pi$ and $0 < y < \frac{\pi}{2}$, in the values of

(*i*) $\sin(x+y)$, (*ii*) $\cos(x+y)$, (*iii*) $\tan(x-y)$. **[Ans.** (i) $\frac{16}{65}$ (*ii*) $\frac{-63}{65}$ (*iii*) $\frac{56}{33}$]

16. Prove that $\cos 130° \cos 40° + \sin 130° \sin 40° = 0$.

17. Prove that $\sin 40° \cos 20° + \cos 40° \sin 20° = \sqrt{3}/2$.

18. Prove that $\cos 75° \cos 15° + \sin 75° \sin 15° = 1/2$.

19. Prove that $\cos 45° \cos 15° - \sin 45° \sin 15° = 1/2$.

20. Prove that $\sin 80° \cos 20° - \cos 80° \sin 20° = \sqrt{3}/2$.

21. Prove that $\cos(n+2)x \cos(n+1)x + \sin(n+2)x \sin(n+1)x = \cos x$.

22. Prove that $\cos\left(\frac{\pi}{4} - x\right)\cos\left(\frac{\pi}{4} - y\right) - \sin\left(\frac{\pi}{4} - x\right)\sin\left(\frac{\pi}{4} - y\right) = \sin(x-y)$.

23. Prove that $\cos(70° + \theta)\cos(10° + \theta) + \sin(70° + \theta)\sin(10° + \theta) = 1/2$.

24. Prove that $\sin(50° + \theta)\cos(20° + \theta) - \cos(50° + \theta)\sin(20° + \theta) = 1/2$.

25. Find the value of (*i*) $\cos(495°)$, (*ii*) $\sin(-1230°)$, (*iii*) $\cot(-315°)$, (*iv*) $\tan(-225°)$, (*v*) $\operatorname{cosec}(-690°)$, (*vi*) $\sec(-420°)$, (*vii*) $\tan(-120°)$, (*viii*) $\sin(870°)$, (*ix*) $\cos(840°)$,

[Ans. (*i*) $-\frac{1}{\sqrt{2}}$, (*ii*) $-\frac{1}{2}$, (*iii*) 1, (*iv*) 1, (*v*) 2, (*vi*) 2, (*vii*) $\sqrt{3}$, (*viii*) $\frac{1}{2}$, (*ix*) $-\frac{1}{2}$]

26. Show that

(*a*) $\frac{\sin^3\alpha - \cos^3\alpha}{\sin\alpha - \cos\alpha} = 1 + \sin\alpha\cos\alpha$

(*b*) $\frac{1-\cos\alpha}{1+\cos\alpha} = \left(\frac{\sin\alpha}{1+\cos\alpha}\right)^2$

(*c*) $\sin^2\alpha = 1/(\cot^2\alpha + 1)$

(*d*) $\cos^2\alpha = 1/(1 + \tan^2\alpha)$

(*e*) $\frac{\tan x}{\sin x} = \frac{1}{\cos x}$

(*f*) $\cos^2\alpha - \sin^2\alpha = 2\cos^2\alpha - 1$

(*g*) $\frac{\sin\alpha}{1+\cos\alpha} = \frac{1-\cos\alpha}{\sin\alpha}$,

(*h*) $(\sin^2\alpha + 2\cot^2\alpha - 1)/\cot^2\alpha = \sin^2\alpha$.

27. For what value of α is the value of the expression $(\sin\alpha + \cos\alpha)^2 + (\sin\alpha - \cos\alpha)^2$ as large as possible.

28. If $\tan\alpha = 2/5$, and a, b, c, d are arbitrary rational numbers, with $5c + 2d \neq 0$, show that $\frac{a\sin\alpha + b\cos\alpha}{c\cos\alpha + d\sin\alpha}$ is a rational number.

29. (*a*) If $\tan\alpha = 2/5$, find the numerical value of $\cos^2\alpha - \sin^2\alpha$.

(*b*) If $\tan\alpha = r$, write an expression in terms of r that represents the value of $\cos^2\alpha - \sin^2\alpha$.

30. If tan $\alpha = 2/5$, find the numerical value of $\dfrac{\sin\alpha - 2\cos\alpha}{\cos\alpha - 3\sin\alpha}$

31. If tan $\alpha = 2/5$, find the numerical value of 2 sin α cos 2 α.

32. (*a*) For which angles α is $\sin^4\alpha - \cos^4\alpha > \sin^2\alpha - \cos^2\alpha$?

(*b*) For which angles α is $\sin^4\alpha - \cos^4\alpha \geq \sin^2\alpha - \cos^2\alpha$?

33. Check the identity $\sin^2\alpha - \sin^2\beta = (\sin\alpha + \sin\beta)(\sin\alpha - \sin\beta)$ in using $\alpha = 30°$, $\beta = 60°$, on your calculator. You will find that, despite Pheobe's specious reasoning, the identity is true for these values. Is this a coincidence?

34. Use the law of cosines and the figure drawn for the proof to give a direct derivation of the formula for cos ($\alpha + \beta$).

35. For any acute angles α and β for which $\cos\alpha \cos\beta \neq 0$, show that $\dfrac{\cos(\alpha+\beta)}{\cos\alpha\cos\beta} = 1 - \tan\alpha$ $\tan\beta$.

36. Prove that $\sin^2\alpha - \sin^2\beta = \sin(\alpha + \beta)\sin(\alpha - \beta)$.

37. Prove that $\cos^2\beta - \cos^2\alpha = \sin(\alpha + \beta)\sin(\alpha - \beta)$.

38. Without using your calculator, find the numerical value of sin 18° cos 12° + cos 18° sin 12°.

39. (*a*) Without using your calcular, try to find the numerical value of sin 113° cos 307° + cos 113° sin 307°.

(*b*) Now use your calculator to check the result.

(*c*) Did you see the addition formulas in part (*a*)? Remember that we have proved the addition formulas only for positive acute angles. Doesn't it look like work for larger angles as well.

40. Simplify the expression $\sin 2\alpha \cos\alpha - \cos 2\alpha \sin\alpha$.

41. Simplify the expression $\sin(\alpha + \beta)\sin\beta + \cos(\alpha + \beta)\cos\beta$.

42. Simplify the expression $\dfrac{\sin(\alpha+\beta) - \cos\alpha\sin\beta}{\cos(\alpha+\beta) + \sin\alpha\sin\beta}$.

43. For any anlge $\alpha < \pi/4$, show that $\sin\left(\alpha + \dfrac{\pi}{4}\right) = \dfrac{\sqrt{2}}{2}(\sin\alpha + \cos\alpha)$.

44. We know that $\cos\dfrac{\pi}{6} = \dfrac{\sqrt{3}}{2}$. Show that :

(*a*) $\cos\dfrac{\pi}{12} = \dfrac{1}{2}\sqrt{2+\sqrt{3}}$ (*b*) $\cos\dfrac{\pi}{24} = \dfrac{1}{2}\sqrt{2+\sqrt{2+\sqrt{3}}}$

(*c*) $\cos\dfrac{\pi}{48} = \dfrac{1}{2}\sqrt{2+\sqrt{2+\sqrt{2+\sqrt{3}}}}$. (*d*) $\cos\dfrac{\pi}{96} = \dfrac{1}{2}\sqrt{2+\sqrt{2+\sqrt{2+\sqrt{2+\sqrt{3}}}}}$.

By evaluating these expressions (with a calculator or computer), observe that they are approaching 1. Can you explain why?

45. We can find another approximation to π by finding nested radical expressions for sin π/12, sin π/24, sin π/48, sin π/96, etc. Using a calculator or computer, find the values of the expressions.

(*a*) $12\sin\frac{\pi}{12} = 6\sqrt{2-\sqrt{3}}$. (*b*) $24\sin\frac{\pi}{24} = 12\sqrt{2-\sqrt{2+\sqrt{3}}}$.

(*c*) $48\sin\frac{\pi}{48} = 24\sqrt{2-\sqrt{2\sqrt{2+\sqrt{3}}}}$. (*d*) $96\cos\frac{\pi}{96} = 48\sqrt{2-\sqrt{2+\sqrt{2+\sqrt{2+\sqrt{3}}}}}$.

46. Show that cos 70° + sin 40° = cos 10°.

47. Find an acute angle α such that cos 55° + cos 65° = cosα.

48. Show that cos 20° + cos 100° + cos 140° = 0.

49. Show that sin 78° + cos 132° = sin 18°.

50. Show that $\frac{\cos 15° + \sin 15°}{\cos 15° - \sin 15°} = \sqrt{3}$.

51. If α + β + γ = π, show that

(*a*) sin (α + β) = sin γ (*b*) cos (α + β) = – cos γ

(*c*) sin 2α + sin 2β + sin 2γ = 2 sinα sinβ sinγ

52. For any angle α, show that, sinα + sin (α + 2π/3) + sin (α + 4π/3) = 0.

53. For any angle α, show that, sinα + 2sin 3α + sin5α = $4\cos^2\alpha$ sin 3α

54. For any three angles, α, β, γ, show that $\frac{\sin(\beta-\gamma)}{\sin\beta\sin\gamma} + \frac{\sin(\gamma-\alpha)}{\sin\gamma\sin\alpha} + \frac{\sin(\alpha-\beta)}{\sin\alpha\sin\beta} = 0$.

55. For any three angles α, β, γ, show that sin (α – β) + sin (α – γ) + sin (β – γ)

$$= 4\cos\frac{(\alpha-\beta)}{2}\sin\frac{(\alpha-\gamma)}{2}\cos\frac{(\beta-\gamma)}{2}.$$

56. For any three angles, α, β, γ, show that,

sin (α + β + γ) + sin (α – β – γ) + sin (α + β – γ) + sin (α – β + γ) = 4 sin α cos β cos γ

57. Prove that $\sin^2 72 - \cos^2 30° = \frac{\sqrt{5}-1}{8}$

58. If α + β = π/4. Proove that $\frac{(1+\tan\alpha)}{(1+\tan\beta)} = 2$.

59. Find the expression for tan (α + β + γ) which involves only tan α, tan β, and tan γ

60. Prove that tanα tan2α tan3α = tan3α – tan2α – tanα when ever all these expressions are defined. For what values of α are some of these expression not defined.

61. Prove that (*i*) 2cos 45° cos 15° = $\frac{\sqrt{3}+1}{2}$, (*ii*) 2sin 75° sin 15° = 1/2, (*iii*) cos 15° – sin 15° = $1/\sqrt{2}$

62. Prove that $2\cos 45° \cos 15° = \frac{\sqrt{3}+1}{2}$

63. Prove that $2 \sin 75° \sin 15° = 1/2$

64. Prove that $\cos 15° - \sin 15° = \frac{1}{\sqrt{2}}$

65. Prove that $\frac{\sin x + \sin 3x}{\cos x - \cos 3x} = \cot x$

66. Prove that $\frac{\sin 7x - \sin 5x}{\cos 7x + \cos 5x} = \tan x$

67. Prove that $(\sin 3x + \sin x) \sin x + (\cos 3x - \cos x) \cos x = 0$.

68. Prove that $\frac{\sin 2x - \sin 2y}{\cos 2y - \cos 2x} = \cot (x + y)$

69. Prove that $\cot 4x (\sin 5x + \sin 3x) = \cot x (\sin 5x - \sin 3x)$

70. Prove that $\frac{\sin x + \sin y}{\sin x - \sin y} = \tan\left(\frac{x+y}{2}\right)\cot\left(\frac{x-y}{2}\right)$

71. If $\cos x + \cos y = 1/3$ and $\sin x + \sin y = 1/4$, Prove that $\left(\frac{x+y}{2}\right) = \frac{3}{4}$

72. Express each of the following as product: (*i*) $\cos 2x - \cos 4x$ (*ii*) $\cos 7x + \cos 5x$ (*iii*) $\sin 7x - \sin 3x$ (*iv*) $\sin 10x + \sin 6x$

[**Ans.** (*i*) $2 \sin 3x \sin x$ (*ii*) $2\cos 6x \cos x$ (*iii*) $2\cos 5x \sin 2x$ (*iv*) $2 \sin 8x \cos 2x$]

73. If $\tan\theta = a/b$. Prove that $a\sin 2\theta + b\cos 2\theta = b$

74. $\sin^2 24° - \sin^2 6 = \frac{\sqrt{5}-1}{8}$.

75. Prove that $\sin^2 72° - \cos^2 30° = \frac{\sqrt{5}-1}{8}$

Prove the following questions:

76. $2 \sin 22½° - \cos 22½° = \frac{1}{\sqrt{2}}$

77. $\cos x \cos 2x \cos 4x \cos 8x = \frac{\sin 16x}{16 \sin x}$.

78. $(\cos^4 x + \sin^4 x) = 1/2\,(2 - \sin^2 2x)$

79. $\frac{\tan 2x}{1 + \sec 2x} = \tan x$.

80. $\sin 2x (\tan x + \cot x) = 2$.

81. $\dfrac{\cos^3 x - \sin^3 x}{\cos x - \sin x} = \dfrac{1}{2}(2 + \sin 2x).$

82. $\dfrac{\cos 2x}{\cos x - \sin x} = \cos x + \sin x$

83. Prove that $\sin \dfrac{\pi}{5} \sin \dfrac{2\pi}{5} \sin \dfrac{3\pi}{5} \sin \dfrac{4\pi}{5} = \dfrac{5}{16}$

84. If $\sin x = \dfrac{\sqrt{5}}{3}$ and $0 < x < \dfrac{\pi}{2}$, find the value of (*i*) sin $2x$ (*ii*) cos $2x$ (*iii*) tan $2x$

[**Ans.** (*i*) $\dfrac{4\sqrt{5}}{9}$, (*ii*) $\dfrac{-1}{9}$, (*iii*) $\dfrac{-4}{\sqrt{5}}$]

85. Prove that $(\cos^4 x + \sin^4 x) = \dfrac{1}{2}\ (2 \sin^2 2x)$.

86. Prove that cosec $2x$ + cot $2x$ = cot x

87. Prove that $3 \sin 40° - 4\sin^3 40° = \dfrac{\sqrt{3}}{2}$.

88. Prove that $\dfrac{\cos^3 x - \sin^3 x}{\cos x - \sin x} = \dfrac{1}{2}(2 + \sin 2x).$

89. Prove that $2 \cos^2 15° - 1 = \dfrac{\sqrt{3}}{2}$.

90. Calculate $\tan \dfrac{13\pi}{12}$ [**Ans.** $2 - \sqrt{3}$]

Unit—III

Complex Number

1.1. DEFINITION OF COMPLEX NUMBER

A complex number can be representd by an expression of the form $a + bi$, where a and b are real numbers and i is a symbol with the property that $i^2 = -1$. The complex number $a + bi$ can also be represented by the ordered pair (a, b) and plotted as a point in a plane (called the Argand plane) as in Figure 1. Thus, the complex number $i = 0 + 1 \cdot i$ is identified with the point (0, 1).

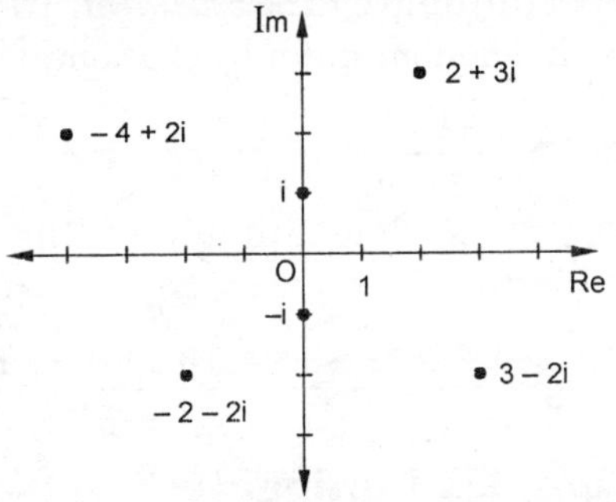

Fig. 1. Complex numbers as points in the Argand plane.

1.1.1. Real and Imaginary Parts

The **real part** of the complex number $a + bi$ is the real number a and the **imaginary part** is the real number b. Thus, the real part of $4 - 3i$ is 4 and the imaginary part is –3. Two complex number $a + bi$ and $c + di$ are equal if $a = c$ and $b = d$, that is, their real parts are equal and their imaginary parts are equal. In the Argand plane the horizontal axis is called the real axis and the vertical axis is called the imaginary axis.

The sum and difference of two complex numbers are defined by *adding* or *subtracting* their real parts and their imaginary parts:

$$(a + bi) + (c + di) = (a + c) + (b + d)i$$
$$(a + bi) - (c + di) = (a - c) + (b - d)i$$

For instance,

$$(1 - i) + (4 + 7i) = (1 + 4) + (-1 + 7)i = 5 + 6i$$

The product of complex numbers is defined so that the usual commutative and distributive laws hold:

$$(a + bi)(c + di) = a(c + di) + (bi)(c + di)$$

$$= ac + adi + bci + bdi^2$$

Since $i^2 = -1$, this becomes

$$(a + bi)(c + di) = (ac - bd) + (ad + bc)i$$

1.1.2. Polar and Cartesian Form and there Conversion

We call $a + bi$ the cartesian term of a complex number.

We know that any complex number $z = a + bi$ can be considered as a point (a, b) and that any such point can be represented by polar coordinates (r, θ) with $r \geq 0$. In fact,

$$a = r \cos\theta \quad \text{and} \quad b = r \sin\theta$$

as in Figure 2. Therefore, we have

$z = a + bi = (r \cos\theta) + (r \sin\theta)i$

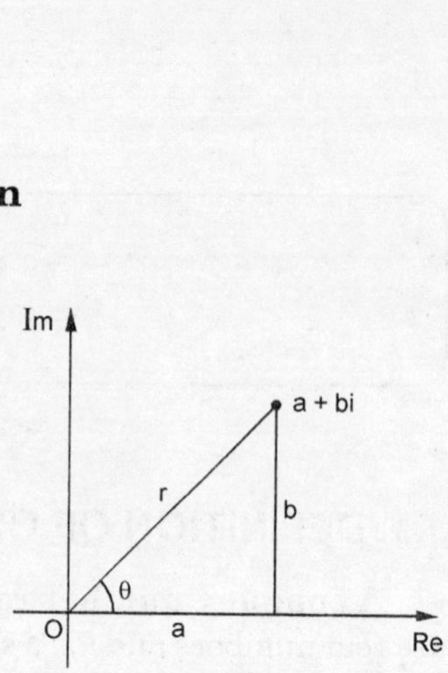

Fig. 2.

1.1.3. Complex Conjugate

Division of complex numbers is much like rationalizing the denominator of a rational expression. For the complex number $z = a + bi$, we define its **complex conjugate** to be $\bar{z} = a - bi$. To find the quotient of two complex number we multiply numerator and denominator by the complex conjugate of the denominator.

Example 1. *Express the number* $\dfrac{-1+3i}{2+5i}$ *in the form* $a + bi$.

Sol. We multiply numerator and denominator by the complex conjugate of $2 + 5i$, namely $2 - 5i$.

$$\frac{-1+3i}{2+5i} = \frac{-1+3i}{2+5i} \cdot \frac{2-5i}{2-5i} = \frac{13+11i}{2^2+5^2} = \frac{13}{29} + \frac{11}{29}i$$

- **Interpretation of the Complex Conjugate**

The geometric interpretation of the complex conjugate is shown in Fig. 3, $\bar{z}$ is the reflection of z in the real axis. We list some of the properties of the complex conjugate in the following box.

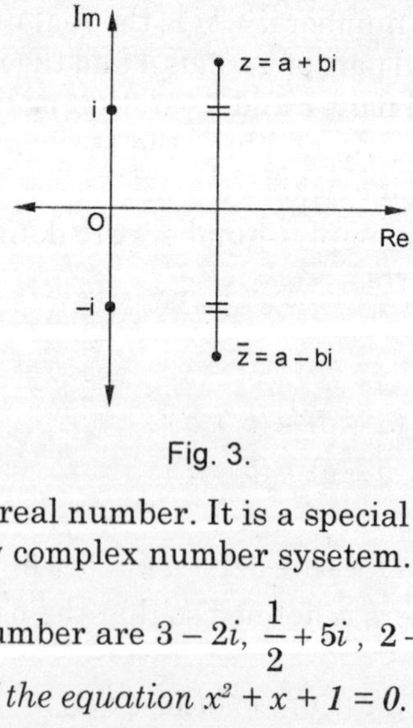

Fig. 3.

The imaginary unit i is not a real number. It is a special symbol *iota* used in the representation of the elements in this new complex number sysetem.

Some examples of complex number are $3 - 2i$, $\frac{1}{2} + 5i$, $2 - \frac{1}{3}i$, $0 + 3i$, $5 + 0i$, $0 + 0i$,

Example 2. *Find the roots of the equation* $x^2 + x + 1 = 0$.

Solution. Using the quadratic formula, we have

$$x = \frac{-1 \pm \sqrt{1^2 - 4 \cdot 1}}{2} = \frac{-1 \pm \sqrt{-3}}{2} = \frac{-1 \pm \sqrt{-3}i}{2}$$

We observe that the solutions of the equation in above example complex conjugates of each other. In general, the solutions of any quadratic equation $ax^2 + bx + c = 0$ with real coefficients a, b, and c are always complex conjugates. (If z is real, $\bar{z} = z$, so z is its own conjugate.)

We have seen that if we allow complex numbers as solutions, then every quadratic equation has solution. More generally, it is true that every polynomial equation

$$a_n x^n + a_{n-1} x^{n-1} + \ldots + a_1 x + a_0 = 0$$

of degree at least one has a solution among the complex numbers. This fact is know as the Fundamental Theorem of Algebra and was proved by Gauss.

Names for Particular Kinds of Complex Number

Standard Form	$a + bi$	
Imaginary Unit:	i	
Complex Number:	$a + bi$	a and b real numbers
Imaginary Number:	$a + bi$	$b \neq 0$
Pure Imaginary Number	$0 + bi = bi$	$b \neq 0$
Real Number:	$a + 0i = a$	
Zero:	$0 + 0i = 0$	
Conjugate of $a + bi$:	$a - bi$	

where a and b are real numbers and i is called the imaginary unit.

1.1.4. Modulus and Argument of a Complex Number

The **modulus** and **argument** of a complex numbers are quantities which can be recognised by looking at an Argand diagram. Recall that any complex number, z, can be represented by a point in the complex plane as shown in Figure 4.

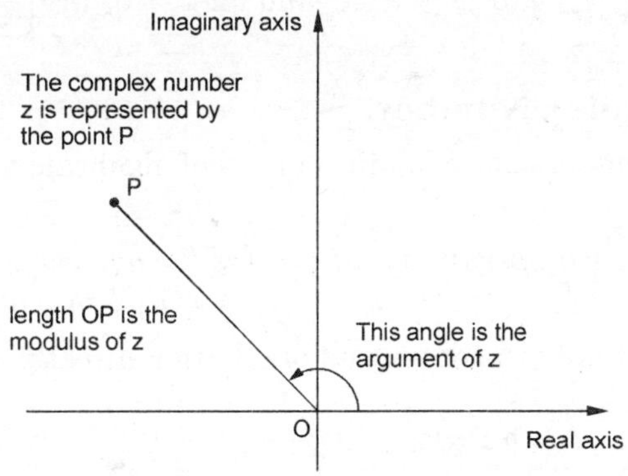

Fig. 4. The complex number z is represented by point P. Its modulus and argument are shown.

Join point P to the origin with a line segment. We associate with this line segment two important quantities. The length of the line segment, that is OP, is called the modulus of the complex number. The angle from the positive axis to the line segment is called the argument of the complex number, z.

• A complex number is written in the form

$$z = x + iy$$

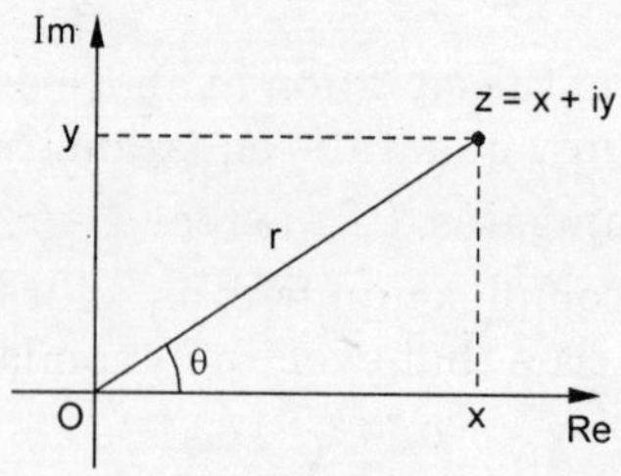

Fig. 5.

• The **modulus** of z is

$$|z| = r = \sqrt{x^2 + y^2}.$$

• The **argument** of z is

$$\arg z = \theta = \arctan\left(\frac{y}{x}\right)$$

When calculating θ, it must take account of the quadrant in which z lies-if in doubt draw an Argand diagram.

• The *Principle value of the argument* is denoted by Arg z, and is the unique value of arg z such that $-\pi < \arg z \leq \pi$. Arg z in obtained by adding or subtracting integer multiples of 2π from arg z.

• Writing a complex number in terms of **polar coordinates** r and θ:

$$z = x + iy = r\cos\theta + ir\sin\theta = r(\cos\theta + i\sin\theta) = re^{i\theta},$$

• For any two complex numbers z_1 and z_2

$$\arg(z_1 z_2) = \arg z_1 + \arg z_2 \text{ and for } z_2 \neq 0,\ \arg\left(\frac{z_1}{z_2}\right) = \arg z_1 + \arg z_2.$$

1.1.5. Addition of Complex Number.

Addition of complex numbers obeys addition laws of algebraic binomials.

Example 3. *Carry out each operation and express the answer in standard from:*

(i) $(2 - 3i) + (6 + 2i)$ *(ii)* $(-5 + 4i) + (0 + 0i)$

Solution. *(i)* We could apply the definition of addition directly, but it is easier to use complex number properties.

$$\begin{aligned}(2 - 3i) + (6 + 2i) &= 2 - 3i + 6 + 2i \\ &= (2 + 6) + (-3 + 2)i \\ &= 8 - i\end{aligned}$$

(ii) $(-5 + 4i) + (0 + 0i) = -5 + 4i$

1.1.6. Subtraction of Complex Numbers

A subtraction of complex numbers obeys subtraction laws as algebraic binomials.

Example 4. (*i*) $-4(-5i) = (-1)(4-5i) = -4+5i$

(*ii*) $(7-3i)-(6+2i) = 7-3i-6-2i = 1-5i$

(*iii*) $(-2+7i)+(2-7i) = -2+7i+2-7i = 0$

1.1.7. Multiplication of Complex Numbers

The multiplication of complex numbers obeys the same laws as multiplication of algebraic binomials, with the exception that $i^2 = -1$.

Example 5.

$$(2-3i)(4+2i) = 8+4i-12i-6i^2, \qquad (i^2=-1),$$
$$= 8+4i-12i+6,$$
$$= 14-8i.$$

1.1.8. Division of Complex Numbers

Division of complex numbers such as $\dfrac{2+3i}{3-4i}$ is performed by multiplying the numerator and the denominator by the conjugate of the denominator.

Example 6.

$$\frac{2+3i}{3-4i} = \frac{2+3i}{3-4i}\times\frac{3+4i}{3+4i},$$
$$= \frac{(2+3i)(3+4i)}{(3-4i)(3+4i)},$$
$$= \frac{6+8i+9i+12i^2}{9+12i-12i-16i^2},$$
$$= \frac{6-12+17i}{9+16}$$
$$= \frac{-6+17i}{25},$$

> **Do you know?**
>
> If z_1 and z_2 are two unrelated (not the conjugate of each other) complex numbers then the result of $z_1 \times z_2$ will always be a complex number. If, on the other hand, you multiply z_1 by its conjugate the result is a real number, *i.e.* $z_1 \times \bar{z}_1$ = real number.

Example 7. *$(-1+3i)(2-5i) = (-1)(2-5i)+3i(2-5i)$*

Sol. $= -2+5i+6i-15(-1) = 13+11i$

Example 8. *Find the modulus and argument of $z = 4+3i$.*

Sol. The complex number $z = 4+3i$ is shown in Figure 6. It has been represented by the point Q which has coordinates (4, 3). The modulus of z is the length of the line OQ which we can find Pythagoras theorem.

$$(OQ)^2 = 4^2+3^2 = 16+9 = 25$$

and hence $\quad OQ = 5.$

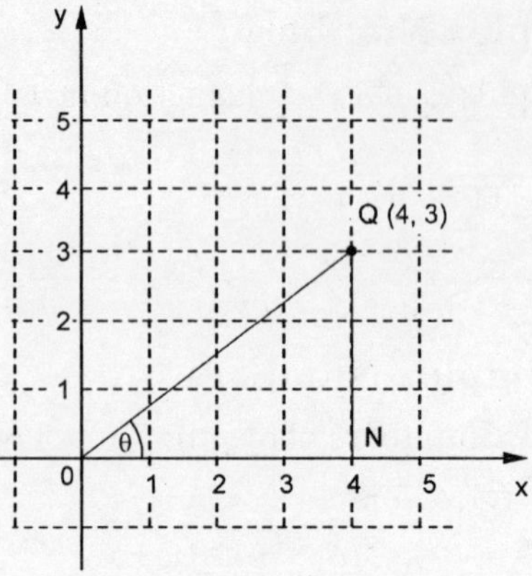

Fig. 6.

Hence the modulus of $z = 4 + 3i$ is 5. To find the argument we must calculate the angle between the x axis and the line segment OQ. We have labelled this θ in Figure 6.

By referring to the right-angled triangle OQN in Figure 6 we see that

$$\tan\theta = \frac{3}{4}$$

$$\theta = \tan^{-1}\frac{3}{4} = 36.97^\circ$$

To summarise, the modulus of $z = 4 + 3i$ is 5 and its argument is $\theta = 36.97^\circ$. There is a special symbol for the modulus of z; this is $|z|$. So, in this example, $|z| = 5$. We also have an abbreviation for argument: we write $\arg(z) = 36.97^\circ$.

Example 9. Find the modulus and argument of $z = 3 - 2i$.

Solution. The Argand diagram is shown in Figure 7. The point P with coordinates $(3, -2)$ represents $z = 3 - 2i$

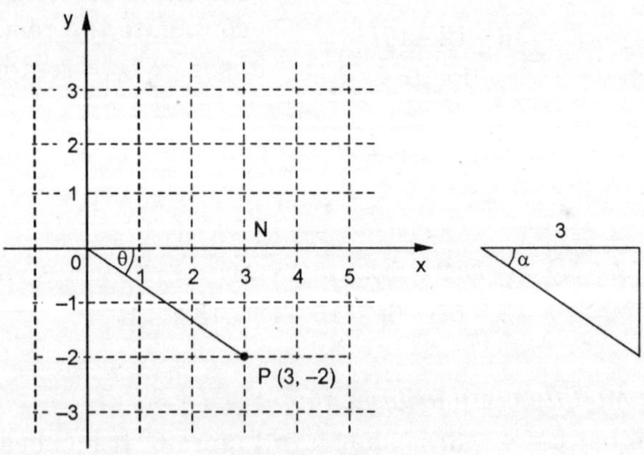

Fig. 7. The complex number $z = 3 - 2i$

We use Pythagoras' theorem in triangle ONP to find the modulus of z:

$$(OP)^2 = 3^2 + 2^2 = 13$$

$$OP = \sqrt{13}$$

Using the symbol for modulus, we see that in this example $|z| = \sqrt{13}$.

We must be more careful with the argument. When the angle θ shown in Figure 7 is measured in a clockwise sense convention dictates that the angle is negative. We can find the size of the angle by referring to the right-angled triangle shown. In that triangle $\tan \alpha = \frac{2}{3}$ so that $\alpha = \tan^{-1}\frac{2}{3} = 33.67°$. This is not the argument of z. The argument of z is $\theta = -33.67°$. This is not the argument of z. The argument of z is $\theta = -33.67°$, arg(z) = –33.67°

Example 10. *Calculate arg z and arg z where*

$$z = 1 - \sqrt{3}i.$$

Sol. Let $\alpha = \arctan\left|\frac{y}{x}\right|$, and use the Argand diagram to determine θ.

So $$\alpha = \arctan\left|\frac{-\sqrt{3}}{1}\right| = \pi/3,$$

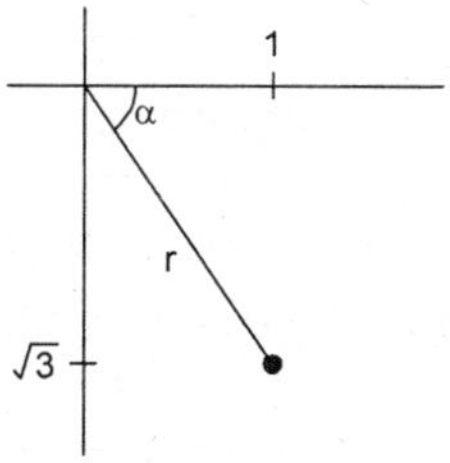

Fig. 8.

where the acute angle is chosen for α.

Here z lies in the 4th quadrant, therefore for any integer n

$$\arg z = \theta = 2n\pi - \pi/3,$$

but

$$\text{Arg } z = -\pi/3.$$

Example 11. *Write $z = 2 + 2i$ in the form $re^{i\theta}$, where $r, \theta \in \mathbb{R}$ with $r > 0$ and $\theta \in (-\pi, \pi]$.*

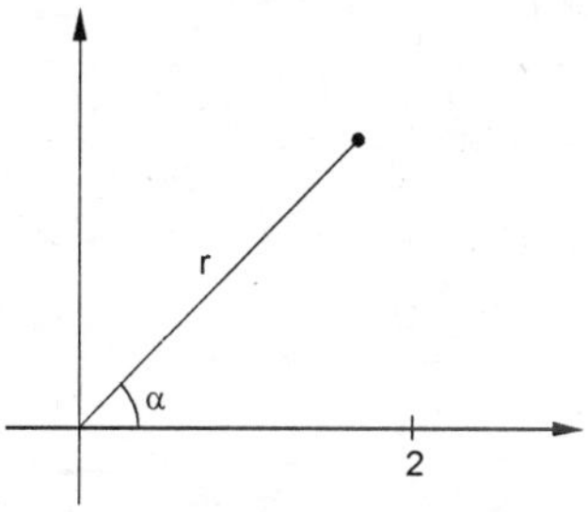

Fig. 9.

Sol. Here $x = 2$ and $y = 2$, so that $r = \sqrt{2^2 + 2^2} = \sqrt{8} = 2\sqrt{2}$ and

$$\alpha = \arctan\left|\frac{2}{2}\right| = \pi/4.$$

Here z lies in the 1st quadrant, therefore arg $z = \theta = \pi/4$; hence $z = 2\sqrt{2}e^{i\pi/4}$.

Example 12. *Write* $z = -2 + 2\sqrt{3}i$ *in the form* $re^{i\theta}$, *where* $r, \theta \in \mathbb{R}$ *with* $r > 0$ *and* $\theta \in (-\pi, \pi]$.

Sol. Here $x = -2$ and $y = 2\sqrt{3}$, so that $r = \sqrt{(-2)^2 + (2\sqrt{3})^2} = 4$ and

$$\alpha = \arctan\left|\frac{2\sqrt{3}}{-2}\right| = \pi/3.$$

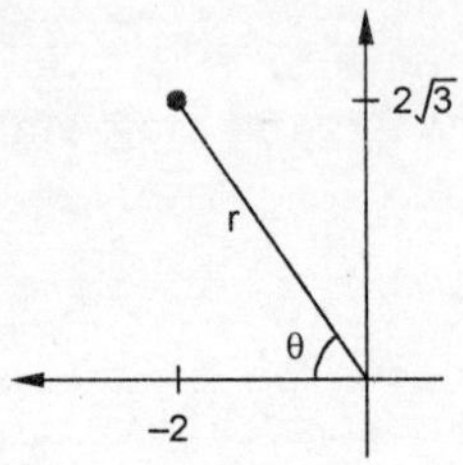

Fig. 10.

Here z lies in the 2nd quadrant, therefore arg $z = \theta = \pi - \pi/3 = 2\pi/3$: hence $z = 4e^{2i\pi/3}$.

Example 13. *Let* $arg(z_1) = 5\pi/6$ *and* $arg(z_2) = -\pi/3$. *Calculate (a) Arg* (z_1z_2).

(b) Arg$\left(\frac{z_1}{z_2}\right)$.

Sol. (*a*) $\arg(z_1z_2) = \arg z_1 + \arg z_2 = 5\pi/6 + (-\pi/3) = \pi/2 \Rightarrow \text{Arg}(z_1z_2) = \pi/2$.

(*b*) $\arg(z_1/z_2) = \arg z_1 - \arg z_2 = 5\pi/6 - (-\pi/3) = 7\pi/6 \Rightarrow \text{Arg}(z_1/z_2) = 7\pi/6 - 2\pi = -5\pi/6$

Example 14. *Given the list of complex numbers:* $3 - 2i$, $\frac{1}{2} + 5i$, $2 - \frac{1}{3}i$ $0 + 3i$, $5 + 0i = 5$, $0 + 0i = 0$

(i) List all the imaginary numbers, pure imaginary numbers, real numbers, and zero.

(ii) Write the conjutgate of each.

Sol. (*i*) Imaginary numbers: $3 - 2i$, $\frac{1}{2} + 5i$, $2 - \frac{1}{3}i$, $3i$

Pure imaginary numbers: $0 + 3i = 3i$

Real numbers: $5 + 0i = 5$, $= 0 + 0i = 0$

Zero: $0 + 0i = 0$

(*ii*) $3 + 2i$, $\frac{1}{2} - 5i$, $2 + \frac{1}{3}i$, $0 - 3i = -3i$, $5 - 0i = 5$, $0 - 0i = 0$

Example 15. *(– 5 + 4i) + (0 + 0i)*

Sol. $(-5 + 4i) + (0 + 0i) = -5 + 4i + 0 + 0i = -5 + 4i$

Example 16. *Carry out each operation and express the answer in standard form:*

(i) $(3 + 2i) + (6 - 4i)$ *(ii)* $(0 + 0i) + (7 - 5i)$

Sol. *(i)* $(3 + 2i) + (6 - 4i) = 3 + 2i + 6 - 4i = 9 - 2i$

(ii) $(0 + 0i) + (7 - 5i) = 0 + 0i + 7 - 5i = 7 + (-5i) = 7 - 5i$

Example 17. *Carry out each operation and express the answer in standard form:*

(a) $-(4 - 5i)$ *(b)* $(7 - 3i) - (6 + 2i)$

(c) $(-2 + 7i) + (2 - 7i)$

Sol.

(a) $-(4 - 5i) = (-1)(4 - 5i) = -4 + 5i$ *(b)* $(7 - 3i) - (6 + 2i) = 7 - 3i - 6 - 2i = 1 - 5i$

(c) $(-2 + 7i) + (2 - 7i) = -2 + 7i + 2 - 7i = 0$

Example 18. *Carry out each operation and express the answer in standard form:*

(a) $-(-3 + 2i)$ *(b)* $(3 - 5i) - (1 - 3i)$

(c) $(-4 + 9i) + (4 - 9i)$

Sol. *(a)* $-(-3 + 2i) = +3 - 2i$

(b) $(3 - 5i) - (1 - 3i) = 3 - 5i - 1 + 3i = 2 - 2i$

(c) $(-4 + 9i) + (4 - 9i) = -4 + 9i + 4 - 9i = 0$

Example 19. *Carry out each operation and express the answer in standard form:*

(a) $(2 - 3i)(6 + 2i)$ *(b)* $1(3 - 5i)$

(c) $i(1 + i)$ *(d)* $(3 + 4i)(3 - 4i)$

Solution. *(a)* $(2 - 3i)(6 + 2i) = 2(6 + 2i) - 3i(6 + 2i) = 12 + 4i - 18i - 6i^2$

$= 12 - 14i - 6(-1)$ Replace i^2 with -1.

$= 18 - 14i$

(b) $1(3 - 5i) = 1 \cdot 3 - 1 \cdot 5i = 3 - 5i$

(c) $i(1 + i) = i + i^2 = i - 1 = -1 + i$

(d) $(3 + 4i)(3 - 4i) = 9 - 12i + 12i - 16i^2 = 9 + 16 = 25$

Example 20. *Carry out each operation and express the answer in standard form:*

(a) $(5 + 2i)(4 - 3i)$ *(b)* $3(-2 + 6i)$

(c) $i(2 - 3i)$ *(d)* $(2 + 3i)(2 - 3i)$

Sol.

(a) $(5 + 2i)(4 - 3i) = 20 - 15i + 8i - 6i^2 = 20 - 15i + 8i + 6 = 26 - 7i$

(b) $3(-2 + 6i) = -6 + 18i$

(c) $i(2 - 3i) = 2i - 3i^2 = 2i - 3(-1) = 2i + 3 = 3 + 2i$

(d) $(2 + 3i)(2 - 3i) = (2)^2 - (3i)^2 = 4 - 9(-1) = 13$

Example 21. *Carry out each operation and express the answer in standard form:*

(a) $\dfrac{1}{2 + 3i}$ *(b)* $\dfrac{7 - 3i}{1 + i}$

Sol. *(a)* Multiply numberator and denominator by the conjugate of the denominator.

$$\frac{1}{2 + 3i} = \frac{1}{2 + 3i} \cdot \frac{2 - 3i}{2 - 3i} = \frac{2 - 3i}{4 - 9i^2} = \frac{2 - 3i}{4 + 9}$$

$$= \frac{2-3i}{13} = \frac{2}{13} - \frac{3}{13}i$$

This answer can be checked by multiplication:

Check $(2+3i)\left(\frac{2}{13}-\frac{3}{13}i\right) = \frac{4}{13} - \frac{6}{13}i + \frac{6}{13}i - \frac{9}{13}i^2$

$$= \frac{4}{13} + \frac{9}{13} = 1$$

(*b*) $\frac{7-3i}{1+i} = \frac{7-3i}{1+i}\cdot\frac{1-i}{1-i} = \frac{7-7i-3i+3i^2}{1-i^2} = \frac{4-10i}{2} = 2-5i$

check $(1+i)(2-5i) = 2 - 5i + 2i - 5i^2 = 7 - 3i$

Example 22. *Carry out each operation and express the answer in standard form:*

(*a*) $\frac{1}{4+2i}$ (*b*) $\frac{6+7i}{2-i}$

Sol. (*a*) $\frac{1}{4+2i} = \frac{1}{4+2i} \times \frac{4-2i}{4-2i} = \frac{4-2i}{16+4}$

$$= \frac{4}{20} - \frac{2i}{20} = \frac{1}{5} - \frac{i}{10}$$

(*b*) $\frac{6+7i}{2-i} = \frac{6+7i}{2-i} \times \frac{2+i}{2+i} = \frac{(6+7i)(2+i)}{4+1} = \frac{12+6i+14i+7i^2}{5} = \frac{12+20i-7}{5} = \frac{5+20i}{5}$

$$= 1 + 4i$$

Example 23. *Carry out the indicated operations and write each answer in standard from:*

(*a*) $(3 - 2i)^2 - 6(3 - 2i) + 13$ (*b*) $\frac{2-3i}{2i}$

Sol.

(*a*) $(3 - 2i)^2 - 6(3 - 2i) + 13 = 9 - 12i + 4i^2 - 18 + 12i + 13 = 9 - 12i - 4 - 18 + 12i + 13 = 0$

(*b*) If a complex number is divided by a pure imaginary number, we can make the denominator real by multiplying numerator and denominator by *i*.

$$\frac{2-3i}{2i}\cdot\frac{i}{i} = \frac{2i-3i^2}{2i^2} = \frac{2i+3}{-2} = -\frac{3}{2} - i$$

Example 24. *Carry out the indicated operations and write each answer in standard form:* $(3 + 2i)^2 - 6(3 + 2i) + 13$

Sol. $(3 + 2i)^2 - 6(3 + 2i) + 13 = 9 + (2i)^2 + 12i - 18 - 12i + 13 = 9 - 4 + 12i - 18 - 12i + 13$

$$= 5 - 18 + 12i - 12i + 13 = -3 + 12i - 12i + 3 = 0$$

Principal Square Root of a Negative Real Number

The principal square root of a negative real number, denoted by $\sqrt{-a}$ where a is positive, is defined by

$$\sqrt{-a} = i\sqrt{a}, \qquad \sqrt{-3} = i\sqrt{3} \quad \sqrt{-9} = i\sqrt{9} = 3i$$

The other square root of $-a$, $a > 0$, is $-\sqrt{-a} = -i\sqrt{a}$.

Now, we write $i\sqrt{a}$ and $i\sqrt{3}$ in place of the standard forms $\sqrt{a}i$ and $\sqrt{3}i$. We follow this convention whenever it appears that i might accidentally slip under a radical sign

$$(\sqrt{ai} \neq \sqrt{a}i, \quad \text{but} \quad \sqrt{a}i = i\sqrt{a}).$$

Thus, $\qquad (i\sqrt{a})^2 = i^2a = -a$

Example 25. *Write in standard form:*

(*i*) $\sqrt{-4}$ $\qquad$ (*ii*) $4 + \sqrt{-5}$

(*iii*) $\dfrac{-3-\sqrt{-5}}{2}$ $\qquad$ (*iv*) $\dfrac{1}{1-\sqrt{-9}}$

Sol.

(*i*) $\sqrt{-4} = i\sqrt{4} = 2i$ $\qquad$ (*ii*) $4 + \sqrt{-5} = 4 + i\sqrt{5}$

(*iii*) $\dfrac{-3-\sqrt{-5}}{2} = \dfrac{-3-i\sqrt{5}}{2} = -\dfrac{3}{2} - \dfrac{\sqrt{5}}{2}i$ $\quad$ (*iv*) $\dfrac{1}{1-\sqrt{-9}} = \dfrac{1}{1-3i} = \dfrac{1\cdot(1+3i)}{(1-3i)\cdot(1+3i)}$

$$= \frac{1+3i}{1-9i^2} = \frac{1+3i}{10} = \frac{1}{10} + \frac{3}{10}i$$

Example 26. *Write in standard form:*

(*i*) $\sqrt{-16}$ $\qquad$ (*ii*) $5 + \sqrt{-7}$

(*iii*) $\dfrac{-5-\sqrt{-2}}{2}$ $\qquad$ (*iv*) $\dfrac{1}{3-\sqrt{-4}}$

Solution. (*i*) $\sqrt{-16} = 4i$ $\qquad$ (*ii*) $5 + \sqrt{-7}$ Thus, $5 + i\sqrt{7}$

(*iii*) $\dfrac{-5-\sqrt{-2}}{2} = -\dfrac{5}{2} - \left(\dfrac{\sqrt{2}}{2}\right)i$ $\qquad$ (*iv*) $\dfrac{1}{3-\sqrt{-4}} = \dfrac{3}{13} + \dfrac{2}{13}i$

EXERCISE 1.1.1.

In problems 1-26, perform in standard form.

1. $(3 + 5i) + (2 + 4i)$ $\qquad$ **2.** $(4 + i) + (5 + 3i)$

3. $(8 - 3i) + (-5 + 6i)$ $\qquad$ **4.** $(-1 + 2i) + (4 - 7i)$

5. $(9 + 5i) - (6 + 2i)$ $\qquad$ **6.** $(3 + 7i) - (2 + 5i)$

7. $(3 - 4i) - (-5 + 6i)$ $\qquad$ **8.** $(-4 - 2i) - (1 + i)$

9. $2 + (3i + 5)$ $\qquad$ **10.** $(2i + 7) - 4i$

11. $(2i)(4i)$ $\qquad$ **12.** $(3i)(5i)$

13. $-2i(4 - 6i)$ $\qquad$ **14.** $(-4i)(2 - 3i)$

15. $(1 + 2i)(3 - 4i)$

16. $(2 - i)(-5 + 6i)$

17. $(3 - i)(4 + i)$

18. $(5 + 2i)(4 - 3i)$

19. $(2 + 9i)(2 - 9i)$

20. $(3 + 8i)(3 - 8i)$

21. $\dfrac{1}{2+4i}$

22. $\dfrac{i}{3+i}$

23. $\dfrac{4+3i}{1+2i}$

24. $\dfrac{3-5i}{2-i}$

25. $\dfrac{7+i}{2+i}$

26. $\dfrac{-5+10i}{3+4i}$

In problems 27-36, Express answers in standard form.

27. $(2-\sqrt{-4})+(5-\sqrt{-9})$

28. $(3-\sqrt{-4})+(-8+\sqrt{-25})$

29. $(9-\sqrt{-9})-(12-\sqrt{-25})$

30. $(-2-\sqrt{-36})-(4+\sqrt{-49})$

31. $(3-\sqrt{-4})(-2+\sqrt{-49})$

32. $(2-\sqrt{-1})(5+\sqrt{-9})$

33. $\dfrac{5-\sqrt{-4}}{7}$

34. $\dfrac{6-\sqrt{-64}}{2}$

35. $\dfrac{1}{2-\sqrt{-9}}$

36. $\dfrac{1}{3-\sqrt{-16}}$

Write problem 37-42 in standard form.

37. $\dfrac{2}{5i}$

38. $\dfrac{1}{3i}$

39. $\dfrac{1+3i}{2i}$

40. $\dfrac{2-i}{3i}$

41. $(2 - 3i)^2 - 2(2 - 3i) + 9$

42. $(2 - i)^2 + 3(2 - i) - 5$

43. Evaluate $x^2 - 2x + 2$ for $x = 1 - i$

44. Evaluate $x^2 - 2x + 2$ for $x = 1 + i$.

45. Simplify: i^{18}, i^{32} and i^{67}

46. Simplify: i^{21}, i^{43} and i^{52}

47. For what real values of x and y will the following equation be a true statement?
$(2x - 1) + (3y + 2)i = 5 - 4i$

48. For what real values of x and y will the following equation be a true statement?
$3x + (y - 2)i = (5 - 2x) + (3y - 8)i$

• Representation of Complex Number

We can write any complex number z in the form

$$\boxed{z = r(\cos\theta + i\sin\theta)}$$

where $r = |z| = \sqrt{a^2+b^2}$ and $\tan\theta = \dfrac{b}{a}$

The angle θ is called the **argunment** of z and we write $\theta = \arg(z)$. Note that $\arg(z)$ is not unique, any two arguments of z differ by an integer multiple of 2π.

The numbers z and w are shown in Figure 11.

The polar form of complex numbers gives insight into multiplication and division. Let

$$z_1 = r_1(\cos\theta_1 + i\sin\theta_1)$$
$$z_2 = r_2(\cos\theta_2 + i\sin\theta_2)$$

be two complex numbers written in polar form. Then

$$z_1z_2 = r_1r_2(\cos\theta_1 + i\sin\theta_1)(\cos\theta_2 + i\sin\theta_2)$$
$$= r_1r_2[(\cos\theta_1\cos\theta_2 - \sin\theta_1\sin\theta_2) + i(\sin\theta_1\cos\theta_2 + \cos\theta_1\sin\theta_2)]$$

Therefore, using the addition formulas for cosine and sine, we have

$$\boxed{z_1z_2 = r_1r_2[\cos(\theta_1+\theta_2) + i\sin(\theta_1+\theta_2)]}$$

This formula says that to multiply two complex numbers we multiply the moduli and add the arguments. (See Figure 11.)

A similar argument using the subtraction formulas for sine and cosine shows that to divide two complex numbers we divide the moduli and subtract the arguments.

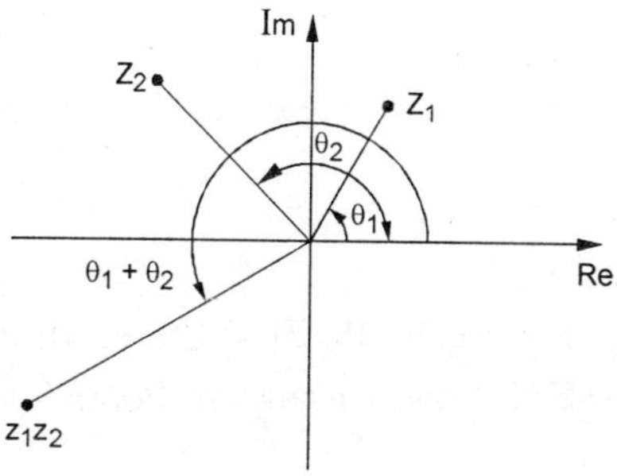

Fig. 11

$$\boxed{\frac{z_1}{z_2} = \frac{r_1}{r_2}[\cos(\theta_1-\theta_2) + i\sin(\theta_1-\theta_2)] \quad z_2 \neq 0}$$

In particular, taking $z_1 = 1$ and $z_2 = z$, (and therefore $\theta_1 = 0$ and $\theta_2 = 0$), we have the following, which is illustrated in Figure 12.

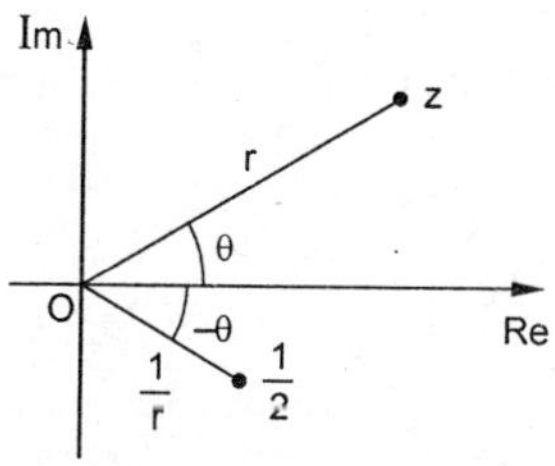

Fig. 12.

$$\boxed{\text{If } z = r(\cos\theta + i\sin\theta), \quad \textit{then } \frac{1}{z} = \frac{1}{r}(\cos\theta - i\sin\theta).}$$

Example 27. *Write the following numbers in polar form.*

(*a*) $z = 1 + i$ (*b*) $w = \sqrt{3} - i$

Sol. (*a*) We have $r = |z| = \sqrt{1^2+1^2} = \sqrt{2}$ and $\tan\theta = 1$, so we can take $\theta = \pi/4$. Therefore, the polar form is

$$z = \sqrt{2}\left(\cos\frac{\pi}{4} + i\sin\frac{\pi}{4}\right)$$

(*b*) Here we have $r = |w| = \sqrt{3+1} = 2$ and $\tan\theta = -1/\sqrt{3}$. Since w lies in the fourth quadrant, we take $\theta = -\pi/6$ and

$$w = 2\left[\cos\left(-\frac{\pi}{6}\right) + i\sin\left(-\frac{\pi}{6}\right)\right]$$

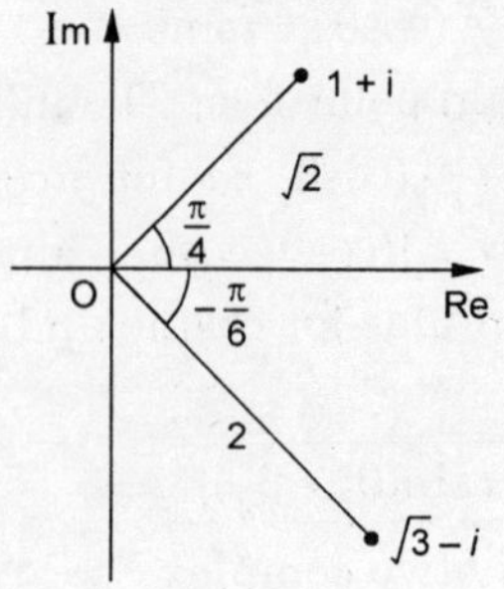

Fig. 13.

Example 28. Find the product of the complex numbers $1 + i$ and $\sqrt{3} - i$ in polar form.

Sol. Now, we take in Polar form

$$1 + i = \sqrt{2}\left(\cos\frac{\pi}{4} + i\sin\frac{\pi}{4}\right)$$

and $$\sqrt{3} - i = 2\left[\cos\left(-\frac{\pi}{6}\right) + i\sin\left(-\frac{\pi}{6}\right)\right]$$

so, $$(1 + i)(\sqrt{3} - i) = 2\sqrt{2}\left[\cos\left(\frac{\pi}{4} - \frac{\pi}{6}\right) + i\sin\left(\frac{\pi}{4} - \frac{\pi}{6}\right)\right]$$

$$= 2\sqrt{2}\left[\cos\frac{\pi}{12} + i\sin\frac{\pi}{12}\right]$$

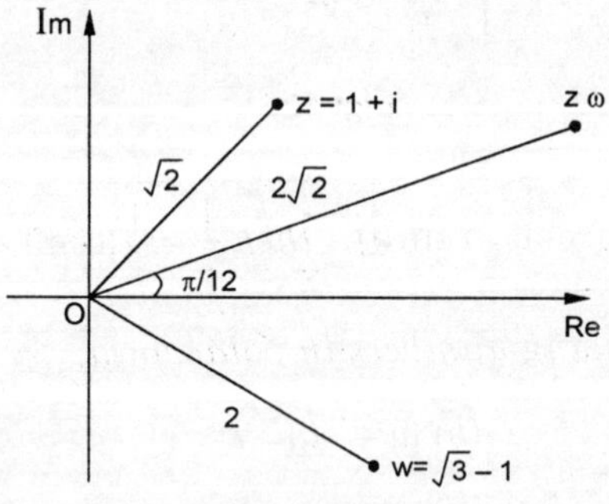

Fig. 14.

1.2. DeMOIVER'S THEOREM

Powers of a complex number.

$$z = r(\cos\theta + i\sin\theta)$$

then $$z^2 = r^2(\cos 2\theta + i\sin 2\theta)$$

and $$z^3 = zz^2 = r^3(\cos 3\theta + i\sin 3\theta)$$

In general, we obtain the followin result, which is named after the French mathematician Abraham De Movire (1667-1754).

If $z = r(\cos\theta + i\sin\theta)$ and n is a positive integer, then $z^n = [r\cos\theta + i\sin\theta)]^n = r^n(\cos n\theta + i\sin n\theta)$.

This says that to take the nth power of a complex number we take the nth power of the modulus and multiply the argument by n.

⇒ **Roots of a Complex Number** Let $z = r(\cos\theta + i\sin\theta)$ and let n be a positive integer. Then z has the n distinct nth roots

$$w_k = r^{1/n}\left[\cos\left(\frac{\theta + 2k\pi}{n}\right) + i\sin\left(\frac{\theta + 2k\pi}{n}\right)\right]$$

where $k = 0, 1, 2, ..., n - 1$.

Notice that each of the nth roots of z has modulus $|w_k| = r^{1/n}$. Thus, all the nth roots of z lie on the circle of radius $r^{1/n}$ in the complex plane. Also, since the argument of each successive nth root exceeds the argument of the previous root by $2\pi/n$, we see that the nth roots of z are equally spaced on this circle.

• Complex Exponentials

We also need to give a meaning to the expression e^z when $z = x + iy$ is a complex number. now,

$$e^z = \sum_{n=0}^{\infty}\frac{z^n}{n!} = 1 + z + \frac{z^2}{2!} + \frac{z^3}{3!} + \cdots$$

and it turns out that this complex exponential function has the same properties as the real exponential function. In particular, it is true that

$$e^{z1 + z2} = e^{z1}e^{z2}$$

If we put $z = iy$, where y is a real number.

$$i^2 = -1,\ i^3 = i^2 i = -i,\ i^4 = 1,\ i^5 = i, ...$$

we get,
$$e^{iy} = 1 + iy + \frac{(iy)^2}{2!} + \frac{(iy)^3}{3!} + \frac{(iy)^4}{4!} + \frac{(iy)^5}{5!} + \cdots$$

$$= 1 + iy - \frac{y^2}{2!} - i\frac{y^3}{3!} + \frac{y^4}{4!} + i\frac{y^5}{5!} + \cdots$$

$$= \left(1 - \frac{y^2}{2!} + \frac{y^4}{4!} - \frac{y^6}{6!} + \cdots\right) + i\left(y - \frac{y^3}{3!} + \frac{y^5}{5!} - \cdots\right)$$

$$= \cos y + i\sin y$$

Here we have used the Taylor series for $\cos y$ and $\sin y$. The result is a famous formula called **Euler's formula**:

$$e^{iy} = \cos y + i\sin y$$

Combining Euler's formula,

$$e^{x + iy} = e^x e^{iy} = e^x(\cos y + i\sin y)$$

Example 29. Find $\left(\frac{1}{2}+\frac{1}{2}i\right)^{10}$.

Sol. Since $\frac{1}{2}+\frac{1}{2}i = \frac{1}{2}(1+i)$, it follows, that $\frac{1}{2}+\frac{1}{2}i$ has the polar form $\frac{1}{2}+\frac{1}{2}i$ = $\frac{\sqrt{2}}{2}\left(\cos\frac{\pi}{4}+i\sin\frac{\pi}{4}\right)$

so by DeMoivre's Theorem,

$$\left(\frac{1}{2}+\frac{1}{2}i\right)^{10} = \left(\frac{\sqrt{2}}{2}\right)^{10}\left(\cos\frac{10\pi}{4}+i\sin\frac{10\pi}{4}\right)$$

$$= \frac{2^5}{2^{10}}\left(\cos\frac{5\pi}{2}+i\sin\frac{5\pi}{2}\right) = \frac{1}{32}i$$

DeMoivre's Theorem can also be used to find the *nth* roots of complex numbers. An *nth* root of the complex number z is a complex number ω such that

$$\omega^n = z$$

1.2.1. The Cube Roots of Unity

Let us assume that the cube root of 1 is z *i.e.*, $\sqrt[3]{1} = z$.

Then, cubing both sides we get, $z^3 = 1$

or, $z^3 - 1 = 0$

or, $(z-1)(z^2+z+1) = 0$

Therefore,either $z - 1 = 0$ *i.e.*, $z = 1$ or , $z^2 + z + 1 = 0$

So, $$z = \frac{-1\pm\sqrt{1^2-4\cdot1\cdot1}}{2\cdot1} = \frac{-1\pm\sqrt{-3}}{2} = -\frac{1}{2}\pm i\frac{\sqrt{3}}{2}$$

the three cube roots of unity are

$1,\ -\frac{1}{2}+i\frac{\sqrt{3}}{2}$ and $-\frac{1}{2}-i\frac{\sqrt{3}}{2}$

among them 1 is real number and the other two are conjugate complex numbers and they are also known as imaginary cube roots of unity.

• Properties of the Cube Roots of Unity

Property I: Among the three cube roots of unity one of the cube roots is real and the other two are conjugate complex numbers.

The three cube roots of unity are 1, $-\frac{1}{2}+i\frac{\sqrt{3}}{2}$ and $-\frac{1}{2}-i\frac{\sqrt{3}}{2}$.

Hence, we conclude that from the cube roots of unity we get 1 is real and the other two i.e., $\frac{1}{2}+i\frac{\sqrt{3}}{2}$ and $-\frac{1}{2}-i\frac{\sqrt{3}}{2}$ are conjugate complex numbers.

Property II: Square of any one imaginary cube root of unity is equal to the other imaginary cube root of unity.

$$\left(\frac{-1+\sqrt{3}i}{2}\right)^2 = \frac{1}{4}[(-1)\cdot 2 - 2\cdot 1\cdot \sqrt{3}i + (\sqrt{3}i)^2] = \frac{1}{4}[1 - 2\sqrt{3}i - 3] = \frac{-1-\sqrt{3}i}{2}$$

And
$$\left(\frac{-1-\sqrt{3}i}{2}\right)^2 = \frac{1}{4}[(1\cdot 2 + 2\cdot 1\cdot \sqrt{3}i + (\sqrt{3}i)^2)]$$

$$= \frac{1}{4}[1 + 2\sqrt{3}i - 3] = \frac{-1+\sqrt{3}i}{2},$$

Thus, we conclude that square of any cube root of unity is equal to the other.

Therefore, suppose ω^2 is one imaginary cube root of unity then the other would be ω.

Property III. The product of the two imaginary cube roots is 1 or, the product of three cube roots of unity is 1.

Let us assume that, $\omega = \dfrac{-1-\sqrt{3}i}{2}$; then , $\omega^2 = \dfrac{-1+\sqrt{3}i}{2}$

Therefore, the product of the two imaginary or complex cube roots $= \omega.\omega^2$

$$= \frac{-1-\sqrt{3}i}{2} \times \frac{-1+\sqrt{3}i}{2}$$

or,
$$\omega^3 = \frac{1}{4}[(-1)^2 - (\sqrt{3}i)^2] = \frac{1}{4}[1 - 3i^2] = \frac{1}{4}[1+3] = \frac{1}{4}\times 4 = 1.$$

Again, the cube roots of unity are $1, \omega, \omega^2$. So, product of cube roots of unity

$$= 1\cdot\omega\cdot\omega^2 = \omega^3 = 1.$$

Therefore, product of the three cube roots of unity is 1.

Property IV. $\omega^3 = 1$

We know that ω is a root of the equation $z^3 - 1 = 0$.

Therefore, ω satisfies the equation $z^3 - 1 = 0$.

Consequently, $\omega^3 - 1 = 0$

or, $\omega = 1$.

Note. Since $\omega^3 = 1$, hence, $\omega^n = \omega^m$, where m is the least non-negative remainder obtained by dividing n by 3.

Property V. The sum of the three cube roots of unity is zero *i.e.*, $1 + \omega + \omega^2 = 0$.

We know that, the sum of the three cube roots of unity $= 1 + \dfrac{-1-\sqrt{3}i}{2} + \dfrac{-1+\sqrt{3}i}{2}$

Or,
$$1 + \omega + \omega^2 = 1 - \frac{1}{2} + \frac{\sqrt{3}}{2}i - \frac{1}{2} - \frac{\sqrt{3}}{2}i = 0.$$

Application.

(*i*) The cube roots of 1 are $1 + \omega + \omega^2$ where, $= \dfrac{-1-\sqrt{3}i}{2}$ or, $\dfrac{-1+\sqrt{3}i}{2}$

(*ii*) $1 + \omega + \omega^2 = 0 \Rightarrow 1 + \omega = -\omega^2$, $1 + \omega^2 = -\omega$ and $\omega + \omega^2 = -1$

(*iii*) $\omega^4 = \omega^3 \cdot \omega = 1 \cdot \omega = \omega$;

$$\omega^5 = \omega^3 \cdot \omega^2 = 1 \cdot \omega^2 = \omega^2;$$
$$\omega^6 = (\omega^3)^2 = (1)^2 = 1.$$

In general, if n be a positive integer then,

$$\omega^{3n} = (\omega^3)^n = 1^n = 1;$$
$$\omega^{3n+1} = \omega^{3n} \cdot \omega = 1 \cdot \omega = \omega;$$
$$\omega^{3n+2} = \omega^{3n} \cdot \omega^2 = 1 \cdot \omega^2 = \omega^2.$$

Property VI. The reciprocal of each imaginary cube roots of units is the other.

The imaginary cube roots of unity are ω and ω^2, where $\omega = \dfrac{-1+\sqrt{3}i}{2}$.

Therefore, $\omega \cdot \omega^2 = \omega^3 = 1$

$$\Rightarrow \quad \omega = \frac{1}{\omega^2} \text{ and } \omega^2 = \frac{1}{\omega}$$

Hence, we conclude that the reciprocal of each imaginary cube roots of unity is the other.

Property VII. If ω and ω^2 are the roots of the equation $z^2 + z + 1 = 0$ then $-\omega$ and $-\omega^2$ are the roots of the equation $z^2 - z + 1 = 0$.

Property VIII. Cube roots of -1 are -1, $-\omega$ and $-\omega^2$.

1.2.2. *n*th Root of Unity

Let n is a positive integer then 'x' is said to be an nth root of unity if satisfies the equation $x^n = 1$. So the equation has n roots which also termed on the nth roots of unity.

Procedure To find n^{th} root of unity

It x is an nth roots of unity then it satisfied the relation $x^n = 1$

'1' can also written $\cos 0 + i \sin 0$

we have $x^n = 1$

$$= \cos 0 + i\sin 0$$
$$= \cos 2k\pi + i\sin 2k\pi \qquad [k \text{ is an integer}]$$

taking the nth root on both side

$$x = (\cos k\pi + i\sin k2\pi)^{1/n}$$
$$= \left\{\cos\left(\frac{2k\pi}{n}\right) + i\sin\left(\frac{2k\pi}{n}\right)\right\} \quad [k = 0, 1, 2, \ldots (n-1)]$$
$$x = \cos\left[\frac{2k\pi}{n}\right] + i\sin\left[\frac{2k\pi}{n}\right], \text{ where } 0 \le k \ldots\ldots n-1$$

The complex numbers are in the form $x + iy$ and plotted on the argand or the complex place.

Also since the roots of unity are in the form $\cos\left[\dfrac{2k\pi}{n}\right] + i\sin\left[\dfrac{2k\pi}{n}\right]$.

Now, we see the value of x and y satisfy the equation of unit circle with the centre *i.e.* $x^2 + y^2 = 1$

• Results of nth Roots of Unity

The cube roots of unity are 1, $\frac{(-1,+i\sqrt{3})}{2}, \frac{(-1-i\sqrt{3})}{2}$

- If ω is one of the imaginary cube root of unity then $1 + \omega + \omega^2 = 0$

  In general, $1 + \omega^n + \omega^{2n} = 0$; where $n \in I$ but is not the multiple of 3.
- In polor form the cube roots of unity are

$$\cos 0 + i\sin 0;\ \cos\frac{2\pi}{3} + i\sin\frac{2\pi}{3};\cos\frac{4\pi}{3} + i\sin\frac{4\pi}{3}$$

- Product of the nth roots of any complex number Z is $Z(-1)^{n-1}$

Three cube roots of unity when plotted on the argand plane constitute the vertices of an equilateral triangle.

The flollowing factorization should be remembered ($a, b, c \in R$ and ω is cube root of unity)

$$a^3 - b^3 = (a - b)\ (a - \omega b)(a - x^2b);$$
$$x^2 + x + 1 = (x - \omega)(x - \omega^2);$$
$$a^3 + b^3 = (a + b)(a + \omega b)\ (a + x^2b)$$
$$a^2 + ab + b^2 = (a - b\omega)(a - b\omega^2)$$
$$a^3 + b^3 + c^3 - 3abc = (a + b + c)(a + \omega b + \omega^2c)(a + \omega^2b + \omega c);$$

Let, 1, $a_1, a_2, \ldots a_{n-1}$, n^{th} roots of unity then

- They are in G.P with common ratio $i\left(\frac{2\pi}{n}\right)$

$$1P + a_1P + a_2P + \ldots a_{n-1}P = 0 \text{ if } P \text{ is not an integral multiple of } n$$
$$(i - a_1)(1 - a_2)\ldots(1 - a_{n-1}) = n \text{ and}$$
$$(1 + a_1)(1 + a_2)\ldots(1 + a_{n-1}) = 0 \text{ if } n \text{ is even and}$$
$$= 1 \text{ if } n \text{ is 0 odd.}$$

- $a_1, a_2, \ldots a_{n-1}$, 1 or –1 according as n is odd or even
- **Logarithm of a Complex Quantity**

(*i*) $\text{Log}_e(\alpha + i\beta) = \frac{1}{2}\log_e(\alpha^2 + \beta^2) + i\left(2n\pi + \tan^{-1}\left(\frac{\beta}{a}\right)\right)$ where $n \in I$.

(*ii*) it represents a set of positive real numbers given by $e^{-(2n\pi + \pi/2)}$, $n \in$ I.

Example 30. *Find the smallest positive integer n for which* $[(1 + i)/(1 - i)]^n = 1$

Sol. The given expression is $\left[\frac{(1+i)}{1-i}\right]^n = 1.$

First, we rationalize the given expresion by multiplying and dividing the given expression by the conjugate of denominator,

$$\left[\frac{(1+i)}{(1-i)}\cdot\frac{(1+i)}{(1+i)}\right]^n = 1$$

This means $\left(\frac{2i}{2}\right)^n = 1$

This gives $i^n = 1.$

The smallest positive integer n for which $R = 1$ is 4.

Hence, the value of n is 4.

Example 31. Let z_1 and z_2 be nth roots of unity which subtend a right angle at the origin, then what should be the form of n if k is an integer?

Sol. We know that by the formula of argument,

$$\text{org } \frac{z_1}{z_2} = \frac{\pi}{2}$$

Hence, $$\frac{z_1}{z_2} = \cos\frac{\pi}{2} + i\sin\frac{\pi}{2} = i$$

So, $$\frac{z_1{}^n}{z_2{}^n} = (i)^n = 1$$

Example 32. Find the six sixth roots of $z = -8$.

Sol. In trigonometric form, $z = 8(\cos \pi + i \sin \pi)$. Applying $n = 6$, we get

$$w_k = 8^{1/6}\left(\cos\frac{\pi + 2k\pi}{6} + i \sin\frac{\pi + 2k\pi}{6}\right)$$

We get the six sixth root of -8 by taking $k = 0, 1, 2, 3, 4, 5$ in this formula:

$$w_0 = 8^{1/6}\left(\cos\frac{\pi}{6} + i \sin\frac{\pi}{6}\right) = \sqrt{2}\left(\frac{\sqrt{3}}{2} + \frac{1}{2}i\right)$$

$$w_1 = 8^{1/6}\left(\cos\frac{\pi}{2} + i \sin\frac{\pi}{2}\right) = \sqrt{2}i$$

$$w_2 = 8^{1/6}\left(\cos\frac{5\pi}{6} + i \sin\frac{5\pi}{6}\right) = \sqrt{2}\left(-\frac{\sqrt{3}}{2} + \frac{1}{2}i\right)$$

$$w_3 = 8^{1/6}\left(\cos\frac{7\pi}{6} + i \sin\frac{7\pi}{6}\right) = \sqrt{2}\left(-\frac{\sqrt{3}}{2} - \frac{1}{2}i\right)$$

$$w_4 = 8^{1/6}\left(\cos\frac{3\pi}{2} + i \sin\frac{3\pi}{2}\right) = -\sqrt{2}i$$

$$w_5 = 8^{1/6}\left(\cos\frac{11\pi}{6} + i \sin\frac{11\pi}{6}\right) = \sqrt{2}\left(\frac{\sqrt{3}}{2} - \frac{1}{2}i\right)$$

Example 33. *Evaluate : (a)* $e^{i\pi}$ *(b)* $e^{-1 + i\pi/2}$

Sol.

(*a*) From Euler's equation, we have

$$e^{i\pi} = \cos \pi + i \sin \pi = -1 + i(0) = -1$$

(*b*) Using Euler's Equation we get

$$e^{-1 + i\pi/2} = e^{-1}\left(\cos\frac{\pi}{2} + i \sin\frac{\pi}{2}\right) = \frac{1}{e}[0 + i(1)] = \frac{i}{e}$$

Finally, we note that Euler's equation provides us with an easier method of proving De Moivre's Theorem:

$$[r(\cos\theta + i\sin\theta)]^n = (re^{i\theta})^n = r^n e^{in\theta} = r^n(\cos n\theta + i\sin n\theta)$$

EXERCISES 1.1.2.

1–14 Evaluate the expression and write your answer in the form $a + bi$.

1. $(5 - 6i) + (3 + 2i)$
2. $\left(4 - \frac{1}{2}i\right) - \left(9 + \frac{5}{2}i\right)$
3. $(2 + 5i)(4 - i)$
4. $(1 - 2i)(8 - 3i)$
5. $\overline{12 + 7i}$
6. $\overline{2i\left(\frac{1}{2} - i\right)}$
7. $\dfrac{1 + 4i}{3 + 2i}$
8. $\dfrac{3 + 2i}{1 - 4i}$
9. $\dfrac{1}{1 + i}$
10. $\dfrac{3}{4 - 3i}$
11. i^3
12. i^{100}
13. $\sqrt{-25}$
14. $\sqrt{-3}\sqrt{-12}$

15–17 Find the complex conjugate and the modulus of the number.

15. $12 - 5i$
16. $-1 + 2\sqrt{2}i$
17. $-4i$
18. Prove the following properties of complex numbers.

    (a) $\overline{z + w} = \overline{z} + \overline{w}$ (b) $\overline{zw} = \overline{z}\,\overline{w}$

    (c) $\overline{z^n} = \overline{z}^n$, where n is a positive integer

    [**Hint:** Write $z = a + bi$, $w = c + di$.]

19–24 Find all solutions of the equation.

19. $4x^2 + 9 = 0$
20. $x^4 = 1$
21. $x^2 + 2x + 5 = 0$
22. $2x^2 - 2x + 1 = 0$
23. $z^2 + z + 2 = 0$
24. $z^2 + \frac{1}{2}z + \frac{1}{4} = 0$

25–28 Write the number in polar form with argument between 0 and 2π.

25. $-3 + 3i$
26. $1 - \sqrt{3}i$
27. $3 + 4i$
28. $8i$

29–32 Find polar forms for zw, z/w, and $1/z$ by first putting z and w into polar form.

29. $z = \sqrt{3} + i$, $w = 1 + \sqrt{3}i$
30. $z = 4\sqrt{3} - 4i$, $w = 8i$

31. $z = 2\sqrt{3} - 2i$, $w = -1 + i$ **32.** $z = 4(\sqrt{3} + i)$, $w = -3 - 3i$

33–36 Find the indicated power using De Moivre's Theorem.

33. $(1 + i)^{20}$ **34.** $(1 - \sqrt{3}i)^5$ **35.** $(2\sqrt{3} + 2i)^5$ **36.** $(1 - i)^8$

37–40 Find the indicated roots. Sketch the roots in the complex plane.

37. The eighth roots of 1 **38.** The fifth roots of 32

39. The cube roots of i **40.** The cube roots of $1 + i$

41–46 Write the number in the form $a + bi$.

41. $e^{i\pi/2}$ **42.** $r^{2\pi i}$ **43.** $e^{i\pi/3}$ **44.** $e^{-i\pi}$

45. $e^{2 + i\pi}$ **46.** $e^{\pi + i}$

47. Use De Moivre's Theorem with $n = 3$ to express $\cos 3\theta$ and $\sin 3\theta$ in terms of $\cos\theta$ and $\sin\theta$.

48. Use Euler's formula to prove the following formulas for $\cos x$ and $\sin x$:

$$\cos x = \frac{e^{ix} + e^{-ix}}{2} \quad \sin x = \frac{e^{ix} - e^{-ix}}{2i}$$

49. If $u(x) = f(x) + ig(x)$ is a complex-valued function of a real variable x and the real and imaginary parts $f(x)$ and $g(x)$ are differentiable fuctions of x, then the derivative of u is defined to be $u'(x) = f'(x) + ig'(x)$. Use this together to prove that if $F(x) = e^{rx}$, then $F'(x) = re^{rx}$ when $r = a + bi$ is a complex number.

50. (*a*) If u is a complex-valued function of a real variable, its indefinite integral $\int u(x)\,dx$ is an antiderivative of u. Evaluate

$$\int e^{(1+i)x}dx$$

(*b*) By considering the real and imaginary parts of the integral in part (*a*), evaluate the real integrals

$$\int e^x \cos x dx \text{ and } \int e^x \sin x dx$$

ANSWERS

1. $8 - 4i$ **3.** $13 + 18i$ **5.** $12 - 7i$ **7.** $\frac{11}{13} + \frac{10}{13}i$

9. $\frac{1}{2} - \frac{1}{2}i$ **11.** $-i$ **13.** $5i$ **15.** $12 + 5i$; 13

17. $4i$, 4 **19.** $\pm\frac{3}{2}i$ **21.** $-1 \pm 2i$ **23.** $-\frac{1}{2} \pm (\sqrt{7}/2)i$

25. $3\sqrt{2}[\cos(3\pi/4) + i\sin(3\pi/4)]$ **27.** $5\left\{\cos\left[\tan^{-1}\left(\frac{4}{3}\right)\right] + i\sin\left[\tan^{-1}\left(\frac{4}{3}\right)\right]\right\}$

29. $4[\cos(\pi/2) + i\sin(\pi/2)]$, $\cos(-\pi/6) + i\sin(-\pi/6)$, $\frac{1}{2}[\cos(-\pi/6) + i\sin(-\pi/6)]$

31. $4\sqrt{2}\left[\cos(7\pi/12) + i\sin(7\pi/12)\right]$

$(2\sqrt{2})\left[\cos(13\pi/12) + i\sin(13\pi/12)\right], \frac{1}{4}\left[\cos(\pi/6) + i\sin(\pi/6)\right]$

33. -1024

35. $-512\sqrt{3} + 512i$

37. $\pm 1, \pm i, (1/\sqrt{2})(\pm 1 \pm i)$

39. $\pm(\sqrt{3}/2) + \frac{1}{2}i, -i$

41. i

43. $\frac{1}{2} + (\sqrt{3}/2)i$

45. $-e^2$

47. $\cos 3\theta = \cos^3\theta - 3\cos\theta\sin^2\theta$, $\sin 3\theta = 3\cos^2\theta\sin\theta - \sin^3\theta$.

• Roots of a Complex Number

We have seen that $\cos\frac{\theta}{q} + i\sin\frac{\theta}{q}$ is one of the values of $(\cos\theta + i\sin\theta)^{1/q}$. Here we shall show that $(\cos\theta + i\sin\theta)^{1/q}$ has q different values.

Since $\cos\theta$ and $\sin\theta$ are periodic functions with period $2n\pi$ (n being an integer).

$$\therefore \quad \cos\theta + i\sin\theta = \cos(2n\pi + \theta) + i\sin(2n\pi + \theta)$$

$$(\cos\theta + i\sin\theta)^{1/q} = [\cos(2n\pi + \theta) + i\sin(2\pi n + \theta)]^{1/q}$$

$$= \cos\left(\frac{2n\pi + \theta}{q}\right) + i\sin\left(\frac{2\pi n + \theta}{q}\right) \quad ...(1)$$

Giving n the values 0, 1, 2, 3, ... $(q-1)$ successively, we get following q values of $(\cos + i\sin)^{1/q}$.

$$\cos\frac{\theta}{q} + i\sin\frac{\theta}{q} \quad (\text{for } n = 0)$$

$$\cos\frac{2\pi + \theta}{q} + i\sin\frac{2\pi + \theta}{q} \quad (\text{for } n = 1)$$

$$\cos\frac{4\pi + \theta}{q} + i\sin\frac{4\pi + \theta}{q} \quad (\text{for } n = 2)$$

... ...

... . ..

$$\cos\frac{2(q-1)\pi + \theta}{q} + i\sin\frac{2(q-1)\pi + \theta}{q} \quad (\text{for } n = q-1)$$

Each of the above value is a root of $(\cos\theta + i\sin\theta)^{1/q}$.

Note 1. If we give to n values greater than $q - 1$, no new roots are obtained but the same values will repeat. For example, when we put $n = q$ in (1), we get $(q + 1)$th root as

$$\cos\left(\frac{2q\pi + \theta}{q}\right) + i\sin\left(\frac{2q\pi + \theta}{q}\right)$$

$$= \cos\left(2\pi + \frac{\theta}{q}\right) + i\sin\left(2\pi + \frac{\theta}{q}\right)$$

$$= \cos\frac{\theta}{q} + i\sin\frac{\theta}{q},$$

Which is same as the value obtained by putting $n = 0$.

Note 2. The roots of $(\cos\theta + i\sin\theta)^{1/q}$ are in G.P., as shown below. The roots of $(\cos\theta + i\sin\theta)^{1/q}$ can be written as

$$\cos\left(\frac{2n\pi + \theta}{q}\right) + i\sin\left(\frac{2n\pi + \theta}{q}\right) \qquad \text{for } n = 0, 1, 2, 3, ..., q - 1.$$

Further,

$$\cos\left(\frac{2n\pi + \theta}{q}\right) + i\sin\left(\frac{2n\pi + \theta}{q}\right)$$

$$= \left(\cos\frac{2n\pi}{q} + i\sin\frac{2n\pi}{q}\right)\left(\cos\frac{\theta}{q} + i\sin\frac{\theta}{q}\right)$$

$$= \left(\cos\frac{2\pi}{q} + i\sin\frac{2\pi}{q}\right)^n\left(\cos\frac{\theta}{q} + i\sin\frac{\theta}{q}\right)$$

$$= \alpha^n \cdot \beta \qquad \text{for } n = 0, 1, 2, 3, ..., q - 1$$

where $\alpha = \cos\frac{2\pi}{q} + i\sin\frac{2\pi}{q}$

and $\beta = \cos\frac{\theta}{q} + i\sin\frac{\theta}{q}$.

Therefore, the roots of $(\cos\theta + i\sin\theta)^{1/q}$ can be written as $\beta, \alpha\beta, \alpha^2, ..., \alpha^{q-1}\beta$.

Thus we see the roots are in G.P. with common ratio

$$\alpha = \cos\frac{2\pi}{q} + i\sin\frac{2\pi}{q}.$$

Example 34. If $\cos\alpha + 2\cos\beta + 3\cos\gamma = 0$

$= \sin\alpha + 2\sin\beta + 2\sin\gamma$, prove that

$\cos 3\alpha + 8\cos 3\beta + 27\cos 3\gamma = 18\cos(\alpha + \beta + \gamma)$,

and $\sin 3\alpha + 8\sin 3\beta + 27\sin 3\gamma = 18\sin(\alpha + \beta + \gamma)$.

Sol. Let $a = \cos\alpha + i\sin\alpha$, $b = 2\cos\beta + 2i\sin\beta$, $c = 3\cos\gamma + 3i\sin\gamma$.

$$\therefore \quad a + b + c = (\cos\alpha + i\sin\alpha) + (2\cos\beta + 2i\sin\beta) + (3\cos\gamma + 3i\sin\gamma)$$

$$= (\cos\alpha + 2\cos\beta + 3\cos\beta) + i(\sin\alpha + 2\sin\beta + 3\sin\gamma) = 0$$

Since $\quad a + b + c = 0$

$$a^3 + b^3 + c^3 = 3abc.$$

i.e., $(\cos\alpha + i\sin\alpha)^3 + (2\cos\beta + 2i\sin\beta)^3 + (3\cos\gamma + 3i\sin\gamma)^3$

$$= 3(\cos\alpha + i\sin\alpha)(2\cos\beta + 2i\sin\beta)(3\cos\gamma + 3i\sin\gamma)$$

$\therefore$ $(\cos 3\alpha + i\sin 3\alpha) + (8\cos 3\beta + 8i\sin 3\beta) + (27\cos 3\gamma + 27i\sin 3\gamma)$

$$= 18[\cos(\alpha+\beta+\gamma) + i\sin(\alpha+\beta+\gamma)]$$

or $(\cos 3\alpha + 8\cos 3\beta + 27\cos 3\gamma) + i(\sin 3\alpha + 8\sin 3\beta + 27\cos 3\gamma)$

$$= 18[\cos(\alpha+\beta+\gamma) + i\sin(\alpha+\beta+\gamma)]$$

Equating real and imaginary parts, we get

$$\cos 3\alpha + 8\cos 3\beta + 27\cos 3\gamma = 18\cos(\alpha+\beta+\gamma)$$

$$\sin 3\alpha + 8\sin 3\beta + 27\sin 3\gamma = 18\sin(\alpha+\beta+\gamma)$$

Example 35. *Find general value of θ, which satisfies the following equation, $(\cos\theta + i\sin\theta)(\cos 2\theta + i\sin 2\theta)\ldots(\cos n\theta + i\sin n\theta) = 1$.*

Sol. Here $(\cos\theta + i\sin\theta)(\cos 2\theta + i\sin 2\theta)\ldots(\cos n\theta + i\sin n\theta) = 1$.

or $(\cos\theta + i\sin\theta)^1 \cdot (\cos\theta + i\sin\theta)^2 \ldots (\cos\theta + i\sin\theta)^n$ [By De-Moivre's Theorem]

$$= \cos 2r\pi + i\sin 2r\pi \qquad [\because \cos 2r\pi + i\sin 2r\pi = 1]$$

or $(\cos\theta + i\sin\theta)^{1+2+3+\ldots+n} = \cos 2r\pi + i\sin 2r\pi$

$$(\cos\theta + i\sin\theta)^{\frac{n(n+1)}{2}} = \cos 2r\pi + i\sin 2r\pi \qquad \left[\because \sum n = \frac{n(n+1)}{2}\right]$$

Applying De-Moivre's theorem, we get

$$\cos\frac{n(n+1)}{2}\theta + i\sin\frac{n(n+1)}{2}\theta = \cos 2r\pi + i\sin 2r\pi.$$

Thus we have

$$\frac{n(n+1)}{2}\theta = 2r\pi$$

$$\theta = \frac{4r\pi}{n(n+1)}$$

Example 36. *Find all the values of $(1)^{1/6}$*

Sol.
$$1 = \cos 0 + i\sin 0 = \cos(2n\pi) + i\sin(2n\pi)$$

$\therefore$
$$(1)^{1/6} = [\cos(2n\pi) + i\sin(2n\pi)]^{1/6}$$

$$= \cos\frac{n\pi}{3} + i\sin\frac{n\pi}{3} \qquad \ldots(1)$$

where $n = 0, 1, 2, 3, 4, 5$ [By De-Moivre's theorem]

Putting $n = 0, 1, 2, 3, 4, 5$ successively in (1), we get the required roots as follows,

(*i*) $\cos 0 + i\sin 0 = 1$ [for $n = 0$]

(*ii*) $\cos\frac{\pi}{3} + i\sin\frac{\pi}{3} = \frac{1}{2} + \frac{i\sqrt{3}}{2}$ [for $n = 1$]

(*iii*) $\cos\frac{2\pi}{3} + i\sin\frac{2\pi}{3} = -\frac{1}{2} + \frac{i\sqrt{3}}{2}$ [for $n = 2$]

(*iv*) $\cos\pi + i\sin\pi = -1$ [for $n = 3$]

(*v*) $\cos\frac{4\pi}{3} + i\sin\frac{4\pi}{3} = -\frac{1}{2} - \frac{i\sqrt{3}}{2}$ [for $n = 4$]

(*vi*) $\cos\frac{5\pi}{3} + i\sin\frac{5\pi}{3} = \frac{1}{2} - \frac{i\sqrt{3}}{2}$ [for $n = 5$]

Thus the roots are 1, –1, $\frac{1}{2} \pm \frac{i\sqrt{3}}{2}$, $-\frac{1}{2} \pm \frac{i\sqrt{3}}{2}$

Example 37. *Find all the values of* $(1+\sqrt{3}i)^{1/3}$.

Sol. Let
$$1 = r\cos\theta$$
$$\sqrt{3} = r\sin\theta$$
$$\therefore \quad r = 2$$
and
$$\theta = \frac{\pi}{3}$$
$$\therefore \quad 1+\sqrt{3}i = \left(\cos\frac{\pi}{3} + i\sin\frac{\pi}{3}\right)$$
$$= 2\left[\cos\left(2n\pi + \frac{\pi}{3}\right) + i\sin\left(2n\pi + \frac{\pi}{3}\right)\right]$$
$$= 2\left[\cos(6n+1)\frac{\pi}{3} + i\sin(6n+1)\frac{\pi}{3}\right]$$
$$\therefore \quad (1+\sqrt{3}i)^{1/3} = (2)^{1/3}\left[\cos(6n+1)\frac{\pi}{3} + i\sin(6n+1)\frac{\pi}{3}\right]^{1/3}$$
$$= (2)^{1/3}\left[\cos(6n+1)\frac{\pi}{9} + i\sin(6n+1)\frac{\pi}{9}\right] \quad \text{for } n = 0, 1, 2$$

Hence the required values are,
$$(2)^{1/3}\left(\cos\frac{\pi}{9} + i\sin\frac{\pi}{9}\right), (2)^{1/3}\left(\cos\frac{7\pi}{9} + i\sin\frac{7\pi}{9}\right), (2)^{1/3}\left(\cos\frac{13\pi}{9} + i\sin\frac{13\pi}{9}\right)$$

Thus the required values are
$$(2)^{1/3}\left(\cos\frac{r\pi}{9} + i\sin\frac{r\pi}{9}\right), \text{ for } r = 1, 7, 13.$$

Example 38. *Prove that the roots of the equation*
$$x^{10} + 11x^5 - 1 = 0 \text{ are } \left(\frac{\pm\sqrt{5}-1}{2}\right)\left(\cos\frac{2n\pi}{5} + i\sin\frac{2n\pi}{5}\right).$$

where $n = 0, 1, 2, 3, 4$

Sol. Here
$$x^{10} + 11x^5 - 1 = 0 \quad \ldots(i)$$
Put $x^5 = t$ in (*i*),
$$\therefore \quad t^2 + 11t - 1 = 0$$
$$t = \frac{-11 \pm \sqrt{(11)^2 - 4\times 1(-1)}}{2} = \frac{-11 \pm 5\sqrt{5}}{2}$$

$\because \quad x^5 = \frac{-11 \pm 5\sqrt{5}}{2} = \frac{-176 \pm 80\sqrt{5}}{32}$

(Multiplying N^r, and D^r. by 16 so as to make the D^r fifth ower of 2.)

$$\therefore \quad x^5 = \left(\frac{-176 \pm 80\sqrt{5}}{32}\right)(\cos 2n\pi + i \sin 2n\pi)$$

$$\therefore \quad x^5 = \left(\frac{\pm\sqrt{5}-1}{2}\right)^5 (\cos 2n\pi + i \sin 2n\pi)$$

$$\therefore \quad x = \left(\frac{\pm\sqrt{5}-1}{2}\right)(\cos 2n\pi + i \sin 2n\pi)^{1/5}$$

or $$x = \left(\frac{\pm\sqrt{5}-1}{2}\right)\left(\cos\frac{2n\pi}{5} + i \sin\frac{2n\pi}{5}\right)$$

where $n = 0, 1, 2, 3, 4.$

Example 39. *Prove that* $\sqrt[n]{a+ib} + \sqrt[n]{a-ib}$ *has n real values and find those of* $\sqrt[3]{1+\sqrt{3}i} + \sqrt[3]{1-\sqrt{3}i}$.

Sol. Let $\quad a = r\cos\theta$ and $b = r\sin\theta$

$\therefore \quad r = (a^2 + b^2)^{1/2}$

and $\quad \theta = \tan^{-1}\frac{b}{a}$

Now $\quad (a+ib)^{1/n} + (a-ib)^{1/n} = [r(\cos\theta + i\sin\theta)]^{1/n} + [r(\cos\theta - i\sin\theta)]^{1/n}$

$= r^{1/n}[\cos(2k\pi + \theta) + i\sin(2k\pi + \theta)]^{1/n} + r^{1/n}[\cos(2k\pi + \theta) - i\sin(2k\pi + \theta)]^{1/n}$

[By De-Movire's theorem]

$$= 2r^{1/n}\cos\left(\frac{2k\pi + \theta}{n}\right) \quad \text{for } k = 0, 1, 2, ..., n-1.$$

Thus there are n real values of the given expression.

Putting $\quad a = 1, b = \sqrt{3}$ and $n = 3$,

We have $\quad r = 2, \theta = \frac{\pi}{3}.$

$$\therefore \quad \sqrt[3]{1+\sqrt{3}i} + \sqrt[3]{1-\sqrt{3}i} = 2.2^{1/3}.\left[\cos\left(\frac{2k\pi + \frac{\pi}{3}}{3}\right)\right] \quad \text{for } k = 0, 1, 2$$

$$= 2^{4/3}\left[\cos(6k+1)\frac{\pi}{9}\right]$$

Giving k the vaues 0, 1, 2 we get the required value as

$$2^{4/3}\cos\frac{\pi}{9},\ 2^{4/3}\cos\frac{7\pi}{9},\ 2^{4/3}\cos\frac{13\pi}{9}$$

Example 40. *Find the 7th root of unity and prove that the sum of their nth power always vanishes unless n be a multiple of 7, n being an integer, and then the sum is 7.*

Sol. Let
$$1 = \cos 0 + i\sin 0$$
$$= \cos 2n\pi + i\sin 2n\pi$$
$$(1)^{1/7} = (\cos 2n\pi + i\sin 2n\pi)^{1/7}$$

Applying De-Moivre's theorem, we have

$$(1)^{1/7} = \cos\frac{2n\pi}{7} + i\sin\frac{2n\pi}{7}$$

Therefore the roots are

$$\cos 0 + i\sin 0 = 1$$
$$\cos\frac{2\pi}{7} + i\sin\frac{2\pi}{7} = \alpha(\text{say})$$
$$= \cos\frac{4\pi}{7} + i\sin\frac{4\pi}{7}\ \alpha^2 \text{ and so on up to } \alpha^6$$

Let the sum of nth powers of the roots be S.

$$S = 1 + \alpha^n + (\alpha^2)^n + \ldots + (\alpha^6)^n$$
$$= 1 + \alpha^n + (\alpha^n)^2 + \ldots + (\alpha^n)^6 \quad [\text{A G.P. with common ratio } \alpha^n]$$
$$S = \frac{1-(\alpha^n)^7}{1-\alpha^n} \quad \ldots(1)$$

$$\left[\text{Sum of G.P.} = \frac{a(1-r^n)}{1-r}\right]$$

When n is not a multiple of 7

$$(\alpha^n)^7 = \left(\cos\frac{2\pi}{7} + i\sin\frac{2\pi}{7}\right)^{7n}$$
$$= \cos\frac{2\pi.7n}{7} + i\sin\frac{2\pi.7n}{7} \quad [\text{By De-Moivre's theroem}]$$
$$= \cos 2n\pi + i\sin 2n\pi = 1$$
$$\therefore \quad 1-(\alpha^n)^7 = 1 - 1 = 0$$

Also
$$1-\alpha^n = 1-\left(\cos\frac{2n\pi}{7} + i\sin\frac{2n\pi}{7}\right) \neq 0$$

as n is not a multiple of 7.

$$\therefore \quad S = 0 \text{ from (1).}$$

When n is a multiple of 7, $n = 7m$ (say)

$$\therefore \qquad \alpha^n = \left(\cos\frac{2\pi}{7} + i\sin\frac{2\pi}{7}\right)^{7m} = \cos 2\pi m + i \sin 2\pi m = 1$$

Similarly α^{2n}, α^{4n} etc. each term is 1.

$$\therefore \qquad S = 1 + \alpha^n + \alpha^{2n} + ... + \alpha^{6n}$$
$$= 1 + 1 + 1 + ... + 1 = 7.$$

Example 41. *Solve the equation $x^4 - x^3 + x^2 - x + 1 = 0$.*

Sol. To solve the given equation, multiple both sides by $(x + 1)$.

$$\therefore \qquad (x + 1)(x^4 - x^3 + x^2 - x + 1) = 0$$

or $$x^5 + 1 = 0$$

or $$x^5 = -1 = \cos\pi \pm i \sin\pi$$
$$= \cos(2n\pi + \pi) \pm i \sin(2n\pi + \pi)$$

$$\therefore \qquad x = [\cos(2n + 1)\pi \pm i \sin(2n + 1)\pi]^{1/3} \text{ for } n = 0, 1, 2.$$

Applying De-Moivre's theoerm, we get

$$x = \cos(2n+1)\frac{\pi}{5} \pm i\sin(2n+1)\frac{\pi}{5}$$

Giving n the values 0, 1, 2 the roots are

$$\cos\frac{\pi}{5} \pm i\sin\frac{\pi}{5}, \cos\frac{3\pi}{5} \pm i\sin\frac{3\pi}{5}, \cos\pi + i\sin\pi.$$

The root $\cos\pi + i \sin\pi = -1$, is due to multiplication by $(x + 1)$.
Hence rejecting $x = -1$, the other four roots are the required roots.

EXERCISE 1.1.3.

1. Simplify (*i*) $\dfrac{(\cos 2\theta - i\sin 2\theta)^7 (\cos 3\theta + i\sin 3\theta)^{-5}}{(\cos 4\theta - i\sin 4\theta)^{-4} (\cos 6\theta + i\sin 6\theta)^6}$

   (*ii*) $\dfrac{(\cos\theta + i\sin\theta)^3}{(\sin\phi + i\cos\phi)^4}$

2. Show that $\{(\cos\theta - \cos\phi) + i(\sin\theta - \sin\phi)\}^n + \{(\cos\theta - \cos\phi) - i(\sin\theta - \sin\phi)\}^n$
   $= 2^{n+1} \sin n\dfrac{\theta - \phi}{2} \cos n\left(\dfrac{\pi + \theta + \phi}{2}\right)$

3. If $x = \cos\theta + i \sin\theta$, show that

   (*i*) $x^m + \dfrac{1}{x^m} = 2 \cos m\theta$

   (*ii*) $x^m - \dfrac{1}{x^m} = 2\, i \sin m\theta.$

4. Prove that

(i) $(a+bi)^{\frac{m}{n}} + (a-bi)^{\frac{m}{n}} = 2(a^2+b^2)^{\frac{m}{2n}} \cos\left(\frac{m}{n}\tan^{-1}\frac{b}{a}\right)$

(ii) $(1+i)^n + (1-i)^n = 2^{\frac{n+2}{2}} \cos\frac{n\pi}{4}$.

5. (i) $x_r = 1, \cos\left(\frac{\pi}{2^r}\right) + i\sin\left(\frac{\pi}{2^r}\right)$, show that

$x_1 . x_2 . x_3 \ldots \infty = -1$.

(ii) Prove that $(1 + \cos\theta + i \sin\theta)^n + (1 + \cos\theta - i \sin\theta)^n = 2^{n+1} \cos^n \frac{\theta}{2} \cos\frac{n\theta}{2}$.

6. (i) If α, β be the roots of $x^2 - 2x + 4 = 0$,

Prove that $\alpha^n + \beta^n = 2^{n+1} \cos\frac{n\pi}{3}$

(ii) Show that the equation whose roots are the nth powers of the roots of $x^2 - 2x \cos\theta + 1 = 0$ is given by $x^2 - 2x \cos n\theta + 1 = 0$.

7. If $\sin\alpha + \sin\beta + \sin\gamma = 0 = \cos\alpha + \cos\beta + \cos\gamma$, prove that

(i) $\cos3\alpha + \cos3\beta + \cos3\gamma = 3 \cos(\alpha + \beta + \gamma)$

(ii) $\sin3\alpha + \sin3\beta + \sin3\gamma = 3 \sin(\alpha + \beta + \gamma)$

(iii) $\cos(\beta + \gamma) + \cos(\gamma + \alpha) + \cos(\alpha + \beta) = \sin(\beta + \gamma) + \sin(\gamma + \alpha) + \sin(\alpha + \beta) = 0$

(iv) $\cos2\alpha + \cos2\beta + \cos2\gamma = \sin2\alpha + \sin2\beta + \sin2\gamma = 0$

8. If $(a_1 + ib_1)(a_2 + ib_2)(a_3 + ib_3) \ldots (a_n + ib_n) = A + iB$.

Prove that

(i) $(a_1^2 + b_1^2)(a_2^2 + b_2^2) \ldots (an^2 + bn^2) = A^2 + B^2$

(ii) $\tan^{-1}\left(\frac{b_1}{a_1}\right) + \tan^{-1}\left(\frac{b_2}{a_2}\right) + \ldots + \tan^{-1}\left(\frac{b_n}{a_n}\right) = \tan^{-1}\left(\frac{B}{A}\right)$

9. If n is a positive integer show that

$(\sqrt{3}+i)^n + (\sqrt{3}-i)^n = 2^{n+1} \cos\frac{n\pi}{6}$.

10. If $x = \cos\theta + i \sin\theta$ and $\sqrt{1-c^2} = nc - 1$, show that $1 + c \cos\theta = \frac{c}{2n}(1+nx)\left(1+\frac{n}{x}\right)$.

11. Find the value of x such that $\dfrac{(x+\alpha)^n-(x+\beta)^n}{\alpha-\beta}=\dfrac{\sin n\theta}{\sin^n\theta}$

where α and β are roots of $t^2 - 2t + 2 = 0$.

12. Prove that if n is a positive integer and $(1 + x)^n = p_0 + p_1x + p_2x^2 + \ldots + p_nx^n$ then

(i) $p_0 - p_2 + p_4 - \ldots = 2^{n/2}\cos\dfrac{n\pi}{4}$. (ii) $p_0 + p_4 + p_8 + \ldots = 2^{n-3} + 2^{\frac{n-2}{2}}\cos\dfrac{n\pi}{4}$.

13. Show that the sum of the infinite series

$1-\dfrac{1.3}{2.4}+\dfrac{1.3.5.7}{2.4.6.8}-\dfrac{1.3.5.7.9.11}{2.4.6.8.10.12}+\ldots$ is $1/2\ (1+\sqrt{2})^{1/3}$.

14. Find all the values of

(i) $(16)^{1/4}$ (ii) $(-1)^{1/5}$

(iii) $(1-i\sqrt{3})^{1/4}$ (iv) $(i)^{1/3}$

(v) $(1 + i)^{2/3}$.

15. Find the continued product of the four values of $\left(\cos\dfrac{\pi}{3}+i\sin\dfrac{\pi}{3}\right)^{3/4}$

16. Show that if α denotes any n^{th} root of unity, then $1 + \alpha + \alpha^2 + \ldots + \alpha^{n-1} = 0$

17. Solve the equation $x^{12} - 1 = 0$ and find the roots that satisfy $x^4 + x^2 + 1 = 0$.

18. Solve (i) $x^6 + x^5 + x^4 + x^3 + x^2 + x + 1 = 0$ (ii) $x^5 - x^4 + x^3 - x^2 + x - 1 = 0$.

19. Solve using De Moivre's theorem the equation $(x - 1)^n = x^n$ (n being a +ve integer) and show that its roots are given by $\dfrac{1}{2}\left\{1+i\cot\left(\dfrac{r\pi}{n}\right)\right\}$. where $r = 0, 1, 2, \ldots (n - 1)$.

20. Solve $x^5 = 1$ by De Moivre's theorem and prove that the sum of nth powers of the roots of the equation, n being an integer not divisible by 5, is zero.

21. Express $p=\dfrac{(\sqrt{3}-1)+i(\sqrt{3}+1)}{2\sqrt{2}}$ in the form $r(\cos\theta + i\sin\theta)$ and derive all the value of $p^{1/6}$.

22. Show that the roots of the equation $(1 + x)^{2n} + (1 - x)^{2n} = 0$ are given by $\pm i\tan\left(\dfrac{2r-1}{4n}\right)\pi$

where $r = 1, 2, \ldots n$.

ANSWERS

1. (*i*) $\cos 81\theta - i \sin 81\theta$ (*ii*) $\cos(3\theta + 4\phi) + i \sin(3\theta + 4\phi)$.

11. $\cot\theta - 1$.

14. (*i*) $\pm 2, \pm 2i$. (*ii*) $\cos\frac{\pi}{5} \pm i\sin\frac{\pi}{5}, \cos\frac{3\pi}{5} \pm i\sin\frac{3\pi}{5}, -1$

(*iii*) $\pm 2^{1/4}\left[\cos\frac{r\pi}{12} - i\sin\frac{r\pi}{12}\right]$, where $r = 1, 7$

(*iv*) $\frac{\pm\sqrt{3}+i}{2}; -i$ (*v*) $2^{1/3}\left(\frac{\pm\sqrt{3}+i}{2}\right), -2^{1/2}\, i$.

15. 1

17. $\pm\left(\cos\frac{\pi}{3} \pm i\sin\frac{\pi}{3}\right)$.

18. (*i*) $\cos\frac{r\pi}{7} \pm i\sin\frac{r\pi}{7}$, where $r = 2, 4, 6$ (*ii*) $\frac{1}{2} \pm \frac{i\sqrt{3}}{2}, -\frac{1}{2} \pm \frac{i\sqrt{3}}{2}, 1$.

21. $\cos\left(\frac{5\pi}{12} + i\sin\frac{5\pi}{12}\right)$. All the values are $\pm\left[\cos\frac{r\pi}{72} + i\sin\frac{r\pi}{72}\right]; r = 5, 29, 53$.

Unit—IV

Differential Calculus

PART 1 : FUNCTION AND LIMITS

1.1. FUNCTION

A variable quantity takes up different value during a set of operations, while a constant quantity remains unchanged. We usually denote variables by x, y, z, u, v, w etc. and constants by a, b, c, d etc.

1.1.1. Concept of Constant and Variable

Two or more quantities may be related. If A is area and 'r' is radius, then area, $A = \pi r^2$. Here, π is constant if we change r, the value of A changes. Thus, area A of the circle is a function of its radius r and it is denoted by $A = f(r)$, where A is dependent variable and r is independent variable.

- If volume 'V' of cube, $V = a^3$, here is a side of cube, and thus volume V is a function of one variable 'a', here we denote it by, $V = f(a)$.
- If volume 'V' of cylinder is the formula $V = \pi r^2 h$, here r, h is radius and height respectively and π is constant, thus function.

$$V = f(r, h).$$

When variables are so related that the value of one of the depends upon the values of the others, the first variable is said to be a function of the others variables.

1.1.2. Open Interval

This interval is an interval that does not include its end points. The open interval $(x : a < x < b)$ is denoted (a, b) although the non-standard notation. $]a, b[$ is used

Open interval]a, b[

1.1.3. Closed Interval

Closed interval [a, b]

A closed interval that include all of its limit points. If the end points of the interval are finite numbers a and b then the interval $\{x : a \leq x \leq b\}$ is denoted $[a, b]$. If one of the end point is $\pm \infty$. Then the interval still contains all of its limit points (although not all of its end points), so $[a, \infty)$ and $(-\infty, b]$ are also closed intervals, as is interval $(-\infty . \infty)$.

Example 1. Let some intervals are:

1. $a \leq x < b$ or $a < x \leq b$
2. $-\infty < x < \infty$ (all real number)
3. $x \geq a$ or $x \leq a$
4. $x > a$ or $x < a$

Thus *closed interval* is an interval that includes all of its end points and *open interval* is an interval that does not contain its end points.

Interval Notation

Interval	*Interval Notation*
$a \leq x \leq b$	$[a, b]$
$a \leq x < b$	$[a, b)$
$a < x \leq b$	$(a, b]$
$x \geq a$	$[a, \infty)$
$x \leq a$	$(-\infty, a]$
$x > a$	(a, ∞)
$x < a$	$(-\infty, a)$
$-\infty < x < \infty$	$(-\infty, \infty)$

1.2. DEFINITION OF FUNCTION

Function is a relationship between two variables is a called a *function.* A function is a rule which maps a number to another unique number also, an input, of the function, we get an output of the function. If a function that added 3 to any number. So, if we apply this function to the number 2, we get the number 5. If we apply this function to the number 8, we get the number 11. If we apply this function to the number x, we get the number $x + 3$.

So, $$f(x) = x + 3$$

The number x that the input of the function is called the *argument* of the function. So if we choose an argument of 2, we get

$$f(2) = 2 + 3 = 5.$$

If we choose an argument of 8, we get

$$f(8) = 8 + 3 = 11.$$

If we choose an argument of –6, we get

$$f(-6) = -6 + 3 = -3$$

If we choose an argument of z, we get

$$f(z) = z + 3.$$

If we choose an argument of x^2, we get

$$f(x^2) = x^2 + 3.$$

Do you know?

- Function is a rule that counts a number to another unique number.
- The input to the function is called the *independent variable,* and is called the *argument* of the function and also the output of the function is called the *dependent variable.*

At first sight, it seems that we can pick any number we choose for the argument. However, that is not the case, as we shall see later. But because we do have some choice in the number we can pick, we call the argument the *independent variable.* The output of the function, *e.g.*, $f(x)$, $f(5)$, etc. depends upon the argument, and so this is called the *dependent variable.*

• Value of a Function at a Point

Function is a compute the value $f(x)$, of the function for a given real number x and say for which real numbers x the rule may be a applied.

Domain. The set of numbers for which a function is defined.

Range. The set of all possible numbers $f(x)$ as x runs over the domain is called the range of the function.

• **Limit of a Function-Informal Approach.** Now, we assume the function

$$f(x) = \frac{16 - x^2}{4 + x} \qquad ...(1)$$

Now, domain is the set of all real numbers except –4. Although f cannot be evaluated at –4 because substituting –4 for x results in the underlined quantity 0/0, $f(x)$ can be calculated at any number x that is very *close* to –4.

x	–4.1	–4.01	–4.001
$f(x)$	8.1	8.01	8.001

x	–3.9	–3.99	–3.999
$f(x)$	7.9	7.99	7.999

...(2)

The two table show that as x approaches –4 from either the left or right, the function values $f(x)$ appear to be approaching 8, in other words, when x is near –4, $f(x)$ is near 8. To interpret the numerical information in (1) graphically, observe that for every number $x \neq -4$. The function f can be simplified by cancellation:

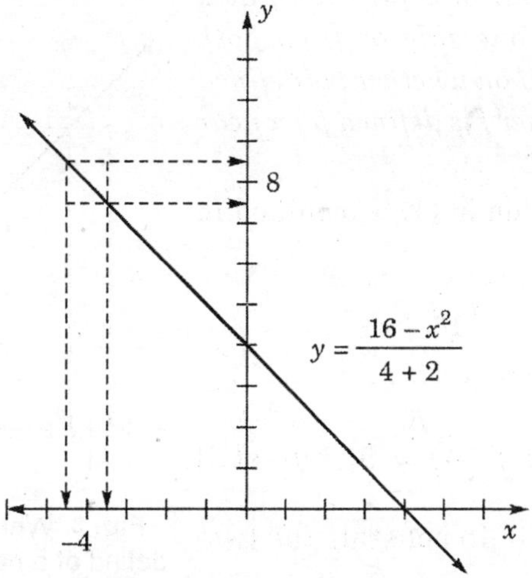

Fig. 1. When x is near –4, $f(x)$ is near 8.

$$f(x) = \frac{16 - x^2}{4 + x} = \frac{(4 + x)(4 - x)}{4 + x} = 4$$

The number L is said to be the left-hand limit of $f(x)$ as x approaches a. Similarly, if $f(x)$ can be made arbitrarily close to a number L_2 by taking x sufficiently close to, but not equal to, a number a from the *right*, then L_2 is the right-hand limit of $f(x)$ as x approaches a and we write

$$f(x) \to L_2 \text{ as } x \to a^+$$

or
$$\lim_{x \to a^+} f(x) = L_2 \qquad ...(3)$$

The quantities in (3) and (4) are also referred to as one-sided limits:

Two-Sided Limits. If both the left-hand limit $\lim_{x \to a} f(x)$ and the right-hand limit $\lim_{x \to a} f(x)$ exist and have a common value L.

$$\lim_{x \to a^-} f(x) = L \text{ and } \lim_{x \to a^+} f(x) = L,$$

then we say that L is the limit of $f(x)$ as x approaches a and write

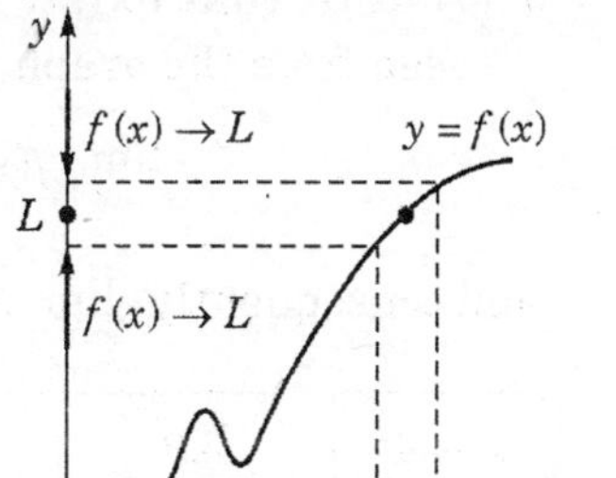

Fig. 2. $f(x) \to L$ as $x \to a^+$ if and only if $f(x) \to L$ as $x \to a^-$ and $f(x) \to L$.

$$\lim_{x \to a^+} f(x) = L \qquad ...(4)$$

A limit such as (4) is said to be a two-sided limit (See Fig. 2). Since the numerical tables in (2) suggest that

$$f(x) \to 8 \text{ as } x \to -4^- \quad \text{and } f(x) \to 8 \text{ as } x \to -4^+, \qquad ...(5)$$

We can replace the two symbolic statement in (6) by the statement

$$f(x) \to 8 \text{ as } x \to -4 \text{ or equivalently } \lim_{x \to 4} \frac{16 - x^2}{4 + x} = 8 \qquad ...(6)$$

(i) **Existence and Non-existence.** Of course a limit (one-sided or two-sided) does not have to exist. But it is important that you keep firmly in mind:

- *The existence of a limit of a function f as x approaches a (from one side or from both sides), does not depend on whether f is defined at a but only on whether f is defined for x near the number a.*

For example, if the function in (1) is modified in the following manner

$$f(x) = \begin{cases} \dfrac{16 - x^2}{4 + x} & x \neq -4 \\ 5. & x = -4 \end{cases}$$

then $f(-4)$ is defined and $f(-4) = 5$, but still $\lim_{x \to -4} \frac{16 - x^2}{4 + x} = 8$. (See Fig. 3). In general, the two-sided limit $\lim_{x \to a} f(x)$ does not exist.

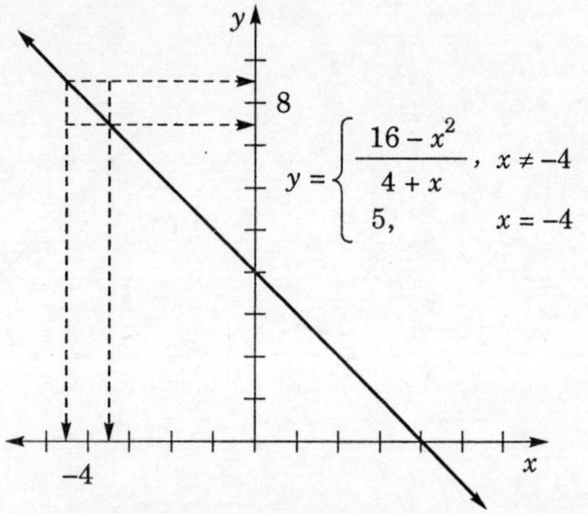

Fig. 3. Whether *f* is defined at *a* or is not defind at a has no bearing on the existence of the limit of *f*(*x*) as $x \to a$.

- If either of the one-sided limits $\lim_{x \to a^-} f(x)$ or $\lim_{x \to a^+} f(x)$ fails to exist, or
- If $\lim_{x \to a^-} f(x) = L_1$ and $\lim_{x \to a^+} f(x) = L_2$, but $L_1 \neq L_2$.

(ii) **A limit that exist.** The graph of the function $f(x) = -x^2 + 2x + 2$ is shown in Fig. 4. As seen from the graph and the company tables, it seems that

$$\lim_{x \to 4^-} f(x) = -6 \text{ and } \lim_{x \to 4^+} f(x) = -6$$

and consequently $\lim_{x \to 4} f(x) = -6$.

$x \to 4^-$	3.9	3.99	3.999
$f(x)$	–5.41000	–5.94010	–5.99400

$x \to 4^+$	4.1	4.01	4.001
$f(x)$	–6.61000	–6.06010	–6.06010

Now function is certainly defined at 4, but at not time did we substitute $x = 4$ into the function to find the value of $\lim_{x \to 4} f(x)$.

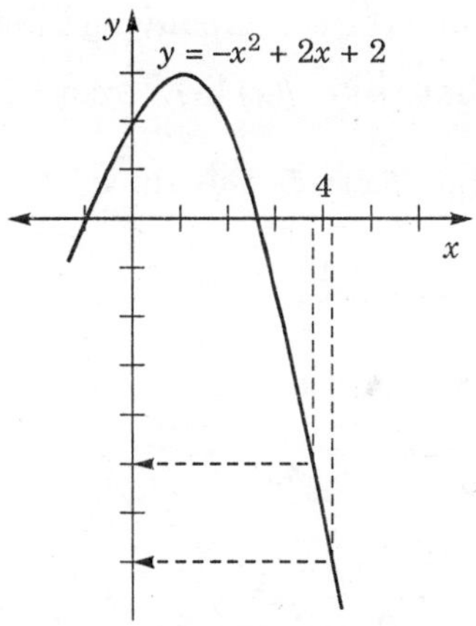

Fig. 4. Graph of function.

The graph of the piecewise-defined function.

$$f(x) = \begin{cases} x^2, & x < 2 \\ -3 + 6 & x > 2 \end{cases}$$

Now that $f(2)$ is not defined, but that is no consequence when considering $\lim_{x \to 2} f(x)$. From the graph and the accompanying tables,

$x \to 2^-$	1.9	1.99	1.999
$f(x)$	3.61000	3.96010	3.99600

$x \to 2^+$	2.1	2.01	2.001
$f(x)$	3.90000	3.99000	3.99900

we see that when we make x close to 2, we can make $f(x)$ arbitrarily close to 4, and so

$$\lim_{x \to 2^-} f(x) = 4 \text{ and } \lim_{x \to 2^+} f(x) = 4.$$

That is, $$\lim_{x \to 2} f(x) = 4.$$

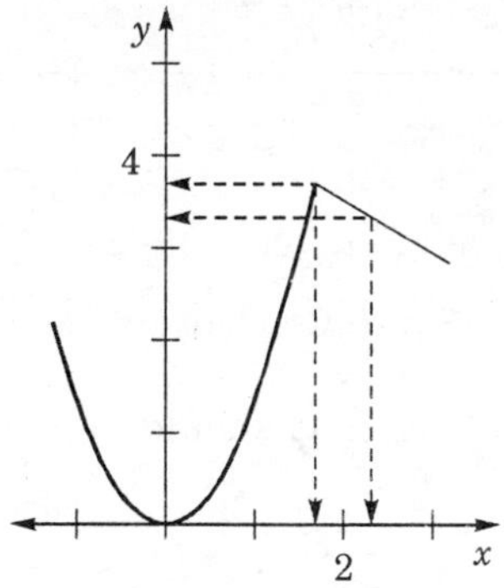

Fig. 5. Graph of function.

(*iii*) (*a*) **A limit that does not exist.** The graph of the piecewise-defined function

$$f(x) = \begin{cases} x + 2 & x \leq 2 \\ -x + 10 & x > 5 \end{cases}$$

is given in Fig. 6. From the graph and the accompanying tables, it appears that as x approaches 5 through numbers less than 5 that $\lim_{x \to 5^-} f(x) = 7$. Then as x approaches 5 through numbers greater than 5 it appears that $\lim_{x \to 5^+} f(x) = 5$. But since

$$\lim_{x \to 5^-} f(x) \neq \lim_{x \to 5^+} f(x),$$

we conclude that $\lim_{x \to 5^+} f(x)$ does not exist.

$x \to 5^-$	4.9	4.99	4.999
$f(x)$	6.90000	6.99000	6.99900

$x \to 5^+$	5.1	5.01	5.001
$f(x)$	4.90000	4.99000	4.99900

(*b*) **A limit that does not exist.** Recall, the greatest integer function of floor function $f(x) = [x]$ is defined to be the greatest integer that is less than or equal to x. The domain of f is the set of real numbers $(-\infty, \infty)$. From the graph in Fig. 7, we see that $f(n)$ is defined for every integer n; for each integer n, $\lim_{x \to n} f(x)$ does not exist. For example, as x approaches, say, the number 3, the two one-sided limits exists but have different values.

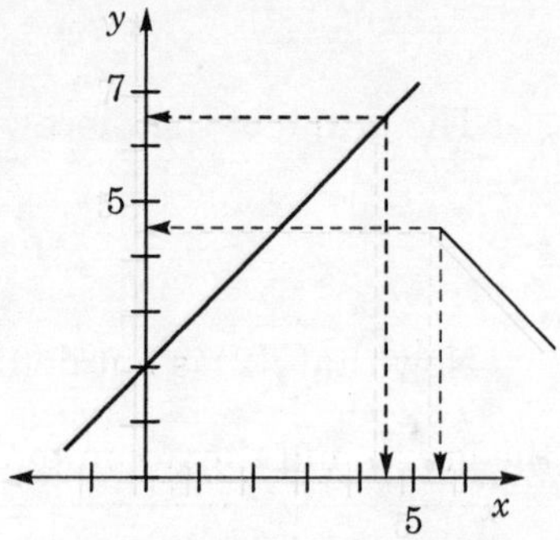

Fig. 6. Graph of function

$$\lim_{x \to 3^-} f(x) = 2 \text{ whereas } \lim_{x \to 3^+} f(x) = 3.$$

In general, for an integer n.

$$\lim_{x \to n^-} f(x) = n - 1 \text{ whereas } \lim_{x \to n^+} f(x) = n.$$

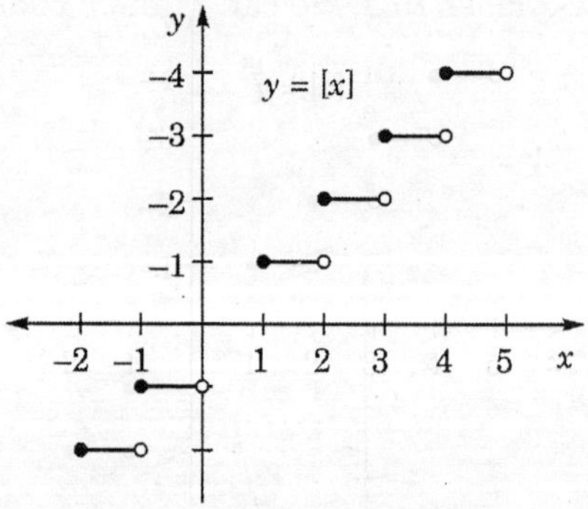

Fig. 7. Graph of function

(*iv*) **A right-hand limit.** From Fig. 8 it should be clear that $f(x) = \sqrt{x} \to 0$ as $x \to 0^+$ that is

$$\lim_{x \to 0^+} \sqrt{x} = 0,$$

It would be incorrect to write $\lim_{x \to 0} \sqrt{x} = 0$ since this notation carries with it the connotation that the limits from the left and from the right exist and are equal to 0. In this case $\lim_{x \to 0^+} \sqrt{x}$ does not exist since $f(x) = \sqrt{x}$ is not defined for $x < 0$.

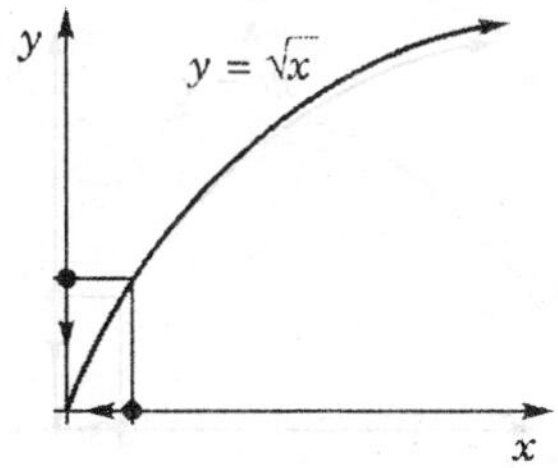

Fig. 8. Graph of function.

If $x = a$ is a vertical asymptote for the graph of $y = f(x)$, then $\lim_{x \to a^-} f(x)$ will always fail to exist because the function values $f(x)$ must become unbounded from at least one side of the line $x = a$.

(*v*) **A limit that does not exist.** A vertical asymptote always corresponds to an infinite break in the graph of a function f. In Fig. 9 we see that the y-axis or $x = 0$ is a vertical asymptote for the graph of $f(x) = 1/x$. The tables

$x \to 0^-$	-0.1	-0.01	-0.001
$f(x)$	-10	-100	-1000

$x \to 0^+$	0.1	0.01	0.001
$f(x)$	10	100	1000

clearly show that the function values $f(x)$ become unbounded in absolute value as we get close to 0. In other words, $f(x)$ is not approaching a real number as $x \to 0^-$ nor as $x \to 0^+$. Therefore, neither the left-hand nor the right-hand limit exists as x approaches 0. Thus we conclude that $\lim_{x \to 0^-} f(x)$ does not exist.

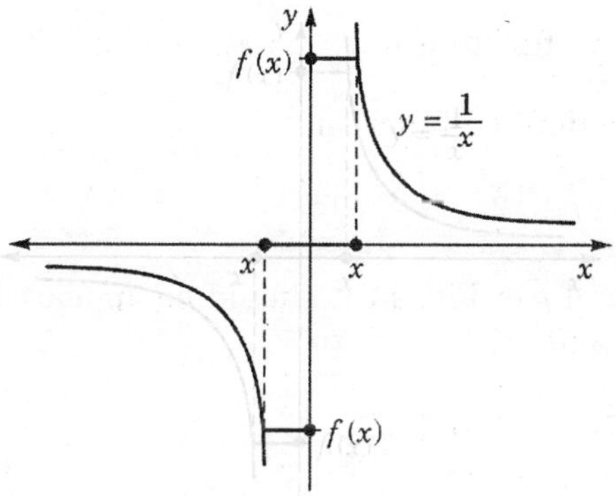

Fig. 9. Graph of function.

(*vi*) **A trigonometric limit.** To do the calculus of the trigonometric functions sin x, cos x, tan x, and so on, it is important to realize that the variable x is either a real number or an angle measured in radians. With that in mind, consider the numerical values of $f(x) = (\sin x)/x$ as $x \to 0^+$ given in the table that follow.

$x \to 0^+$	0.1	0.001	0.001	0.0001
$f(x)$	0.99833416	0.99998333	0.99999983	0.99999999

Because sin x is an odd function, for $x > 0$ and $-x < 0$, we have $\sin(-x) = -\sin x$ and as a consequences

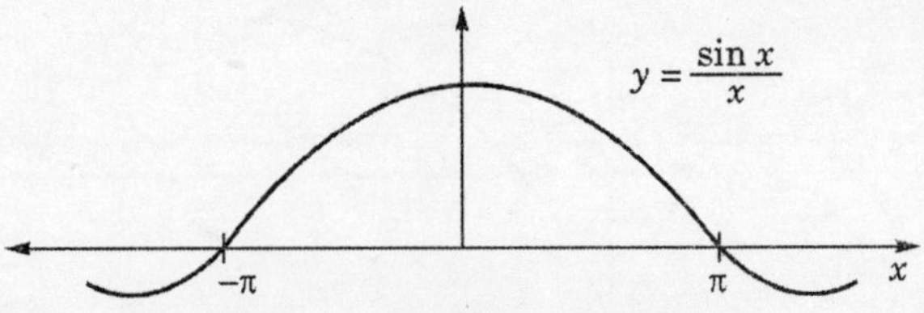

Fig. 10. Graph of function.

$$f(-x) = \frac{\sin(-x)}{-x} = \frac{\sin x}{x} = f(x).$$

f is an even function. The table of numerical values as well as the graph of f strongly suggest the following result:

$$\lim_{x \to 0} \frac{\sin x}{x} = 1.$$

$$\lim_{x \to 0} \frac{1 - \cos x}{x} = 0.$$

(*vii*) **An indeterminate form.** A limit of a quotient $f(x)/g(x)$, where both the numerator and the denominator approach 0 as $x \to a$, is said to have the indeterminate form 0/0.

$$\lim_{x \to 0} \frac{f(x+h) - f(x)}{h},$$

We have the indeterminate form 0/0.

The limit $\lim_{x \to 0} |x|$ has the indeterminate form 0/0, but unlike fails to exist. To see why, let us examine the graph of the function $f(x) = |x|/x$. For $x \neq 0$. $|x| = \begin{cases} x, & x > 0 \\ -x, & x < 0 \end{cases}$ and so we recognize f as the piecewise-defined function.

$$f(x) = \frac{|x|}{1} = \begin{cases} 1, & x > 0 \\ -1, & x < 0 \end{cases} \quad ...(i)$$

From (*i*) and the graph of f in Fig. 11 it should be appeared then both the left-hand and right-hand limits of f exist and

$$\lim_{x \to 0^-} \frac{|x|}{x} = -1 \text{ and } \lim_{x \to 0^+} \frac{|x|}{x} = 1.$$

Because these one-sided limits are different, we conclude that $\lim_{x \to 0^+} |x|/x$ does not exist.

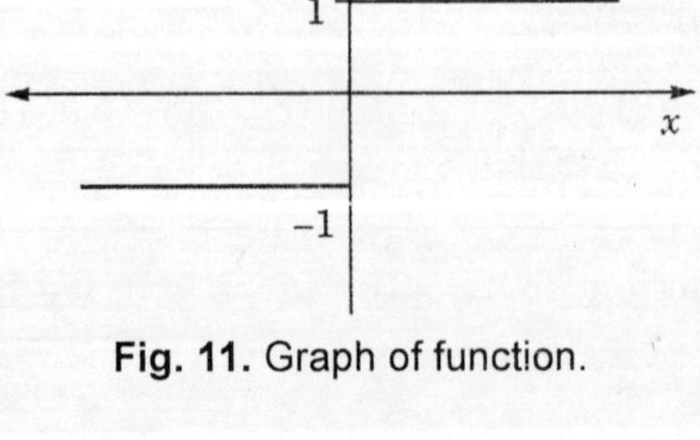

Fig. 11. Graph of function.

1.2.1 Types of Function

In general, we denote a function of x and y, we can write it is a function of x

$$y = f(x)$$

A function $f(x) = a_0 + a_1x + a_2x^2 + \ldots a_nx^n$, $a_n \neq 0$, where n is positive integer, is called polynomial function of degree n.

1. **Algebraic functions:** (*a*) The quotient of two polynomials defines *rational algebraic function.*

   For example: $$f(x) = \frac{(2x^4 - 3x^2 + 2x + 3)}{(x^4 + 2x^2 - 3x + 2)}$$

   (*b*) If the functions involving radicals of one or more polynomials, are called *irrational algebraic functions.*

   For example: $\sqrt{(2x^2 - 4x + 3)}$, $\sqrt{(5x^2 + 5x + 2)}$.

2. **Transcendental Functions:** Non algebraic functions are called *transcendental functions.*

   Example: $\sin x$, $\cos x$, $\tan x$, $\cos^{-1} x$, e^x, $\log x$ ...

3. **Even Function:** If the function which satisfies the condition $f(-x) = f(x)$ is called *even function.*

   For example: $f(x) = 30x^4 - 35x^4 + 2$

   Replacing x by $-x$ in $f(x)$, we get

   $$f(-x) = 30\,(-x)^4 - 35\,(-x)^4 + 2$$

   Both get same, thus it is called *even function.*

4. **Odd Functions:** A function which satisfies the condition

   $f(-x) = -f(x)$ is called an *odd function* of x.

5. **Explicit Function:** When y is expressed directly in terms of x, as in the examples given above, it is said to be an explicit function of x.

6. **Implicit Function:** Relation between x and y is given in a form such as

   $$x^3 + y^3 + 3\,axy = 0.$$

   In general $f(x, y) = K$

   Thus, y is implicit function of x.

Example 2. *If* $$f(x) = \log\left(\frac{1-x}{1+x}\right),$$

prove that $$f(a) + f(b) = f\left(\frac{a+b}{1+ab}\right).$$

Solution. Consider

$$f(a) + f(b) = \log\left(\frac{1-a}{1+a}\right) + \log\left(\frac{1-b}{1+b}\right)$$

$$= \log\left(\frac{1-a}{1+a}\right) + \log\left(\frac{1-b}{1+b}\right)$$

$$= \log \left\{ \left(\frac{1-a}{1+a} \right) \times \left(\frac{1-b}{1+b} \right) \right\}$$

$$= \log \left\{ \frac{(1+ab)-(a+b)}{(1+ab)+(a+b)} \right\}$$

$$= \log \left\{ \frac{1 - \left(\frac{a+b}{1+ab} \right)}{1 + \left(\frac{a+b}{1+ab} \right)} \right\} = f \left(\frac{a+b}{1+ab} \right).$$

1.3. CONCEPT AND DEFINITION OF LIMIT

Let f be a function defined on some open interval that contains the number a, except possible at a itself. We say *the limit of f(x) as x approaches a is L,* and we write

$$\lim_{x \to a} f(x) = L$$

If for every number $\in > 0$ there is a corresponding number $\delta > 0$ such that

$$|f(x) - L| < \in$$

where ever

$$0 < |x - a| < \delta.$$

Let us consider the function $f(x) = \dfrac{x^4 + 64}{x^2 - 4}$ where, $f(x)$ takes a finite value for every value of x except when $x = 2$, is of the meaningless from ∞. Whenever a function $f(x)$ assumes the form $\dfrac{0}{0}, \dfrac{\infty}{\infty}, 0 \times \infty, \infty - \infty, 1^{\infty}$ etc.

For a particular value of $x = a$, it is not defined for that value of x and then it is said to be indeterminate at $x = a$,

If we take a value of x very near to a when $x = a - h$ or $x = a + h$. Where h is very small, we get value of function $f(x)$, if we take

$$x = 3 \Rightarrow 3, 3.5, 3.6, 3.7 \ldots 3.999$$
$$f(x) = 6 \Rightarrow 6, 6.5, 6.6., 6.7 \ldots 6.999.$$

This number 6 is said to be limit of function

$$f(x) = \frac{x^2 - 16}{x - 4}$$

When x tends to 4 and we write as,

$$\lim_{x \to 4} \left(\frac{x^2 - 16}{x - 4} \right) = 0$$

In general, as $x \to a, f(x) \to l$, then l is called limit of the functions $f(x)$ which is symbolically written as $\lim_{x \to a} f(x) = l$. In case $x \to a$, through value of $x > a$, we call it is right hand Limit or R.H.L.

• When $x = a + h$ and $h \to 0$, so that $x \to a$

$$\lim_{x \to a^+} f(x) = \lim_{h \to 0^+} f(a+h)$$

where 0^+ indicates that it is R.H.L.

• When $x \to a$, through $x < a$, it gives the left hand limit of L.H.L. Here, we take $x = a - h$, then $h \to 0$ so that $x \to a$,

Thus, $$\boxed{\lim_{x \to 0^-} f(x) = \lim_{h \to 0^-} f(a-h)}$$

where 0^- indicates that it is L.H.L.

Let $x = 4 + h$. Now, R.H.L is given by

$$\underset{x \to a^+}{\text{Lt}} f(x) = \underset{x \to 4^+}{\text{Lt}} \left(\frac{x^2 - 16}{x - 4}\right)$$

$$= \underset{x \to 4^{+0}}{\text{Lt}} \left\{\frac{(4+h)^2 - 16}{(4+h) - 4}\right\}$$

$$f(x) = \underset{h \to 0}{\text{Lt}} = 8 + h = 8$$

For L.H.L, Let $x = 4 - h$.

Now $$\underset{x \to a^-}{\text{Lt}} f(x) = \underset{x \to 4}{\text{Lt}} \left(\frac{x^4 - 16}{x - 4}\right)$$

$$= \underset{h \to 0}{\text{Lt}} \left\{\frac{(4-h)^2 - 16}{(4-h) - 4}\right\}$$

$$= 8 - h = 8$$

Thus, R.H.L = L.H.L = 8.

• Algebra of Limits

Algebra of limits: We have observed that the limiting process respects addition, subtraction, multiplication and division as long as the limits and functions under consideration are well defined. This is not a coincidence. In fact below, we formalise these as a theorem without proof.

Theorem. Let f and g be two functions such that both $\lim_{x \to a} f(x)$ and $\lim_{x \to a} g(x)$ exist then

(*i*) Limit of sum of two functions is sum of the limits of the functions, *i.e.*,

$$\lim_{x \to a} [f(x) + g(x)] = \lim_{x \to a} f(x) + \lim_{x \to a} g(x).$$

(*ii*) Limit of difference of two functions is difference of the limits of the functions, *i.e.*,

$$\lim_{x \to a} [f(x) - g(x)] = \lim_{x \to a} f(x) - \lim_{x \to a} g(x).$$

(*iii*) Limit of product of the functions is product of the limits of the functions, *i.e.*,

$$\lim_{x \to a} [f(x) . g(x)] = \lim_{x \to a} f(x) . \lim_{x \to a} g(x).$$

(*iv*) Limit of quotient of two functions is quotient of the limits of the functions (wherever the denominator is non zero), *i.e.*,

$$\lim_{x \to a} \frac{f(x)}{g(x)} = \frac{\lim_{x \to 0} f(x)}{\lim_{x \to 0} g(x)}$$

Note: In particular as a special case of (*iii*), when $g(x)$ is the constant function such that $g(x) = \lambda$, for real number λ, we have

$$\lim_{x \to a} [(\lambda . f)(x)] = \lambda . \lim_{x \to a} f(x).$$

1.4. STANDARD LIMITS OF ALGEBRAIC, TRIGONOMETRIC, EXPONENTIAL AND LOGARITHMIC FUNCTIONS

If $f_1(x)$ and $f_2(x)$ be two functions of x such that

$$\operatorname*{Lt}_{x \to a} f_1(x) = l_1 \text{ and } \operatorname*{Lt}_{x \to a} f_2(x) = l_2, \text{ then}$$

(*a*) $\lim_{x \to a} \{f_1(x) \pm f_2(x)\} = \lim_{x \to a} f_1(x) \pm \lim_{x \to a} f_2(x)$

$= l_1 \pm l_2$

(*b*) $\lim_{x \to a} \{f_1(x) \times f_2(x)\} = \lim_{x \to a} f_1(x) \times \lim_{x \to a} f_2(x)$

$= l_1 \times l_2$

(*c*) $\lim_{x \to a} \{f_1(x)/f_2(x)\} = \dfrac{\lim_{x \to a} f_1(x)}{\lim_{x \to a} f_2(x)}$

$= \dfrac{l_1}{l_2}$, where $l_2 \neq 0$.

Limit of Algebraic Functions

This can be proven by dividing both the numerator and denominator by x^n. If the numerator is a polynomial of higher degree, the limit does not exist. If the denominator is of higher degree, the limit is 0.

- $\lim_{x \to \infty} (2x^4 - x^2 - 8x)$
- $\lim_{t \to 0} \left(\frac{1}{3} t5 + 2t^3 - t^2 + 8\right)$

Limits of Trigonometric Functions

- $\lim_{x \to 0} \dfrac{\sin x}{x} = 1$
- $\lim_{x \to 0} \dfrac{1 - \cos x}{x} = 0$
- $\lim_{x \to 0} x \sin\left(\dfrac{1}{x}\right) = 1$

Do you know?

(*i*) $\operatorname*{Lt}_{x \to 0} \left(\dfrac{\sin x}{x}\right) = 1$

(*ii*) $\operatorname*{Lt}_{x \to 0} (1 + x)^{1/x} = e$

(*iii*) $\operatorname*{Lt}_{x \to \infty} x^{1/x} = 1$

(*iv*) $\operatorname*{Lt}_{x \to \infty} \left(\dfrac{x^n}{n!}\right) = 0 \quad (x > 0 \text{ and fixed})$

(*v*) $\operatorname*{Lt}_{x \to 0} \dfrac{a^x - 1}{x} = \log_e a$

(*vi*) $\operatorname*{Lt}_{x \to \infty} (x)^{1/n} = 1 \quad (x > 0 \text{ and fixed})$

(*vii*) $\operatorname*{Lt}_{n \to \infty} \left(\dfrac{\log n}{n}\right) = 0$

(*viii*) $\operatorname*{Lt}_{x \to a} \dfrac{x^n - a^n}{x - a} = na^{n-1}$

$\log_e = 1, \log 1 = 0, \log 0 = \infty$
$\log \infty = \infty, e^0 = 1, e^\infty = \infty, e^{-0} = 0$

(*ix*) $\lim_{x \to \infty} \left(1 + \dfrac{a}{n}\right)^n = e^a.$

Limits of Exponential Functions

- $\lim_{x \to 0} (1+x)^{\frac{1}{x}} = \lim_{x \to \infty} \left(1 + \frac{1}{r}\right)^r = e$
- $\lim_{x \to 0} \frac{e^x - 1}{x} = 1$
- $\lim_{x \to 0} \frac{e^{ax} - 1}{bx} = \frac{a}{b}$
- $\lim_{x \to 0} \frac{c^{ax} - 1}{bx} = \frac{a}{b} \ln c$
- $\lim_{x \to 0} x^x = 1$

Limits of Logarithmic Functions

- $\lim_{x \to 0} \frac{\ln(1+x)}{x} = 1$
- $\lim_{x \to 0} \frac{\ln(1+ax)}{bx} = \frac{a}{b}$
- $\lim_{x \to 0} \frac{\log_c(1+ax)}{bx} = \frac{a}{b \ln c}$

Example 3.

(*i*) $\lim_{x \to \infty} \left\{\frac{(3x+1)(10x+7)}{(2x-5)(3x+8)}\right\}$ (*ii*) $\lim_{x \to 0} \left(\frac{1-\cos x}{x^2}\right)$

(*iii*) $\lim_{h \to 0} \left\{\frac{\sqrt{(x+h)} - \sqrt{x}}{h}\right\}$ (*iv*) $\lim_{x \to 0} \left(\frac{\tan x - \sin x}{\sin^3 x}\right)$.

Solution. (*i*) $\lim_{x \to \infty} \left\{\frac{(3x+1)(10x+7)}{(2x-5)(3x+8)}\right\}$

$$= \lim_{x \to \infty} \left\{\frac{\left(3 + \frac{1}{x}\right)\left(10 + \frac{7}{x}\right)}{\left(2 - \frac{5}{x}\right)\left(3 + \frac{8}{x}\right)}\right\} = \frac{(3+0)(10+0)}{(2-0)(3+0)} = 5.$$

(*ii*) $$\lim_{x \to 0} \left(\frac{1-\cos x}{x^2}\right) = \lim_{x \to 0} \left(\frac{1 - \left(1 - 2\sin^2 \frac{x}{2}\right)}{x^2}\right) = \lim_{x \to 0} \left(\frac{2\sin^2 \frac{x}{2}}{x^2}\right)$$

$$= \lim_{x \to 0} \left(\frac{2\sin^2 y}{4y^2}\right), \text{ where } x = 2y$$

$$= \frac{1}{2} \lim_{y \to 0} \left(\frac{\sin y}{y}\right)^2 = \frac{1}{2} . 1^2 = \frac{1}{2}.$$

(*iii*) $$\lim_{h \to 0} \left[\frac{\sqrt{x+h} - \sqrt{x}}{h}\right] = \lim_{h \to 0} \left[\left(\frac{\sqrt{x+h} - \sqrt{x}}{h}\right) \times \left(\frac{\sqrt{x+h} + \sqrt{x}}{\sqrt{x+h} + \sqrt{x}}\right)\right]$$

$$= \lim_{h \to 0} \left[\frac{(x+h) - x}{h(\sqrt{x+h} + \sqrt{x})}\right] = \lim_{h \to 0} \left(\frac{1}{\sqrt{x+h} + \sqrt{x}}\right)$$

$$= \frac{1}{2\sqrt{x}}.$$

$$(iv)\ \lim_{x\to 0}\left(\frac{\tan x - \sin x}{\sin^3 x}\right) = \lim_{x\to 0}\left[\frac{\sin x\,(1-\cos x)}{\cos x \sin^3 x}\right] = \lim_{x\to 0}\left[\frac{(1-\cos x)}{\cos x\,(1-\cos^2 x)}\right]$$

$$= \lim_{x\to 0}\frac{1}{\cos x\,(1+\cos x)} = \frac{1}{1\,(1+1)} = \frac{1}{2}$$

$$= \lim_{x\to 0}\frac{\sin x\,(1-\cos x)}{\sin 3x \cos x}$$

$$= \lim_{x\to 0}\left[\frac{x^2}{\sin^2 x}\cdot\frac{2\sin^2 x/2}{x^2}\cdot\frac{1}{\cos x}\right]$$

$$= \frac{1}{2}\left(\lim_{x\to 0}\cdot\frac{\sin x}{x}\right)^{-2}\cdot\left(\lim_{x\to 0}\frac{\sin x/2}{x/2}\right)^{2}\cdot\left(\lim_{x\to 0}\frac{1}{\cos x}\right)$$

$$= \frac{1}{2}\times(1)^{-2}\times(1)^2\times 1 = \frac{1}{2}.$$

Example 4. $\lim\limits_{x\to 0}\dfrac{\tan x}{x}$

$$\lim_{x\to 0}\frac{\sin x}{x\cos x} = \lim_{x\to 0}\frac{\sin x}{x}\lim_{x\to 0}\frac{1}{\cos x} = 1.1 = 1.$$

Example 5. $\lim\limits_{x\to 2}\dfrac{x^3-2x^2}{x^2-5x+6} = \lim\limits_{x\to 2}\dfrac{x^2\,(x-2)}{(x-2)\,(x-3)}\lim\limits_{x\to 2}\dfrac{x^2}{(x-3)}$

$$= \frac{(2)^2}{2-3} = \frac{4}{-1} = -4.$$

Example 6. $\lim\limits_{x\to 1}\left[\dfrac{x-2}{x^2-x} - \dfrac{1}{x^3-3x^2+2x}\right] = \left[\dfrac{x-2}{x\,(x-1)} = \dfrac{1}{x\,(x^2-3x+2)}\right]$

$$= \left[\frac{x-2}{x\,(x-1)} - \frac{1}{x\,(x-1)\,(x-2)}\right] = \left[\frac{x^2-4x+4-1}{x\,(x-1)\,(x-2)} = \frac{x^2-4x+3}{x\,(x-1)\,(x-2)}\right].$$

Evaluating the function at 1, we get

$$\lim_{x\to 1}\left[\frac{x^2-2}{x^2-x} - \frac{1}{x^3-3x^2+2x}\right] = \lim_{x\to 1}\frac{x^2-4x+3}{x\,(x-1)\,(x-2)}$$

$$= \lim_{x\to 1}\frac{(x-3)\,(x-1)}{x\,(x-1)\,(x-2)} = \lim_{x\to 1}\frac{x-3}{x\,(x-2)} = \frac{1-3}{1\,(1-2)} = 2.$$

Example 7. $\lim_{x \to 1} \left[\frac{x^2+1}{x+100}\right] = \frac{1^2+1}{1+100} = \frac{2}{101}.$

Example 8. $\lim_{x \to 2} \left[\frac{x^2-4}{x^3-4x^2+4x}\right] = \lim_{x \to 2} \frac{(x+2)(x-2)}{x(x-2)^2}$

$$= \lim_{x \to 2} \frac{(x+2)}{x(x-2)} = \frac{2+2}{2(2-2)} = \frac{4}{0} \text{ which is not defined.}$$

Example 9. $\lim_{x \to 0} \frac{\sin 4x}{\sin 2x} = \lim_{x \to 0} \left[\frac{\sin 4x}{4x} . \frac{2x}{\sin 2x} . 2\right]$

$$= 2 \lim_{x \to 0} \left[\frac{\sin 4x}{4x}\right] \div \left[\frac{\sin 2x}{2x}\right]$$

$$= 2 \lim_{4x \to 0} \left[\frac{\sin 4x}{4x}\right] \div \lim_{2x \to 0} \left[\frac{\sin 2x}{2x}\right] = 2.1.1 = 2.$$

Example 10. *Evaluate* $\lim_{x \to 2} \left\{\frac{(x^2-4)}{\sqrt{3x-2}-\sqrt{x+2}}\right\}.$

Sol. $\lim_{x \to 2} \left\{\frac{(x^2-4)}{\sqrt{3x-2}-\sqrt{x+2}}\right\}$

$$= \lim_{x \to 2} \left\{\frac{(x^2-4)}{\sqrt{3x-2}-\sqrt{x+2}} \times \frac{(\sqrt{3x-2}+\sqrt{x-2})}{(\sqrt{3x-2}+\sqrt{x-2})}\right\}$$

$$= \lim_{x \to 2} \left\{\frac{(x^2-4)(\sqrt{3x-2}+\sqrt{x-2})}{(3x-2)-(x+2)}\right\}$$

$$= \lim_{x \to 2} \left\{\frac{(x^2-4)(\sqrt{3x-2}+\sqrt{x-2})}{2(x-2)}\right\}$$

$$= \lim_{x \to 2} \frac{(x+2)(\sqrt{3x-2}+\sqrt{x-2})}{2}$$

$= 8$ [putting $x = 2$]

Example 11. *Evaluate* $\lim_{x \to 0} \left(\frac{e^{3x}-1}{x}\right).$

Sol. $\lim_{x \to 0} \left(\frac{e^{3x}-1}{x}\right) = \lim_{3x \to 0} \left\{\left(\frac{e^{3x}-1}{3x}\right) \times 3\right\}$ $[\because \ (x \to 0) \Rightarrow (3x \to 0)]$

$$= 3 \times \lim_{y \to 0} \left(\frac{e^y - 1}{y} \right), \text{ where } y = 3x$$

$$= (3 \times 1) = 3. \qquad \left[\because \lim_{y \to 0} \left(\frac{e^y - 1}{y} \right) = 1 \right]$$

Example 12. *Evaluate* $\lim_{x \to 2} \left(\frac{e^x - e^2}{x - 2} \right)$.

Sol. Put $(x - 2) = y$, so that when $x \to 2$ then $y \to 0$.

$$\therefore \quad \lim_{x \to 2} \left(\frac{e^x - e^2}{x - 2} \right) = \lim_{y \to 0} \left(\frac{e^{y+2} - e^2}{y} \right)$$

$$= \lim_{y \to 0} \left\{ e^2 . \left(\frac{e^y - 1}{y} \right) \right\}$$

$$= e^2 . \lim_{y \to 0} \left(\frac{e^y - 1}{y} \right) = (e^2 \times 1) = e^2. \qquad \left[\because \lim_{y \to 0} \left(\frac{e^y - 1}{y} \right) = 1 \right]$$

Example 13. *Evaluate:*

(*i*) $\lim_{x \to 0} \left(\frac{e^{-x} - 1}{x} \right)$ (*ii*) $\lim_{x \to 0} \left(\frac{e^x - e^{-x}}{x} \right)$ (*iii*) $\lim_{x \to 0} \left(\frac{e^x + e^{-x} - 2}{x^2} \right)$.

Sol. (*i*) $\lim_{x \to 0} \left(\frac{e^{-x} - 1}{x} \right) = \lim_{y \to 0} \left(\frac{e^y - 1}{-y} \right)$, where $-x = y$

$$= - \lim_{y \to 0} \left(\frac{e^y - 1}{y} \right) = -1 \qquad \left[\because \lim_{y \to 0} \left(\frac{e^y - 1}{y} \right) = 1 \right]$$

(*ii*) $\lim_{x \to 0} \left(\frac{e^x - e^{-x}}{x} \right) = \left\{ \frac{(e^x - 1) - (e^{-x} - 1)}{x} \right\}$

$$= \lim_{x \to 0} \left\{ \left(\frac{(e^x - 1)}{x} \right) - \left(\frac{(e^{-x} - 1)}{x} \right) \right\}$$

$$= \lim_{x \to 0} \left(\frac{e^x - 1}{x} \right) - \lim_{x \to 0} \left(\frac{e^{-x} - 1}{x} \right)$$

$$= 1 - \lim_{y \to 0} \left(\frac{e^y - 1}{-y} \right), \quad \text{where } -x = y \quad \left[\because \lim_{x \to 0} \left(\frac{e^x - 1}{x} \right) = 1 \right]$$

$$= 1 + \lim_{y \to 0} \left(\frac{e^y - 1}{y} \right) = (1 + 1) = 2 \qquad \left[\because \lim_{x \to 0} \left(\frac{e^y - 1}{y} \right) = 1 \right]$$

$$(iii)\ \lim_{x\to 0}\left(\frac{e^x+e^{-x}-2}{x^2}\right)=\lim_{x\to 0}\left(\frac{e^{2x}+1-2e^x}{x^2e^x}\right)$$

$$=\lim_{x\to 0}\left\{\left(\frac{e^x-1}{x}\right)^2\cdot\frac{1}{e^x}\right\}$$

$$=\lim_{x\to 0}\left(\frac{e^x-1}{x}\right)^2\cdot\lim_{x\to 0}\frac{1}{e^x}$$

$$=\left(1^2\times\frac{1}{e^0}\right)=(1\times 1)=1. \qquad \left[\because\ \lim_{x\to 0}\left(\frac{e^x-1}{x}\right)=1\right]$$

Example 14. *Evaluate* $\lim_{x\to 0}\left(\frac{3^x-2^x}{x}\right)$.

Sol. $\lim_{x\to 0}\left(\frac{3^x-2^x}{x}\right)=\lim_{x\to 0}\left\{\frac{(3^x-1)-(2^x-1)}{x}\right\}$

$$=\lim_{x\to 0}\left(\frac{3^x-1}{x}\right)-\lim_{x\to 0}\left(\frac{2^x-1}{x}\right)$$

$$=(\log 3-\log 2)=\log\frac{3}{2}. \qquad \left[\because\ \lim_{x\to 0}\left(\frac{a^x-1}{x}\right)=\log a\right]$$

Example 15. *Evaluate* $\lim_{x\to a}\left\{\frac{x^{12}-a^{12}}{x-a}\right\}$.

Sol. $\lim_{x\to a}\left\{\frac{x^{12}-a^{12}}{x-a}\right\}=12a^{(12-1)}=12a^{11}$ $\qquad \left[\because\ \lim_{x\to a}\left(\frac{x^n-a^n}{x-a}\right)=na^{n-1}\right]$

Example 16. *Evaluate* $\lim_{x\to 2}\left(\frac{x^5-32}{x^3-8}\right)$.

Sol. $\lim_{x\to 2}\left(\frac{x^5-32}{x^3-8}\right)=\lim_{x\to 2}\left(\frac{x^5-2^5}{x^3-2^3}\right)$

$$=\lim_{x\to 2}\left\{\left(\frac{x^5-2^5}{x^3-2}\right)\div\left(\frac{x^3-2^3}{x-2}\right)\right\}$$

$$=\lim_{x\to 2}\left(\frac{x^5-2^5}{x-2}\right)\div\lim_{x\to 2}\left(\frac{x^3-2^3}{x-2}\right)$$

$$=[(5\times 2^4)\div(3\times 2^2)] \qquad \left[\because\ \lim_{x\to a}\left(\frac{x^n-a^n}{x-a}\right)=na^{n-1}\right]$$

$$= \left(\frac{80}{12}\right) = \frac{20}{3}.$$

Example 17. *Evaluate* $\lim\limits_{x \to 1} \dfrac{x^{15} - 1}{x^{10} - 1}$.

Sol. $$\lim_{x \to 1} \frac{x^{15} - 1}{x^{10} - 1} = \lim_{x \to 1} \left\{\left(\frac{x^{15} - 1}{x - 1}\right) \div \left(\frac{x^{10} - 1}{x - 1}\right)\right\}$$

$$= \lim_{x \to 1} \left(\frac{x^{15} - 1}{x - 1}\right) \div \lim_{x \to 1} \left(\frac{x^{10} - 1}{x - 1}\right)$$

$$= (15 \times 1^{14}) \div (10 \times 1^{9}) = \frac{15}{10} = \frac{3}{2}.$$

Example 18. *Evaluate* $\lim\limits_{x \to a} \left(\dfrac{x^m - a^m}{x^n - a^n}\right)$.

Sol. $$\lim_{x \to a} \left(\frac{x^m - a^m}{x^n - a^n}\right) = \lim_{x \to a} \left\{\left(\frac{x^m - a^m}{x - a}\right) \div \left(\frac{x^n - a^n}{x - a}\right)\right\}$$

$$= \lim_{x \to a} \left(\frac{x^m - a^m}{x - a}\right) \div \lim_{x \to a} \left(\frac{x^n - a^n}{x - a}\right)$$

$$= (ma^{m-1}) \div (na^{n-1})$$

$$= \frac{ma^{m-1}}{na^{n-1}} = \frac{m}{n} a^{m-n}.$$

Example 19. *Evaluate* $\lim\limits_{x \to a} \dfrac{(x + 2)^{3/2} - (a + 2)^{3/2}}{x - a}$.

Sol. $$\lim_{x \to a} \frac{(x + 2)^{3/2} - (a + 2)^{3/2}}{x - a}$$

$$= \lim_{(x+2) \to (a+2)} \frac{(x + 2)^{3/2} - (a + 2)^{3/2}}{(x + 2) - (a + 2)}$$

$$= \frac{3}{2} \cdot (a + 2)^{\left(\frac{3}{2} - 1\right)} = \frac{3}{2} (a+2)^{1/2} = \frac{3}{2} \sqrt{(a + 2)} \qquad \left[\because \lim_{x \to a} \left(\frac{x^n - a^n}{x - a}\right) = na^{n-1}\right]$$

Example 20. *Evaluate* $\lim\limits_{x \to 0} \dfrac{(1 + x)^n - 1}{x}$.

Sol. Put $(1 + x) = y$, so that when $x \to 0$ then $y \to 1$.

$$\therefore \quad \lim_{x \to 0} \frac{(1 + x)^n - 1}{x} = \lim_{y \to 1} \left(\frac{y^n - 1}{y - 1}\right)$$

$$= \lim_{y \to 1} \left(\frac{y^n - 1^n}{y - 1} \right) = n \times 1^{(n-1)} = (n \times 1) = n.$$

Example 21. *Evaluate* $\lim\limits_{x \to 1} \left(\dfrac{1 - x^{-1/3}}{1 - x^{-2/3}} \right)$.

Sol. $\lim\limits_{x \to 1} \left(\dfrac{1 - x^{-1/3}}{1 - x^{-2/3}} \right) = \lim\limits_{x \to 1} \left[\dfrac{(1 - x^{-1/3})}{(1)^2 - (x^{-1/3})^2} \right]$

$$= \lim_{x \to 1} \frac{(1 - x^{-1/3})}{(1 - x^{-1/3})(1 + x^{-1/3})}$$

$$= \lim_{x \to 1} \frac{1}{(1 + x^{-1/3})} = \frac{1}{\{1 + (1)^{-1/3}\}} = \frac{1}{2}.$$

Example 22. *Evaluate* $\lim\limits_{x \to 0} \dfrac{(\sqrt{1 + 3x} - \sqrt{1 - 3x})}{x}$.

Sol. $\lim\limits_{x \to 0} \dfrac{(\sqrt{1 + 3x} - \sqrt{1 - 3x})}{x}$

$$= \lim_{x \to 0} \left\{ \frac{(\sqrt{1 + 3x} - \sqrt{1 - 3x})}{x} \times \frac{(\sqrt{1 + 3x} + \sqrt{1 - 3x})}{(\sqrt{1 + 3x} + \sqrt{1 - 3x})} \right\}$$

$$= \lim_{x \to 0} \frac{\{(1 + 3x) - (1 - 3x)\}}{x(\sqrt{1 + 3x} + \sqrt{1 - 3x})} = \lim_{x \to 0} \frac{6x}{x(\sqrt{1 + 3x} + \sqrt{1 - 3x})}$$

$$= \lim_{x \to 0} \frac{6x}{(\sqrt{1 + 3x} + \sqrt{1 - 3x})} = \frac{6}{(\sqrt{1} + \sqrt{1})} = \frac{6}{2} = 3.$$

Example 23. *Find the* $\lim\limits_{x \to 0} \dfrac{(1 - \cos x \sqrt{\cos 2x})}{x^2}$.

Sol. $\lim\limits_{x \to 0} \dfrac{(1 - \cos x \sqrt{\cos 2x})}{x^2}$

$$= \lim_{x \to 0} \left\{ \frac{(1 - \cos x \sqrt{\cos 2x})}{x^2} \times \frac{(1 + \cos x \sqrt{\cos 2x})}{(1 + \cos x \sqrt{\cos 2x})} \right\}$$

$$= \lim_{x \to 0} \left\{ \frac{(1 - \cos^2 x \cos 2x)}{x^2} \times \frac{1}{(1 + \cos x \sqrt{\cos 2x})} \right\}$$

$$= \lim_{x \to 0} \frac{(1 - \cos^2 x \cos 2x)}{x^2} \times \lim_{x \to 0} \frac{1}{(1 + \cos x \sqrt{\cos 2x})}$$

$$= \lim_{x \to 0} \left\{ \frac{1 - \cos^2 x\,(1 - 2\sin^2 x)}{x^2} \right\} \times \frac{1}{(1 + 1\sqrt{1})}$$

$$= \lim_{x \to 0} \frac{(1 - \cos^2 x + 2\sin^2 x \cos^2 x)}{x^2} \times \frac{1}{2}$$

$$= \frac{1}{2} \times \lim_{x \to 0} \frac{(\sin^2 x + 2\sin^2 x \cos^2 x)}{x^2}$$

$$= \frac{1}{2} \times \lim_{x \to 0} \frac{\sin^2 x\,(1 + 2\cos x^2)}{x^2}$$

$$= \frac{1}{2} \times \lim_{x \to 0} \left(\frac{\sin x}{x} \right)^2 \times \lim_{x \to 0} (1 + 2\cos^2 x)$$

$$= \frac{1}{2} \times 1^2 \times (1 + 2 \times 1^2) = \frac{3}{2}.$$

Example 24. *Find the* $\displaystyle \lim_{x \to \frac{\pi}{4}} \frac{(\sin x - \cos x)}{\left(x - \frac{\pi}{4} \right)}$.

Sol. Put $\left(x - \frac{\pi}{4} \right) = y$, so that when $x \to \frac{\pi}{4}$ then $y \to 0$.

$$\therefore \quad \lim_{x \to \frac{\pi}{4}} \frac{(\sin x - \cos x)}{\left(x - \frac{\pi}{4} \right)}$$

$$= \lim_{y \to 0} \frac{\left\{ \sin\left(\frac{\pi}{4} + y \right) - \cos\left(\frac{\pi}{4} + y \right) \right\}}{y} \qquad \text{[putting } (x = y)\text{]}$$

$$= \lim_{y \to 0} \frac{\left\{ \left(\sin\frac{\pi}{4} \cos y + \cos\frac{\pi}{4} \sin y \right) - \left(\cos\frac{\pi}{4} \cos y - \sin\frac{\pi}{4} \sin y \right) \right\}}{y}$$

$$= \frac{2}{\sqrt{2}} \times \lim_{y \to 0} \left(\frac{\sin y}{y} \right) = (\sqrt{2} \times 1) = \sqrt{2}.$$

Example 25. *Calculate* $\displaystyle \lim_{y \to \frac{\pi}{2}} \frac{\cos x}{\left(\frac{\pi}{2} - x \right)}$.

Sol. Put $\left(x = \frac{\pi}{2} \right) = y$, so that when $x \to \frac{\pi}{2}$ then $y \to 0$.

$$= \lim_{y \to \frac{\pi}{2}} \frac{\cos x}{\left(\frac{\pi}{2} - x\right)} = \lim_{y \to 0} \frac{\cos\left(\frac{\pi}{2} + y\right)}{-y}$$

$$= \lim_{y \to 0} \left(\frac{-\sin y}{-y}\right) = \lim_{y \to 0} \frac{\sin y}{y} = 1.$$

Example 26. *Evaluate* $\lim_{x \to \frac{\pi}{6}} \frac{(\sqrt{3}\sin x - \cos x)}{\left(x - \frac{\pi}{6}\right)}$.

Sol. $\lim_{x \to \frac{\pi}{6}} \frac{(\sqrt{3}\sin x - \cos x)}{\left(x - \frac{\pi}{6}\right)}$

$$= \lim_{x \to \frac{\pi}{6}} \frac{2\left(\frac{\sqrt{3}}{2}\sin x - \frac{1}{2}\cos x\right)}{\left(x - \frac{\pi}{6}\right)} = \lim_{x \to \frac{\pi}{6}} \frac{2\left(\sin x \cos\frac{\pi}{6} - \cos x \sin\frac{\pi}{6}\right)}{\left(x - \frac{\pi}{6}\right)}$$

$$= \lim_{x \to \frac{\pi}{6}} \frac{\sin\left(x - \frac{\pi}{6}\right)}{\left(x - \frac{\pi}{6}\right)} = 2\lim_{y \to 0} \frac{\sin y}{y} = 2. \qquad \left[\text{putting } \left(x - \frac{\pi}{6}\right) = y\right]$$

Example 27. *Evaluate:*

(*i*) $\lim_{x \to 0} \left(\frac{1 - \cos 4x}{1 - \cos 5x}\right)$ (*ii*) $\lim_{x \to 0} \left(\frac{1 - \cos mx}{1 - \cos nx}\right)$.

Sol. (*i*) $\lim_{x \to 0} \left(\frac{1 - \cos 4x}{1 - \cos 5x}\right) = \lim_{x \to 0} \frac{2\sin^2 2x}{2\sin^2 (5x/2)}$

$$= \lim_{x \to 0} \frac{\sin^2 2x}{\sin^2 (5x/2)} = \lim_{x \to 0} \left\{\frac{\left\{\frac{\sin 2x}{2x}\right\}^2 . 4x^2}{\left\{\frac{\sin (5x/2)}{(5x/2)}\right\} . \frac{25x^2}{4}}\right\}$$

$$= \frac{16}{25} . \frac{\lim_{2x \to 0} \left(\frac{\sin 2x}{2x}\right)^2}{\lim_{\frac{5x}{2} \to 0} \left\{\frac{\sin (5x/2)}{(5x/2)}\right\}^2} = \left(\frac{16}{25} \times \frac{1^2}{1^2}\right) = \frac{16}{25}.$$

$$(ii)\ \lim_{x\to 0}\left(\frac{1-\cos mx}{1-\cos nx}\right) = \lim_{x\to 0}\left\{\frac{2\sin^2(mx/2)}{2\sin^2(nx/2)}\right\} = \lim_{x\to 0}\left\{\frac{\sin^2(mx/2)}{\sin^2(nx/2)}\right\}$$

$$= \lim_{x\to 0}\left\{\frac{\dfrac{\sin^2(mx/2)}{(mx/2)^2}\times\dfrac{m^2x^2}{4}}{\dfrac{\sin^2(nx/2)}{(nx/2)^2}\times\dfrac{n^2x^2}{4}}\right\} = \frac{m^2}{n^2}\cdot\frac{\lim\limits_{\frac{mx}{2}\to 0}\left\{\dfrac{\sin(mx/2)}{(mx/2)}\right\}^2}{\lim\limits_{\frac{nx}{2}\to 0}\left\{\dfrac{\sin(nx/2)}{nx/2}\right\}^2}$$

$$= \left(\frac{m^2}{n^2}\times\frac{1^2}{1^2}\right) = \frac{m^2}{n^2}.$$

Example 28. *Evaluate* $\lim\limits_{x\to 0}\left(\dfrac{\sin 2x+\sin 6x}{\sin 5x-\sin 3x}\right)$.

Sol. $\lim\limits_{x\to 0}\left(\dfrac{\sin 2x+\sin 6x}{\sin 5x-\sin 3x}\right)$

$$= \lim_{x\to 0}\left(\frac{2\sin 4x\cos 2x}{2\cos 4x\sin x}\right)$$

$$= \lim_{x\to 0}\left\{\frac{\sin 4x\cos 2x}{\cos 4x\sin x}\right\}$$

$$= \lim_{x\to 0}\left\{\frac{\sin 4x}{4x}\times\frac{x}{\sin x}\times\cos 2x\times\frac{1}{\cos 4x}\times 4\right\}$$

$$= 4\times\lim_{x\to 0}\frac{\sin 4x}{4x}\times\lim_{x\to 0}\left(\frac{x}{\sin x}\right)\times\lim_{x\to 0}\cos 2x\times\frac{1}{\lim\limits_{x\to 0}\cos 4x}$$

$$= \left(4\times 1\times 1\times 1\times\frac{1}{1}\right) = 4.$$

Example 29. *Evaluate* $\lim\limits_{x\to \pi}\left(\dfrac{1+\cos x}{\tan^2 x}\right)$.

Sol. $\lim\limits_{x\to \pi}\left(\dfrac{1+\cos x}{\tan^2 x}\right) = \lim\limits_{x\to \pi}\dfrac{\cos^2 x\,(1+\cos x)}{\sin^2 x}$

$$= \lim_{x\to \pi}\frac{\cos^2 x\,(1+\cos x)}{(1-\cos^2 x)}$$

$$= \lim_{x\to \pi}\frac{\cos^2\pi}{(1-\cos\pi)} = \frac{1}{2}. \qquad \text{[putting } x=\pi\text{]}$$

Example 30. *Evaluate* $\lim\limits_{x\to 0}\dfrac{x\tan x}{(1-\cos x)}$.

Sol. $\lim_{x \to 0} \frac{x \tan x}{(1 - \cos x)} = \lim_{x \to 0} \frac{x \sin x}{\cos x (1 - \cos x)}$

$$= \lim_{x \to 0} \frac{x \sin^2 x}{\sin x \cos x (1 - \cos x)}$$

$$= \lim_{x \to 0} \frac{x (1 - \cos^2 x)}{\sin x \cos x (1 - \cos x)} = \lim_{x \to 0} \frac{x (1 + \cos x)}{\sin x \cos x}$$

$$= \lim_{x \to 0} \frac{x}{\sin x} . \lim_{x \to 0} \frac{(1 + \cos x)}{\cos x} = \left(1 \times \frac{2}{1}\right) = 2.$$

Example 31. *Evaluate:*

(*i*) $\lim_{x \to \frac{\pi}{2}} (\sec x - \tan x)$ (*ii*) $\lim_{x \to 0} \frac{(\text{cosec } x - \cot x)}{x}$

Sol. (*i*) $\lim_{x \to \frac{\pi}{2}} (\sec x - \tan x) = \lim_{x \to \frac{\pi}{2}} \left(\frac{1}{\cos x} - \frac{\sin x}{\cos x} \right)$

$$= \lim_{x \to \frac{\pi}{2}} \frac{(1 - \sin x)}{\cos x} = \lim_{x \to \frac{\pi}{2}} \frac{(1 - \sin x) \cos x}{\cos^2 x}$$

$$= \lim_{x \to \frac{\pi}{2}} \frac{(1 - \sin x) \cos x}{(1 - \sin^2 x)}$$

$$= \lim_{x \to \frac{\pi}{2}} \frac{\cos x}{(1 + \sin x)} = 0. \qquad \left[\text{putting } x = \frac{\pi}{2}\right]$$

(*ii*) $\lim_{x \to 0} \frac{(\text{cosec } x - \cot x)}{x} = \lim_{x \to 0} \frac{\left(\frac{1}{\sin x} - \frac{\cos x}{\sin x} \right)}{x}$

$$= \lim_{x \to 0} \frac{(1 - \cos x)}{x \sin x} = \lim_{x \to 0} \frac{(1 - \cos x) \sin x}{x \sin^2 x}$$

$$= \lim_{x \to 0} \frac{(1 - \cos x) \sin x}{x (1 - \cos^2 x)} = \lim_{x \to 0} \frac{\sin x}{x (1 + x \cos x)}$$

$$= \lim_{x \to 0} \frac{\sin x}{x} . \lim_{x \to 0} \frac{1}{(1 + \cos x)} = \left(1 \times \frac{1}{2}\right) = \frac{1}{2}.$$

Example 32. *Evaluate:*

(*i*) $\lim_{x \to 0} \left(\frac{\tan x - \sin x}{\sin^3 x} \right)$ (*ii*) $\lim_{x \to 0} \left(\frac{\tan x - \sin x}{x^3} \right)$.

Sol. (*i*) $\lim_{x \to 0} \left(\frac{\tan x - \sin x}{\sin^3 x} \right) = \lim_{x \to 0} \frac{(\sin x - \sin x \cos x)}{\cos x \sin^3 x}$

$$= \lim_{x \to 0} \frac{\sin x\,(1 - \cos x)}{\cos x \sin^3 x} = \lim_{x \to 0} \frac{(1 - \cos x)}{\cos x \sin^2 x}$$

$$= \lim_{x \to 0} \frac{(1 - \cos x)}{\cos x\,(1 + \cos^2 x)}$$

$$= \lim_{x \to 0} \frac{1}{\cos x\,(1 + \cos x)} = \frac{1}{2} \qquad \text{[putting } x = 0\text{]}$$

$$(ii)\ \lim_{x \to 0} \frac{(\tan x - \sin x)}{x^3} = \lim_{x \to 0} \left(\frac{\sin x - \sin x \cos x}{x^3 \cos x} \right)$$

$$= \lim_{x \to 0} \frac{\sin x\,(1 - \cos x)}{x^3 \cos x} = \lim_{x \to 0} \left\{ \frac{\sin x}{x} \cdot \frac{2\sin^2 (x/2)}{x^2} \cdot \frac{1}{\cos x} \right\}$$

$$= \lim_{x \to 0} \left\{ \frac{\sin x}{x} \cdot \frac{2\sin^2 (x/2)}{(x/2)^2} \cdot \frac{1}{4} \cdot \frac{1}{\cos x} \right\}$$

$$= \frac{1}{2} \cdot \lim_{x \to 0} \frac{\sin x}{x} \cdot \lim_{x \to 0} \left\{ \frac{\sin (x/2)^2}{(x/2)} \right\} \cdot \lim_{x \to 0} \frac{1}{\cos x}$$

$$= \left(\frac{1}{2} \times 1 \times 1^2 \times 1 \right) = \frac{1}{2}.$$

Example 33. *Evaluate* $\lim_{x \to 0} \left(\frac{x^3 \cot x}{1 - \cos x} \right)$.

Sol. $\lim_{x \to 0} \left(\frac{x^3 \cot x}{1 - \cos x} \right)$

$$= \lim_{x \to 0} \left\{ \frac{x^3 \cot x}{(1 - \cos x)} \times \frac{(1 + \cos x)}{(1 + \cos x)} \right\}$$

$$= \lim_{x \to 0} \frac{x^3 \cos x\,(1 + \cos x)}{\sin x\,(1 - \cos^2 x)} = \lim_{x \to 0} \frac{x^3 \cos x\,(1 + \cos x)}{\sin^3 x}$$

$$= \lim_{x \to 0} \left(\frac{x}{\sin x} \right)^3 \times \lim_{x \to 0} \cos x \times \lim_{x \to 0} (1 + \cos x)$$

$$= (1^3 \times 1 \times 2) = 2.$$

Example 34. *Evaluate* $\lim_{x \to 0} \left(\frac{\sin 3x + 7x}{4x + \sin 2x} \right)$.

Sol. $\lim_{x \to 0} \left(\frac{\sin 3x + 7x}{4x + \sin 2x} \right) = \lim_{x \to 0} \left(\frac{\frac{\sin 3x}{x} + 7}{4 + \frac{\sin 2x}{x}} \right)$ [dividing num. and denom. by x]

$$= \lim_{x \to 0} \left[\frac{\left(\frac{\sin 3x}{3x}\right).3 + 7}{4 + \left(\frac{\sin 2x}{2x}\right).2} \right] = \lim_{x \to 0} \frac{\left[3 \times \lim\limits_{3x \to 0} \frac{\sin 3x}{3x}\right] + 7}{4 + \left[2 \times \lim\limits_{2x \to 0} \left(\frac{\sin 2x}{2x}\right)\right]}$$

$$= \frac{(3 \times 1) + 7}{4 + (2 \times 1)} = \frac{10}{6} = \frac{5}{3}.$$

Example 35. *Evaluate* $\lim\limits_{x \to 0} \left(\frac{\tan 3x - 2x}{3x - \sin^2 x}\right)$.

Sol. $\lim\limits_{x \to 0} \left(\frac{\tan 3x - 2x}{3x - \sin^2 x}\right)$

$$= \lim_{x \to 0} \frac{\left(\frac{\tan 3x}{3x} - \frac{2}{3}\right)}{\left(1 - \frac{1}{3} . \frac{\sin x}{x} . \sin x\right)} \qquad \text{[dividing num. and denom. by } 3x]$$

$$= \frac{\left(\lim\limits_{3x \to 0} \frac{\tan 3x}{3x} - \frac{2}{3}\right)}{\left\{1 - \frac{1}{3} \lim\limits_{x \to 0} \frac{\sin x}{x} . \lim\limits_{x \to 0} \sin x\right\}} = \frac{\left(1 - \frac{2}{3}\right)}{1 - \left(\frac{1}{3} \times 1 \times 0\right)} = \frac{\left(\frac{1}{3}\right)}{1} = \frac{1}{3}.$$

Example 36. *Evaluate* $\lim\limits_{x \to 0} \frac{x \tan 4x}{1 - \cos 4x}$.

Sol. $\lim\limits_{x \to 0} \frac{x \tan 4x}{1 - \cos 4x} = \lim\limits_{x \to 0} \frac{x \sin 4x}{\cos 4x \, (2 \sin^2 2x)}$

$$= \lim_{x \to 0} \frac{2x \sin 2x \cos 2x}{\cos 4x \, (2 \sin^2 2x)} = \lim_{x \to 0} \left(\frac{\cos 2x}{\cos 4x} . \frac{2x}{\sin 2x} \times \frac{1}{2}\right)$$

$$= \frac{1}{2} \times \frac{\lim\limits_{2x \to 0} \cos 2x}{\lim\limits_{4x \to 0} \cos 4x} \times \lim_{2x \to 0} \left(\frac{2x}{\sin 2x}\right) = \left(\frac{1}{2} \times \frac{1}{1} \times 1\right) = \frac{}{2}$$

Example 37. *Evaluate:*

(*i*) $\lim\limits_{x \to 0} \frac{\sin 2x}{x}$ (*ii*) $\lim\limits_{x \to 0} \frac{\sin 3x}{5x}$ (*iii*) $\lim\limits_{x \to 0} \left(\frac{\sin ax}{\sin bx}\right)$.

Sol. (*i*) $\lim\limits_{x \to 0} \frac{\sin 2x}{x} = \lim\limits_{x \to 0} \left(\frac{2 \times \sin 2x}{2x}\right)$

$$= 2. \lim_{2x \to 0} \frac{\sin 2x}{2x} \qquad [\because \text{ when } x \to 0 \text{ then } 2x \to 0]$$

$$= (2 \times 1) = 2 \qquad \left[\because \lim_{\theta \to 0} \frac{\sin \theta}{\theta} = 1\right]$$

(ii) $\lim_{x \to 0} \frac{\sin 3x}{5x}$

$$= \lim_{x \to 0} \left(\frac{3}{5} \cdot \frac{\sin 3x}{3x}\right)$$

$$= \frac{3}{5} \cdot \lim_{3x \to 0} \frac{\sin 3x}{3x} \qquad [\because \text{ when } x \to 0 \text{ then } 3x \to 0]$$

$$= \left(\frac{3}{5} \times 1\right) = \frac{3}{5} \qquad \left[\because \lim_{\theta \to 0} \frac{\sin \theta}{\theta} = 1\right]$$

(iii) $\lim_{x \to 0} \left(\frac{\sin ax}{\sin bx}\right)$

$$= \lim_{x \to 0} \frac{ax.\left(\frac{\sin ax}{ax}\right)}{bx.\left(\frac{\sin bx}{bx}\right)}$$

$$= \frac{a}{b} \cdot \lim_{x \to 0} \frac{\left(\frac{\sin ax}{ax}\right)}{\left(\frac{\sin bx}{bx}\right)}$$

$$= \frac{a}{b} \cdot \frac{\lim_{ax \to 0} \frac{\sin ax}{ax}}{\lim_{bx \to 0} \frac{\sin bx}{bx}} \qquad [\because x \to 0 \Rightarrow ax \to 0 \text{ and } bx \to 0]$$

$$= \left(\frac{a}{b} \times \frac{1}{1}\right) = \frac{a}{b}. \qquad \left[\because \lim_{\theta \to 0} \frac{\sin \theta}{\theta} = 1\right]$$

Example 38. *Evaluate* $\lim_{x \to 0} \frac{\sin 5x}{\tan 3x}$.

Sol. $\lim_{x \to 0} \frac{\sin 5x}{\tan 3x}$

$$= \lim_{x \to 0} \frac{5x.\left(\frac{\sin 5x}{5x}\right)}{3x.\left(\frac{\tan 3x}{3x}\right)}$$

$$= \frac{5}{3} \cdot \frac{\lim_{5x \to 0}\left(\frac{\sin 5x}{5x}\right)}{\lim_{3x \to 0}\left(\frac{\tan 3x}{3x}\right)} = \left(\frac{5}{3} \times \frac{1}{1}\right) = \frac{5}{3}.$$

Example 39. *Evaluate* $\lim_{x \to 0} \frac{(1 - \cos x)}{x^2}$.

Sol. $\lim_{x \to 0} \frac{(1 - \cos x)}{x^2}$

$$= \lim_{x \to 0} \frac{2 \sin^2 (x/2)}{\left(\frac{x}{2}\right)^2 \times 4}$$

$$= \frac{1}{2} \lim_{x \to 0} \left\{\frac{\sin (x/2)}{(x/2)}\right\}^2 = \left(\frac{1}{2} \times 1^2\right) = \frac{1}{2}.$$

Example 40. *Evaluate* $\lim_{x \to 0} \left(\frac{3^{2x} - 2^{3x}}{x} \right)$.

Sol. $$\lim_{x \to 0} \left(\frac{3^{2x} - 2^{3x}}{x} \right) = \lim_{x \to 0} \left\{ \frac{(3^{2x} - 1) - (2^{3x} - 1)}{x} \right\}$$

$$= \lim_{x \to 0} \left(\frac{3^{2x} - 1}{x} \right) - \lim_{x \to 0} \left(\frac{2^{3x} - 1}{x} \right)$$

$$= \lim_{x \to 0} \left\{ \left(\frac{3^{2x} - 1}{2x} \right) . 2 \right\} - \lim_{x \to 0} \left\{ \left(\frac{2^{3x} - 1}{3x} \right) . 3 \right\}$$

$$= 2. \lim_{2x \to 0} \left(\frac{3^{2x} - 1}{2x} \right) - 3 . \lim_{3x \to 0} \left(\frac{2^{3x} - 1}{3x} \right)$$

$$= (2 \log 3 - 3 \log 2) \qquad \left[\because \lim_{y \to 0} \left(\frac{a^y - 1}{y} \right) \log e \right]$$

$$= [\log (3)^2 - \log (2)^3] = (\log 9 - \log 8) = \log \left(\frac{9}{8} \right).$$

Example 41. *Evaluate* $\lim_{x \to 0} \left(\frac{3^{2x} - 1}{2^{3x} - 1} \right)$.

Sol. $$\lim_{x \to 0} \left(\frac{3^{2x} - 1}{2^{3x} - 1} \right) = \lim_{x \to 0} \frac{\left\{ \left(\frac{3^{2x} - 1}{2x} \right) . 2x \right\}}{\left\{ \left(\frac{2^{3x} - 1}{3x} \right) . 3x \right\}} = \frac{2}{3} . \frac{\lim_{2x \to 0} \left(\frac{3^{2x} - 1}{2x} \right)}{\lim_{3x \to 0} \left(\frac{2^{3x} - 1}{3x} \right)}$$

$$= \frac{2}{3} . \frac{\log 3}{\log 2} \qquad \left[\because \lim_{y \to 0} \left(\frac{a^y - 1}{y} \right) = \log e \right]$$

$$= \frac{2 \log 3}{3 \log 2} = \frac{\log (3^2)}{\log (2^3)} = \frac{\log 9}{\log 8}.$$

Example 42. *Evaluate* $\lim_{x \to 1} \frac{x^2 - \sqrt{x}}{\sqrt{x} - 1}$.

Sol. $$\lim_{x \to 1} \frac{x^2 - \sqrt{x}}{\sqrt{x} - 1} = \lim_{x \to 1} \left\{ \frac{(x^2 - \sqrt{x})}{(\sqrt{x} - 1)} \times \frac{(\sqrt{x} + 1)}{(\sqrt{x} + 1)} \times \frac{(x^2 + \sqrt{x})}{(x^2 + \sqrt{x})} \right\}$$

$$= \lim_{x \to 1} \frac{(x^4 - x)(\sqrt{x} + 1)}{(x - 1)(x^2 + \sqrt{x})} = \lim_{x \to 1} \frac{x (x^3 - 1)(\sqrt{x} + 1)}{(x - 1)(x^2 + \sqrt{x})}$$

$$= \lim_{x \to 1} \frac{x(x-1)(x^2+x+1)(\sqrt{x}+1)}{(x-1)(x^2+\sqrt{x})}$$

$$= \lim_{x \to 1} \frac{x(x^2+x+1)(\sqrt{x}+1)}{(x^2+\sqrt{x})}$$

$$= \frac{1 \times (1^2+1+1)(\sqrt{1}+1)}{(1^2+\sqrt{1})} = \left(\frac{1 \times 3 \times 2}{2}\right) = 3. \qquad \text{[putting } x = 1\text{]}$$

Example 43. *Evaluate* $\lim\limits_{x \to a} \dfrac{\sqrt{a+2x}-\sqrt{3x}}{\sqrt{3a+x}-2\sqrt{x}}$.

Sol. $\lim\limits_{x \to a} \dfrac{(\sqrt{a+2x}-\sqrt{3x})}{(\sqrt{3a+x}-2\sqrt{x})}$

$$= \lim_{x \to a} \frac{(\sqrt{a+2x}-\sqrt{3x})}{(\sqrt{3a+x}-2\sqrt{x})} \times \frac{(\sqrt{3a+x}+2\sqrt{x})}{(\sqrt{3a+x}+2\sqrt{x})} \times \frac{(\sqrt{a+2x}+\sqrt{3x})}{(\sqrt{a+2x}+\sqrt{3x})}$$

$$= \lim_{x \to a} \frac{[(a+2x)-3x] \times (\sqrt{3a+x}+2\sqrt{x})}{[(3a+x)-4x] \times (\sqrt{a+2x}+\sqrt{3x})}$$

$$= \lim_{x \to a} \frac{(a-x) \times (\sqrt{3a+x}+2\sqrt{x})}{3(a-x) \times (\sqrt{a+2x}+\sqrt{3x})} = \lim_{x \to a} \frac{(\sqrt{3a+x}+2\sqrt{x})}{3(\sqrt{a+2x}+\sqrt{3x})}$$

$$= \frac{(\sqrt{4a}+2\sqrt{a})}{3(\sqrt{3a}+\sqrt{3a})} = \frac{4\sqrt{a}}{6\sqrt{3}\sqrt{a}} = \frac{2}{3\sqrt{3}}.$$

Example 44. *Evaluate* $\lim\limits_{x \to 1} \dfrac{(2x-3)(\sqrt{x}-1)}{(2x^2+x-3)}$.

Sol. $\lim\limits_{x \to 1} \dfrac{(2x-3)(\sqrt{x}-1)}{(2x^2+x-3)} = \lim\limits_{x \to 1} \dfrac{(2x-3)(\sqrt{x}-1)}{(2x+3)(x-1)}$

$$= \lim_{x \to 1} \frac{(2x-3)(\sqrt{x}-1)(\sqrt{x}+1)}{(2x+3)(x-1)(\sqrt{x}+1)}$$

$$= \lim_{x \to 1} \frac{(2x-3)(x-1)}{(2x+3)(x-1)(\sqrt{x}+1)}$$

$$= \lim_{x \to 1} \frac{(2x-3)}{(2x+3)(\sqrt{x}+1)} = \frac{(2 \times 1 - 3)}{(2 \times 1 + 3)(\sqrt{1}+1)} = \frac{-1}{10}.$$

Example 45. *Evaluate* $\lim\limits_{x \to 0} \dfrac{(1+x)^6-1}{(1+x)^5-1}$.

Sol. Put $(1+x) = y$, so that when $x \to 0$ then $y \to 1$.

$\therefore \quad \lim\limits_{x \to 0} \dfrac{(1+x)^6-1}{(1+x)^5-1}$

$$= \lim_{y \to 1} \left(\frac{y^6 - 1}{y^5 - 1} \right) = \frac{\lim_{y \to 1} \left(\frac{y^6 - 1}{y - 1} \right)}{\lim_{y \to 1} \left(\frac{y^5 - 1}{y - 1} \right)}$$

$$= \frac{\lim_{y \to 1} \left(\frac{y^6 - 1^6}{y - 1} \right)}{\lim_{y \to 1} \left(\frac{y^5 - 1^5}{y - 1} \right)} = \frac{6 \times 1^{(6-1)}}{5 \times 1^{(5-1)}}$$

$$= \frac{6 \times 1^5}{5 \times 1^4} = \frac{6}{5}. \qquad \left[\because \lim_{y \to a} \left(\frac{x^n - a^n}{x - a} \right) = na^n \right]$$

EXERCISE 1.1.1.

1. If $f(x) = x^3 - 2x^3 + 5x + 9$, find the value of $f(0)$, $f(2)$, $f(-3)$, $(x + 1)$ and $f(2/x)$.
   [**Ans.** $f(0) = 9$, $f(2) = 19$, $f(-3) = -51$, $f(x + 1) = x^3 + x^2 + 4n + 8$]

2. If $f(x) = \dfrac{2x - 3}{5x - 2}$, show that $x = f(y)$.

3. If $f(x) = \dfrac{1 + x}{1 - x}$, show that $f[f(\tan \theta)] = -\cot \theta$.

4. If $\phi(x) = \dfrac{1}{2}(b^x + b^{-x})$ and $\theta(x) = 1/2\,(b^x - b^{-x})$.
   Show that
   (*i*) $\phi(x + y) = \phi(x)\,\phi(y) + \theta(x)\,\theta(y)$ (*ii*) $\phi(x + y) = \phi(x)\,\theta(y) + \phi(y)\,\theta(x)$.

5. Find $f(-1)$, $f(0)$, $f(1)$, when

$$f(x) = \begin{cases} \text{arc sin } x & \text{for } -1 \le x \le 0 \\ \text{arc tan } x & \text{for } 0 < x < \infty \end{cases}$$

   $\left[\textbf{Ans. } \dfrac{-\pi}{2}; 0; \dfrac{\pi}{4} \right]$

Evaluate the following limits:

6. $\displaystyle \operatorname*{Lt}_{x \to 1} \frac{x - 1}{\sqrt{x^2 + 3} - 2}$

7. $\displaystyle \operatorname*{Lt}_{x \to 0} \frac{\sqrt{1 + x} - \sqrt{1 - x}}{x}$

8. $\displaystyle \operatorname*{Lt}_{x \to 3} \frac{x^3 + 2x^2 - 14x - 3}{x^2 - x - 6}$

9. $\displaystyle \operatorname*{Lt}_{x \to \infty} \frac{\sqrt{x^4 + 1} - 2x^2 - 1}{x^2}$

10. $\displaystyle \operatorname*{Lt}_{x \to 0} \frac{x^2 \sin(1/x)}{\sin x}$

11. $\displaystyle \operatorname*{Lt}_{x \to 0} \frac{\sqrt{(1 + x)} - 1}{\sqrt{x}}$

12. $\displaystyle \operatorname*{Lt}_{x \to 0} \frac{\sin mx - \sin nx}{x} \; (m \ne n)$

13. $\displaystyle \operatorname*{Lt}_{x \to 0} \frac{1 - \cos^2 x}{2x^2}$

14. $\displaystyle \frac{\tan x - \sin x}{x^3}$

15. $\displaystyle \operatorname*{Lt}_{x \to 0} \frac{\operatorname{cosec} x - \cot x}{x}$

16. $\displaystyle \operatorname*{Lt}_{x \to 0} \frac{\cos ax - \cos bx}{x^2}$

17. $\displaystyle \operatorname*{Lt}_{x \to 0} \frac{x^2 \sin(1/x)}{\sin x}$

18. $\underset{x \to 0}{\text{Lt}} \dfrac{\sqrt{(1+x)}-1}{\sqrt{x}}$

19. $\underset{x \to 0}{\text{Lt}} (1+\sin x)^{\cot x} (1+\sin x)^{\cot x}$

20. $\underset{x \to 0}{\text{Lt}} \dfrac{e^{\sin x}-1}{x}$

21. $\underset{x \to 0}{\text{Lt}} \dfrac{\sqrt{(1+x+x^2)}-1}{\sin 4x}$

22. $\underset{x \to 0}{\text{Lt}} \left\{\dfrac{\sqrt{(1+\sin x)}-\sqrt{(1-\sin x)}}{x}\right\}$

23. $\lim_{x \to a} \left(\dfrac{xe^{-x}-ae^{-a}}{x-a}\right) = (1-a)\, e^{-a}$

24. $\lim_{x \to 0} \left[\dfrac{x^2+3x^3+(2x-\sin x)^2}{2x^2\cos^2 x-5x^3}\right] = 1.$

25. $\lim_{t \to 0} \left[\dfrac{\sqrt{(1+t)}-\sqrt{1-t}}{t}\right] = \lim_{x \to 0} \left[\dfrac{(1+x)^n-1}{nx}\right]$

Evaluate the following limits

26. $\lim_{x \to 2} (5-x)$ [**Ans.** 3]

27. $\lim_{x \to 1} (6x^2-4x+3)$ [**Ans.** 5]

28. $\lim_{x \to 3} \left(\dfrac{x^2+9}{x+3}\right)$ [**Ans.** 3]

29. $\lim_{x \to 3} \left(\dfrac{x^2-4x}{x-2}\right)$ [**Ans.** –3]

30. $\lim_{x \to 5} \left(\dfrac{x^2-25}{x-5}\right)$ [**Ans.** 10]

31. $\lim_{x \to 1} \left(\dfrac{x^3-1}{x-1}\right)$ [**Ans.** 3]

32. $\lim_{x \to -2} \left(\dfrac{x^3+8}{x+2}\right)$ [**Ans.** 12]

33. $\lim_{x \to 3} \left(\dfrac{x^4-81}{x-3}\right)$ [**Ans.** 108]

34. $\lim_{x \to 3} \left(\dfrac{x^2-4x+3}{x^2-2x-3}\right)$ [**Ans.** 1/2]

35. $\lim_{x \to a} \left(\dfrac{x^{5/2}-a^{5/2}}{x-a}\right)$ $\left[\textbf{Ans. } \dfrac{5}{2} a^{3/2}\right]$

36. $\lim_{x \to a} \left\{\dfrac{x+2^{5/3}-(a+2)^{5/3}}{x-a}\right\}$ $\left[\textbf{Ans. } \dfrac{5}{3}(a+3)^{2/3}\right]$

37. $\lim_{x \to 1} \left(\dfrac{x^n-1}{x-1}\right)$ [**Ans.** n]

38. $\lim_{x \to 2} \left(\dfrac{x^5-32}{x^3-8}\right)$ $\left[\textbf{Ans. } \dfrac{20}{3}\right]$

39. $\lim_{x \to \frac{1}{2}} \left(\dfrac{4x^2-1}{2x-1}\right)$ [**Ans.** 2]

40. $\lim_{x \to a} \left(\dfrac{\sqrt{x}-\sqrt{a}}{x-a}\right)$ $\left[\textbf{Ans. } \dfrac{1}{2\sqrt{a}}\right]$

41. $\lim_{x \to 0} \left(\dfrac{\sqrt{x+h}-\sqrt{x}}{h}\right)$ $\left[\textbf{Ans. } \dfrac{1}{2\sqrt{x}}\right]$

42. $\lim_{h \to 0} \dfrac{1}{h}\left\{\dfrac{1}{\sqrt{x+h}}-\dfrac{1}{\sqrt{x}}\right\}$ [**Ans.** $-1/2x^{3/2}$]

43. $\lim_{x \to 0} \left(\dfrac{\sqrt{1+x}-1}{x}\right)$ [**Ans.** 1/2]

44. $\lim_{x \to 0} \left(\dfrac{\sqrt{2-x}-\sqrt{2+x}}{x}\right)$ [**Ans.** $-1/\sqrt{2}$]

45. $\lim_{x \to 0} \left(\dfrac{\sqrt{1+x+x^2}-1}{x}\right)$ [**Ans.** 1/2]

46. $\lim_{x \to 0} \left(\dfrac{\sqrt{3+x}-1}{2-x}\right)$ $\left[\textbf{Ans. } \dfrac{1}{\sqrt{3}+1}\right]$

47. $\lim_{x \to 0} \left(\dfrac{2x}{\sqrt{a+x}-\sqrt{a-x}}\right)$ [**Ans.** $2\sqrt{a}$]

48. $\lim_{x \to 1} \left(\dfrac{\sqrt{3+x}-\sqrt{5-x}}{x^2-1}\right)$ [**Ans.** 1/4]

49. $\lim_{x \to 1} \left(\frac{x^2 - 4}{\sqrt{x+2} - \sqrt{3x-2}} \right)$ **[Ans.** –8] **50.** $\lim_{x \to 4} \left(\frac{3 - \sqrt{5+x}}{1 - \sqrt{5-x}} \right)$ **[Ans.** –1/3]

51. $\lim_{x \to 0} \left(\frac{\sqrt{a+x} - \sqrt{a}}{x\sqrt{a(a+x)}} \right)$ $\left[\textbf{Ans. } \frac{1}{2a^{3/2}} \right]$ **52.** $\lim_{x \to 0} \left(\frac{\sqrt{1+x^2} - \sqrt{1+x}}{\sqrt{1+x^3} - \sqrt{1+x}} \right)$ **[Ans.** 1]

53. $\lim_{x \to 1} \left(\frac{x^4 - 3x^2 + 2}{x^3 - 5x^2 + 3x + 1} \right)$ **[Ans.** 1/2] **54.** $\lim_{x \to 2} \left(\frac{3^x + 3^{3-x} - 12}{3^{3-x} - 3^{x/2}} \right)$ **[Ans.** –4/3]

55. $\lim_{x \to 0} \left(\frac{e^{4x} - 1}{x} \right)$ **[Ans.** 4] **56.** $\lim_{x \to 0} \left(\frac{e^{2+x} - e^2}{x} \right)$ **[Ans.** e^2]

57. $\lim_{x \to 1} \left(\frac{e^x - e^4}{x - 4} \right)$ **[Ans.** e^4] **58.** $\lim_{x \to 0} \left(\frac{e^{3x} - e^{2x}}{x} \right)$ **[Ans.** 1]

59. $\lim_{x \to 0} \left(\frac{e^x - x - 1}{x} \right)$ **[Ans.** 0] **60.** $\lim_{x \to 0} \left(\frac{e^{bx} - e^{ax}}{x} \right)$, $0 < a < b$ **[Ans.** $(b - a)$]

61. $\lim_{x \to 0} \left(\frac{a^x - b^x}{x} \right)$ **[Ans.** $(\log a - \log b)$] **62.** $\lim_{x \to 0} \left(\frac{a^x - a^{-x}}{x} \right)$ **[Ans.** $2 \log a$]

63. $\lim_{x \to 0} \left(\frac{2^x - 1}{x} \right)$ **[Ans.** log 2] **64.** $\lim_{x \to 0} \left(\frac{3^{2+x} - 9}{x} \right)$ **[Ans.** 9 log 3]

PART II : CONTINUITY AND DISFERENTIATION AND ITS APPLICATIONS

1.1. EXAMINE THE CONTINUITY OF A FUNCTION AT ANY POINT

A function $f(x)$ is continuous at $x - a$, if corresponding to a positive number, however, small but not zero, there exists a positive number δ, such that $| f(x) - f(a) | < \varepsilon$ for all value of x in the interval $(a - \delta, a + \delta)$.

In other words, $f(x)$ is continuous at $x = a$, if $\lim_{x \to a} f(x)$ exists and is equal to $f(a)$. Thus $f(x)$ is continuous at $x = a$, if its value does not change suddenly around $x = a$.

A function $f(x)$ is said to be continuous in an interval (a, b) if it is continuous for every value of x in (a, b).

Every polynomial function of x is continuous in any arbitrary given interval.

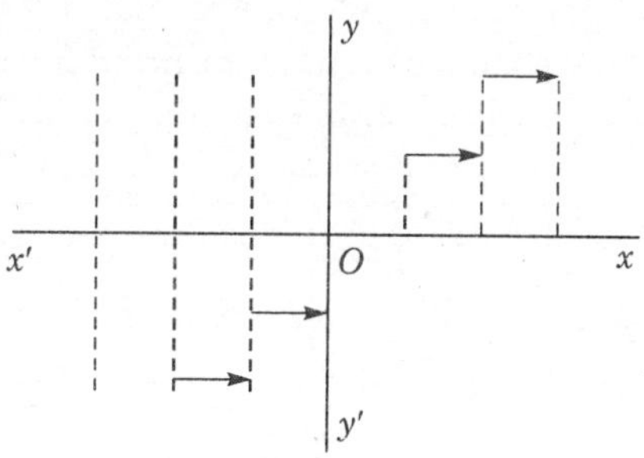

Fig. 12

Example 1. $|x|$

$x > 0, x \leq 0$

Continuity of Function of Point

This is a function to find that point on x-axis remain close to each other.

Let the function

$$f(x) = \begin{cases} 1, \text{ if } x \leq 0 \\ 2, \text{ if } x > 0 \end{cases} \quad \text{(See in Fig. 13.)}$$

Example 2. *Draw* $|x|$ *and check continuity?*

Solution. $$f(x) = \begin{cases} -x, \text{ if } x < 0 \\ x, \text{ if } x \geq 0 \end{cases}$$

Clearly, the function is defined at 0 and $f(0) = 0$

$$f(x) = |x| \text{ at } x = 0 \quad \text{(See in Fig. 14)}$$

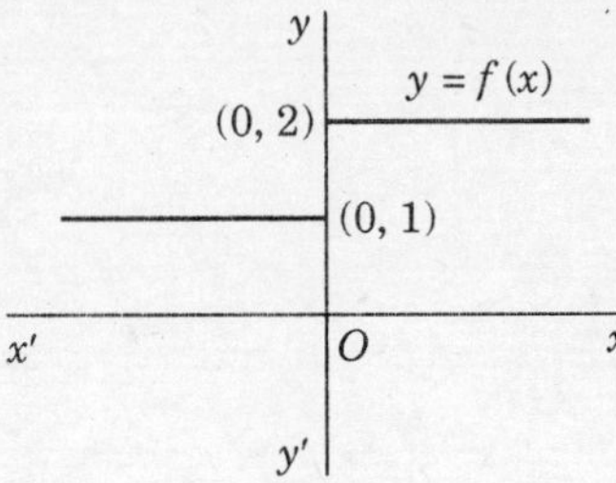

Fig. 13

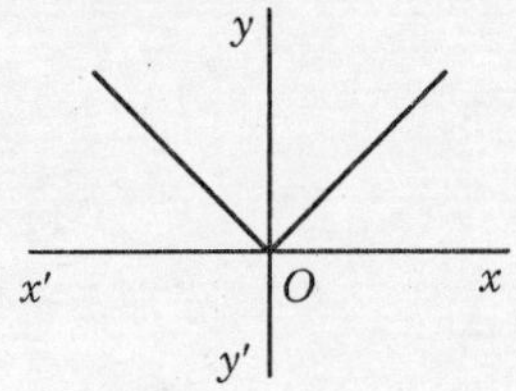

Fig. 14

Example 3. *Discuss the continuity of the function f defined by* $f(x) = \begin{Bmatrix} x+2 & x \leq 1 \\ x-2 & \text{if } x > 1 \end{Bmatrix}$.

Sol. The function f is defined at all points.

Case (*i*) if $C < 1$ $\quad f(x) = C + 2 = x + 2$

Therefore, $\lim\limits_{x \to C} f(x) = \lim\limits_{x \to C} (x + 2) = C + 2$, $f(x)$ is continuous at all points $x < 1$.

Case (*ii*) if $C > 1$, then $f(C) = C - 2$, therefore

$$\lim_{x \to C} f(x) = \lim_{x \to C} (x - 2) = (C - 2) = f(C)$$

Thus, $f(x)$ is continuous at all points $x > 1$.

Case (*iii*) if $C = 1$, then the left hand limit of f at $x = 1$

$$\lim_{x^+ \to 1} f(x) = \lim_{x \to 1} (x + 2) = 1 + 2 = 3$$

The right hand limit of f at $x = 1$, $\lim\limits_{x^- \to 1} f(x) = \lim\limits_{x \to 1} (x - 2) = 1 - 2 = -1$.

Since the left and right hand limit of $f(x)$ at $x = 1$, do not coincide, $f(x)$ is not continuous at $x = 1$. Hence $x = 1$, is the only point of discontinuity of $f(x)$. (See in Fig. 15.)

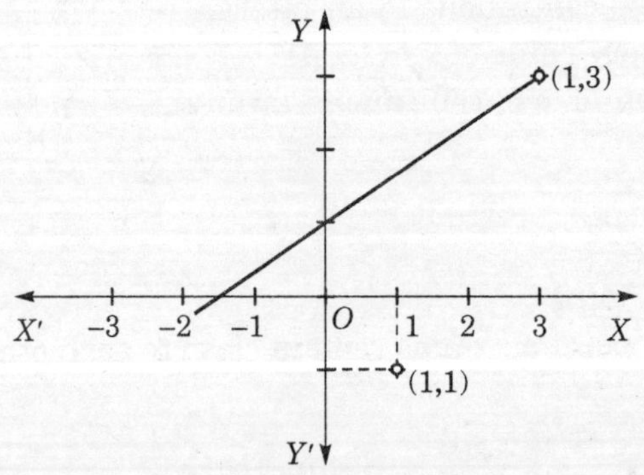

Fig. 1.15

Example 4. *Discuss the continuity of the function defined by*

$$f(x) = \begin{cases} x + 2 & \text{if } x < 0 \\ -x + 2, & \text{if } x > 0 \end{cases}.$$

Sol. Observe that the function is defined at all real numbers except at 0. Domain of definition of this function is

$D_1 \cup D_2$ where $D_1 = \{x \in R : x < 0\}$

and $D_2 = \{x \in \mathbf{R} : x > 0\}$

If $c \in D_1$, then $\lim_{x \to c} f(x) = \lim_{x \to c} (x + 2) = c + 2 = f(c)$ and hence f is continuous in D_1.

If $c \in D_2$, then $\lim_{x \to c} f(x) = \lim_{x \to c} (-x + 2) = -c + 2 = f(c)$ and hence f is continuous in D_2.

Since f is continuous at all points in the domain of f, we deduce that f is continuous. Graph of this function is given the Fig. 16. Note that to graph this function we need to lift the pen from the plane of the paper, but we need to do that only for those points where the function is not defined.

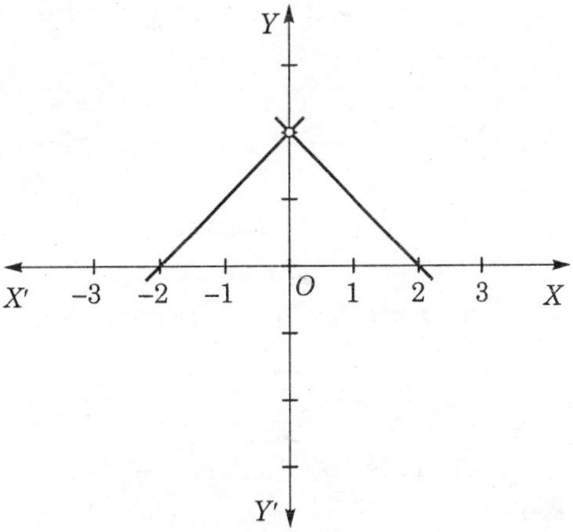

Fig. 16.

Example 5. *Discuss the continuity of the function f given by*

$$f(x) = \begin{cases} x, & \text{if } x \geq 0 \\ x^2, & \text{if } x < 0 \end{cases}.$$

Sol. Clearly the function is defined at every real number. Graph of the function is given in Fig. 17. By inspection, it seems prudent to partition the domain of definition of f into three disjoint subsets of the real line.

Let $D_1 = \{x \in \mathbf{R} : x < 0\}$, $D_2 = \{0\}$ and $D_3 = \{x \in \mathbf{R} : x > 0\}$.

Case 1. Let c be real number which is not equal to any integer. It is evident from graph that for all real numbers *close* to c the value of function is equal to c $\lim_{x \to c} f(x) = \lim_{x \to c} [x] = [c]$. Also $f(c) = [c]$ and hence the function is continuous numbers not equal to integers.

Case 2. Let c be an integer. Then we can find a sufficiently small real number $r > 0$ such that $[c - r] = c - 1$ whereas $[c + r] = c$.

This, in terms of limits mean that

$$\lim_{x \to c^-} f(x) = c - 1, \quad \lim_{x \to c^-} f(x) = c$$

Since these limits cannot be equal to each other for any c, the function discontinuous at every integral point.

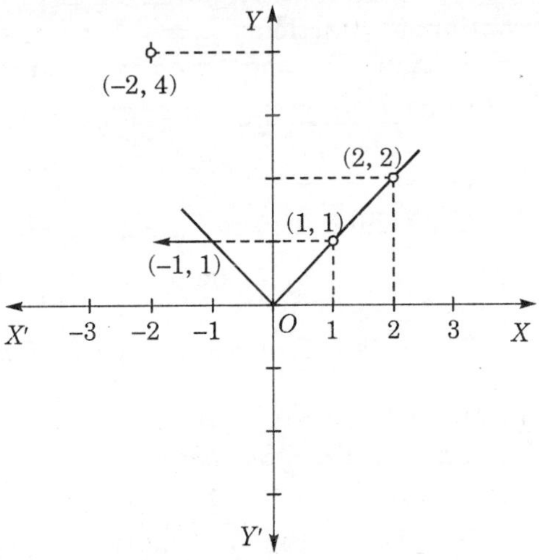

Fig. 17.

Algebraic Continuous Functions

Continuity of a function at a point is entirely dictated by the limit of the function at that point, and it is reasonable to expect result analogous to the case of limits.

Theorem. Suppose f and g be two real functions continuous at a real number 'c'.

Then

(1) $f + g$ is continuous at $x = c$. (2) $f - g$ is continuous at $x = c$.

(3) $f \cdot g$ is continuous at $x = c$. (4) $\left(\dfrac{f}{g}\right)$ is continuous at $x = c$, [provided $g(c) \neq 0$].

Proof. We are investigating continuity of $(f + g)$ at $x = c$. Clearly it is defined at $x = c$. We have

$$\lim_{x \to c} (f + g)(x) = \lim_{x \to c} [f(x) + g(x)] \qquad \text{(by definition of } f + g)$$
$$= \lim_{x \to c} f(x) + \lim_{x \to c} g(x) \qquad \text{(by the theorem on limit)}$$
$$= f(c) + g(c) \qquad \text{(as } f \text{ and } g \text{ are common)}$$
$$= (f + g)(c) \qquad \text{(by definition of } f - g)$$

Hence, $f + g$ is continuous at $x = c$.

Proofs for the remaining parts are similar.

Example 6. *Check the continuity of the function of given by*

$$f(x) = \begin{cases} x, & \text{if } x \geq 0 \\ x^2, & \text{if } x < 0 \end{cases}.$$

Sol. This function is defined at every real number. Graph of function is given is this Fig. 18.

Thus prudent to partition the domain of definition of f into three disjoint subsets of real line.

Let $D_1 = \{x \in \mathbf{R} : x < 0\}$, $D_2 = \{0\}$ and $D_3 = \{x \in \mathbf{R} : x > 0\}$,

Case 1. At any point in D_1, $f(x) = x^2$, it is continuous function.

Case 2. At any point in D_3, $f(x) = x$, it is continuous function.

Case 3. At $x = 0$, $f(x) = x$, $f(0) = 0$.

The left hand limit of f at 0 is

$$\lim_{x \to 0^-} f(x) = \lim_{x \to 0^-} x^2 = 0^2 = 0$$

The right hand limit of $f(x)$ at 0 is

$$\lim_{x \to 0^+} f(x) = \lim_{x \to 0^+} x = 0$$

Thus $\lim\limits_{x \to 0} f(x) = 0 = f(0)$ and hence $f(x)$ is continuous function.

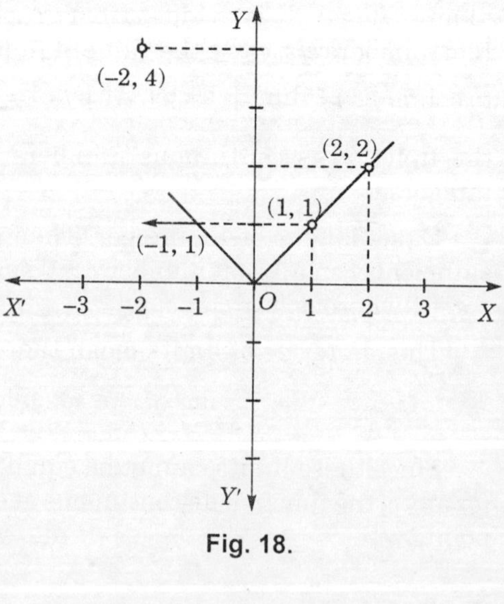

Fig. 18.

Example 7. *Show that every polynomial function is continuous.*

Sol. Recall that a function p is a polynomial function if it is defined by $p(x) = a_0 + a_1 x + ... + a_n x^n$

for some natural number n, $a_n \neq 0$ and $a_i \in \mathbf{R}$. Clearly this function is defined for every real number. For a fixed real number c, we have

$$\lim_{x \to c} p(x) = p(c)$$

By definition, p is continuous at c. Since c is any real number, p is continuous at every number and hence p is a continuous function.

Example 8. *Find all the points of discontinuity of the greatest integer function defined by $f(x) = [x]$ denotes the greatest integer less than or equal to x.*

Sol. First observe that f is defined for all real numbers. Graph of the function is given Fig. 19. From the graph it looks like that f is discontinuous at every integral point. Below we explore, if this is true.

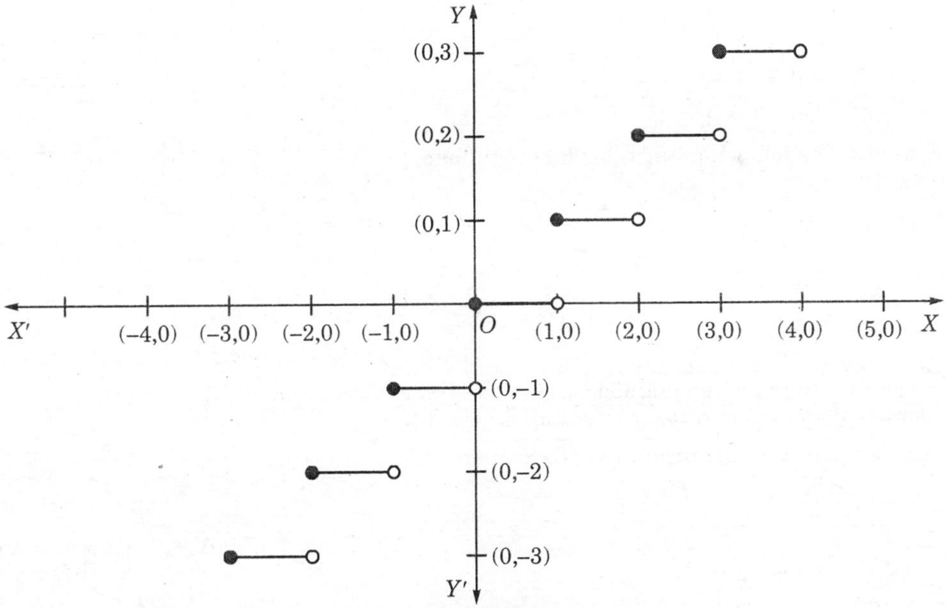

Fig. 19.

EXERCISE 1.1.2.

1. Function $f(x)$ defined by $f(x) = \begin{cases} x, & \text{if } x \le 1 \\ 5, & \text{if } x > 1 \end{cases}$ continuous at $x = 0$? At $x = 1$? at $x = 2$?

Find all points of discontinuity of $f(x)$, where $f(x)$ is defined by.

[**Ans.** $f(x)$ is continuous at $x = 0$, $x = 2$ Not continuous at $x = 1$]

2. $f(x) = \begin{cases} 2x + 3, & \text{if } x \le 2 \\ 2x - 3, & \text{if } x > 2 \end{cases}$ [**Ans.** Discontinuous at $x = 2$]

3. $f(x) = \begin{cases} x^3 - 3, & \text{if } x \le 2 \\ x^2 + 1, & \text{if } x > 2 \end{cases}$ [**Ans.** No point of discontinuity]

4. $f(x) = \begin{cases} |x| + 3, & \text{if } x \le -3 \\ -2x, & \text{if } -3 < x < 3 \\ 6x + 2 & \text{if } x \ge 3 \end{cases}$ [**Ans.** Discontinuous at $x = 3$]

5. $f(x) = \begin{cases} \dfrac{|x|}{x}, & \text{if } x \neq 0 \\ 0 & \text{if } x = 0 \end{cases}$ [**Ans.** Discontinuous at $x = 0$]

6. $f(x) = \begin{cases} x+1, & \text{if } x \geq 1 \\ x^2+1, & \text{if } x < 1 \end{cases}$ [**Ans.** Discontinuous at $x = 2$]

7. $f(x) = \begin{cases} x & \text{if } x \geq 0 \\ x^2 & \text{if } x < 0 \end{cases}$ [**Ans.** Discontinuous at $x = 0$]

8. $f(x) = \begin{cases} x^{10} - 1 & \text{if } x \leq 1 \\ x^2, & \text{if } x > 1 \end{cases}$ [**Ans.** $f(x)$ is discontinuous at $x = 1$]

9. $f(x) = \begin{cases} -2, & \text{if } x \leq -1 \\ 2x, & \text{if } -1 < x \leq 1 \\ 2, & \text{if } x > 1 \end{cases}$ [**Ans.** Continuous]

10. Examine the following functions for continuously

(*a*) $f(x) = x - 5$ (*b*) $f(x) = \dfrac{1}{x-5}, x \neq 5$

(*c*) $f(x) = \dfrac{x^2 - 25}{x+5}, x \neq -5$ (*d*) $f(x) = |x - 5|$.

[**Ans.** (*a*), (*b*), (*c*) and (*d*) are all continuous function]

11. Show that the function defined by $g(x) = x - [x]$ is discontinuous at all integral points. Here $[x]$ denotes the greatest integer less than or equal to x.

12. Find all points of discontinuity of $f(x)$, where

$f(x) = \begin{cases} \dfrac{\sin x}{x} & \text{if } x < 0 \\ x+1 & \text{if } x \geq 0 \end{cases}$ [**Ans.** No discontinuous point]

13. Determine if f defined by $f(x) = \begin{cases} x^2 \sin 1/x, & \text{if } x \neq 0 \\ 0 & \text{if } x = 0 \end{cases}$ is a continuous function.

[**Ans.** Yes, $f(x)\ x \in \mathbf{R}$]

14. $f(x) = \begin{cases} \dfrac{k \cos x}{\pi - 2\pi}, & \text{if } x \neq \dfrac{\pi}{2} \\ 3 & \text{if } x = \dfrac{\pi}{2} \end{cases}$ at $x = \pi/2$. [**Ans.** K = 6]

15. $f(x) = \begin{cases} kx+1, & \text{if } x \leq \pi \\ \cos x & \text{if } x > \pi \end{cases}$ at $x = \pi$. [**Ans.** $x = 1$ is the only point of discontinuity]

16. Show that the function defined by $f(x) = \cos(x^2)$ in a continuous function.

17. Show that the function by $f(x) = |\cos x|$ is continuous function.

18. Find that all the points of discontinuity of f defined by $f(x) = |x| - |x + 1|$.

[**Ans.** No discontinuous point]

19. Show that the function defined by $g(x) = x - [x]$ is discontinuous at all integral points, $[x]$ denotes the greatest integer less or equal to x.

20. Find the values of a and b such that the function defined by

$$f(x) = \begin{cases} 5, & \text{if } x \le 2 \\ ax - b, & \text{if } 2 < x > 10 \\ 21, & \text{if } x \ge 10 \end{cases}$$ [**Ans.** $a = 2, b = 1$]

21. $f(x) = \begin{cases} kx + 1, & \text{if } x \le 5 \\ 3x - 5, & \text{if } x > 5 \end{cases}$ at $x = 5$. [**Ans.** $k = 9/5$]

22. $f(x) = \begin{cases} kx^2 + 1, & \text{if } x \le 2 \\ 3, & \text{if } x > 2 \end{cases}$ at $x = 2$. $\left[\textbf{Ans. } k = \frac{3}{4}\right]$

23. Prove that the function $f(x) = 5x - 3$ is continuous at $x = 0$, at $x = -3$ and at $x = 3$.

24. Prove that the function $f(x) = x^n$ is continuous at $x = n$, where n is a positive integer.

25. Show that the function defined by $f(x) = \sin(x^3)$ is a continuous function.

26. Evaluate:

(*i*) $\lim\limits_{x \to \infty} \left(\dfrac{x - 1}{x + 3}\right)^{x+2}$ (*ii*) $\lim\limits_{x \to \infty} \left(\dfrac{x^2 + 2}{2x^2 + 1}\right)^{x^2}$

(*iii*) $\lim\limits_{x \to 0} \left\{\dfrac{1}{x} \log \sqrt{\dfrac{1 + x}{1 - x}}\right\}$ (*iv*) $\lim\limits_{x \to 0} \left(\dfrac{\sinh x}{x}\right)$

(*v*) $\lim\limits_{x \to 0} \left(\dfrac{\cosh x - 1}{x^2}\right)$ (*vi*) $\lim\limits_{x \to 0} \dfrac{|\sin x|}{x}$.

27. If (x) $\dfrac{x^2 - 16}{x - 4}$, for $x \ne 4$

$f(x) = 8$, for $x = 4$

Show that $f(x)$ is continuous at $x = 4$.

28. A function is defined as follows

$$\begin{aligned} f(x) &= 2x + 1, && 0 < x \le 2 \\ f(x) &= -x^2 + 3x + 3, && 2 < x < 3 \\ f(x) &= 3, && 3 < x \le 5. \end{aligned}$$

Discuss the continuity of $f(x)$ at $x = 2$ and $x = 3$. [**Ans.** Continuous at $x = 2$ and $x = 3$]

29. Discuss the continuity of the function $f(x)$ defined by $f(x) = |x| + |x - 1|$ at $x = 0, 1$.

30. Discuss the continuity at $x = 0$ of the function defined below

$f(x) = \dfrac{|x|}{x}$, $x \ne 0$

$f(x) = 1, x = 0$. [**Ans.** Discontinuous at $x = 0$]

31. A function $f(x)$ is defined as follows:

$$\begin{aligned} f(x) &= x \sin \frac{1}{x}, && \text{for } x \ne 0 \\ &= 0, && \text{for } x = 0. \end{aligned}$$

Show that $f(x)$ is continuous at $x = 0$.

32. Prove that the function

$$f(x) = \frac{\sin^2(ax)}{x^2}, \text{ for } x \ne 0. = 1, \quad \text{for } x = 0$$

is discontinuous at $x = 0$, unless $a = 1$.

• Definition of Derivative of a Function

Differentiation is the action of computing a derivative.

The derivative of a function $y = f(x)$ of a variable x is measure of the rate at which the value of y of the function changes w.r.t. the change of the variable x.

A function $f(x)$ is said to be differentiable at $x = a$, if $\lim\limits_{h \to 0} \dfrac{f(a+h)-f(a)}{h}$ and $\lim\limits_{h \to 0} \dfrac{f(a-h)-f(a)}{-h}$ exist and also they are equal and finite. This limit is called the derivative of f at a.

The above function $f(x)$ is said to be non-differentiable at $x = a$ if either

(*i*) both the above limits do not exist

or (*ii*) they exist but are not equal to the same finite number.

Every function $f(x)$ which possesses a finite derivative for a given value of x must be a continuous of x at that point. The converse of this is not true *i.e.*, a function may be continuous but not differentiable.

Example 9. *Show that the function*

$$\begin{aligned} f(x) &= x && \text{for} \quad 0 \le x \le 1 \\ &= -2-x && \text{for} \quad x \ge 1 \end{aligned}$$

is continuous but non-differentiable at $x = 1$.

Sol. $$\text{L.H.L.} = \lim_{h \to 1 \to 0} f(x) = \lim_{x \to 1} f(x) = 1$$

$$\text{R.H.L.} = \lim_{x \to 1+0} f(x) = \lim_{x \to 1} (2-x) = 2-1 = 1$$

Also, $$f(1) = 2 - 1 = 1.$$

$\therefore$ $$\lim_{x \to 1} f(x) = f(1) = 1.$$

Thus, $f(x)$ is continuous at $x = 1$.

Next, $$\lim_{h \to 0} \frac{f(1+h)-f(1)}{h} = \lim_{h \to 0} \frac{\{2-(1+h)\}-1}{h}$$

since, $$f(1) = 1.$$

$$= \lim_{h \to 0} \left(\frac{-h}{h}\right) = \lim_{h \to 0} (-1) = 1.$$

Also, $$\lim_{h \to 0} \frac{f(1-h)-f(1)}{-h} = \lim_{h \to 0} \frac{(1-h)-(1)}{-h}$$

$$= \lim_{h \to 0} (1) = 1.$$

Since the two limits are unequal, $f(x)$ is not differentiable at $x = 1$.

Example 10. *Show that*

$$\begin{aligned} f(x) &= x^2 \sin \frac{1}{x}, \quad x \ne 0 \\ &= 0, \qquad x = 0 \end{aligned}$$

is continuous and differentiable at $x = 0$.

Sol. Here, $\lim\limits_{x \to 0} f(x) = \lim\limits_{x \to 0} \left(x^2 \sin \dfrac{1}{x}\right) = 0,$

Since $\sin \dfrac{1}{x}$ is finite for every value of x.

$\therefore \qquad \lim_{x \to 0} f(x) = f(0) = 0.$

Thus, $f(x)$ is continuous at $x = 0$.

Further, $\displaystyle \lim_{h \to 0} \frac{f(0+h) - f(0)}{h} = \lim_{h \to 0} \frac{f(h) - (0)}{h}$

$$= \lim_{h \to 0} \frac{h^2 \sin \frac{1}{h} - 0}{h} = \lim_{h \to 0} \left(h \sin \frac{1}{h} \right) = 0$$

Also, $\displaystyle \lim_{h \to 0} \frac{f(0-h) - f(0)}{-h} = \lim_{h \to 0} \frac{f(-h) - f(0)}{-h}$

$$= \lim_{h \to 0} \frac{(-h)^2 \sin\left(\frac{1}{-h}\right) - 0}{-h} = \lim_{h \to 0} \left(h \sin \frac{1}{h} \right) = 0$$

Since both the limits are finite and equal, $f(x)$ is differentiable at $x = 0$.

Example 11. *If* $y = 3x^2 + 5x + 8$, *find* $\dfrac{dy}{dx}$.

Sol. Let Δx and Δy be the increments in x and y respectively, then

$$y + \Delta y = 3(x + \Delta x)^2 + 5(x + \Delta x) + 8$$
$$= (3x^2 + 5x + 8) + (6x \,.\, \Delta x + 5\Delta x + 3\Delta x^2)$$

$$\Rightarrow \qquad \Delta y = \Delta x\,(6x + 5 + 3\Delta x)$$

$$\Rightarrow \qquad \frac{\Delta y}{\Delta x} = (6x + 5 + 3\Delta x) = \lim_{\Delta x \to 0} (6x + 5 + 3\Delta x) = 6x + 5.$$

• Derivative by the Method of First Principle

We can obtain derivatives of functions *ab-initio*, *i.e.*, from the fundamental definition *viz.*

$$\frac{dy}{dx} = \underset{h \to 0}{\text{limit}} \left\{ \frac{f(x+h) - f(x)}{h} \right\}.$$

Now we differentiate some more functions from the first principles.

Example 12. *Differentiate from the first principles:*

(*i*) *tan 3x* (*ii*) *sec 2x.*

Sol. (*i*) Let $y = f(x) = \tan 3x$,

then $f(x + h) = \tan 3\,(x + h)$.

$$\therefore \qquad \frac{dy}{dx} = \underset{h \to 0}{\text{limit}} \left\{ \frac{\tan 3(x+h) - \tan 3x}{h} \right\}$$

$$= \underset{h \to 0}{\text{limit}} \left\{ \frac{\sin 3(x+h) \cos 3x - \sin 3x \cos 3(x+h)}{h \,.\, \cos 3(x+h) \,.\, \cos 3x} \right\}$$

$$= \underset{h \to 0}{\text{limit}} \left\{ \frac{\sin (3x + 3h - 3x)}{h \cos 3(x+h) \cos 3x} \right\}$$

$$= \lim_{h \to 0} \left\{ 3 \left(\frac{\sin 3h}{3h} \right) \frac{1}{\cos(3x + 3h)\cos 3x} \right\}$$

$$= 3 \times \frac{1}{\cos^2 3x} = 3 \sec^2 3x$$

(*ii*) Let $y = f(x) = \sec 2x,$

then, $f(x + h) = \sec 2(x + h).$

Now, $$\frac{dy}{dx} = \lim_{h \to 0} \left\{ \frac{\sec 2(x+h) - \sec 2x}{h} \right\}$$

$$= \lim_{h \to 0} \left\{ \frac{\cos 2x - \cos 2(x+h)}{h \cos 2x \cos 2(x+h)} \right\}$$

$$= \lim_{h \to 0} \left\{ \frac{2 \sin(2x + h) \sin h}{h \cos 2x \cos 2(x+h)} \right\}$$

$$= \lim_{h \to 0} \left\{ 2 \left(\frac{\sin h}{h} \right) \frac{\sin(2x+h)}{\cos 2x \cos 2(x+h)} \right\}$$

$$= \frac{2.1.\sin 2x}{\cos 2x . \cos 2x} = 2 \tan 2x \sec 2x.$$

Example 13. *Differentiate from the first principle:*

(*i*) $\sin^{-1} x$ (*ii*) $\cot^{-1} x$.

Sol. (*i*) Let $y = \sin^{-1} x$

or $x = \sin y.$

Then, $x + h = \sin(y + k).$

Now, $$\frac{dx}{dy} = \lim_{\Delta y \to 0} \frac{\Delta x}{\Delta y}$$

$$= \lim_{k \to 0} \frac{h}{k}$$

$$= \lim_{k \to 0} \left\{ \frac{(x+h) - x}{k} \right\}$$

$$= \lim_{k \to 0} \left\{ \frac{\sin(y+k) - \sin y}{k} \right\}$$

$$= \lim_{k \to 0} \left\{ \frac{2 \cos\left(y + \frac{1}{2}k\right) \sin \frac{k}{2}}{k} \right\}$$

$$= \underset{k \to 0}{\text{limit}} \left\{ \cos\left(y + \frac{1}{2}k\right) \times \left(\frac{\sin \frac{1}{2}k}{\frac{1}{2}k} \right) \right\}$$

$$= \cos y \times 1 = \cos y.$$

$$\therefore \qquad \frac{dy}{dx} = \frac{1}{\cos y}$$

$$= \frac{1}{\sqrt{1 - \sin^2 y}} = \frac{1}{\sqrt{(1 - x^2)}}$$

(*ii*) Let $\qquad y = \cot^{-1} x$

or $\qquad x = \cot y$

Then, $\qquad x + h = \cot (y + k)$

$$\frac{dx}{dy} = \underset{k \to 0}{\text{limit}} \left\{ \frac{\cot (y + k) - \cot y}{k} \right\}$$

$$= \underset{k \to 0}{\text{limit}} \left\{ \frac{\sin y \cot (y + k) - \sin (y + k) \cot y}{k \sin (y + k) \sin y} \right\}$$

$$= \underset{k \to 0}{\text{limit}} \left\{ \frac{\sin (y - y - k)}{k \sin (y + k) \sin y} \right\}$$

$$= \underset{k \to 0}{\text{limit}} \left\{ \left(- \frac{\sin k}{k} \right) \frac{1}{\sin (y + k) \sin y} \right\}$$

$$= - \text{cosec}^2 y.$$

$$\therefore \qquad \frac{dy}{dx} = -\frac{1}{\text{cosec}^2 y} = -\frac{1}{1 + \cot^2 y} = -\frac{1}{(1 + x^2)}$$

1.2. DIFFERENTIATION BY DEFINITION OF x^n, $\sin x$, $\cos x$, $\tan x$, e^x, $\log_a x$

(*a*) Differentiation of x^n.

Let $\qquad y = f(x)\, x^n$, then $f(x + h) = (x + h)^n$

$$\therefore \qquad \frac{dy}{dx} = \underset{h \to 0}{\text{limit}} \left\{ \frac{(x + h)^n - x^n}{h} \right\}$$

$$= \underset{h \to 0}{\text{limit}} \frac{x^n}{h} \left\{ \left(1 + \frac{h}{x}\right)^n - 1 \right\}$$

$$= \underset{h \to 0}{\text{limit}} \frac{x^n}{h} \left[\left\{ 1 + n \,.\, \frac{h}{x} + \frac{n(n-1)}{2\,!} \frac{h^2}{x^2} + \ldots \right\} - 1 \right]$$

$$= \underset{h \to 0}{\text{limit}} \frac{x^n}{h} \left[\frac{nh}{x} + \frac{n(n-1)}{2!} \frac{h^2}{x^2} + \ldots \right]$$

$$= \underset{h \to 0}{\text{limit}}\ x^n \left[\frac{n}{x} + \frac{n(n-1)h}{2x^2} + \ldots \right]$$

$$= x^n \cdot \frac{n}{x} = nx^{n-1}$$

where n is a real number. This result is useful for obtaining derivatives of polynomial and other algebraic functions of x.

(*b*) **Differentiation of sin *x*.**

Let $y = f(x) = \sin x$, then $f(x+h) = \sin(x+h)$

$$\therefore \quad \frac{dy}{dx} = \underset{h \to 0}{\text{limit}} \left\{ \frac{\sin(x+h) - \sin x}{h} \right\}$$

$$= \underset{h \to 0}{\text{limit}} \left\{ \frac{2\cos\left(x + \frac{h}{2}\right) \sin\left(\frac{h}{2}\right)}{h} \right\}$$

$$= \underset{h \to 0}{\text{limit}} \cos\left(x + \frac{h}{2}\right) \times \underset{h \to 0}{\text{limit}} \left(\frac{\sin \frac{h}{2}}{\frac{h}{2}} \right)$$

$$= \cos x \times 1 = \cos x.$$

(*c*) **Differentiation of cos *x*.**

Let $y = f(x) = \cos x$, then $f(x+h) = \cos(x+h)$

$$\therefore \quad \frac{dy}{dx} = \underset{h \to 0}{\text{limit}} \left\{ \frac{\cos(x+h) - \cos x}{h} \right\}$$

$$= \underset{h \to 0}{\text{limit}} \left[\frac{2\sin\left(x + \frac{h}{2}\right) \sin\left(-\frac{h}{2}\right)}{h} \right]$$

$$= \underset{h \to 0}{\text{limit}} \sin\left(x + \frac{h}{2}\right) \times \left\{ -\underset{h \to 0}{\text{limit}} \left(\frac{\sin \frac{h}{2}}{\frac{h}{2}} \right) \right\}$$

$$= \sin x \times (-1) = -\sin x.$$

(*d*) **Differentiation of tan*x***

Let $f(x) = e^x$

$$f(x+h) = \tan(x+h)$$

$$\frac{d}{dx}(\tan x) = \lim_{h \to 0} \frac{f(x+h) - f(x)}{h} = \lim_{h \to 0} \frac{\tan(x+h) - \tan x}{h}$$

$$= \lim_{h\to 0}\frac{\dfrac{\sin(x+h)}{\cos(x+h)}-\dfrac{\sin x}{\cos x}}{h} = \lim_{h\to 0}\frac{\sin(x+h)\cos x-\sin x\cos(x+h)}{h\cos x.\cos(x+h)}$$

$$= \lim_{h\to 0}\frac{\sin(x+h-x)}{h\cos x.\cos(x+h)} = \lim_{h\to 0}\frac{\sin h}{h}\lim_{h\to 0}\frac{1}{\cos x.\cos(x+h)}$$

$$= 1\times\lim_{h\to 0}\frac{1}{\cos x.\cos(x+h)} = \lim_{h\to 0}\frac{1}{\cos x(\cos(x+h))} = \frac{1}{\cos^2 x} = \sec^2 x$$

(*e*) Differentiation of e^x

Let $f(x) = e^x$

$$f'(x) = \lim_{h\to 0}\frac{f(x+h)-f(x)}{h} = \lim_{h\to 0}\frac{e^x(e^x-1)}{h} = e^x\times\lim_{h\to 0}\frac{e^h-1}{h}$$

From first principle $\dfrac{d}{dx}(e^x) = e^x$

$$\lim_{h\to 0}\frac{e^h-1}{h} = 1 = e^x\times 1 = e^x$$

(*f*) Differentiation of $\log_a x$.

Let $y = f(x) = \log_a x$, then $f(x+h) = \log_a(x+h)$

$$\therefore \quad \frac{dy}{dx} = \underset{h\to 0}{\text{limit}}\left\{\frac{f(x+h)-f(x)}{h}\right\} = \underset{h\to 0}{\text{limit}}\left\{\frac{\log_a(x+h)-\log_a x}{h}\right\}$$

$$= \underset{h\to 0}{\text{limit}}\left\{\frac{1}{h}\log_a\left(\frac{x+h}{x}\right)\right\} = \underset{h\to 0}{\text{limit}}\left\{\frac{1}{x}.\frac{x}{h}\log_a\left(1+\frac{h}{x}\right)\right\}$$

$$= \frac{1}{x}\underset{h\to 0}{\text{limit}}\left\{\log_a\left(1+\frac{h}{x}\right)^{\frac{x}{h}}\right\} = \frac{1}{x}\log_a e.$$

(*g*) **Differentiation of cos a^x.**

Let $y = f(x) = a^x$, then $f(x+h) = a^{x+h}$.

$$\therefore \quad \frac{dy}{dx} = \underset{h\to 0}{\text{limit}}\left\{\frac{a^{x+h}-a^x}{h}\right\} = \underset{h\to 0}{\text{limit}}\left\{a^x\,\frac{(a^h-1)}{h}\right\}$$

$$= a^x\underset{h\to 0}{\text{limit}}\left\{\frac{(a^h-1)}{h}\right\} = a^x . \log_e a.$$

Cor. When $a = e$, $y = e^x$ and then

$$\frac{dy}{dx} = e^x\log_e e = e^x . 1 = e^x.$$

1.3. DIFFERENTIATION OF SUM, PRODUCT AND QUOTIENT OF FUNCTION, DIFFERENTIATION OF FUNCTION OF A FUNCTION

I. Derivative of sum or difference. *If u, v, w be functions of x, then*

$$\frac{d}{dx}(u + v - w) = \frac{du}{dx} + \frac{dv}{dx} - \frac{dw}{dx}$$

II. Derivative of a product. *If u, v be functions of x, then*

$$\frac{d}{dx}(uv) = u\frac{dv}{dx} + v\frac{du}{dx}$$

i.e., derivative of the product of two functions

= *1st function × derivative of 2nd + 2nd function × derivative of 1st.*

In particular, $\frac{d}{dx}(cu) = c\frac{du}{dx}$.

III. Derivative of a quotient. *If u, v, be functions of x, then*

$$\frac{d}{dx}\left(\frac{u}{v}\right) = \frac{v\frac{du}{dx} - u\frac{dv}{dx}}{v^2}$$

i.e., derivative of the quotient of two functions

$$= \frac{\text{Denom.} \times \text{derivative of Num.} - \text{Num.} \times \text{derivative of denominator}}{(\text{Denom.})^2}$$

IV. Derivative of a function of a function. *If $y = f(t)$ and $t = \phi(x)$, then*

$$\frac{dy}{dx} = \frac{dy}{dt} \cdot \frac{dt}{dx}$$

i.e., derivative of y w.r.t. x

= *(derivative of 1st function w.r.t. 2nd) × (derivative of 2nd function w.r.t. x)*

• Derivatives of Composite Functions

Definition:

If f and g are two functions defined by $y = f(u)$ and $u = g(x)$ respectively then a function defined by $y = f[g(x)]$ or $fog(x)$ is called a composite function or a function of a function.

The theorem for finding the derivative of a composite function is known as the **CHAIN RULE.**

Theorem:

If f and g are differentiable and are defined by $y = f(u)$ and $u = g(x)$, then the composite function $y = f[g(x)]$ is differentiable and we have $\frac{dy}{dx} = \frac{dy}{du} \times \frac{du}{dx}$.

Corollary:

If $y = f(u)$, $u = g(v)$ and $v = h(x)$ where f, g and h are differentiable functions of u, v and x respectively,

then $$\frac{dy}{dx} = \frac{dy}{du} \times \frac{du}{dv} \times \frac{dv}{dx}$$

Let the function as being the result of combining two functions.

If y is a function of u and u is a function of x, then

$$\frac{dy}{dx} = \frac{dy}{du} \times \frac{du}{dx}$$

The expressions $\frac{dy}{du}$ and $\frac{du}{dx}$ are not really fractions but rather they stand for the derivative of a function with respect to a variable. So, the chain rule we can think of them as fractions, so that the du cancels from the top and the bottom, leaving just $\frac{dy}{dx}$.

Example 14. *Differentiate* $y = (x^2 + 1)^{17}$.

Sol. Put $u = x^2 + 1$, so that $y = u^{17}$. The alternative formulation of the chain rules says that

$$\frac{dy}{dx} = \frac{dy}{du} \times \frac{du}{dx}$$
$$= 17u^{16} \times 2x$$
$$= 17(x^2 + 1)^{16} \times 2x$$

which is the same result as before. Again, if $y = (z^3 + 4z^2 - 3z - 3)^{-6}$, then set $u = z^3 + 4z^2 - 3z - 3$ so that $y = u^{-6}$ and

$$\frac{dy}{dx} = \frac{dy}{du} \times \frac{du}{dx}$$
$$= -6u^{-7} \times (3z^2 + 8z - 3).$$

Example 15. *Differentiate* $(3x^2 - 5)^3$.

Sol. We have, $\frac{d}{dx}(3x^2 + 5)^3 = 3(3x^2 - 5)^2 \times 6x = 18x(3x^2 - 5)^2$

Alternatively, we could first let $u = 3x^2 - 5$ and then $y = u^3$. So,

$$\frac{dy}{dx} = \frac{dy}{du} \times \frac{du}{dx} = 3u^2 \times 6x = 18x(3x^2 - 5)^2.$$

Example 16. *Find* $\frac{dy}{dx}$, *if* $y = \sqrt{x^2 + 1}$.

Sol. The outside function is $\sqrt{\cdot} = (\cdot)^{\frac{1}{2}}$ which has derivative $\frac{1}{2}(\cdot)^{-\frac{1}{2}}$, and the inside function is $x^2 + 1$ so that

$$y' = \frac{1}{2}(x^2 + 1)^{-\frac{1}{2}} \times 2x$$

Alternatively, if $u = x^2 + 1$, we have

$$y = \sqrt{u} = u^{\frac{1}{2}}.$$

So, $$\frac{dy}{dx} = \frac{1}{2}u^{-\frac{1}{2}} \times 2x = \frac{1}{2}(x^2 + 1)^{-\frac{1}{2}} \times 2x.$$

EXERCISE 1.1.3.

A. *Differentiate the following using the composite function rule:*

1. $(2x + 3)^2$ **[Ans.** $8x + 12$]
2. $(x^2 + 2x + 1)^{12}$ **[Ans.** $12(x^2 + 2x + 1)^{12}(2x + 2)$]
3. $(3 - x)^{21}$ **[Ans.** $-21(3 - x)^{20}$]
4. $(x^3 - 1)^5$ **[Ans.** $15x^2(x^3 - 1)^4$]
5. $f(t) = \sqrt{t^2 - 5t + 7}$ $\left[\textbf{Ans. } \frac{1}{2}(t^2 - 5t + 7)^{\frac{1}{2}}(2t - 5)\right]$
6. $g(z) = \dfrac{1}{\sqrt{2 - z^4}}$ **[Ans.** $2z^3(2 - z^4)^{-3/2}$]
7. $y = (t^3 - \sqrt{t})^{-3.8}$ $\left[\textbf{Ans. } -3.8(t^3 - \sqrt{t})^{-4.8}\left(3t^2 - \frac{1}{2\sqrt{t}}\right)\right]$
8. $z = \left(x + \dfrac{1}{x}\right)^{\frac{3}{7}}$ $\left[\textbf{Ans. } \frac{3}{7}\left(x - \frac{1}{x}\right)^{-\frac{4}{7}}\left(1 - \frac{1}{x^2}\right)\right]$

B. *Differentiate the following functions using the composite function rule and the product or quotient rule.*

1. $\dfrac{x}{\sqrt{1 - x^2}}$ $\left[\textbf{Ans. } \frac{\sqrt{1 - x^2} + x^2(1 - x^2)^{-\frac{1}{2}}}{1 - x^2}\right]$
2. $x^{1/3}(1 - x)^{2/3}$ $\left[\textbf{Ans. } \frac{1}{3}x^{-\frac{1}{3}}(1 - x)^{\frac{2}{3}} - \frac{2}{3}x^{\frac{1}{3}}(1 - x)^{-\frac{1}{3}}\right]$
3. $x\sqrt{(1 - x)}$ $\left[\textbf{Ans. } \sqrt{1 - x} - \frac{x}{2\sqrt{1 - x}}\right]$
4. $(2x - 1)^2(x + 3)^3$ **[Ans.** $4(2x - 1)(x + 3)^3 + 3(2x - 1)^2(x + 3)^2$]
5. $(x + 2)(x + 3)^2$ **[Ans.** $(x + 3)^2 + 2(x + 2)(x + 3)$]

• Derivatives of Explicit Functions

An explicit function is a function is which one variable is defined only in forms of the other variable.

Ex.

(*i*) $y = x^2 + 7, \dfrac{dy}{dx} = \dfrac{d(x^2 + 7)}{dx}$ (*ii*) $y = \sqrt{9 - x^2}, \dfrac{dy}{dx} = \dfrac{d}{dx}\sqrt{9 - x^2}$

(*iii*) $y = -\sqrt{9 - x^2}, \dfrac{dy}{dx} = -\dfrac{d}{dx}\sqrt{9 - x^2}$

1.4. DIFFERENTIATION

1.4.1. Differentiation of Trigonometric Inverse Function

The basic trigonometric functions have corresponding inverse functions when appropriate restrictions are placed on the domain of the original functions. All the trigonometric functions have derivatives, which are summarized as follows:

(*a*) Derivatives of $\sin^{-1} x$ and $\cos^{-1} x$.

Let $\sin^{-1} x = y$ then ; $x = \sin y$

and $$\frac{dx}{dy} = \cos y$$

Thus, $$\frac{dy}{dx} = \frac{1}{\cos y}$$

$$= \frac{1}{\sqrt{1 - \sin^2 y}} = \frac{1}{\sqrt{1 - x^2}}$$

$\therefore$ $$\frac{d}{dx}(\sin^{-1} x) = \frac{1}{\sqrt{1 - x^2}}.$$

Similarly, $$\frac{d}{dx}(\cos^{-1} x) = -\frac{1}{\sqrt{1 - x^2}}$$

(*b*) Derivatives of $\tan^{-1} x$ and $\cot^{-1} x$.

Let $$\tan^{-1} x = y, \text{ then}$$
$$x = \tan y$$

and $$\frac{dx}{dy} = \sec^2 y$$

or $$\frac{dy}{dx} = \frac{1}{\sec^2 y} = \frac{1}{1 + \tan^2 y} = \frac{1}{1 + x^2}$$

$\therefore$ $$\frac{d}{dx}(\tan^{-1} x) = \frac{1}{(1 + x^2)}$$

Similarly, $$\frac{d}{dx}(\cot^{-1} x) = \frac{1}{(1 + x^2)}$$

(*c*) Derivatives of $\sec^{-1} x$ and $\text{cosec}^{-1} x$.

Let $\sec^{-1} x = y$, then, $x = \sec y$

and $$\frac{dx}{dy} = \sec y \tan y$$

$\because$ $$\frac{dy}{dx} = \frac{1}{\sec y \tan y} = \frac{1}{\sec y\sqrt{\sec^2 y - 1}} = \frac{1}{x\sqrt{x^2 - 1}}$$

$\therefore$ $$\frac{d}{dx}(\sec^{-1} x) = \frac{1}{x\sqrt{x^2 - 1}}$$

Similarly, $$\frac{d}{dx}(\text{cosec}^{-1} x) = \frac{1}{x\sqrt{x^2 - 1}}.$$

1.4.2. Logarithmic Differentiation

Let us now briefly get the derivatives for logarithms. In this case we need to start with the following fact about functions that are inverse of each other.

If $f(x)$ and $g(x)$ are inverse of each other then,

$$g'(x) = \frac{1}{f'(g(x))} = \frac{1}{e^{g(x)}} = \frac{1}{e^{\ln x}} = \frac{1}{x}$$

Putting this all together gives,

$$\frac{dx}{x}(\ln x) = \frac{1}{x} \qquad x > 0$$

$$\frac{dx}{x}(\ln|x|) = \frac{1}{x} \qquad x \neq 0$$

Using this all we need to avoid is $x = 0$.

$$\log_a x = \frac{\ln x}{\ln a}$$

Differentiation is then fairly simple.

$$\frac{d}{dx}(\log_a x) = \frac{d}{dx}\left(\frac{\ln x}{\ln a}\right) = \frac{1}{\ln a}\frac{d}{dx}(\ln x) = \frac{1}{x \ln a}$$

We took advantages of the fact that a was a constant and so $\ln a$ is also a constant and can be factored out of the derivative. Putting all this together gives,

$$\frac{d}{dx}(\log_a x) = \frac{1}{x \ln a}$$

Here is a summary of the derivatives in this section

(i) $\frac{d}{dx}(e^x) = e^x$ (ii) $\frac{d}{dx}(a^x) = a^x \ln a$

(iii) $\frac{d}{dx}(\ln x) = \frac{1}{x}$ (iv) $\frac{d}{dx}(\log_a x) = \frac{1}{x \ln a}$

Now, we have the derivations of the formulas out of the way let's compute a couple of derivations.

1.4.3. Exponential Differentiation

Let $f(x)$ is the exponential function,

$$f(x) = ax$$

Now, differentiate this, the power rule that we looked at required the exponent to be a fixed number and the base to be a variable. That is exactly the opposite from what we have got with this function. So, we are going to have to start with the definition of the derivative.

$$f'(x) = \lim_{h \to 0} \frac{f(x+h) - f(x)}{h} = \lim_{h \to 0} \frac{a^{x+h} - a^x}{h} = \lim_{h \to 0} \frac{a^x a^h - a^x}{h}$$

$$= \lim_{h \to 0} \frac{a^x(a^h - 1)}{h}$$

Now, we take differentiate, $f'(x) = a^x \lim_{h \to 0} \frac{(a^h - 1)}{h}$

Now let us notice that the limit we have got above is exactly the definition of the derivative of $f(x) = a^x$ at $x = 0$, *i.e.*, $f'(0)$. Therefore, the derivative becomes,

$$f'(x) = f'(0)\, a^x$$

So, we are kind of stuck. We need to know the derivative in order to get the derivative.

There are one value of a that we can deal with at this point. Back in the *exponential functions* section of the review chapter we stated that e = 2. 71828182845905... What we did not do however is actually define where e comes fro. There are infact a variety of ways to define e. Here are three of them.

Some definitions of e.

1. $e = \lim_{x\to\infty}\left(1+\frac{1}{n}\right)^n$ 2. e is the unique positive number for which $\lim_{x\to\infty}\frac{e^h-1}{h}=1$

3. $e=\sum_{n=0}^{\infty}\frac{1}{n!}$ For the natural exponential function, $f(x) = e^x$ we have $f'(0)=\lim_{h\to 0}\frac{e^h-1}{h}=1.$

So, provided we are using the natural exponential function we get the following

$$f(x) = e^x \quad \Rightarrow \quad f'(x) = e^x$$

$$f(x) = a^x \quad \Rightarrow \quad f'(x) = a^x \ln (a)$$

- **Derivatives of Implicit Functions**

It is a function in which one variable is not expressed only in terms the independent variable.

(*i*) $y^4 + 7y^2x - y^2x^3 - 9x^7 = 3$. It cannot be solved explicitly for y in terms of x.

(*ii*) $y^2 + x^2 = 9$. It can be solved for either the top or the bottom of the circle.

(*iii*) $y - x^2 = 7$. It is parabola.

To find $\frac{dy}{dx}$ in case of implicit function $f(x, y) = c$, it is not necessary to solve this equation for y in terms of x.

We can differentiate both sides of $f(x, y) = c$, w.r.t. x and then collect the terms involving $\frac{dy}{dx}$ on one side and rest of the terms on the other side.

$$\frac{d}{dx}(y^n) = ny^{n-1}\frac{dy}{dx}$$

$$\frac{d}{dx}(x^n) = nx^{n-1}\frac{dy}{dx} = nx^{n-1}$$

Thus,

- y as a differentiable function of x.
- Differentiate both sides of the equation with respect to x.

- **Derivatives of Parametric Functions**

Let the parametric equations of a function be $x = f(t)$, $y = \phi(t)$, t being the parameter.

Then, $$\frac{dx}{dt} = f'(t), \quad \frac{dy}{dt} = \phi'(t)$$

$$\therefore \quad \frac{\mathbf{dy}}{\mathbf{dx}} = \frac{\mathbf{dy}}{\mathbf{dt}}\Big/\frac{\mathbf{dx}}{\mathbf{dt}}.$$

Example 17. *If* $x = a (\cos \theta + \log \tan \theta/2)$, $y = a \sin \theta$, *find* dy/dx.

Solution. We have, $\frac{dx}{d\theta} = a\left\{-\sin\theta + \frac{1}{\tan\theta/2}\cdot\frac{d}{d\theta}(\tan\theta/2)\right\}$

$$= a\left\{-\sin\theta + \frac{\cos\theta/2}{\sin\theta/2} \cdot \sec^2\theta/2 \cdot \frac{1}{2}\right\}$$

$$= a\left\{-\sin\theta + \frac{1}{2\sin\theta/2 \cdot \cos\theta/2}\right\} = a\,(-\sin\theta + 1/\sin\theta)$$

$$= a\left(\frac{1-\sin^2\theta}{\sin\theta}\right) = \frac{a\cos^2\theta}{\sin\theta}$$

and $$\frac{dy}{d\theta} = a\cos\theta$$

$$\therefore \quad \frac{dy}{dx} = \frac{dy}{d\theta}\bigg/\frac{dx}{d\theta} = a\cos\theta \cdot \frac{\sin\theta}{a\cos^2\theta} = \tan\theta.$$

• Applications of Derivatives of Parametric Functions

When Cartesian coordinates of a curve is represented as functions of the same variable (usually written t), they are called the **parametric equations.**

Thus, parametric equations in the xy-plane

$$x = x(t) \text{ and } y = y(t) \quad \text{or} \quad x = f(t) \text{ and } y = g(t).$$

denote the x and y coordinate of the graph of a curve in the plane.

Assume that f and g are differentiable and $f'(t)$ is not 0 then, given parametric curve can be expressed as $y = y(x)$ and this function is differentiable at x, that is

$$x = f(t) \qquad \text{or} \qquad t = f^{-1}(x)$$

by plugging into $\quad y = g(t) \quad$ obtained is $\quad y = g[f^{-1}(x)]$

Therefore, we use the chain rule and the derivative of the inverse function to find the derivative of the parametric functions.

$$y'(x) = g'[f^{-1}(x)]\,[f^{-1}(x)]' = g'[f^{-1}(x)] \cdot \frac{1}{f'[f^{-1}(x)]'},$$

$$y'(x) = \frac{g'(t)}{f'(t)} \qquad \text{or} \qquad \frac{dy}{dx} = \frac{\frac{dy}{dt}}{\frac{dx}{dt}}.$$

Examples of Derivative of Parametric Functions

Example 18. *Write equation of the line tangent to the curve $x = t + 1$ and $y = -t^2 + 4$ at the point $t = 1$.*

Solution. The equation $x = t + 1$ solve for t and put into $y = -t^2 + 4$, thus

$$t = x - 1 \;\Rightarrow\; y = -t^2 + 4, \quad y = -(x-1)^2 + 4$$

i.e., $y - 4 = -(x-1)^2$ or $y = -x^2 + 2x + 3$ translated parabola with the vertex $V(x_0, y_0)$, so $V(1, 4)$.

When plotting points of a parametric curve by increasing t, the graph of the function is traced out in the direction of motion.

x	y	t
⋮	⋮	⋮
– 2	– 5	– 3
– 1	0	– 2
0	3	– 1
1	4	0
2	3	1
3	0	2
4	– 5	3
⋮	⋮	⋮

$x(t) = t + 1, \quad t = 1$

$y(t) = -t^2 + 4,$

$x(1) = 1 + 1 = 2,$

$y(1) = -1^2 + 4 = 3,$

$P_1 = (2, 3)$

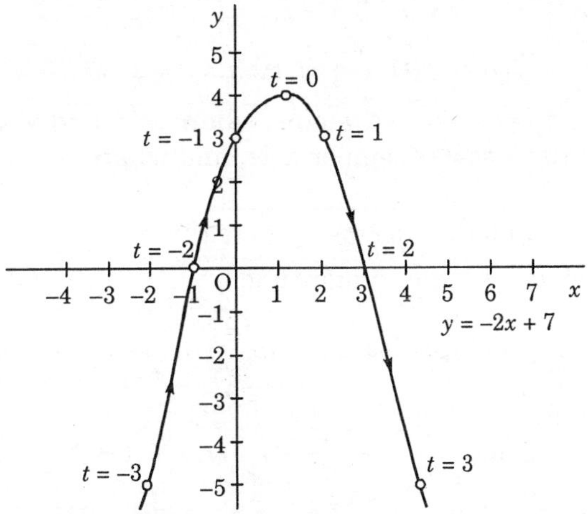

Fig. 1.20.

The derivative of the given parametric equations at $t = 1$ is the slope of the tangent line,

$$\left.\begin{array}{ll} x(t) = t + 1, & x'(t) = 1 \\ y(1) = -t^2 + 4, & y'(t) = -2t \end{array}\right\} \Rightarrow y'(x) = \frac{y'(t)}{x'(t)} = \frac{-2t}{1} = -2t$$

Since $t = x - 1$ then $m = y'(x) = -2(x - 1) = -2(2 - 1) = -2$, $m = -2$.

Therefore, the equation of the line tangent to the given parametric curve at $t = 1$ or the point P_1 (2, 3) is

$$y - y_1 = m(x - x_1), y - 3 = -2(x - 2) \Rightarrow y = -2x + 7.$$

Derivative of Vector-Valued Functions

If the radius vector r of a point in a plane depends on a parameter t, say t represents time, so that its magnitude and direction change continuously while t changes, then its arrow sweeping out a curve.

Let $r(t)$ denotes its value at the moment t and $r(t + h)$ represents its value at $t + h$, and P and P_1 are the corresponding points of the curve, as is shown in the figure below.

The increment $\Delta r = r(t + h) - r(t)$ is the vector that falls in the direction of the secant line PP_1 and points from P to P_1.

The difference quotient $\frac{\vec{r}(t + h) - \vec{r}(t)}{h}$, obtained by division with the scalar h, is the vector of the same direction but of different length.

The limit of the difference quotient as $h \to 0$.

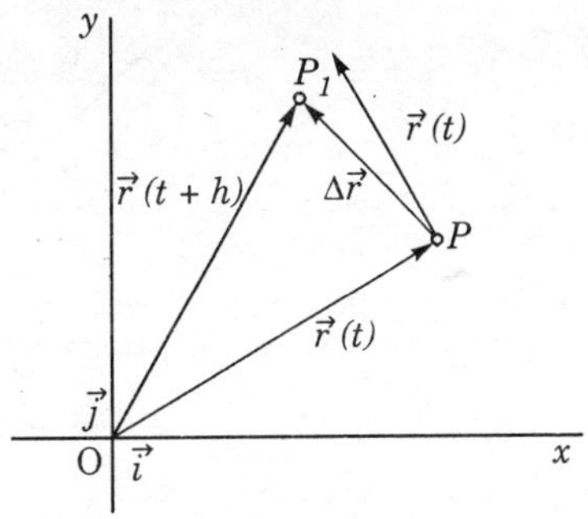

Fig. 1.21

$$\vec{r}(t) = \lim_{h \to 0} \frac{\vec{r}(t + h) - \vec{r}(t)}{h}$$

is the derivative vector of the vector-valued function that falls in the direction of the line tangent to the curve at P.

If $x(t)$ and $y(t)$ are the scalar components of the vector $r(t)$ then, according to rules of vector algebra,

$\Delta r = [x(t + h) - x(t)]\, \boldsymbol{i} + [y(t + h) - y(t)]\, \boldsymbol{j}$

and by use of the definition of the derivative

$$\vec{r}(t) = x'(t)\ \vec{t} + y'(t)\ \vec{j}$$

is the derivative vector, where $x'(t)$ and $y'(t)$ are its scalar components, and where

$$|\vec{r}(t)| = \sqrt{[x'(t)]^2 + [y'(t)]^2}$$

is its length or magnitude.

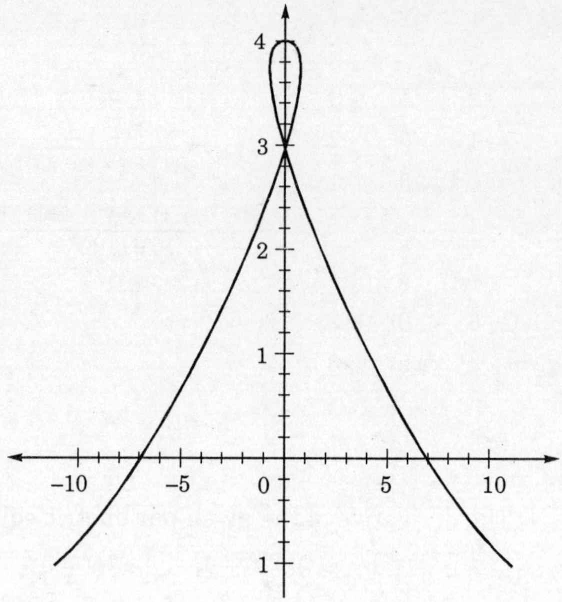

Fig. 1.22

Example 19. *To find* $\dfrac{dy}{dx}$ *when* $x = t^3 - t$

and $y = 4 - t^2$.

Sol. $x = t^3 - t, y = 4 - t^2$

$$\frac{dx}{dt} = 3t^2 - 1, \frac{dy}{dt} = -2t$$

From the chain rule, we have

$$\frac{dy}{dx} = \frac{\frac{dy}{dt}}{\frac{dx}{dt}} = \frac{-2t}{3t^2 - 1}$$

Now we use parametric differentiation.

Example 20. *To find* $\dfrac{dy}{dx}$ *when* $x = t^3$ *and* $y = t^2 - t$.

Sol. Let t is $-2 < t < t^2$,

t	-2	-1	0	1	2
x	-8	-1	0	1	8
y	6	2	0	0	2

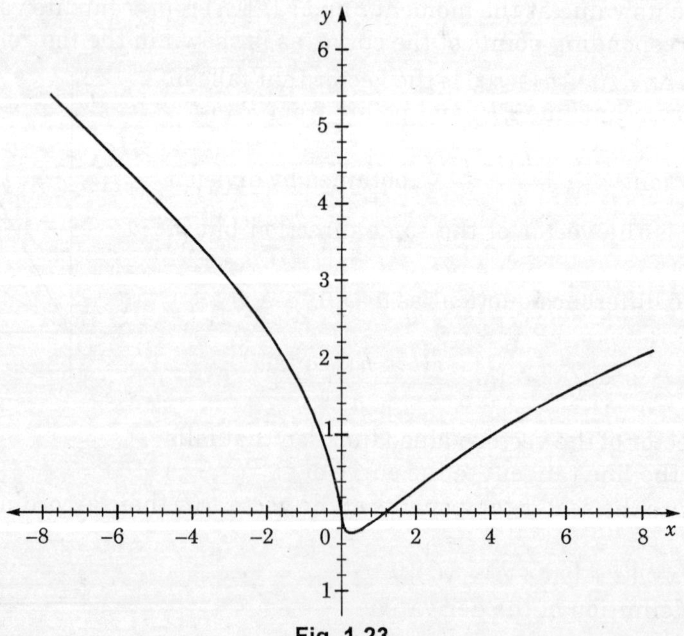

Fig. 1.23

From $\quad x = t^3 \quad y = t^2 - t$

we differentiate with respect to t

$$\frac{dy}{dt} = 3t^2 \qquad \frac{dy}{dt} = 2t - 1$$

Then, using the chain rule,

$$\frac{dy}{dx} = \frac{\frac{dy}{dt}}{\frac{dx}{dt}} \text{ provided } \frac{dx}{dt} \neq 0$$

$$\frac{dy}{dx} = \frac{2t - 1}{3t^2}$$

From this we can see that when $t = \frac{1}{2}, \frac{dy}{dx} = 0$ and so $t = \frac{1}{2}$ is a stationary value. When $t = \frac{1}{2}, x = \frac{1}{8}$ and $y = -\frac{1}{4}$ and these are the coordinates of the stationary point.

We also note that when $t = 0$, $\frac{dy}{dx}$ is infinite and so the y-axis is tangent to the curve at the point (0, 0).

Example 21. Differentiate each of the following functions :

(*a*) $R(w) = 4^w - 5\log_9 w$ (*b*) $f(x) = 3e^x + 10x^3 \ln x$ (*c*) $y = \frac{5e^x}{3e^x + 1}$

Sol. (*a*) This will be the only example that does not involve the natural exponential and natural logarithm functions.

$$R'(w) = 4w \ln 4 - \frac{5}{w \ln 9}$$

(*b*) Not much to this one. Just remember to use the product rule on the second term.

$$f'(x) = 3e^x + 30x^2 \ln x + 10x^3 \left(\frac{1}{x}\right) = 3e^x + 30x^2 \ln x + 10x^2$$

(*c*) We will need to use the quoteint rule on this one.

$$y' = \frac{5e^x(3e^x + 1) - (5e^x)(3e^x)}{(3e^x + 1)^2} = \frac{15e^{2x} + 5e^x - 15e^{2x}}{(3e^x + 1)^2}$$

$$= \frac{5e^x}{(3e^x + 1)^2}$$

Example 22. *Find the derivative of* $(\sin x)^{\tan^{-1} x}$.

Solution. Let $\quad y = (\sin x)^{\tan^{-1} x}$.

Then, $\quad \log y = \tan^{-1} x \,.\, \log \sin x$

$$\therefore \quad \frac{d}{dx}(\log y) = \frac{d}{dx}(\tan^{-1}x \,.\, \log \sin x)$$

$$\frac{d}{dy}(\log y)\frac{dy}{dx} = \log \sin x \,.\, \frac{d}{dx}\tan^{-1}x + \tan^{-1}x.\,\frac{d}{dx}\log \sin x$$

$$\frac{1}{y}\frac{dy}{dx} = \log \sin x \,.\, \frac{1}{1+x^2} + \tan^{-1}x\left(\frac{1}{\sin x}\,.\,\cos x\right)$$

$$\frac{dy}{dx} = y\left(\frac{\log \sin x}{1+x^2} + \cot x \tan^{-1} x\right)$$

$$= (\sin x)^{\tan^{-1}x}\left(\frac{\log \sin x}{1+x^2} + \cot x \tan^{-1} x\right)$$

• SECOND ORDER DERIVATIVES

The differentiation of function *f* is the derivative of *f*, where the last term is the second order derivative expression.

If measures how the rate of change of a quantity is itself changing. The second order derivative of the position of a vehicle w.r.t. time is the instantaneous acceleration of the vehicle, or the rate at which the velocity of the vehicle changing w.r.t. time.

i.e.,
$$a = \frac{du}{dt} = \frac{d^2x}{dt^2}.$$

Example 23. *To find the second order derivative* $\frac{d^2y}{dx^2}$, *when*

$$x = t^2, \quad y = t^3.$$

Solution. We have, $x = t^2$, $y = t^3$ differentiating, we find

$$\frac{dx}{dt} = 2t, \quad \frac{dy}{dt} = 3t^2$$

By chain rule,
$$\frac{dy}{dx} = \frac{\frac{dy}{dt}}{\frac{dx}{dt}} \quad \text{provided } \frac{dx}{dt} \neq 0$$

Thus,
$$\frac{dy}{dx} = \frac{3t^2}{2t} = \frac{3t}{2}$$

We can apply the chain rule a second time in order to find the second order derivative, $\frac{d^2y}{dx^2}$.

$$\frac{d^2y}{dx^2} = \frac{d}{dx}\left(\frac{dy}{dx}\right) = \frac{\frac{d}{dt}\frac{dy}{dx}}{\frac{dx}{dt}} = \frac{\frac{3}{2}}{2t} = \frac{3}{4t}.$$

Example 24. *Find* $\frac{d^2y}{dx^2}$, *when*

$$x = t^3 + 3t^2 \quad \text{and} \quad y = t^4 - 8t^2$$

Solution. We have $x = t^3 + 3t^2$ and $y = t^4 - 8t^2$ differentiating

$$\frac{dx}{dt} = 3t^2 = 6t, \quad \frac{dy}{dt} = 4t^3 - 16t$$

By chain rule, $$\frac{dy}{dx} = \frac{\frac{dy}{dt}}{\frac{dx}{dt}} \text{ provided } \frac{dx}{dt} \neq 0$$

so that $$\frac{dy}{dx} = \frac{4t^3 - 16t}{3t^2 + 6t}$$

This can be simplified as follows

$$\frac{dy}{dx} = \frac{4t(t^2 - 4)}{3t(t + 2)} = \frac{4t(t + 2)(t - 2)}{3t(t + 2)} = \frac{4(t - 2)}{3}$$

We can apply the chain rule a second time in order to find the second derivative, $\frac{d^2y}{dx^2}$.

$$\frac{d^2y}{dx^2} = \frac{d}{dx}\left(\frac{dy}{dx}\right) = \frac{\frac{d}{dt}\frac{dy}{dx}}{\frac{dx}{dt}} = \frac{\frac{4}{3}}{3t^2 + 6t} = \frac{4}{9t(t + 2)}$$

EXERCISE 1.1.4.

1. For each of the following functions determine $\frac{dy}{dx}$.

(*a*) $x = t^2 + 1, y = t^3 - 1$ $\left[\textbf{Ans. } \frac{3t}{2}\right]$

(*b*) $x = 3 \cos t, y = 3 \sin t$ [**Ans.** $-\cot t$]

(*c*) $x = t + \sqrt{t}, y = t - \sqrt{t}$ $\left[\textbf{Ans. } \frac{2\sqrt{t} - 1}{2\sqrt{t} + 1}\right]$

(*d*) $x = 2t^3 + 1, y = t^2 \cot t$ $\left[\textbf{Ans. } \frac{2 \cos t - t \sin t}{6t}\right]$

(*e*) $x = te^{-t}, y = 2t^2 + 1$ $\left[\textbf{Ans. } \frac{4te^t}{1 - t}\right]$

2. Determine the co-ordinates of the stationary points of each of the following functions:

(*i*) $x = 2t^2 + 1, y = te^{-2t}$ $\left[\textbf{Ans. } \frac{5}{4}, \frac{1}{2e}\right]$

(*ii*) $x = \sqrt{t} + 1, y = t^3 - 12t$ for $t > 0$ [**Ans.** $(1 + \sqrt{2} - 16)$]

(*iii*) $x = 5t^4, y = 5t^6 - t^5$ for $t > 0$ $\left[\textbf{Ans. } \frac{5}{1296}, \frac{-1}{46656}\right]$

(*iv*) $x = t + t^2, y = \sin t$ for $0 < t < \pi$ $\left[\textbf{Ans. } \frac{\pi}{2} + \frac{\pi^2}{4}, 1\right]$

(*v*) $x = te^{2t}, y = 2^2e^{-t}$ for $t > 0$ [**Ans.** $2e^4, 4e^{-2}$)]

3. For each of the following functions determine $\frac{d^2y}{dx^2}$ in terms of t:

(*i*) $x = 3t^2 + 4t, y = \sin 2t$ $\left[\textbf{Ans. } \frac{-2(3t+2)\sin 2t - 3\cos 3t}{2(3t+2)^3}\right]$

(*ii*) $x = e^{-t}, y = t^3 + t + 1$ [**Ans.** 4. $(3t^2 + 6t + 1)e^{2t}$]

(*iii*) $x = \frac{1}{2}t^2 + 2, y = \sin(t+1)$ $\left[\textbf{Ans. } \frac{-t\sin(t+1) - \cos(t+1)}{t^3}\right]$

(*iv*) $x = 3t^2 + 1, y = t^3 - 2t^2$ $\left[\textbf{Ans. } \frac{1}{12t}\right]$

(*v*) $x = \sin t, y = \cos t$ [**Ans.** 7. $-\sec^3 t$]

4. If $x = 2\cos\theta - \cos 2\theta$, $y = 2\sin\theta - \sin 2\theta$, find $\frac{dy}{dx}$, when $\theta = \frac{\pi}{2}$.

[**Hint.**

$$\frac{dy}{d\theta} = -2\sin\theta + \sin 2\theta$$

$$\frac{dy}{d\theta} = 2\cos\theta - 2\cos 2\theta$$

$$\therefore \quad \frac{dy}{dx} = \frac{dy}{d\theta}\Big/\frac{dx}{d\theta} = \frac{2(\cos\theta - \cos 2\theta)}{2(\sin 2\theta - \sin\theta)} = \left(\frac{\cos\theta - \cos 2\theta}{\sin 2\theta - \sin\theta}\right)$$

Now at $\theta = \frac{\pi}{2}$, $$\frac{dy}{dx} = \frac{\cos\frac{\pi}{2} - \cos\pi}{\sin\pi - \sin\frac{\pi}{2}} = -1.$$]

5. $y = \sin^{-1}(3x - 4x^3)$.

[**Hint.** Let $x = \sin\theta$, then

$$y = \sin^{-1}(3\sin\theta - 4\sin^3\theta)$$
$$= \sin^{-1}(\sin^3\theta) = 3\theta = 3\sin^{-1}x$$

$$\therefore \quad \frac{dy}{dx} = \frac{3}{\sqrt{(1-x^3)}}.$$]

6. Differentiate $\tan^{-1}\left(\frac{2x}{1-x^2}\right)$ with respect to $\sin^{-1}\left(\frac{2x}{1+x^2}\right)$.

[**Hint.** Let $y = \tan^{-1}\left(\frac{2x}{1-x^2}\right)$ and $z = \sin^{-1}\left(\frac{2x}{1+x^2}\right)$.

Now we have to find $\frac{dy}{dz}$.

Put $x + \tan\theta$. Then

$$y = \tan^{-1}\left(\frac{2\tan\theta}{1-\tan^2\theta}\right) = \tan^{-1}(\tan 2\theta) = 2\theta$$

and $z = \sin^{-1}\left(\frac{2\tan\theta}{1+\tan^2\theta}\right) = \sin^{-1}(\sin 2\theta) = 2\theta$

$$\therefore \quad \frac{dy}{dz} = \frac{d(2\theta)}{d(2\theta)} = 1 \quad \Big]$$

7. Differentiate

$$\tan^{-1}\left\{\frac{\sqrt{(1+x^2)}-\sqrt{(1+x^2)}}{\sqrt{(1+x^2)}+\sqrt{(1-x^2)}}\right\}.$$

[**Hint.** Let $x^2 = \cos 2\theta$, then

$$\sqrt{(1+x^2)} = \sqrt{(1+\cos 2\theta)} = \sqrt{2}\cos\theta$$

$$\sqrt{(1-x^2)} = \sqrt{(1-\cos 2\theta)} = \sqrt{2}\sin\theta$$

Thus the given function simplifies to

$$\tan^{-1}\left\{\frac{\sqrt{2}\cos\theta-\sqrt{2}\sin\theta}{\sqrt{2}\cos\theta+\sqrt{2}\sin\theta}\right\}$$

$$= \tan^{-1}\left(\frac{1-\tan\theta}{1+\tan\theta}\right)$$

$$= \tan^{-1}\left\{\tan\left(\frac{\pi}{4}-\theta\right)\right\}$$

$$= \frac{\pi}{4} - \theta = \frac{\pi}{4} - \frac{1}{2}\cos^{-1}x^2$$

$$\therefore \quad \frac{d}{dx}\left\{\frac{\pi}{4}-\frac{1}{2}\cos^{-1}x^2\right\} = \frac{x}{\sqrt{(1-x^4)}} \quad \Big]$$

8. Differentiate $\frac{\sinh^{-1}x}{x^2}$ w.r.t. x.

Do you know?

- $\frac{d}{dx}(\sin^{-1}x) = \frac{1}{\sqrt{x^2+1}}$
- $\frac{d}{dx}(\cosh^{-1}x) = \frac{1}{\sqrt{x^2-1}}$
- $\frac{d}{dx}(\tanh^{-1}x) = \frac{1}{1-x^2}$
- $\frac{d}{dx}(\coth^{-1}x) = \frac{1}{x^2-1}, x > 1$
- $\frac{d}{dx}(\text{sech}^{-1}x) = \frac{1}{x\sqrt{(1-x^2)}}$
- $\frac{d}{dx}(\text{cosech}^{-1}x) = -\frac{1}{x\sqrt{(x^2-1)}}$.

Following result may be found useful.

- $\sinh^{-1}x + \log_e(x + \sqrt{x^2+1})$
- $\cosh^{-1}x = \log_e(x + \sqrt{(x^2-1)})$
- $\tanh^{-1}x = \frac{1}{2}\log\left(\frac{1+x}{1-x}\right)$

[**Hint.** Let $y = x^{-2} . \sinh^{-1} x$,

then $$\frac{dy}{dx} = x^{-2} . \frac{d}{dx} (\sinh^{-1} x) + \sinh^{-1} x . \frac{d}{dx} (x^{-2})$$

$$= x^{-2} . \frac{1}{\sqrt{(1+x^2)}} + \sinh^{-1} x . (-2x^{-3})$$

$$= \frac{1}{x^3} \left\{ \frac{x}{\sqrt{(1+x^2)}} - \sinh^{-1} x \right\} \Bigg]$$

9. If $y = e^{\cosh^{-1} x^2}$, find $\frac{dy}{dx}$.

[**Hint.** We have $$y = e^{\cosh^{-1} x^2} = e^{\log \{(x^3 + \sqrt{(x^4 - 1)}\}}$$

$$= \{x^2 + (x^4 - 1)^{1/2}\}$$

$$\therefore \quad \frac{dy}{dx} = \left\{ 2x + \frac{1}{2} (x^4 - 1)^{-1/2} \times 4x^3 \right\}$$

$$= 2x \left\{ 1 + \frac{x^2}{\sqrt{(x^4 - 1)}} \right\} \Bigg]$$

10. Differentiate $\tanh^{-1} \left\{ \frac{x^2}{\sqrt{(1+x^2)}} \right\}$.

[**Hint.** Let $x = \sinh \theta$ and let,

$$y = \tanh^{-1} \left\{ \frac{x^2}{\sqrt{(1+x^2)}} \right\}$$

$$= \tanh^{-1} \left\{ \frac{\sin \theta}{\sqrt{(1 + \sinh^2 \theta)}} \right\}$$

$$= \tanh^{-1} (\tan \theta) = \theta = \sinh^{-1} x$$

$$\frac{dy}{dx} = \frac{x}{\sqrt{(1+x^2)}} . \Bigg]$$

Some Derivatives by the Method of First Principle

Example 25. *Differentiate from the first principles:*

(*i*) $\cos^2 x$, (*ii*) $\sin x^2$, (*iii*) $\sqrt{(\sin x)}$.

Solution. (*i*) Let $y = f(x) = \cos^2 x$, then $f(x+h) = \cos^2 (x+h)$

$\because \quad \dfrac{dy}{dx} = \lim\limits_{h \to 0} \left\{ \dfrac{\cos^2 (x+h) - \cos^2 x}{h} \right\}$

$$= \lim_{h \to 0} \left[\frac{\{\cos (x+h) + \cos x\} \{\cos (x+h) - \cos x\}}{h} \right]$$

$$= \lim_{h \to 0} \left[\left\{ 2 \cos \left(x + \frac{h}{2} \right) \cos \frac{h}{2} \right\} \left\{ - \sin \left(x + \frac{h}{2} \right) \times \frac{\sin \frac{h}{2}}{\frac{h}{2}} \right\} \right]$$

$$= [\{2 \cos x \,.\, 1\} \{- \sin x \,.\, 1\}] = -2 \sin x \cos x = - \sin 2x.$$

(*ii*) Let $y = f(x) = \sin x^2$, then $f(x+h) = \sin (x+h)^2$.

Now, $\dfrac{dy}{dx} = \lim\limits_{h \to 0} \left\{ \dfrac{\sin (x+h)^2 - \sin x^2}{h} \right\}$

$$= \lim_{h \to 0} \frac{\left[2 \cos \left\{ \frac{(x+h)^2 + x^2)}{2} \right\} \sin \left\{ \frac{(x+h)^2 - x^2}{2} \right\} \right]}{h}$$

$$= \lim_{h \to 0} \left[2 \cos \left(x^2 + xh + \frac{1}{2} h^2 \right) \frac{\sin \left(xh + \frac{h^2}{2} \right)}{h} \right]$$

$$= \lim_{h \to 0} \left[2 \cos \left(x^2 + xh + \frac{h^2}{2} \right) . \frac{\sin \left\{ \left(x + \frac{h}{2} \right) h \right\}}{\left(x + \frac{h}{2} \right) h} \times \left(x + \frac{h}{2} \right) \right]$$

$$= 2 \cos x^2 \,.\, 1 \,.\, x = 2x \cos x^2.$$

(*iii*) Let $y = f(x) = \sqrt{\sin x}$. Then $f(x+h) = \sqrt{\sin (x+h)}$

Now, $\dfrac{dy}{dx} = \lim\limits_{h \to 0} \left\{ \dfrac{\sqrt{\sin (x+h)} - \sqrt{\sin x}}{h} \right\}$

$$= \lim_{h \to 0} \left\{ \frac{\sqrt{\sin (x+h)} - \sqrt{\sin x}}{h} \right\} \times \left[\frac{\sqrt{\sin (x+h)} + \sqrt{\sin x}}{\sqrt{\sin (x+h)} + \sqrt{\sin x}} \right]$$

$$= \lim_{h \to 0} \left[\frac{\sin (x+h) - \sin x}{h \{\sqrt{\sin (x+h)} + \sqrt{\sin x}\}} \right]$$

$$= \lim_{h \to 0} \left[\frac{2 \cos \left(x + \frac{h}{2} \right) \sin \frac{h}{2}}{h\{\sqrt{\sin (x+h)} + \sqrt{\sin x}\}} \right]$$

$$= \lim_{h \to 0} \left[\frac{\cos\left(x + \frac{h}{2}\right)}{\sqrt{\sin(x+h)} + \sqrt{\sin x}} \cdot \frac{\sin \frac{h}{2}}{\frac{h}{2}} \right]$$

$$= \frac{\cos x}{\sqrt{\sin x} + \sqrt{\sin x}} . 1 = \frac{\cos x}{2\sqrt{\sin x}}.$$

Example 26. *Differentiate the functions:*

(*i*) $\dfrac{\sqrt{(x+1)} + \sqrt{(x-1)}}{\sqrt{(x+1)} - \sqrt{(x-1)}}$ (*ii*) $\log \sqrt{(x^2 + y^2)} = \tan^{-1} \dfrac{x}{y}$.

Solution. (*i*) Let $y = \dfrac{\sqrt{(x+1)} + \sqrt{(x-1)}}{\sqrt{(x+1)} - \sqrt{(x-1)}}$

Rationalising the denominator, we have

$$y = \left[\frac{\sqrt{(x+1)} + \sqrt{(x-1)}}{\sqrt{(x+1)} - \sqrt{(x-1)}} \times \frac{\sqrt{(x+1)} + \sqrt{(x-1)}}{\sqrt{(x+1)} + \sqrt{(x-1)}} \right]$$

$$= \left[\frac{(x+1) + (x-1) + 2\sqrt{(x+1)}\sqrt{(x-1)}}{(x+1) - (x-1)} \right]$$

$$= [x + (x^2 - 1)^{1/2}]$$

$$\therefore \quad \frac{dy}{dx} = \left[1 + \frac{1}{2}(x^2 - 1)^{\frac{1}{2}} \times 2x \right]$$

$$= \left[1 + \frac{x}{\sqrt{(x^2 - 1)}} \right]$$

(*ii*) The given implicit function is

$$\frac{1}{2} \log (x^2 + y^2) = \tan^{-1} \frac{x}{y}$$

Differentiating w.r.t., we get

$$\frac{1}{2} \cdot \frac{1}{(x^2 + y^2)} \left(2x + 2y \frac{dy}{dx} \right) = \left(\frac{1}{1 + \frac{x^2}{y^2}} \right) \left(\frac{y - x \frac{dy}{dx}}{y^2} \right)$$

or
$$\frac{x}{(x^2 + y^2)} + \frac{y}{(x^2 + y^2)} \frac{dy}{dx} = \frac{y}{(x^2 + y^2)} - \frac{x}{x^2 + y^2} \cdot \frac{dy}{dx}$$

Multiplying by $(x^2 + y^2)$ and simplifying, we get

$$\frac{dy}{dx} = \frac{y - x}{y + x}.$$

Example 27. *Differentiate*

(i) $\frac{1}{2}x\sqrt{(a^2-x^2)}+\frac{1}{2}a^2\sin^{-1}\left(\frac{x}{a}\right)$ *(ii)* $\log\sqrt{\left(\frac{1+t+t^2}{1-t+t^2}\right)}$.

Solution. *(i)* Let $y=\frac{1}{2}x(a^2-x^2)^{\frac{1}{2}}+\frac{a^2}{2}\sin^{-1}\left(\frac{x}{a}\right)$

$$\therefore \quad \frac{dy}{dx}=\frac{1}{2}\left\{1.(a^2-x^2)^{\frac{1}{2}}+x.\frac{1}{2}(a^2-x^2)^{-\frac{1}{2}}\times(-2x)\right\}$$

$$+\frac{1}{2}a^2\frac{1}{\sqrt{\left(1-\frac{x^2}{a^2}\right)}}\times\frac{1}{a}$$

$$=\frac{1}{2}\left[\left\{\sqrt{(a^2-x^2)}-\frac{x^2}{\sqrt{a^2-x^2}}\right\}+\frac{a^2}{\sqrt{(a^2-x^2)}}\right]$$

$$=\frac{1}{2}\left[\frac{a^2-x^2-x^2+a^2}{\sqrt{(a^2-x^2)}}\right]=\sqrt{(a^2-x^2)}.$$

(ii) Let $y=\frac{1}{2}\{\log(1+t+t^2)-\log(1-t+t^2)\}$

$$\therefore \quad \frac{dy}{dt}=\frac{1}{2}\left[\left(\frac{1+2t}{1+t+t^2}\right)-\left(\frac{-1+2t}{1-t+t^2}\right)\right]$$

$$=\frac{1}{2}\left\{\frac{(2t+1)(t^2+1-t)-(2t-1)(t^2+1+t)}{(t^2+1+t)(t^2+1-t)}\right\}$$

$$=\frac{1-t^2}{(t^4+t^2+1)}.$$

Example 28. *Find* $\frac{dy}{dx}$, *when*

(i) $x=a\left(\cos\theta+\log\tan\frac{\theta}{2}\right); y=a\sin\theta$ *(ii)* $x=a\cos^{-1}\sqrt{\left(\frac{y}{a}\right)}+\sqrt{(ay-y^2)}$

(iii) $x=\frac{3at}{1+t^3}; y=\frac{3at}{1+t^3}$ *at* $t=1$.

Solution. *(i)* Differentiating x and y w.r.t. θ, we have

$$\frac{dx}{d\theta}=a\left(-\sin\theta+\frac{1}{\tan\theta/2}\cdot\frac{1}{2}\sec^2\frac{\theta}{2}\right)$$

$$=a\left(-\sin\theta+\frac{1}{\sin\theta}\right)=a\cos^2\theta\sin\theta$$

and $$\frac{dy}{d\theta} = a \cos \theta$$

$$\therefore \quad \frac{dy}{dx} = \frac{dy}{d\theta} \cdot \frac{d\theta}{dx} = a \cos \theta \times \frac{\sin \theta}{a \cos^2 \theta} = \tan \theta.$$

(ii) Differentiating w.r.t. y, we have

$$x = a \cos^{-1} \sqrt{\left(\frac{y}{a}\right)} + \sqrt{ay - y^2}$$

$$\frac{dx}{dy} = a \,.\, \frac{1}{\sqrt{(1 - y\, a)}} \times \frac{d}{dy}\left(\frac{\sqrt{y}}{\sqrt{a}}\right) + \frac{1}{2}(ay - y^2)^{-\frac{1}{2}} \times (a - 2y)$$

$$= \frac{-a\sqrt{a}}{\sqrt{(a - y)}} \times \frac{1}{2\sqrt{ay}} + \frac{a - 2y}{2\sqrt{(ay - y^2)}}$$

$$= \frac{-a + a - 2y}{2\sqrt{y}\,\sqrt{(a - y)}} = \frac{y}{\sqrt{(ay - y^2)}}$$

$$\therefore \quad \frac{dy}{dx} = -\sqrt{(ay - y^2)}/y\,.$$

(iii) $$x = \frac{3at}{1+t^3}\,;\; y = \frac{3at}{1+t^3} \text{ at } t = 1$$

$$\frac{dx}{dt} = \frac{3a(1 + t^3) - 3at \times 3t^2}{(1 + t^2)^2} = \frac{3at(2 - t^3)}{(1 + t^3)^2}$$

and $$\frac{dy}{dt} = \frac{6at\,(1 + t^3) - 3at^2 . 3t^2}{(1 + t^3)^2} = \frac{3at(2 - t^3)}{(1 + t^3)^2}.$$

$$\therefore \quad \frac{dy}{dx} = \frac{dy}{dt}\Big/\frac{dx}{dt} = \frac{t(2 - t^3)}{(1 - 2t^3)}$$

Now at $t = 1$, $$\frac{dy}{dx} = \frac{1(2 - 1)}{(1 - 2)} = -1.$$

Example 29. *If* $y = log\left(\dfrac{1 + \sqrt{2x} + x^2}{1 - \sqrt{2x} + x^2}\right) + 2\, tan^{-1}\left(\dfrac{x\sqrt{2}}{1 - x^2}\right)$ *show that* $\dfrac{dy}{dx} = \dfrac{4\sqrt{2}}{(1 + x^4)}$.

Solution. We have $y = \log(1 + \sqrt{2x} + x^2) - \log(1 - \sqrt{2x} + x^2) + 2\tan^{-1}\left\{\dfrac{x\sqrt{2}}{1 - x^2}\right\}$

$$\therefore \quad \frac{dy}{dx} = \frac{(\sqrt{2} + 2x)}{(1 + \sqrt{2x} + x^2)} - \frac{(-\sqrt{2} + 2x)}{(1 - \sqrt{2x} + x^2)} + 2\,.\,\frac{1}{1 + \dfrac{2x}{(1 - x^2)^2}} \times \frac{d}{dx}\left\{\frac{x\sqrt{2}}{1 - x^2}\right\}$$

$$= \frac{(2x + \sqrt{2})(1 + x^2 - 2\sqrt{x}) - (2x - \sqrt{2})(1 + x^2 + \sqrt{2x})}{(1 + x^2 + \sqrt{2x})(1 + x^2 - \sqrt{2x})} + \frac{2\sqrt{2}\,(1 + x^2)}{(1 - x^2)^2 + 2x^2}$$

$$= \frac{2\sqrt{2}\,(1 - x^2)}{(1 + x^4)} + \frac{2\sqrt{2}\,(1 + x^2)}{(1 + x^4)} = \frac{4\sqrt{2}}{(1 + x^4)}.$$

Example 30. *If* $y = (2-x)\left(\dfrac{3-x}{1+x}\right)^{1/3}$, *find* $\dfrac{dy}{dx}$.

Solution. Taking logs on both sides, we have

$$\log y = \log (2-x) + \frac{1}{3}\{\log (3-x) - \log (1+x)\}.$$

Now differentiating w.r.t. x, we have

$$\frac{1}{y}\frac{dy}{dx} = \frac{1}{(2-x)} \times (-1) + \frac{1}{3} - \left\{\frac{1}{(3-x)} \times (-1) - \frac{1}{(1+x)}\right\}$$

$$= \frac{(x^2-7)}{(x-2)(x-3)(x+1)} \qquad \text{On simplifying, we get}$$

$$\therefore \quad \frac{dy}{dx} = \frac{(x^2-7)}{(3-x)^{2/3}(1+x^2)^{3/2}}.$$

Example 31. *Find* $\dfrac{dy}{dx}$, *when* $x^3 + y^3 = 3axy$.

Solution. *Differentiating the given equation w.r.t. x, we have*

$$3x^2 + 3y^2\frac{dy}{dx} = 3a\left(y + x\frac{dy}{dx}\right)$$

or
$$3(y^2 - ax)\frac{dy}{dx} = 3(ay - x^2)$$

or
$$\frac{dy}{dx} = \left(\frac{ay - x^2}{y^2 - ax}\right).$$

Example 32. *Differentiate by first principle.*

Solution. (*i*) Let $y = (x + 5)$.

Let δy be an increment in y, corresponding to an increment δx in x.

Then, $\quad y + \delta y = (x + \delta x + 5)^7$

$\Rightarrow \quad \delta y = (x + \delta x + 5)^7 - (x+5)^7$

$$\Rightarrow \quad \frac{\delta y}{\delta x} = \frac{(x+\delta x+5)^7 - (x+5)^7}{\delta x}$$

$$\Rightarrow \quad \frac{dy}{dx} = \lim_{\delta x \to 0}\frac{\delta y}{\delta x}$$

$$= \lim_{\delta x \to 0}\frac{(x+\delta x+5)^7 - (x+5)^7}{\delta x}$$

$$= \lim_{(x+\delta x+5)\to(x+5)}\frac{[(x+dx+5)^7 - (x+5)^7]}{[(x+\delta x+5) - (x+5)]}$$

$$= \lim_{u \to a}\left(\frac{u^7 - a^7}{u - a}\right), \text{ where } (x+\delta x+5) = u \text{ and } (x+5) = a$$

$$= 7a^{(7-1)} = 7(x+5)^6 \qquad \left[\because \lim_{u \to a} \left(\frac{u^n - a^n}{u - a}\right) = na^{n-1}\right]$$

Thus, $\dfrac{d}{dx}(x+5)^7 = 7(x+5)^6.$

Example 33. *Differentiate each of the following:*

(i) $(x+5)^7$ *(ii)* $\left(x - \dfrac{1}{x}\right)$.

Solution. (*i*) Let $y = (x+5)^7$.

Let δy be an increment in y, corresponding to an increment δx in x.

$$\therefore \quad y + \delta y = (x + \delta x + 5)^7$$

$$\Rightarrow \quad \delta y = (x + \delta x + 5)^7 - (x+5)^7$$

$$\Rightarrow \quad \frac{\delta y}{\delta x} = \frac{(x + \delta x + 5)^7 - (x+5)^7}{\delta x}$$

$$\Rightarrow \quad \frac{dy}{dx} = \lim_{\delta x \to 0} \frac{\delta y}{\delta x}$$

$$= \lim_{\delta x \to 0} \frac{(x + \delta x + 5)^7 - (x+5)^7}{\delta x}$$

$$= \lim_{(x+\delta x+5) \to (x+5)} \frac{[(x + \delta x + 5)^7 - (x+5)^7]}{[(x + \delta x + 5) - (x+5)]}$$

$$= \lim_{u \to a} \left(\frac{u^7 - a^7}{u - a}\right), \text{ where } (x + \delta x + 5) = u \text{ and } (x+5) = a$$

$$= 7a^{(7-1)} = 7(x+5)^6 \qquad \left[\because \lim_{u \to a} \left(\frac{u^n - a^n}{u - a}\right) = na^{n-1}\right]$$

Thus, $\dfrac{d}{dx}(x+5)^7 = 7(x+5)^6.$

(*ii*) Let $y = \left(x - \dfrac{1}{x}\right)$.

Let δy be an increment in y, corresponding to an increment δx in x.

$$\therefore \quad y + \delta y = \left[(x + \delta x) - \frac{1}{(x + \delta x)}\right]$$

$$\Rightarrow \quad \delta y = \left[(x + \delta x) - \frac{1}{(x + \delta x)}\right] - \left(x - \frac{1}{x}\right)$$

$$\Rightarrow \quad \delta y = \delta x + \left\{\frac{1}{x} - \frac{1}{(x + \delta x)}\right\}$$

$$\Rightarrow \quad \frac{\delta y}{\delta x} = \frac{\delta x + \left\{\dfrac{(x+\delta x) - x}{x(x+\delta x)}\right\}}{\delta x}$$

$$\Rightarrow \quad \frac{dy}{dx} = \lim_{\delta x \to 0} \frac{\delta y}{\delta x}$$

$$= \lim_{\delta x \to 0} \frac{\left\{\delta x + \dfrac{\delta x}{x(x+\delta x)}\right\}}{\delta x} = \lim_{\delta x \to 0} \left[1 + \frac{1}{x(x+\delta x)}\right] = \left(1 + \frac{1}{x^2}\right).$$

Thus, $\dfrac{d}{dx}\left(x - \dfrac{1}{x}\right) = \left(1 + \dfrac{1}{x^2}\right).$

Example 34. *Find the derivative of each of the following by the method of first principle.*

(i) sin 2x *(ii) cos 3x* *(iii) tan 5x.*

Solution. *(i)* Let $y = \sin 2x$.

Let δy be an increment in y, corresponding to an increment δx in x.

$$\therefore \quad y + dx = \sin 2(x + dx)$$

$$\Rightarrow \quad \delta y = \sin(2x + 2\delta x) - \sin 2x$$

$$\Rightarrow \quad \frac{\delta y}{\delta x} = \frac{\sin(2x + 2\delta x) - \sin 2x}{\delta x}$$

$$\Rightarrow \quad \frac{dy}{dx} = \lim_{\delta x \to 0} \frac{\delta y}{\delta x}$$

$$= \lim_{\delta x \to 0} \frac{[\sin(2x + 2\delta x) - \sin 2x]}{\delta x}$$

$$= \lim_{\delta x \to 0} \frac{2\cos(2x + \delta x)\sin \delta x}{\delta x}$$

$$\left\{\because \sin C - \sin D = 2\cos\left(\frac{C-D}{2}\right)\sin\left(\frac{C-D}{2}\right)\right\}$$

$$= 2 \,.\, \lim_{\delta x \to 0} \cos(2x + \delta x) \,.\, \lim_{\delta x \to 0} \frac{\sin \delta x}{\delta x}$$

$$= (2\cos 2x \times 1) = 2\cos 2x.$$

Thus, $\dfrac{d}{dx}(\sin 2x) = 2\cos 2x.$

(ii) Let $y = \cos 3x$.

Let δy be an increment in y, corresponding to an increment δx in x.

$$\therefore \quad y + \delta y = \cos 3(x + \delta x)$$

$$\Rightarrow \quad \delta y = \cos(3x + 3\delta x) - \cos 3x$$

$$\Rightarrow \quad \frac{\delta y}{\delta x} = \frac{\cos(3x + 3\delta x) - \cos 3x}{\delta x}$$

$$\Rightarrow \quad \frac{dy}{dx} = \lim_{\delta x \to 0} \frac{\delta y}{\delta x}$$

$$= \lim_{\delta x \to 0} \frac{[\cos(3x + 3\delta x) - \cos 3x)]}{\delta x}$$

$$= \lim_{\delta x \to 0} \frac{-2 \sin\left(3x + \frac{3}{2}\delta x\right) \sin\left(\frac{3}{2}\delta x\right)}{\delta x}$$

$$\left\{\because (\cos C - \cos D) = -2 \sin\left(\frac{C+D}{2}\right) \sin\left(\frac{C-D}{2}\right)\right\}$$

$$= \lim_{\delta x \to 0} \sin\left(3x + \frac{3}{2}\delta x\right) . \lim_{\delta x \to 0} \frac{3 \sin\left(\frac{3}{2}\delta x\right)}{\left(\frac{3}{2}\delta x\right)}$$

$$= (-\sin 3x \times 3 \times 1) = -3 \sin 3x$$

Thus, $\quad \frac{d}{dx}(\cos 3x) = -3 \sin 3x.$

(*iii*) Let $y = \tan 5x$.

Let δy be an increment in y, corresponding to an increment δx in x.

$\therefore \quad y + \delta y = \tan 5(x + \delta x)$

$\Rightarrow \quad \delta y = \tan(5x + 5\delta x) - \tan 5x$

$\Rightarrow \quad \frac{\delta y}{\delta x} = \frac{\tan(5x + 5\delta x) - \tan 5x}{\delta x}$

$$\frac{dy}{dx} = \lim_{dx \to 0} \frac{\delta y}{\delta x}$$

$$= \lim_{dx \to 0} \frac{\tan(5x + 5\delta x) - \tan 5x}{\delta x}$$

$$\Rightarrow \quad = \lim_{dx \to 0} \frac{\left\{\frac{\sin(5x + 5\delta x)}{\cos(5x + 5\delta x)} - \frac{\sin 5x}{\cos 5x}\right\}}{\delta x}$$

$$= \lim_{\delta x \to 0} \frac{\sin(5x + 5\delta x)\cos 5x - \cos(5x + 5\delta x)\sin 5x}{\cos(5x + 5\delta x) . \cos 5x . \delta x}$$

$$= \lim_{\delta x \to 0} \frac{\sin(5x + 5\delta x - 5x)}{\cos(5x + 5\delta x) . \cos 5x . \delta x}$$

$$[\because \sin A \cos B - \cos A \sin B = \sin(A - B)]$$

$$= \frac{1}{\cos 5x} . \lim_{\delta x \to 0} \frac{1}{\cos(5x + 5\delta x)} . \lim_{\delta x \to 0} \frac{\sin(5\delta x) \times 5}{(5\delta x)}$$

$$= \left[\frac{1}{\cos 5x} \times \frac{1}{\cos 5x} \times (1 \times 5)\right] = \frac{5}{\cos^2 5x} = 5 \sec^2 5x.$$

Thus, $\quad \frac{d}{dx}(\tan 5x) = 5 \sec^2 5x.$

Example 35. *Differentiate x sin x by the method of first principle.*

Solution. Let $y = x \sin x$.

Let δy be an increment in y, corresponding to an increment δx in x.

Then, $$y + \delta y = (x + \delta x) \sin (x + \delta x)$$

$$\Rightarrow \quad \delta y = (x + \delta x) \sin (x + \delta x) - x \sin x$$

$$\Rightarrow \quad \frac{\delta y}{\delta x} = \frac{(x + \delta x) \sin (x + \delta x) - x \sin x}{\delta x}$$

$$\Rightarrow \quad \frac{dy}{dx} = \lim_{\delta x \to 0} \frac{\delta y}{\delta x}$$

$$= \lim_{\delta x \to 0} \frac{\{(x + \delta x) \sin (x + \delta x) - x \sin x\}}{\delta x}$$

$$= \lim_{dx \to 0} \left[\frac{x \{\sin (x + \delta x) - \sin x\}}{\delta x} + x \sin x \right]$$

$$= \lim_{\delta x \to 0} \frac{x \{\sin (x + \delta x) - \sin x\}}{\delta x} + \lim_{\delta x \to 0} \sin (x + \delta x)$$

$$= \lim_{\delta x \to 0} \left\{ \frac{2x \cos \left(x + \frac{\delta x}{2} \right) \sin \left(\frac{\delta x}{2} \right)}{\delta x} \right\} + \sin x$$

$$= \left\{ x \,.\, \lim_{\delta x \to 0} \cos \left(x + \frac{\delta x}{2} \right) . \lim_{\delta x \to 0} \frac{\sin (\delta x/2)}{(\delta x/2)} \right\} + \sin x$$

$$= (x \,.\, \cos x \,.\, 1) + \sin x = (x \cos x + \sin x).$$

Thus, $$\frac{d}{dx} (x \sin x) = (x \cos x + \sin x).$$

Example 36. *Differentiate* $\frac{\sin x}{x}$ *by the method of first principle.*

Solution. Let $y = \frac{\sin x}{x}$.

Let δy be an increment in y, increment in y, corresponding to an increment δx in x.

Then, $$y + \delta y = \frac{\sin (x + \delta x)}{(x + \delta x)}$$

$$\Rightarrow \quad \delta y = \left\{ \frac{\sin (x + \delta x)}{(x + \delta x)} - \frac{\sin x}{x} \right\} = \frac{\{x \sin (x + \delta x) - (x + \delta x) \sin x\}}{x(x + \delta x)}$$

$$\Rightarrow \quad \frac{\delta y}{\delta x} = \frac{\{x \sin (x + \delta x) - (x + \delta x) \sin x\}}{x(x + \delta x) \,.\, \delta x}$$

$$\Rightarrow \quad \frac{dy}{dx} = \lim_{\delta x \to 0} \frac{\delta y}{\delta x}$$

$$= \lim_{\delta x \to 0} \frac{\{x \sin (x + \delta x) - (x + \delta x) \sin x\}}{x(x + \delta x) \,.\, \delta x}$$

$$= \lim_{\delta x \to 0} \frac{x[\sin(x+\delta x) - \sin x](\delta x)\sin x}{x(x+\delta x).\,\delta x}$$

$$= \lim_{\delta x \to 0} \frac{x\left[2\cos\left(x+\frac{\delta x}{2}\right)\sin\left(\frac{\delta x}{2}\right)\right] - (\delta x)\sin x}{x(x+\delta x).\,\delta x}$$

$$= \left\{\lim_{\delta x \to 0}\left(\cos x + \frac{\delta x}{2}\right).\lim_{\delta x \to 0}\frac{\sin(\delta x/2)}{(\delta x/2)}.\lim_{\delta x \to 0}\frac{1}{(x+\delta x)}\right\} - \lim_{\delta x \to 0}\frac{1}{x(x+\delta x)}$$

$$= \left(\cos x \times 1 \times \frac{1}{x}\right) - \frac{\sin x}{x^2} = \left(\frac{\cos x}{x} - \frac{\sin x}{x^2}\right) = \frac{(x\cos x - \sin x)}{x^2}$$

Thus, $\frac{d}{dx}\left(\frac{\sin x}{x}\right) = \frac{(x\cos x - \sin x)}{x^2}$.

Example 37. *Differentiate cot (2x + 1) by the method of first principle.*

Solution. Let $y = \cot(2x+1)$.

Let δy be an increment in y, corresponding to an increment δx in x.

Then, $y + \delta y = \cot[2(x+\delta x)+1]$

$\Rightarrow \delta y = \cot[2(x+\delta x)+1] - \cot(2x+1)$

$$\Rightarrow \frac{\delta y}{\delta x} = \frac{\cot[2(x+\delta x)+1] - \cot(2x+1)}{\delta x}$$

$$\Rightarrow \frac{dy}{dx} = \lim_{\delta x \to 0}\frac{\delta y}{\delta x}$$

$$= \lim_{\delta x \to 0}\frac{\cot(2x+2\delta x+1) - \cot(2x+1)}{\delta x}$$

$$= \lim_{\delta x \to 0}\frac{\left\{\frac{\cos(2x+2\delta x+1)}{\sin(2x+2\delta x+1)} - \frac{\cos(2x+1)}{\sin(2x+1)}\right\}}{\delta x}$$

$$= \lim_{\delta x \to 0}\frac{\sin(2x+1)\cos(2x+2\delta x+1) - \cos(2x+1)\sin(2x+2\delta x+1)}{\delta x.\sin(2x+2\delta x+1).\sin(2x+1)}$$

$$= \lim_{\delta x \to 0}\frac{\sin[(2x+1)-(2x+2\delta x+1)]}{\delta x.\sin(x\,2\delta x+1).\sin(2x+1)}$$

$[\because \sin A\cos B - \cos A\sin B = \sin(A-B)]$

$$= \lim_{\delta x \to 0}\frac{\sin(-2\delta x)}{\delta x.\sin(2x+2\delta x+1).\sin(2x+1)}$$

$$= \frac{1}{\sin(2x+1)}\lim_{\delta x \to 0}\left\{\frac{-\sin(2\delta x)}{2\delta x}\times 2\right\}.\lim_{\delta x \to 0}\frac{1}{\sin(2x+2\delta x+1)}$$

$$= \frac{1}{\sin(2x+1)} \cdot (-1) \times 2 \times \frac{1}{\sin(2x+1)} = \frac{-2}{\sin^2(2x+1)}$$

$$= -2 \operatorname{cosec}^2(2x+1)$$

Thus, $\frac{d}{dx}[\cot(2x+1)] = -2 \operatorname{cosec}^2(2x+1)$.

Example 38. *Find the derivative of each of the following from the first principle.*

(*i*) $\sqrt{2x+3}$ (*ii*) $\sqrt{4-x}$ (*iii*) $\frac{1}{\sqrt{x}}$.

Solution. (*i*) Let $y = \sqrt{2x+3}$.

Let dy be an increment in y, corresponding to an increment dx in x.

Then, $y + \delta y = \sqrt{2(x+\delta x)+3}$

$$\Rightarrow \quad \delta y = \sqrt{2(x+\delta x)+3} - \sqrt{2x+3}$$

$$\Rightarrow \quad \frac{\delta y}{\delta x} = \frac{\sqrt{2(x+\delta x)+3} - \sqrt{2x+3}}{\delta x}$$

$$\Rightarrow \quad \frac{dy}{dx} = \lim_{\delta x \to 0} \frac{\delta y}{\delta x} = \lim_{\delta x \to 0} \frac{\sqrt{2(x+\delta x)+3} - \sqrt{2x+3}}{\delta x}$$

$$= \lim_{\delta x \to 0} \left\{ \frac{(\sqrt{2x+2\delta x+3} - \sqrt{2x+3})}{\delta x} \times \frac{(\sqrt{2x+2\delta x+3} + \sqrt{2x+3})}{(\sqrt{2x \times 2\delta x+3} + \sqrt{2x+3})} \right\}$$

$$= \lim_{\delta x \to 0} \frac{\{(2x+2\delta x+3) - (2x+3)\}}{\delta x.(\sqrt{2x+2\delta x+3} + \sqrt{2x+3})}$$

$$= \lim_{\delta x \to 0} \frac{2\delta x}{\delta x.(\sqrt{2x+2\delta x+3} + \sqrt{2x+3})}$$

$$= \lim_{\delta x \to 0} \frac{2}{(\sqrt{2x+2\delta x+3} + \sqrt{2x+3})}$$

$$= \frac{2}{2\sqrt{2x+3}} = \frac{1}{\sqrt{2x+3}}$$

Thus, $\frac{d}{dx}(\sqrt{2x+3}) = \frac{1}{\sqrt{2x+3}}$.

(*ii*) Let $y = \sqrt{4-x}$.

Let δy be an increment in y, corresponding to an increment δx in x.

Then, $y + \delta y = \sqrt{4-(x+\delta x)}$

$$\Rightarrow \quad \delta y = \sqrt{4-(x+\delta x)} - \sqrt{4-x}$$

$$\Rightarrow \quad \frac{\delta y}{\delta x} = \frac{\sqrt{4-(x+\delta x)} - \sqrt{4-x}}{\delta x}$$

$$\Rightarrow \quad \frac{dy}{dx} = \lim_{\delta x \to 0} \frac{\delta y}{\delta x}$$

$$= \lim_{\delta x \to 0} \frac{\{\sqrt{4-(x+\delta x)} - \sqrt{4-x}\}}{\delta x}$$

$$= \lim_{\delta x \to 0} \left\{ \frac{(\sqrt{4-x-\delta x} - \sqrt{4-x})}{\delta x} \times \frac{(\sqrt{4-x-\delta x} + \sqrt{4-x})}{(\sqrt{4-x-\delta x} + \sqrt{4-x})} \right\}$$

$$= \lim_{\delta x \to 0} \frac{\{(4-x-\delta x)-(4-x)\}}{\delta x\,(\sqrt{4-x-\delta x} + \sqrt{4-x})}$$

$$= \lim_{\delta x \to 0} \frac{-\delta x}{\delta x\,(\sqrt{4-x-\delta x} + \sqrt{4-x})}$$

$$= \lim_{\delta x \to 0} \frac{-1}{(\sqrt{4-x-\delta x} + \sqrt{4-x})} = \frac{-1}{2\sqrt{4-x}}.$$

$$\therefore \quad \frac{d}{dx}(\sqrt{4-x}) = \frac{-1}{2\sqrt{4-x}}.$$

(*iii*) Let $y = \dfrac{1}{\sqrt{x}}$.

Let δy be an increment in y, corresponding to an increment δx in x.

Then, $$y + \delta y = \frac{1}{\sqrt{x+\delta x}}$$

$$\Rightarrow \quad \delta y = \left(\frac{1}{\sqrt{x+\delta x}} - \frac{1}{\sqrt{x}} \right)$$

$$\Rightarrow \quad \frac{\delta y}{\delta x} = \frac{1}{\delta x} \cdot \left[\frac{1}{\sqrt{x+\delta x}} - \frac{1}{\sqrt{x}} \right]$$

$$\Rightarrow \quad \frac{dy}{dx} = \lim_{\delta x \to 0} \frac{\delta y}{\delta x}$$

$$= \lim_{\delta x \to 0} \frac{1}{\delta x} \cdot \left[\frac{1}{\sqrt{x+\delta x}} - \frac{1}{\sqrt{x}} \right]$$

$$= \lim_{\delta x \to 0} \left\{ \frac{(\sqrt{x} - \sqrt{x+\delta x})}{\delta x \,.\, \sqrt{x+\delta x} \,.\, \sqrt{x}} \times \frac{(\sqrt{x} + \sqrt{x+\delta x})}{(\sqrt{x} + \sqrt{x+\delta x})} \right\}$$

$$= \lim_{\delta x \to 0} \left[\frac{\{x - (x+\delta x)\}}{\delta x . \sqrt{x+\delta x} \,.\, \sqrt{x} \,.\, (\sqrt{x} + \sqrt{x+\delta x})} \right]$$

$$= \lim_{\delta x \to 0} \left[\frac{-\delta x}{\delta x \,.\, \sqrt{x + \delta x} \,.\, \sqrt{x} \,.\, (\sqrt{x} + \sqrt{x + \delta x})} \right\}$$

$$= \lim_{\delta x \to 0} \left[\frac{-1}{\sqrt{x + \delta x} \,.\, \sqrt{x} \,.\, (\sqrt{x} + \sqrt{x + \delta x})} \right\}$$

$$= \frac{1}{\sqrt{x} \,.\, \sqrt{x} \,.\, (2\sqrt{x})} = \frac{-1}{2x^{3/2}}$$

Thus, $\dfrac{d}{dx}\left(\dfrac{1}{\sqrt{x}}\right) = \dfrac{-1}{2x^{3/2}}$.

Example 39. *Find the derivative of* $x^{-3/2}$ *by the method of first principle.*

Solution. $y = x^{-3/2}$.

Consider δy be an increment in y, corresponding to an increment δx in x.

Then, $\quad y + \delta y = (x + \delta x)^{-3/2}$

$\Rightarrow \quad \delta y = (x + \delta x)^{-3/2} - x^{-3/2}$

$\Rightarrow \quad \dfrac{\delta y}{\delta x} = \dfrac{\{(x + \delta x)^{-3/2} - x^{-3/2}\}}{\delta x}$

$\Rightarrow \quad \dfrac{dy}{dx} = \lim\limits_{\delta x \to 0} \dfrac{\delta y}{\delta x}$

$$= \lim_{\delta x \to 0} \frac{\{(x + \delta x)^{-3/2} - x^{-3/2}\}}{\delta x} = \lim_{\delta x \to 0} \frac{\{(x + \delta x)^{-3/2} - x^{-3/2}\}}{\{(x + \delta x) - x\}}$$

$$= \lim_{u \to x} \frac{(u^{-3/2} - x^{-3/2})}{(u - x)}, \text{ where } u = (x + \delta x)$$

$$= -\frac{3}{2} x^{\left(\frac{-3}{2} - 1\right)} \quad \left[\because \lim_{u \to a} \frac{u^n - a^n}{u - a} = na^{n-1}\right]$$

$$= -\frac{3}{2} x^{-5/2}.$$

Thus, $\dfrac{d}{dx}(x^{-3/2}) = -\dfrac{3}{2} x^{-5/2}$.

Example 40. *Find the derivative of* e^{x^2} *by the method of first principle.*

Solution. Let $y = e^{x^2}$.

Let δy be an increment in y, corresponding to an increment δx in x.

$\therefore \quad y + \delta y = e^{(x + \delta x)^2}$

$\Rightarrow \quad \delta y = e^{(x + \delta x)^2} - e^{x^2}$

$\Rightarrow \quad \dfrac{\delta y}{\delta x} = \dfrac{e^{(x + \delta x)^2} - e^{x^2}}{\delta x}$

$\Rightarrow \quad \dfrac{dy}{dx} = \lim\limits_{\delta x \to 0} \dfrac{\delta y}{\delta x}$

$$= \lim_{\delta x \to 0} \frac{e^{(x + \delta x)^2} - e^{x^2}}{\delta x} = \lim_{\delta x \to 0} \frac{e^{(x^2 + 2x\delta x + \delta x^2)} - e^{x^2}}{\delta x}$$

$$= \lim_{\delta x \to 0} \frac{e^{x^2}\,[e^{2x\delta x + \delta x^2)} - 1]}{\delta x}$$

$$= e^{x^2} . \lim_{\delta x \to 0} \frac{(e^{2x\delta x + \delta x^2} - 1)}{(2x\delta x + \delta x^2)} \times \frac{(2x\delta x + \delta x^2)}{\delta x}$$

$$= e^{x^2} . \lim_{\theta x \to 0} \left(\frac{e^{\theta} - 1}{\theta}\right) . \lim_{\delta x \to 0} (2x + \delta x), \text{ where } (2x\delta x + \delta x^2) = \theta$$

[clearly, $\theta \to 0$ when $\delta x \to 0$]

$$= (e^{x^2} \times 1 \times 2x) = 2xe^{x^2} \qquad \left[\because \lim_{\theta \to 0}\left(\frac{e^{\theta} - 1}{\theta}\right) = 1\right]$$

Thus, $\dfrac{d}{dx}(e^{x^2}) = 2xe^{x^2}$.

Example 41. *Find the derivative of $e^{\sin x}$ by the method of first principle.*

Solution. Let $y = e^{\sin x}$.

Let δy be an increment in y, corresponding to an increment δx in x.

$\therefore \quad y + \delta y = e^{\sin (x + \delta x)}$

$\Rightarrow \quad \delta y = e^{\sin (x + \delta x)} - e^{\sin x}$

$$\Rightarrow \quad \frac{\delta y}{\delta x} = \frac{e^{\sin (x + \delta x)} - e^{\sin x}}{\delta x}$$

$$\Rightarrow \quad \frac{dy}{dx} = \lim_{\delta x \to 0} \frac{[e^{\sin (x + \delta x)} - e^{\sin x}]}{\delta x} = \lim_{\delta x \to 0} \frac{e^{\sin x} . [e^{\sin (x + \delta x)} - \sin x - 1]}{\delta x}$$

$$= e^{\sin x} . \lim_{\delta x \to 0} \frac{[e^{\sin (x + \delta x) - \sin x} - 1]}{\{\sin (x + \delta x) - \sin x\}} . \frac{[\sin (x + \delta x) - \sin x]}{\delta x}$$

$$= e^{\sin x} . \lim_{\delta x \to 0} \left(\frac{e^{\theta} - 1}{\theta}\right) . \lim_{\delta x \to 0} \frac{\sin (x + \delta x) - \sin x}{\delta x}$$

where $\theta = \sin (x + \delta x) - \sin$ [clearly, $(\delta x \to 0) \Rightarrow (\theta \to 0)$]

$$= e^{\sin x} . 1 . \lim_{\delta x \to 0} \frac{2 \cos\left(x + \frac{\delta x}{2}\right) \sin\left(\frac{\delta x}{2}\right)}{\delta x}$$

$$= e^{\sin x} . \lim_{\delta x \to 0} \cos\left(x + \frac{\delta x}{2}\right) . \lim_{\delta x \to 0} \frac{\sin\left(\frac{\delta x}{2}\right)}{\left(\frac{\delta x}{2}\right)}$$

$$= (e^{\sin x} . \cos x . 1) = e^{\sin x} \cos x.$$

Thus, $\dfrac{d}{dx}(e^{\sin x}) = e^{\sin x} \cos x$.

Example 42. *Differentiate each of the following by the method of first principle.*

(*i*) $\sqrt{sin\ x}$ (*ii*) $\sqrt{cos\ x}$

(*iii*) $\sqrt{tan\ x}$ (*iv*) $\sqrt{cosec\ x}$.

Solution. (*i*) Let $y = \sqrt{\sin x}$.

Let δy be an increment in y, corresponding to an increment δx in x.

$$\therefore \quad y + \delta y = \sqrt{\sin (x + \delta x)}$$

$$\Rightarrow \quad \delta y = \sqrt{\sin (x + \delta x)}\ \sqrt{\sin x}$$

$$\Rightarrow \quad \frac{\delta y}{\delta x} = \frac{\sqrt{\sin (x + \delta x)} - \sqrt{\sin x}}{\delta x}$$

$$\Rightarrow \quad \frac{dy}{dx} = \lim_{\delta x \to 0} \frac{\delta y}{\delta x}$$

$$= \lim_{\delta x \to 0} \frac{\sqrt{\sin (x + \delta x)} - \sqrt{\sin x}}{\delta x}$$

$$= \lim_{\delta x \to 0} \frac{\{\sqrt{\sin (x + \delta x)} - \sqrt{\sin x}\}}{\delta x} \times \frac{\{\sqrt{\sin (x + \delta x)} + \sqrt{\sin x}\}}{\{\sqrt{\sin (x + \delta x)} + \sqrt{\sin x}\}}$$

$$= \lim_{\delta x \to 0} \frac{\{\sin (x + \delta x) - \sin x\}}{\delta x \,.\, \{\sqrt{\sin (x + \delta x)} + \sqrt{\sin x}\}}$$

$$= \lim_{\delta x \to 0} \frac{2 \cos \left(x + \frac{\delta x}{2}\right) \sin \left(\frac{\delta x}{2}\right)}{\{\sqrt{\sin (x + \delta x)} + \sqrt{\sin x}\} \,.\, \delta x}$$

$$= \lim_{\delta x \to 0} \left(\cos x + \frac{\delta x}{2}\right) \,.\, \lim_{\delta x \to 0} \frac{\sin \left(\frac{\delta x}{2}\right)}{\left(\frac{\delta x}{2}\right)} \,.\, \lim_{\delta x \to 0} \frac{1}{\{\sqrt{\sin (x + \delta x)} + \sqrt{\sin x}\}}$$

$$= \left\{\cos x \,.\, 1 \,.\, \frac{1}{2\sqrt{\sin x}}\right\} = \frac{\cos x}{2\sqrt{\sin x}}.$$

Thus, $\quad \dfrac{d}{dx}\sqrt{\sin x} = \dfrac{\cos x}{2\sqrt{\sin x}}$.

(*ii*) Let $y = \sqrt{\cos x}$.

Let δy be an increment in y, corresponding to an increment δx in x.

$$\therefore \quad y + \delta y = \sqrt{\cos (x + \delta x)}$$

$$\Rightarrow \quad \delta y = \sqrt{\cos (x + \delta x)} - \sqrt{\cos x}$$

$$\Rightarrow \quad \frac{\delta y}{\delta x} = \frac{\sqrt{\cos (x + \delta x)} - \sqrt{\cos x}}{\delta x}$$

$$\Rightarrow \quad \frac{dy}{dx} = \lim_{\delta x \to 0} \frac{\{\sqrt{\cos(x+\delta x)} - \sqrt{\cos x}\}}{\delta x}$$

$$= \lim_{\delta x \to 0} \frac{\{\sqrt{\cos(x+\delta x)} - \sqrt{\cos x}\}}{\delta x} \times \frac{\{\sqrt{\cos(x+\delta x)} - \sqrt{\cos x}\}}{\{\sqrt{\cos(x+\delta x)} + \sqrt{\cos x}\}}$$

$$= \lim_{\delta x \to 0} \frac{\{\cos(x+\delta x) - \cos x\}}{\delta x \,.\, \{\sqrt{\cos(x+\delta x)} + \sqrt{\cos x}\}}$$

$$= \lim_{\delta x \to 0} \frac{-2\left(x+\frac{\delta x}{2}\right)\sin\left(\frac{\delta x}{2}\right)}{\delta x \,.\, \{\sqrt{\cos(x+\delta x)} + \sqrt{\cos x}\}}$$

$$\left[\because \quad \cos C - \cos D = -2\sin\left(\frac{C+D}{2}\right)\sin\left(\frac{C-D}{2}\right)\right]$$

$$= \lim_{\delta x \to 0} \sin\left(x+\frac{\delta x}{2}\right) . \lim_{\delta x \to 0} \frac{\sin(\delta x/2)}{(\delta x/2)} . \lim_{\delta x \to 0} \frac{1}{\{\sqrt{\cos(x+\delta x)} + \sqrt{\cos x}\}}$$

$$= (-\sin x) \times 1 \times \frac{1}{2\sqrt{\cos x}} = \frac{-\sin x}{2\sqrt{\cos x}}.$$

Thus, $\frac{d}{dx}\sqrt{\cos x} = \frac{-\sin x}{2\sqrt{\cos x}}$.

(*iii*) Let $y = \sqrt{\tan x}$.

Let δy be an increment in y, corresponding to an increment δx in x.

$$\therefore \quad y + \delta y = \sqrt{\tan(x+\delta x)}$$

$$\Rightarrow \quad \delta y = \sqrt{\tan(x+\delta x)} - \sqrt{\tan x}$$

$$\Rightarrow \quad \frac{\delta y}{\delta x} = \frac{\sqrt{\tan(x+\delta x)} - \sqrt{\tan x}}{\delta x}$$

$$\Rightarrow \quad \frac{dy}{dx} = \lim_{\delta x \to 0} \frac{\delta y}{\delta x}$$

$$= \lim_{\delta x \to 0} \frac{\sqrt{\tan(x+\delta x)} - \sqrt{\tan x}}{\delta x}$$

$$= \lim_{\delta x \to 0} \left\{\frac{\sqrt{\tan(x+\delta x)} - \sqrt{\tan x}}{\delta x} \times \frac{\sqrt{\tan(x+\delta x)} + \sqrt{\tan x}}{\sqrt{\tan(x+\delta x)} + \sqrt{\tan x}}\right\}$$

$$= \lim_{\delta x \to 0} \frac{\tan(x+\delta x) - \tan x}{\delta x\,[\sqrt{\tan(x+\delta x)} + \sqrt{\tan x}]}$$

$$= \lim_{\delta x \to 0} \frac{\left\{\frac{\sin(x+\delta x)}{\cos(x+\delta x)} - \frac{\sin x}{\cos x}\right\}}{\delta x\,[\sqrt{\tan(x+dx)} + \sqrt{\tan x}]}$$

$$= \lim_{\delta x \to 0} \frac{\sin(x+\delta x)\cos x - \cos(x+\delta x)\sin x}{\cos(x+\delta x)\cos x\,.\,\delta x\,.\,[\sqrt{\tan(x+\delta x)}+\sqrt{\tan x}]}$$

$$= \lim_{\delta x \to 0} \frac{\sin(x+\delta x - x)}{\cos(x+\delta x)\,.\,\cos x\,.\,\delta x\,.\,(\sqrt{\tan(x+\delta x)}+\sqrt{\tan x})}$$

$$= \frac{1}{\cos x}\,.\,\lim_{\delta x \to 0}\frac{1}{\cos(x+\delta x)}\,.\,\lim_{\delta x \to 0}\frac{\sin \delta x}{\delta x}\,.\,\lim_{\delta x \to 0}\frac{1}{\sqrt{\tan(x+\delta x)}+\sqrt{\tan x)}}$$

$$= \left(\frac{1}{\cos x}\,.\,\frac{1}{\cos x}\,.\,1\,.\,\frac{1}{2\sqrt{\tan x}}\right) = \frac{\sec^2 x}{2\sqrt{\tan x}}.$$

Thus, $\dfrac{d}{dx}(\sqrt{\tan x}) = \dfrac{\sec^2 x}{2\sqrt{\tan x}}$.

(*iv*) Let $y = \sqrt{\operatorname{cosec} x}$.

Let δy be an increment in y, corresponding to an increment δx in x.

$\therefore \quad y + \delta y = \sqrt{\operatorname{cosec}(x+\delta x)}$.

$\Rightarrow \quad \delta y = -\sqrt{\operatorname{cosec}(x+\delta x)} - \sqrt{\operatorname{cosec} x}$

$\Rightarrow \quad \dfrac{\delta y}{\delta x} = \dfrac{\sqrt{\operatorname{cosec}(x+\delta x)} - \sqrt{\operatorname{cosec} x}}{\delta x}$

$\Rightarrow \quad \dfrac{\delta y}{\delta x} = \lim\limits_{\delta x \to 0} \dfrac{\delta y}{\delta x}$

$$= \lim_{\delta x \to 0} \frac{\sqrt{\operatorname{cosec}(x+\delta x)} - \sqrt{\operatorname{cosec} x}}{\delta x}$$

$$= \lim_{\delta x \to 0} \frac{\left\{\dfrac{1}{\sqrt{\sin(x+\delta x)}} - \dfrac{1}{\sqrt{\sin x}}\right\}}{\delta x}$$

$$= \lim_{\delta x \to 0} \frac{\{\sqrt{\sin x} - \sqrt{\sin(x+\delta x)}\}}{\delta x\,.\,\sqrt{\sin(x+\delta x)}\,.\,\sqrt{\sin x}} \times \frac{\{\sqrt{\sin x} + \sqrt{\sin(x+\delta x)}\}}{\{\sqrt{\sin x} + \sqrt{\sin(x+\delta x)}\}}$$

$$= \lim_{\delta x \to 0} \frac{\{\sin x - \sin(x+\delta x)\}}{\{\sqrt{\sin(x+\delta x)}\,.\,\sqrt{\sin x}\}} \times \frac{1}{\delta x\,.\,\{\sqrt{\sin x} + \sqrt{\sin(x+\delta x)}\}}$$

$$= \lim_{\delta x \to 0} \frac{-2\cos\left(x+\dfrac{\delta x}{2}\right)\sin\left(\dfrac{\delta x}{2}\right)}{\{\sqrt{\sin(x+\delta x)}\,.\,\sqrt{\sin x}\}\,.\,\delta x\,.\,\{\sqrt{\sin x} + \sqrt{\sin(x+\delta x)}\}}$$

$$= -\lim_{\delta x \to 0} \cos\left(x+\frac{\delta x}{2}\right)\,.\,\lim_{\delta x \to 0}\frac{\sin\left(\dfrac{\delta x}{2}\right)}{\left(\dfrac{\delta x}{2}\right)}$$

$$\lim_{\delta x \to 0}\frac{1}{\sqrt{\sin(x+\delta x)}\,.\,\sqrt{\sin x}}\,.\,\lim_{\delta x \to 0}\frac{1}{\sqrt{\sin x} + \sqrt{\sin(x+\delta x)}\}}$$

$$= -\cos x \times 1 \times \frac{1}{\sqrt{\sin x} \cdot \sqrt{\sin x}} \cdot \frac{1}{(\sqrt{\sin x} + \sqrt{\sin x})}$$

$$= \frac{-\cos x}{\sin x} \cdot \frac{1}{2\sqrt{\sin x}} = -\frac{1}{2}\sqrt{\operatorname{cosec} x}\ \cot x.$$

Thus, $\quad \dfrac{d}{dx}(\sqrt{\operatorname{cosec} x}) = -\dfrac{1}{2}\sqrt{\operatorname{cosec} x}\ \cot x.$

Example 43. *Differentiate each of the following by the method of first principle.*

(i) $\sin \sqrt{x}$ *(ii)* $\cos \sqrt{x}$.

Solution. (*i*) Let $y = \sin \sqrt{x}$.

Let δy be an increment in y, corresponding to an increment δx in x.

$$\therefore \quad y + \delta y = \sin \sqrt{x + \delta x}$$

$$\Rightarrow \quad \delta y = \sin \sqrt{x + \delta x} - \sin \sqrt{x}$$

$$\Rightarrow \quad \frac{\delta y}{\delta x} = \frac{\sin \sqrt{x + \delta x} - \sin \sqrt{x}}{\delta x}$$

$$\Rightarrow \quad \frac{dy}{dx} = \lim_{\delta x \to 0} \frac{\delta y}{\delta x}$$

$$= \lim_{\delta x \to 0} \frac{\{\sin \sqrt{x + \delta x} - \sin \sqrt{x}\}}{\delta x}$$

$$= \lim_{\delta x \to 0} \frac{2 \cos \left(\dfrac{\sqrt{x + \delta x} + \sqrt{x}}{2}\right) \sin \left(\dfrac{\sqrt{x + dx} - \sqrt{x}}{2}\right)}{\{(x + \delta x) - x\}}$$

$$\left[\because \quad \sin C - \sin D = 2 \cos \left(\frac{C + D}{2}\right) \sin \left(\frac{C - D}{2}\right)\right]$$

$$= \lim_{\delta x \to 0} \frac{2 \cos \left(\dfrac{\sqrt{x + \delta x} + \sqrt{x}}{2}\right) \sin \left(\dfrac{\sqrt{x + \delta x} - \sqrt{x}}{2}\right)}{(\sqrt{x + \delta x} + \sqrt{x})(\sqrt{x + \delta x} - \sqrt{x})}$$

$$[\because \quad (x + \delta x) - x = (\sqrt{x + \delta x} + \sqrt{x})(\sqrt{x + \delta x} - \sqrt{x})]$$

$$= \lim_{\delta x \to 0} \frac{\sin \left(\dfrac{\sqrt{x + \delta x} - \sqrt{x}}{2}\right)}{\left(\dfrac{\sqrt{x + \delta x} - \sqrt{x}}{2}\right)} \cdot \lim_{\delta x \to 0} \cos \left(\frac{\sqrt{x + \delta x} + \sqrt{x}}{2}\right)$$

$$\times \lim_{\delta x \to 0} \frac{1}{(\sqrt{x + \delta x} + \sqrt{x})}$$

$$= \lim_{\delta x \to 0} \frac{\sin \theta}{\theta} \cdot \lim_{\delta x \to 0} \cos\left(\frac{\sqrt{x+\delta x}+\sqrt{x}}{2}\right) \cdot \lim_{\delta x \to 0} \frac{1}{(\sqrt{x+\delta x}+\sqrt{x})}$$

$$\text{where } \frac{(\sqrt{x+\delta x}-\sqrt{x})}{2} = \theta \text{ and } \delta x \to 0 \Rightarrow \theta \to 0$$

$$= 1 \times \cos\sqrt{x} \times \frac{1}{2\sqrt{x}} = \frac{\cos\sqrt{x}}{2\sqrt{x}}.$$

Thus, $$\frac{d}{dx}(\sin\sqrt{x}) = \frac{\cos\sqrt{x}}{2\sqrt{x}}.$$

(*ii*) Let $y = \cos\sqrt{x}$.

Let δy be an increment in y, corresponding to an increment δx in x.

$\therefore$ $$y + \delta y = \cos\sqrt{x+\delta x}$$

$\Rightarrow$ $$\delta y = \cos\sqrt{x+\delta x} - \cos\sqrt{x}$$

$\Rightarrow$ $$\frac{\delta y}{\delta x} = \frac{\{\cos\sqrt{x+\delta x} - \cos\sqrt{x}\}}{\delta x}$$

$\Rightarrow$ $$\frac{dy}{dx} = \lim_{\delta x \to 0} \frac{\delta y}{\delta x}$$

$$= \lim_{\delta x \to 0} \frac{\{\cos\sqrt{x+\delta x} - \cos\sqrt{x}\}}{\delta x}$$

$$= \lim_{\delta x \to 0} \frac{-2\sin\left(\frac{\sqrt{x+\delta x}-\sqrt{x}}{2}\right)\sin\left(\frac{\sqrt{x-\delta x}-\sqrt{x}}{2}\right)}{\{(x+\delta x)-x\}}$$

$$= \lim_{\delta x \to 0} \frac{-2\sin\left(\frac{\sqrt{x+\delta x}-\cos\sqrt{x}}{2}\right)\sin\left(\frac{\sqrt{x-\delta x}-\sqrt{x}}{2}\right)}{(\sqrt{x+\delta x}+\sqrt{x})(\sqrt{x+\delta x}-\sqrt{x})}$$

$$= -\lim_{\delta x \to 0} \sin\left(\frac{\sqrt{x+\delta x}+\sqrt{x}}{2}\right) \cdot \lim_{\delta x \to 0} \frac{\sin\left(\frac{\sqrt{x-\delta x}-\sqrt{x}}{2}\right)}{\left(\frac{\sqrt{x-\delta x}-\sqrt{x}}{2}\right)}$$

$$\times \lim_{\delta x \to 0} \frac{1}{(\sqrt{x+\delta x}+\sqrt{x})}$$

$$= -\lim_{\delta x \to 0} \sin\left(\frac{\sqrt{x+\delta x}+\sqrt{x}}{2}\right) \cdot \lim_{\theta \to 0} \frac{\sin\theta}{\theta} \cdot \lim_{\delta x \to 0} \frac{1}{(\sqrt{x+\delta x}+\sqrt{x})}$$

$$\text{where } \frac{(\sqrt{x+\delta x}-\sqrt{x})}{2} = \theta$$

$$= -\sin\left(\frac{2\sqrt{x}}{2}\right) \times 1 \times \frac{1}{2\sqrt{x}} = \frac{-\sin\sqrt{x}}{2\sqrt{x}}.$$

Thus, $\frac{d}{dx}(\cos\sqrt{x}) = \frac{-\sin\sqrt{x}}{2\sqrt{x}}$.

Example 44. *Find the derivative of the following from the first principle.*

(*i*) *sin*^{2} *x* (*ii*) *cos*^{2} *x*.

Solution. (*i*) Let $y = \sin^2 x$.

Let δy be an increment in y, corresponding to an increment δx in x.

$$\therefore \quad y + \delta y = \sin^2(x + \delta x)$$

$$\Rightarrow \quad \delta y = \sin^2(x + \delta x) - \sin^2 x$$

$$\Rightarrow \quad \frac{\delta y}{\delta x} = \frac{\sin^2(x+\delta x) - \sin^2 x}{\delta x}$$

$$\Rightarrow \quad \frac{dy}{dx} = \lim_{\delta x \to 0} \frac{\delta y}{\delta x}$$

$$= \lim_{\delta x \to 0} \frac{\sin^2(x+\delta x) - \sin^2 x}{\delta x}$$

$$= \lim_{\delta x \to 0} \frac{\sin^2(2x+\delta x) - \sin \delta x}{\delta x}$$

$$[\because \sin^2 A - \sin^2 B = \sin(A+B) \,.\, \sin(A-B)]$$

$$= \lim_{\delta x \to 0} \sin(2x + \delta x) \,.\, \lim_{\delta x \to 0} \frac{\sin \delta x}{\delta x}$$

$$= (\sin 2x \times 1) = \sin 2x.$$

Thus, $\frac{d}{dx}(\sin^2 x) = \sin 2x$.

(*ii*) Let $y = \cos^2 x$.

Let δy be an increment in y, corresponding to an increment δx in x.

$$\therefore \quad y + \delta y = \cos^2(x + \delta x)$$

$$\Rightarrow \quad \delta y = \cos^2(x + \delta x) - \cos^2 x$$

$$\Rightarrow \quad \frac{\delta y}{\delta x} = \frac{\cos^2(x+\delta x) - \cos^2 x}{\delta x}$$

$$\Rightarrow \quad \frac{dy}{dx} = \lim_{\delta x \to 0} \frac{\delta y}{\delta x}$$

$$= \lim_{\delta x \to 0} \frac{\cos^2(x+\delta x) - \cos^2 x}{\delta x}$$

$$= \lim_{\delta x \to 0} \frac{[\cos(x+\delta x) + \cos x] \times [\cos(x+\delta x) - \cos x]}{\delta x}$$

$$= \lim_{\delta x \to 0} \frac{\left\{2\cos\left(x + \frac{\delta x}{2}\right)\cos\left(\frac{\delta x}{2}\right)\right\}\left\{-2\sin\left(x+\frac{\delta x}{2}\right)\sin\left(\frac{\delta x}{2}\right)\right\}}{\delta x}$$

$$= -2 \lim_{\delta x \to 0} \cos\left(x + \frac{\delta x}{2}\right) \,.\, \lim_{\delta x \to 0} \cos\left(\frac{\delta x}{2}\right)$$

$$\times \lim_{\delta x \to 0} \sin\left(x + \frac{\delta x}{2}\right) \cdot \lim_{\delta x \to 0} \frac{\sin(\delta x/2)}{(\delta x/2)}$$

$$= (-2 \cos x \times 1 \times \sin x \times 1) = -2 \sin x \cos x = -\sin 2x$$

Thus, $\frac{d}{dx}(\cos^2 x) = -\sin 2x.$

Example 45. *Find the derivative of the following from the first principle.*

(i) $\cos(x^2 + 1)$ *(ii)* $\tan x^2$.

Solution. (*i*) Let $y = \cos(x^2 + 1)$.

Let δy be an increment in y, corresponding to an increment δx in x.

$$\therefore \quad y + \delta y = \cos[(x + \delta x)^2 + 1]$$

$$\Rightarrow \quad \delta y = \cos[(x + \delta x)^2 + 1] - \cos(x^2 + 1)$$

$$\Rightarrow \quad \frac{\delta y}{\delta x} = \frac{\cos[(x + \delta x)^2 + 1] - \cos(x^2 + 1)}{\delta x}$$

$$\Rightarrow \quad \frac{dy}{dx} = \lim_{\delta x \to 0} \frac{\delta y}{\delta x}$$

$$= \lim_{\delta x \to 0} \frac{\{\cos[(x + \delta x)^2 + 1] - \cos(x^2 + 1)\}}{\delta x}$$

$$= \lim_{\delta x \to 0} \frac{-2\sin\left[\frac{\{(x + \delta x)^2 + 1 + (x^2 + 1)\}}{2}\right] \cdot \sin\left[\frac{\{(x + \delta x)^2 + 1 - (x^2 + 1)\}}{2}\right]}{\delta x}$$

$$= -2 \cdot \lim_{\delta x \to 0} \sin\left[x^2 + x \cdot \delta x + 1 + \frac{(\delta x)^2}{2}\right] \cdot \frac{\sin\left[\left(x + \frac{\delta x}{2}\right) \cdot \delta x\right]}{\left[\left(x + \frac{\delta x}{2}\right) \cdot \delta x\right]} \cdot \left(x + \frac{\delta x}{x}\right)$$

$$= -2 \cdot \lim_{\delta x \to 0} \sin\left[x^2 + x \cdot \delta x + 1 + \frac{(\delta x)^2}{2}\right] \cdot \lim_{\theta \to 0} \frac{\sin\theta}{\theta} \cdot \lim_{\delta x \to 0}\left(x + \frac{\delta x}{2}\right),$$

where $\theta = \left(x + \frac{\delta x}{2}\right) \cdot \delta x$; clearly, $[\delta x \to 0] \Rightarrow [\theta \to 0]$

$$= -2 \sin(x^2 + 1) \times 1 \times x$$

$$= -2x \sin(x^2 + 1)$$

Thus, $\frac{d}{dx}[\cos(x^2 + 1)] = -2x \sin(x^2 + 1).$

(*ii*) Let $y = \tan x^2$.

Let δy be an increment in y, corresponding to an increment δx in x.

$$\therefore \quad y + \delta y = \tan(x + \delta x)^2$$

$$\Rightarrow \quad \delta y = \tan(x + \delta x)^2 - \tan x^2$$

$$\Rightarrow \quad \frac{\delta y}{\delta x} = \frac{\tan(x + \delta x)^2 - \tan x^2}{\delta x}$$

$$\Rightarrow \quad \frac{dy}{dx} = \lim_{\delta x \to 0} \frac{\delta y}{\delta x}$$

$$= \lim_{\delta x \to 0} \frac{\tan (x + \delta x)^2 - \tan x^2}{\delta x} = \lim_{\delta x \to 0} \frac{\left\{ \frac{\sin (x + \delta x)^2}{\cos (x + \delta x)^2} - \frac{\sin x^2}{\cos x^2} \right\}}{\delta x}$$

$$= \lim_{\delta x \to 0} \frac{[\sin (x + \delta x)^2 \cos x^2 - \cos (x + \delta x)^2 \sin x^2]}{\delta x \cdot \cos (x + \delta x)^2 \cdot \cos x^2}$$

$$= \lim_{\delta x \to 0} \frac{\sin [(x + \delta x)^2 - x^2]}{\delta x \cdot \cos (x + \delta x)^2 \cdot \cos x^2}$$

$$= \lim_{\delta x \to 0} \frac{\sin [(2x + \delta x) - \delta x]}{[(2x + \delta x) \cdot \delta x]} \cdot \frac{(2x + \delta x)}{\cos (x + dx)^2 \cdot \cos x^2}$$

$$= \left(\lim_{\delta x \to 0} \frac{\sin \theta}{\theta} \right) \cdot \lim_{\delta x \to 0} (2x + \delta x) \cdot \lim_{\delta x \to 0} \frac{1}{\cos (x + \delta x)^2 \cdot \cos x^2},$$

where $(2x + \delta x) \cdot \delta x = \theta$ [clearly, $(\delta x \to 0) \Rightarrow (\theta \to 0)$]

$$= 1 \times 2x \times \frac{1}{\cos x^2 \cdot \cos x^2} = 2x \sec^2 x^2$$

Thus, $\frac{d}{dx} (\tan x^2) = 2x \sec^2 x^2$.

Example 46. *Differentiate xe^x by the method of first principle.*

Solution. Let $y = xe^x$.

Let δy be an increment in y, corresponding to an increment δx in x.

$$\therefore \quad y + \delta y = (x + \delta x)\, e^{x + \delta x}$$

$$\Rightarrow \quad \delta y = (x + \delta x)e^{x + \delta x} - xe^x$$

$$\Rightarrow \quad \frac{\delta y}{\delta x} = \frac{(x + \delta x)\, e^{x + \delta x} - xe^x}{\delta x}$$

$$\Rightarrow \quad \frac{dy}{dx} = \lim_{\delta x \to 0} \frac{\delta y}{\delta x}$$

$$= \lim_{\delta x \to 0} \frac{(x + \delta x)\, e^{x + \delta x} - xe^x}{\delta x}$$

$$= \lim_{\delta x \to 0} \frac{xe^{x + \delta x} + xe^x \;\delta x \cdot e^{x + \delta x}}{\delta x} = \lim_{\delta x \to 0} \left[xe^x \cdot \left(\frac{e^{\delta x} - 1}{\delta x} \right) + e^{x + \delta x} \right]$$

$$= xe^x \cdot \lim_{\delta x \to 0} \left(\frac{e^{\delta x} - 1}{\delta x} \right) + \lim_{\delta x \to 0} e^{x + \delta x}$$

$$= (x^{ex} \times 1) + e^x \qquad \left[\because \lim_{\delta x \to 0} \left(\frac{e^{\delta x} - 1}{\delta x} \right) = 1 \right]$$

$$= (x^{ex} + e^x) = (x + 1)e^x$$

Hence, $\frac{d}{dx}(xe^x) = (x+1)e^x.$

EXERCISE 1.1.5.

1. Differentiate:

(i) $\cos^{-1} x$ $\left[\textbf{Ans. } -\frac{1}{\sqrt{1-x^2}}\right]$ (ii) $\tan^{-1} 2x$ $\left[\textbf{Ans. } \frac{2}{1+4x^2}\right]$

(iii) $\sec^{-1} 3x$ $\left[\textbf{Ans. } \frac{1}{x\sqrt{9x^2-1}}\right]$ (iv) $\text{cosec}^{-1} 4x$ $\left[\textbf{Ans. } -\frac{1}{x\sqrt{16x^2-1}}\right]$

2. Differentiate by the method of first principles.

(i) $\cot 2x$ [**Ans.** $-2\,\text{cosec}^2 2x$] (ii) $\text{cosec}\, 3x$ [**Ans.** $-3\,\text{cosec}\, 3x \cot 3x$]

3. Show that $\frac{d}{dx}\left[\log\left(\frac{1+x\sin x}{1-x\sin x}\right)\right] = \frac{2(\sin x + x\cos x)}{1-x^2\sin^2 x}$

4. If $y = \sqrt{\frac{1-\sin\theta}{1+\sin\theta}}$, show that $(1+\sin\theta)\frac{dy}{d\theta} + 1 = 0$

5. If $y = \sqrt{\frac{1-\cos t}{1+\cos t}}$, show that $\frac{dy}{dt} = \frac{1}{2}\sec^2\frac{t}{2}$

6. If $y = \frac{\sin^{-1} x}{\sqrt{(1+x^2)}}$, show that $(1-x^2)\frac{dy}{dx} - xy = 1$

7. Differentiate:

(i) $a^{\sin x}$ [**Ans.** $\log a \,.\, \cos x \,.\, a^{\sin x}$] (ii) $5^{\sqrt{1+x^2}}$ $\left[\textbf{Ans. } \frac{x\log 5}{\sqrt{1+x^2}}5^{\sqrt{1+x^2}}\right]$

(iii) $e^{\tan^{-1}(\cos\sqrt{x})}$ $\left[\textbf{Ans. } \frac{\sin\sqrt{x}\, e^{\tan^{-1}}(\cos\sqrt{x})}{2\sqrt{x}\,(1+\cos^2\sqrt{x}}\right]$

(iv) $2^x \,.\, e^{\sin^{-1} x}$ $\left[\textbf{Ans. } 2^x e^{\sin^{-1}x}\frac{1}{\sqrt{1-x^2}} + \log 2\right]$

(v) $a^x \,.\, x^x$ [**Ans.** $a^x x^{a-1}(a + x\log a)$]

8. If $y = \log\left(\frac{\sqrt{x^2+1}+x}{\sqrt{x^2+1}-x}\right)$, show that $\frac{dy}{dx} = 2(x^2+1)^{-1/2}$

9. If $y = \sin\left(\log\sqrt{\frac{x}{1+x}}\right)$, show that $\frac{dy}{dx} = \frac{1}{2x(x+1)}\cos\left(\log\sqrt{\frac{x}{x+1}}\right)$

10. Differentiate:

(i) $\log(\sec x + \tan x)$ [**Ans.** $\sec x$] (ii) $\log(x + \sqrt{x^2+a^2})$ $\left[\textbf{Ans. } \frac{1}{\sqrt{x^2+a^2}}\right]$

(iii) $\log\sqrt{\tan x}$ [**Ans.** $\text{cosec}\, 2x$] (iv) $\log\left(\frac{a+b\sin x}{a-b\sin x}\right)$ $\left[\textbf{Ans. } \frac{2ab\cos x}{a^2-b^2\sin^2 x}\right]$

(v) $\log\left(\dfrac{1+\sqrt{x}}{1-\sqrt{x}}\right)$ $\left[\text{Ans. } \dfrac{1}{\sqrt{x}\,(1-x)}\right]$

11. Differentiate:

(i) $\sin^{-1}\sqrt{\cos x}$ $\left[\text{Ans. } -\dfrac{1}{2}\sqrt{1+\sec x}\right]$ (ii) $\tan^{-1}\left(\dfrac{b}{a}\tan x\right)$ $\left[\text{Ans. } \dfrac{ab\sec^2 x}{a^2+b^2\tan^2 x}\right]$

(iii) $\tan^{-1}(\sin x+\cos x)$ $\left[\text{Ans. } \dfrac{\cos x-\sin x}{2+\sin 2x}\right]$ (iv) $\cos^{-1}\left(\dfrac{b+a\cos x}{a+b\cos x}\right)$ $\left[\text{Ans. } \dfrac{\sqrt{a^2-b^2}}{a+b\cos x}\right]$

(v) $\tan^{-1}\left(\dfrac{x\sin\alpha}{1-x\cos\alpha}\right)$ $\left[\text{Ans. } \dfrac{\sin\alpha}{1-2x\cos\alpha+x^2}\right]$

12. Differentiate:

(i) $\sin^n x\cos^m x$ [**Ans.** $n\sin^{n-1}x\cos^{m+1}x - m\sin^{n+1}\cos^{m-1}x$]

(ii) $x \,.\, \sin(ax^2+bx+c)$ [**Ans.** $\sin(ax^2+bx+c)+(2ax+b)x \,.\, x\cos(ax^2+bx+c)$]

(iii) $\sin\left(\dfrac{1+x^2}{1+x^3}\right)$ $\left[\text{Ans. } \dfrac{2x-3x^2-x^4}{(1+x^3)^2}\cos\left(\dfrac{1+x^2}{1+x^3}\right)\right]$

(iv) $\dfrac{a\cos x+b}{a+b\cos x}$ $\left[\text{Ans. } \dfrac{(b^2-a^2)\sin x}{(a+b\cos x)^2}\right]$

(v) $\dfrac{\sec x-\tan x}{\sec x+\tan x}$ $\left[\text{Ans. } -\dfrac{2\sec x}{(\sec x+\tan x)^2}\right]$

13. Differentiate:

(i) $\dfrac{x^3+1}{x^2+x-1}$ $\left[\text{Ans. } \dfrac{(x^4+2x^3-3x^2-2x-1)}{(x^2-x^2)^{5/2}}\right]$

(ii) $\dfrac{a^2+x^2}{\sqrt{a^2-x^2}}$ $\left[\text{Ans. } \dfrac{x(3a^2-x^2)}{(a^2-x^2)^{3/2}}\right]$

(iii) $\dfrac{x}{(a^2-x^2)^{3/2}}$ $\left[\text{Ans. } \dfrac{(a^2+2x^2)}{(a^2-2x^2)^{5/2}}\right]$

(iv) $\sqrt{\dfrac{1+x}{1-x}}$ $\left[\text{Ans. } \dfrac{1}{(1-x)\sqrt{1-x^2}}\right]$

(v) $\dfrac{\sqrt{x+1}+\sqrt{x-1}}{\sqrt{x+1}-\sqrt{x-1}}$ $\left[\text{Ans. } 1+\dfrac{x}{\sqrt{x^2-1}}\right]$

14. (i) Given $y(1+t)=3$, prove that $3t\dfrac{dy}{dx}=y(y-3)$.

(ii) Given $xy+px+q=0$, prove that $x^2\dfrac{dy}{dx}=$ constant.

15. (i) If $y=\sqrt{(1+x^4)}$, show that $y\dfrac{dy}{dx}=2x^3$

(ii) If $y=(x-\sqrt{1+x^2})$, show that $(1+x^2)\left(\dfrac{dy}{dx}\right)^2=y^2$

16. Differentiate the following functions:

(i) $(ax)^m+\left(\dfrac{b}{x}\right)^{5/2}+\dfrac{c}{x}+1$ $\left[\text{Ans. } ma^m x^{m-1}-\dfrac{5}{2}b^{5/2}x^{-7/2}-\dfrac{c}{x^2}\right]$

(ii) $\dfrac{2x^2 + 3x + 4}{\sqrt{x}}$ $\left[\textbf{Ans. } 3\sqrt{x} + \dfrac{3}{2\sqrt{x}} - \dfrac{2}{x\sqrt{x}}\right]$

17. (i) Show that the function $f(x) = x^3 \cos \dfrac{1}{x}$, $x \neq 0 = 0$,

$= 0$, for $x = 0$

is derivable at $x = 0$.

(ii) A function is defined below

$f(x) = 1$, when $x < 0$

$f(x) = 1 + \sin x$, when $0 \le x < \dfrac{\pi}{2}$

$f(x) = 2 + \left(x - \dfrac{\pi}{2}\right)^2$, when $x \ge \dfrac{\pi}{2}$

Show that (a) $f(x)$ is continuous at $x = 0$ and $x = \dfrac{\pi}{2}$ (b) $f(x)$ non-differentiate at $x = 0$ but differentiable at $x = \dfrac{\pi}{2}$.

18. If $f(x) = x^2$ when $x \le 0$; $f(x) = 1$ when $0 < x < 1$; $f(x) = \dfrac{1}{x}$ when $x > 1$, show that $f(x)$ is discontinuous as well as non-differentiable at $x = 0$ as well as at $x = 1$.

19. (i) Show that $f(x) = |x|$ is not differentiable at $x = 0$.

(ii) A function $\phi(x)$ is defined as follows:

$$\phi(x) = 1 + x, \quad \text{for } x \le 2$$
$$= 5 - x, \quad \text{for } x \ge 2$$

Show that $\phi(x)$ is continuous but non-differentiable at $x = 2$.

20. If $y = a + bx + cx^2$, obtain $\dfrac{dy}{dx}$ by the method of first principle. [**Ans.** $b + 2cx$]

Differentiable following questions:

21. $(ax + b)$ [**Ans.** a]

22. $\left(ax^2 + \dfrac{b}{x}\right)$ $\left[\textbf{Ans. } \left(2ax - \dfrac{b}{x^2}\right)\right]$

23. $3x^2 + 2x - 5$ [**Ans.** $6x + 2$]

24. $x^3 - 2x^2 + x + 3$ [**Ans.** $3x^2 - 4x + 1$]

25. x^8 [**Ans.** $8x^7$]

26. $\dfrac{1}{x^3}$ $\left[\textbf{Ans. } \dfrac{-3}{x^4}\right]$

27. $\dfrac{1}{x^5}$ $\left[\textbf{Ans. } \dfrac{-5}{x^6}\right]$

28. $\sqrt{ax + b}$ $\left[\textbf{Ans. } \dfrac{2}{2\sqrt{ax + b}}\right]$

29. $\sqrt{5x - 4}$ $\left[\textbf{Ans. } \dfrac{5}{2\sqrt{5x - 4}}\right]$

30. $\dfrac{1}{\sqrt{x + 2}}$ $\left[\textbf{Ans. } \dfrac{-1}{2(x + 2)^{3/2}}\right]$

31. $\dfrac{1}{\sqrt{2x + 3}}$ $\left[\textbf{Ans. } \dfrac{-1}{(2x + 3)^{3/2}}\right]$

32. $\dfrac{1}{\sqrt{6x - 5}}$ $\left[\textbf{Ans. } \dfrac{-3}{(6x - 5)^{3/2}}\right]$

33. $\dfrac{1}{\sqrt{2 - 3x}}$ $\left[\textbf{Ans. } \dfrac{3}{2(2 - 3x)^{3/2}}\right]$

34. $\dfrac{2x + 3}{3x + 2}$ $\left[\textbf{Ans. } \dfrac{-5}{(3x + 2)^2}\right]$

35. $\dfrac{5-x}{5+x}$ $\left[\textbf{Ans. } \dfrac{-10}{(5+x)^2}\right]$ **36.** $\dfrac{x^2+1}{x}, x \neq 0$ $\left[\textbf{Ans. } \left(1-\dfrac{1}{x^2}\right)\right]$

37. $\sqrt{\cos 3x}$ $\left[\textbf{Ans. } \dfrac{-3}{2} \cdot \dfrac{\sin 3x}{\sqrt{\cos 3x}}\right]$ **38.** $\sqrt{\sec x}$ $\left[\textbf{Ans. } \dfrac{1}{2} \tan x \sqrt{\sec x}\right]$

39. $\tan^2 x$ [**Ans.** $2 \tan x \sec^2 x$] **40.** $\sin (2x+3)$ [**Ans.** $2 \cos (2x+3)$]

41. $\tan (3x+1)$ [**Ans.** $3 \sec^2 (3x+1)$]

42. If $y = \sqrt{\dfrac{\sec x - \tan x}{\sec x + \tan x}}$, show that $\dfrac{dy}{dx} = \sec x (\tan x + \sec x)$.

43. If $y = \sqrt{\dfrac{1-x}{1+x}}$, prove that $(1-x^2) \dfrac{dy}{dx} + y = 0$.

44. If $y = \dfrac{\cos x + \sin x}{\cos x - \sin x}$, show that $\dfrac{dy}{dx} = \sec^2 \left(x + \dfrac{\pi}{4}\right)$.

45. If $y = \dfrac{\cos x - \sin x}{\cos x + \sin x}$, show that $\dfrac{dy}{dx} + y^2 + 1 = 0$

Differentiable following questions:

46. $y = \dfrac{\sin x + x^2}{\cot 2x}$ [**Ans.** $2(\sin x + x^2) \sec^2 2x + (\cos x + 2x) \tan 2x$]

47. $y = \sin \left(\dfrac{1+x^2}{1-x^2}\right)$ $\left[\textbf{Ans. } \dfrac{4x}{(1-x^2)} \cdot \cos \left(\dfrac{1+x^2}{1-x^2}\right)\right]$

48. $y = \cos \left(\dfrac{1-x^2}{1+x^2}\right)$ $\left[\textbf{Ans. } \dfrac{4x}{(1-x^2)^2} \cdot \sin \left(\dfrac{1+x^2}{1-x^2}\right)^2\right]$

49. $y = e^x \log (\sin 2x)$ [**Ans.** $e^x [2\cot 2x + \log (\sin 2x)]$]

50. $y = \sin (\sqrt{\sin x + \cos x})$ $\left[\textbf{Ans. } \dfrac{(\cos x - \sin x) \cos (\sqrt{\sin x + \cos x})}{2\sqrt{\sin x + \cos x}}\right]$

51. $\dfrac{e^{2x} + x^3}{\operatorname{cosec} 2x}$ [**Ans.** $2e^{2x} (\cos 2x + \sin 2x) + 2x^3 \cos 2x = 3x^2 \sin 2x$]

52. $\sqrt{\dfrac{1+e^x}{1-e^x}}$ $\left[\textbf{Ans. } \dfrac{e^x}{(1-e^x)(1-eh2x)^{1/2}}\right]$

53. $\sqrt{\dfrac{1+\sin x}{1-\sin x}}$ $\left[\textbf{Ans. } \dfrac{1}{(1-\sin x)} \text{ or } \sec x (\sec x + \tan x)\right]$

54. $\sqrt{\dfrac{a^2-x^2}{a^2+x^2}}$ $\left[\textbf{Ans. } \dfrac{-2_a^2 x}{(a^2+x^2)(a^2-x^2)^{1/2}}\right]$

55. $\sqrt{\dfrac{1-x^2}{1+x^2}}$ $\left[\textbf{Ans. } \dfrac{-2x}{(1+x^2)^{3/2}(1-x^2)^{1/2}}\right]$

56. $\dfrac{e^{2x} + e^{-2x}}{e^{2x} - e^{-2x}}$ $\left[\textbf{Ans. } \dfrac{-8}{(e^{2x} - e^{-2x})^2}\right]$

57. $\dfrac{e^{x} + e^{-x}}{e^{x} - e^{-x}}$ $\left[\textbf{Ans. } \dfrac{-4}{(e^{x} - e^{-x})^2}\right]$

58. $e^{(x \sin x + \cos x)}$ [**Ans.** $(x \cos x) . e^{(x \sin x + \cos x)}$]

59. $\cos (x^3 . e^x)$ [**Ans.** $- x^2 e^x (x + 3) . \sin (x^3 e^x)$]

60. $e^{-5x} \cot 4x$ [**Ans.** $- e^{-5x}(4 \text{ coses}^2 4x + 5 \cot 4x)$]

61. $e^{3x} \cos 2x$ [**Ans.** $e^{3x} (3 \cos 2x - 2 \sin 2x)$]

62. $e^{2x} \sin 3x$ [**Ans.** $e^{2x} (3 \cos 3x + 2\sin 3x)$]

63. $\cos (\sin \sqrt{ax + b})$ $\left[\textbf{Ans. } \dfrac{-a}{2\sqrt{ax + b}} (\cos \sqrt{ax + b}) . \sin (\sin \sqrt{ax + b})\right]$

64. $\sin x \sin 2x$ [**Ans.** $2 \sin x \cos 2x + \cos x \sin 2x$]

65. $\cos 3x \sin 5x$ [**Ans.** $5 \cos 3x \cos 5x - 3 \sin 3x \sin 5x$]

66. $\sqrt{\cot \sqrt{x}}$ $\left[\textbf{Ans. } \dfrac{-\text{cosec}^2 \sqrt{x}}{4\sqrt{x} . \sqrt{\cos \sqrt{x}}}\right]$

67. $\sqrt{x \sin x}$ $\left[\textbf{Ans. } \dfrac{(x \cos + \sin x)}{2\sqrt{x \sin x}}\right]$

68. $\sqrt{\sin x^3}$ $\left[\textbf{Ans. } \dfrac{3x^2 \cos x^3}{2\sqrt{\sin x^3}}\right]$

69. $\cos^2 x^3$ [**Ans.** $- 3x^2 \sin (2x^3)$]

70. $\sin^2 (2x + 3)$ [**Ans.** $2 \sin (4x + 6)$]

71. $\dfrac{1}{(x^2 - x + 3)^3}$ $\left[\textbf{Ans. } \dfrac{-3 (2x - 1)}{(x^2 - x + 3)^4}\right]$

72. $(ax^2 + bx + c)^n$ [**Ans.** $n(2ax + b) (ax^2 + bx + c)^{n-1}$]

73. $(3x^2 - x + 1)^4$ [**Ans.** $4(6x - 1) (3x^2 - x + 1)^3$]

74. $(3 - 4x)^5$ [**Ans.** $- 20(3 - 4x)^4$]

75. $(5 + 7x)^6$ [**Ans.** $42(5 + 7x)^5$]

76. $\sqrt{\sin x}$ $\left[\textbf{Ans. } \dfrac{\cos x}{2 \sqrt{\sin x}}\right]$

77. $e^{\cot x}$ [**Ans.** $- (\text{cosec}^2 x) (e^{\cot} x)$]

78. e^{x4} [**Ans.** $4x^3 (e^{x4})$]

79. $\tan \sqrt{x}$ $\left[\textbf{Ans. } \dfrac{\sec^2 \sqrt{x}}{2\sqrt{x}}\right]$

80. $\tan^3 x$ [**Ans.** $3 \tan^2 x \sec^2 x$]

81. $\cot^2 x$ [**Ans.** $- 2 \cot x \text{ cosec}^2 x$]

82. $\cos x^3$ [**Ans.** $- 3x^2 \sin x^3$]

83. $\tan 3x$ [**Ans.** $3 \sec^2 3x$]

84. $\cos 5x$ [**Ans.** $- 5 \sin 5x$]

85. $\sin 4x$ [**Ans.** $4 \cos 4x$]

1.4.4. SUCESSIVE DIFFERENTIATION

The derivative of function of x is in general also a function of x. This new function may be differentiable, in which case the derivative of the first derivative is called the second derivative of the original function. Similarly, the derivative of the second derivator is called the third derivate and on the *n-th* derivative.

Thus
$$y = 3x^4$$
$$\frac{dy}{dx} = 12x^3$$
$$\frac{dy}{dx}\left(\frac{dy}{dx}\right) = 36x^2$$
$$\frac{d}{dx}\left[\frac{d}{dx}\left(\frac{dy}{dx}\right)\right] = 72x$$

and so on.

Notation
$$\frac{d}{dx}\left(\frac{dy}{dx}\right) = \frac{d^2y}{dx^2}$$
$$\frac{d}{dx}\left[\frac{d}{dx}\left(\frac{dy}{dx}\right)\right] = \frac{d}{dx}\left(\frac{d^2y}{dx^2}\right) = \frac{d^3y}{dx^3}$$

....

....

....

$$\frac{d}{dx}\left(\frac{d^{n-1}y}{dx^{n-1}}\right) = \left(\frac{d^ny}{dx^n}\right)$$

if $f = f(x)$, successive derivative area also denoted by

$$f'(x), f''(x), f'''(x), f^4(x), \ldots\, f^{(n)}(x);$$
$$y', y'', y''', y^4, \ldots\, y^{(n)}(x);$$

or
$$\frac{d}{dx}f(x), \frac{d^2}{dx^2}f(x), \frac{d^3}{dx^3}f(x), \ldots \frac{d}{dx^n}f(x)$$

The *n*-th derivative. In general expression involving n may be found for nth derivative.

Ex. If $y = e^{ax}$, find $\frac{d^ny}{dx^n}$

$$\frac{dy}{dx} = ae^{ax}, \quad \frac{d^2y}{dx^2} = a^2e^{ax}, \quad \frac{d^ny}{dx^n} = a^n \,.\, e^{ax}$$

If $y = \log_n$, find $\frac{d^ny}{dx^n}$

$$\frac{dy}{dx} = \frac{1}{x}, \frac{d^2y}{dx^2} = -\frac{1}{x^2}, \frac{d^2y}{dx^3} = \frac{1.2}{x^3}, \frac{d^ny}{dx^n} = \frac{(-1)^{n-1}(n-1)}{x^n}$$

- **Standard Terms**

Standard Results

(*a*) If
$$y = (ax + b)^m, \text{ then}$$
$$y_1 = m.a(ax + b)^{m-1}$$

$$y_2 = m(m-1).a^2(ax+b)^{m-2}$$

$$y_3 = m(m-1)(m-2).a^3(ax+b)^{m-3}$$

In general, $y_n = m(m-1)(m-2).(m-n+1).a^n(ax+b)^{m-n}$

In particular (i) $y_n = \dfrac{m!}{(m-n)!}a^n(ax+b)^{m-n}$

If m is a positive integer $> n$

(ii) $y_n = 0$, if m is a positive integer $< n$.

(iii) $y_n = n!\,a^n$, when $m = n$

(iv) $y = (a^x + b)^{-1}$, when $m = -1$,

$$y^n = (-1)(-2)(-3)\ldots(-n).a^n(ax+b)^{-1-n}$$

$$= (-1)^n(n!)a^n.(ax+b)^{-(n+1)}$$

(b) If $y = e^{ax}$, then $y_1 = ae^{ax}$, $y_2 = a^2 = e^{ax}$, $y_3 = a^3 e^{ax}$.

This can be generalised to $y^n = a^n e^{ax}$

(c) If $y = log\,(ax + b)$, then

$$y_1 = a/(ax+b) = a(ax+b)^{-1}$$

$$y_2 = a^2.(-1)(ax+b)^{-2}$$

$$y_3 = a^3.(-1)(-2)(ax+b)^{-3}$$

$$y_4 = a^4.(-1)(-2)(-3)(ax+b)^{-4} = a^4(-1)^3 3!(ax+b)^{-4}$$

$$\therefore \quad y_n = a^n(-1)^{n-1}(n-1)!(ax+b)^{-n}$$

(d) If $y = \sin(ax + b)$, then

$$y_1 = a.\cos(ax+b) = a.\sin\left(ax+b+\frac{x}{2}\right)$$

on using $\cos\theta = \sin\left(\theta+\dfrac{x}{2}\right)$. Again differentiating, we get

$$y_2 = a^2.\cos\left(ax+b+\frac{\pi}{2}\right)$$

$$= a^2.\sin\left(ax+b+\frac{\pi}{2}+\frac{\pi}{2}\right) = a^2.\sin\left(ax+b+\frac{\pi}{2}\right)$$

Thus, $y_3 = a^3.\sin\left(ax+b+3.\dfrac{\pi}{2}\right)$

Continuing this process, we get

$$y_n = a^n\sin\left(ax+b+\frac{n\pi}{2}\right)$$

(e) If $y = \cos(ax + b)$, then proceeding exactly as above we get $y_n = a^n\cos\left(ax+b+\dfrac{n\pi}{2}\right)$

(f) If $y = e^{ax}\sin(bx + c)$, then

$$y_1 = ae^{ax}.\sin(bx+c) + e^{ax}.b\cos(bx+c)$$

$$= e^{ax}\{a\sin(bx+c) + b\cos\{(bx+c)\}$$

Now let $a = r\cos\phi$ and $b = r\sin\phi$, we get

$$y_1 = e^{ax}\,.\,r\{\sin(bx+c)\cos\phi + \cos(bx+c)\sin\phi\}$$
$$= r\,e^{ax}\sin(bx+c+\phi)$$

Similarly, again differentiating and simplifying, we have

$$y_2 = r^2 e^{ax}\sin(bx+c+2\phi).$$

and $$y_3 = r^3 e^{ax}\sin(bx+c+3\phi).$$

If this process is continued n times, we

$$y_n = r^n e^{ax}\sin(bx+c+n\phi).$$

where $$r = \sqrt{(a^2+b^2)} \text{ and } \phi = \tan^{-1}\frac{b}{a}$$

(*g*) If $y = e^{ax}\cos(bx+c)$, then proceeding as above we have

$$y_n = r^n e^{ax}\cos(bx+c+n\phi)$$

where r and ϕ are given.

Example 47. *Find the nth derivative of* $\dfrac{x^4}{(x-1)(x-2)}$.

Sol. Let $$y = \frac{x^4}{(x-1)(x-2)} = x^2+3x+7+\frac{15x-14}{(x-1)(x-2)}$$

Resolving into partial fractions, we have

$$y = x^2+3x+7-\frac{1}{(x-1)}+\frac{16}{(x-2)}$$

Now differentiating n (> 2) times and $m = -1$, we have

$$y_n = (-1)^n(n!)\left\{\frac{16}{(x-2)^{n+1}} - \frac{1}{(x-1)^{n+1}}\right\}$$

Example 48. *If* $y = \sin^4 x$, *find* y_n.

Sol. Here $$y = \sin^4 x = (\sin x^2)^2 = \left(\frac{1-\cos 2x}{2}\right)^2$$

$$= \frac{1}{4}(1-2\cos 2x+\cos^2 2x)$$

$$= \frac{1}{4}(1-2\cos 2x+\frac{1}{2}(1+\cos 4x) = \frac{1}{8}(3-4\cos 2x+\cos 4x)$$

Now differentiating n times w.r.t., we have

$$y_n = \frac{1}{8}\left\{-4.2^n\cos\left(2x+\frac{n\pi}{2}\right)+4^n\cos\left(4x+\frac{n\pi}{2}\right)\right\}$$

$$= 2^{n-1}\left\{2^{n-2}\cos\left(4x+\frac{n\pi}{2}\right)-\cos\left(2x+\frac{n\pi}{3}\right)\right\}$$

Example 49. *If* $y = e^{3x}\cos x\cos 2x\sin x$, *find* y_n.

Sol. Here $$y = \frac{1}{2}e^{3x}(2\cos x \sin x)\cos 2x = \frac{1}{2}e^{3x}\sin 2x \cos 2x$$

$$= \frac{1}{4}e^{3x}\sin 4x$$

We have, $$y_n = \frac{1}{4}.5^n e^{3x}\sin\left(4x + n\tan^{-1}\frac{4}{3}\right)$$

Example 50. *Find y_n when $y = \tan^{-1}\left(\frac{x}{a}\right)$.*

Sol. Differentiating y w.r.t., we have

$$y_1 = \frac{a}{x^2 + a^2} = \frac{a}{(x - i\,a)(x + ia)} = \frac{1}{2i}\left\{\frac{1}{(x - ia)} - \frac{1}{(x + ia)}\right\}$$

Differentiating $(n - 1)$ times, we have

$$y_n = \frac{(-1)^{n-1}(n-1)!}{2i}\left\{\frac{1}{(x - ia)^n} - \frac{1}{(x + ia)^n}\right\}$$

Now y_n can be simplified by using De Moivre's theorm. Let $x = r\cos\phi$ and $a = r\sin\phi$.

Then $$(x - i\,a)^{-n} = (r\cos\phi - i\,r\sin\phi)^{-n}$$

$$= r^{-n}(\cos\phi - i\sin\phi)^{-n} = r^{-n}(\cos n\phi + i\sin n\phi)$$

Similarly $$(x + i\,a)^{-n} = (r^{-n}(\cos n\phi - i\sin n\phi)$$

$\therefore$ $$y_n = \frac{(-1)^{n-1}(n-1)!}{2i}r^{-n}\{(\cos n\phi + i\sin n\phi) - (\cos n\phi - i\sin n\phi)\}$$

$$= (-1)^{n-1}(n-1)!\,r^{-n}\sin n\phi = (-1)^{n-1}(n-1)!\,a^{-n}\sin^n\phi\sin n\phi)\}$$

for $$a = r\sin\phi \Rightarrow \frac{1}{r} = \frac{1}{a}\sin\phi$$

or $$r^{-n} = a^{-n}\sin n\,\phi.$$

Example 51. *Find the nth derivative of $\frac{1}{(1 + x + x^2)}$.*

Sol. Let $$y = \frac{1}{x^2 + x + 1} = \frac{1}{\left(x + \frac{1}{2}\right)^2 + \frac{3}{4}} = \frac{1}{t^2 + a^2}$$

where $t = x + \frac{1}{2}$ and $a = \frac{\sqrt{3}}{2}$

Now, $$y_n = \frac{(-1)^n n!}{a^{n+2}}\sin^{n+1}\theta\sin(n+1)\theta$$

$$= \tan\theta = \frac{\sqrt{3}}{2x+1},\quad a = \frac{\sqrt{3}}{2}$$

EXERCISE 1.1.6.

1. If $y=\left(\frac{1+x}{1-x}\right)^n$, show that $\frac{dy}{dx}=\frac{2ny}{(1-x^2)}$ and $\frac{d^2y}{dx^2}=\frac{2(n+x)}{(1-x^2)}\cdot\frac{dy}{dx}$

2. Find $\frac{d^2y}{dx^2}$, where $x = a(\theta + \sin\theta)$, $y = a(1 - \cos\theta)$. **Ans.** $\frac{1}{4a}\sec^4\frac{\theta}{2}$

3. If $y = a e^{-kt}\cos(pt + c)$, show that $\frac{d^2y}{dt^2}+2k\frac{dy}{dt}+n^2y=0$ where $n^2 = p^2 + k^2$

4. If $p^2 = a^2 + \cos^2\theta + b^2\sin^2\theta$, prove that $p+\frac{d^2p}{d\theta^2}=\frac{a^2b^2}{p^3}$

5. If $y = \sin kx + \cos kx$, prove that $y_n = k^n\{1 + (-1)^n \sin 2kx\}^{1/2}$.

6. Find the nth differential co-efficients of

(i) $\cos 2x \sin 3x$ **Ans.** $\frac{1}{2}\left[5^n\sin\left(5x+\frac{n\pi}{2}\right)+\sin\left(x+\frac{n\pi}{2}\right)\right]$

(ii) $3\cos 5x\cos 3x + x^n$ **Ans.** $\frac{3}{2}\left[8^n\cos\left(8x+\frac{n\pi}{2}\right)+2^n\cos\left(2x+\frac{n\pi}{2}\right)\right]+n!$

(iii) $\cos x\cos 2x\cos 3x$ **Ans.** $\frac{1}{4}\left[6^n\cos\left(6x+\frac{n\pi}{2}\right)+4^n\cos\left(4x+\frac{n\pi}{2}\right)+2^n\cos\left(2x+\frac{n\pi}{2}\right)\right]$

(iv) $\sin^2 x\cos^3 x$ **Ans.** $\frac{1}{16}\left[2\cos\left(x+\frac{n\pi}{2}\right)-3^n\cos\left(3x+\frac{n\pi}{2}\right)-5^n\cos\left(5x+\frac{n\pi}{2}\right)\right]$

(v) $\cos^4 x$ **Ans.** $\frac{1}{8}\left[2^{n+2}\cos\left(2x+\frac{n\pi}{2}\right)+4^n\cos\left(4x+\frac{n\pi}{2}\right)\right]$

(vi) $e^x\sin x\sin 2x$ **Ans.** $\frac{1}{2}e^x\left[2^{n/2}\cos\left(x+\frac{n\pi}{4}\right)-10^{n/2}\cos\left(3x+n\tan^{-1}3\right)\right]$

7. Find the nth derivatives of

(i) $\frac{2x-1}{(x-2)(x+1)}$ **Ans.** $(-1)^n n![(x-2)^{-(n+1)}+(x+1)^{-(n+1)}]$

(ii) $(1-5x+6x^2)^{-1}$ **Ans.** $\left[\frac{n!.3^{n+1}}{(1-3x)^{n+1}}-\frac{n!.2^{n+1}}{(1-2x)^{n+1}}\right]$

(iii) $\frac{4x}{(x-1)^2(x+1)}$ **Ans.** $(-1)^n n!\left[\frac{1}{(x-1)^{n+1}}-\frac{1}{(x+1)^{n+1}}+\frac{2(n+1)}{(x-1)^{n+2}}\right]$

(iv) $\tan^{-1}\left\{\left(\frac{1+x}{1-x}\right)\right\}$ **Ans.** $(-1)^{n-1}(n-1)!\sin^n\theta\sin n\theta$, where $\theta = \cos^{-1}x$

8. If $y = \tan^{-1} x$, show that

$$y_n = (n-1)! \cos\left\{ny + (n-1)\frac{\pi}{2}\right\} \cos^n y.$$

9. If $y = e^{x\cos\beta} \cos(x\sin\beta)$, show that $y_n = e^{x\cos\beta}.\cos(x\sin\beta + n\beta)$

10. Prove that the value of the nth derivative of $\frac{x^3}{x^3-1}$ for $x = 0$, is zero when n is even and ($-n!$) when n is odd and greater than one.

11. If $y = \log\sqrt{\frac{2x+1}{x-2}}$, show that

$$y_n = \frac{1}{2}(-1)^{n-1}(n-1)!\left\{\frac{2^n}{(2x+1)^n} - \frac{1}{(x-2)^n}\right\}$$

12. If $y = \cosh 2x$, show that (*i*) $y_n = 2^n \sinh 2x$ when n is odd and (*ii*) $y_n = 2^n \cosh 2x$ when n is even.

[**Hint.** $y = \cosh 2x = ½ (e^{2x} + e^{-2x})$. Now find y_n.]

13. If $y = e^{ax} \sin bx$, prove that (*i*) $y_n = (a \sec\theta)^n e^{ax} \sin\left(bx + n\tan^{-1}\frac{b}{a}\right)$

(*ii*) $y_{n+1} - 2ay_n + (a^2+b^2)y_{n-1} = 0$

14. If $f(x) = e^{-ax}\cos(bx+c)$ show that

$$f^n(x) = (-1)^n(a^2+b^2)^{n/2}e^{-ax}\cos\left(bx + c + n\tan^{-1}\frac{b}{a}\right)$$

Also find $f^n(0)$ when $a = b = 1$ and $c = 0$

15. If $(1-x^2)\tan y = 2x$, show that

$$y_n = 2(-1)^{n-1}(n-1)!\sin^n w \sin nw, \text{ where } \cot w = x$$

16. If $y = x\log\left(\frac{x-1}{x+1}\right)$, prove that

$$y_n = (-1)^n(n-2)!\left[\frac{x-n}{(x-1)^n} - \frac{x+n}{(x+1)^n}\right]$$

17. If $y = x(a^2+x^2)^{-1}$, show that

$$y_n = (-1)^n n!\, a^{-n-1}\sin^{n+1}\phi\cos(n+1)\phi$$

where $\phi = \tan^{-1}\left(\frac{a}{x}\right)$

18. If $y = x\log(x+1)$, prove that

$$y_n = \frac{(-1)^{n-2}(n-2)!(x+n)}{(x+1)^n}$$

• Leibnitz's Theorem for *n*

If $y = u.v$, where u and v are functions of x having derivatives of nth order, then

$y_n = u_n . v + {}^nC_1 . u_{n-1} . v_1 + {}^nC_2 . u_{n-2} . v_2 + ... + {}^nC_r\, u_{n-r} . v_r + ... + u . v_n$ where suffixes in u and v denote the order of differentiation with respect to x and ${}^nC_1$, ${}^nC_2$, ... have their usual meanings.

This theorem is proved by mathematical induction. There are two functions are ϕ given below:

(*i*) If one of the functions out of the product is a polynomial function of x, we will generally take it as v while the function whose nth derivate is easily known, will be taken as u. For example, in case

$$y = (3x^2 - 7x + 4)\, e^{5x},$$

take $u = e^{5x}$ and $v = 3x^2 - 7x + 4$

Then $\quad u_n = 5^n e^{5x}$

$$u_{n-1} = 5^{n-1} e^{5x}$$

$$u_{n-2} = 5^{n-2} e^{5x} \text{ etc.}$$

and $\quad v_1 = 6x - 7,\ v_2 = 6,\ v_3 = 0 = v_4 = v_5 = ...$ etc.

$\therefore \quad y^n = D^n \{e^{5x} \times (3x^2 - 7x + 4)\}$

$$= (5^n e^{5x}) \times (3x^2 - 7x + 4) + {}^nC_1(5^{n-1}e^{5x}) \times (6x - 7) + {}^nC_2(5^{n-2}e^{5x}) \times 6$$

$$= 5^{n-2} e^{5x} \{25(3x^2 - 7x + 4) + 5n(6x - 7) + 3n(n-1)\}$$

Here we get the first three terms only as all the rest vanish.

(*ii*) The Leibnitz's theorem can also be stated as follows :

$$D^n(uv) = D^n u.v + {}^nC_1.D^{n-1}u.Dv + {}^nC_2 D^{n-2}u.D^2 v + ...$$

$$+ {}^nC_r.D^{n-r}u.D^r v + ... + u.D^n v.$$

Example 52. If $y = x^n/(x+1)$, shows that $y_n = \dfrac{n!}{(x+1)^{n+1}}$

Sol. Let $\quad u = \dfrac{1}{(x+1)} = (x+1)^{-1}$

and $\quad v = x^n$

Then $\quad u_1 = -(x+1)^{-2},$

$$u_2 = (-1)^2\, 2!\, (x+1)^{-3}$$

$$u_n = (-1)^n\, n!\, (x+1)^{-(n+1)}$$

and $\quad v_1 = nx^{n-1},$

$$v_2 = n(n-1)\, x^{n-2}, ..., v_n = n!$$

Now $\quad y_n = D^n (u \times v)$

$$= (-1)^n n!(x+1)^{-(n+1)}.x^n + {}^nC_1(-1)^{n-1} \times (n-1)!(x+1)^{-n} nx^{n-1} +$$

$${}^nC_2.(-1)^{n-2} \times (n-2)!(x+1)^{-(n-1)}.n(n-1)x^{n-2} + ... + \frac{1}{x+1}.n!$$

$$= \frac{(-1)^n n!}{(x+1)^{n+1}} \left\{ x^n - {}^nC_1 x^{n-1}.(x+1) + {}^nC_2.x^{n-2}.(x+1)^2 - ... + (x+1)^n \right\}$$

$$= \frac{(-1)^n n!}{(x+1)^{n+1}}\left\{[x-(x+1)]^n\right\} = \frac{(-1)^n n!}{(x+1)^{n+1}}\left\{(-1)^n\right\} = n!/(x+1)^{n+1}$$

Example 53. *If* $\cos^{-1}\left(\frac{y}{b}\right) = \log\left(\frac{x}{n}\right)^n$ *prove that* $x^2 y_{n+2} + (2n+1)xy_{n+1} + 2n^2\ y_n = 0$

Sol. Simplifying $\cos^{-1}\left(\frac{y}{b}\right) = n\log\left(\frac{x}{n}\right)$

$$\Rightarrow \left(\frac{y}{b}\right) = \cos\left\{n\log\left(\frac{x}{n}\right)\right\}$$

Differentiating $\quad y_1 = -b\sin\left\{n\log\left(\frac{x}{n}\right)\right\}\times\frac{n}{x}$

$\Rightarrow \quad y_1 \,.\, x = -bn\sin\{n\log(x/n)\}.$

Differentiating again and simplifying, we have

$$y_2 \,.\, x + y_1 \,.\, 1 = -bn\cos\{n\log(x/n)\} \,.\, n/x = -\frac{n^2}{x}y$$

or $\quad y_2 \,.\, x^2 + y_1 \,.\, x + n^2 y = 0$

Now differentiating n times by Leibnitz's theorem, we have

$$(y_{n+2}.x^2 + {}^nC_1 y_{n+1}.2x + {}^nC_2 y_n.2) + (y_{n+1}.x + {}^nC_1 y_n.1) + n^2 y_n = 0$$

or $$(x^2 y_{n+2} + 2nx\ y_{n+1} + n(n-1)y_n + (xy_{n+1} ny_n) + n^2 y_n = 0$$

or $$x^2 y_{n+2} + (2n+1)xy_{n+1} + 2n^2 y_n = 0$$

Example 54. If $y = e^{a\sin^{-1}x}$, *prove that*

(i) $(1-x^2)\,y_2 - xy_1 - a^2 y = 0$ $\qquad$ *(ii)* $(1-x^2)\,y_{n+2} - (2n+1)\,x\,y_{n+1} - (n^2+a^2)\,y_n = 0$

Sol. Differentiating y w.r.t. x, we get

$$y_1 = \frac{ae^{a\sin^{-1}x}}{\sqrt{(1-x^2)}} = \frac{ay}{\sqrt{1-x^2}} \qquad \text{...}(i)$$

or $\quad y_1^2(1-x^2) = a^2y^2$

Differentiating again, we have

$$2y_1y_2 \,.\, (1-x^2) + y_1^2 \,.\, (-2x) = a^2 \,.\, 2yy$$

or $$y_2.(1-x^2) - y_1.x - a^2 y = 0 \qquad \text{...}(ii)$$

Differentiating (*ii*), n times by Leibnitz theorem, we have

$$\{y_{n+2}.(1-x^2) + {}^nC_1 y_{n+1}.(-2x) + {}^nC_2.y_n.(-2)\} - (y_{n+1}.x + {}^nC_1 y_n.1) - a^2 y_n = 0$$

or $$(1-x^2)y_{n+2} - (2n+1)xy_{n+1} - (n^2+a^2)y_n = 0 \qquad \text{...}(iii)$$

Now putting $x = 0$ in (*i*), (*ii*) and (*iii*), we get

$$(y_1)_0 = a,$$

$$(y_2)_0 = a^2$$

and $\quad (y_{n+2})_0 = (n^2 + a^2)(y_n)_0 \quad ...(iv)$

Taking $\quad n = 1, 3, 5, 7, \ldots (n-2)$ in (iv), we get

$$(y_3)_0 = (1^2 + a^2)(y_1)_0 = (1 + a^2) \,.\, a$$
$$(y_5)_0 = (3^2 + a^2)(y_3)_0 = (3^2 + a^2)(1 + a^2) \,.\, a$$
$$(y_7)_0 = (5^2 + a^2)(y_5)_0 = (5^2 + a^2)(3^2 + a^2)(1 + a^2) \,.\, a$$
$$\vdots \quad \vdots$$
$$(y_n)_0 = \{(n-2)^2 + a^2\}(y_{n-2})_0$$
$$= \{(n-2)^2 + a^2\} \,.\, \{(n-4)^2 + a^2\}(y_{n-4})_0$$
$$= \{(n-2)^2 + a^2\} \,.\, \{(n-4)^2 + a^2\} \ldots (3^2 + a^2)(1 + a^2)\, a$$

This is when n is odd. When n is even, taking $n = 2, 4, 6, \ldots. (n-2)$ in (iv), we have

$$(y_4)_0 = (2^2 + a^2)(y_2)_0 = (2^2 + a^2) \,.\, a^2$$
$$(y_6)_0 = (4^2 + a^2)(y_4)_0 = (4^2 + a^2) \,.\, (2^2 + a^2) \,.\, a^2$$
$$(y_8)_0 = (6^2 + a^2)(y_6)_0 = (6^2 + a^2) \,.\, (4^2 + a^2)(2^2 + a^2) \,.\, a^2$$
$$\vdots \quad \vdots$$
$$(y_n)_0 = \{(n-2)^2 + a^2\}(y_{n-2})_0$$
$$= \{(n-2)^2 + a^2\}\{(n-4)^2 + a^2\}(y_{n-4})_0$$
$$= \{(n-2)^2 + a^2\}\{(n-4)^2 + a^2\} \ldots (4^2 + a^2)(2^2 + a^2)a^2,$$

when n is even.

EXERCISE 1.1.7.

1. Find the nth derivative of

   (a) $x^2 e^{3x}$

   (b) $x^4 \sin 3x$

   (c) $x^n e^x$

   (d) $e^x \log x$

2. If $y = a \cos(\log x) + b \sin(\log x)$, show that

   (a) $x^2 y_2 + x y_1 + y = 0$

   (b) $x^2 y_{n+2} + (2n+1)x\, y_{n+1} + (n^2 + 1)y_n = 0$

3. If $y^{1/m} + y^{-1/m} = 2x$, prove that

   $(x^2 - 1)y_{n+2} + (2n+1)x\, y_{n+1} + (n^2 - m^2)\, y_n = 0$

Do you know

If $\quad y = f(x),$

then $\quad (y)_0 = (y)_{x=0} = f(0),$

$(y_1)_0 = (y_1)_{x=0} = f'(0),$

$(y_2)_0 = (y_2)_{x=0} = f''(0),$

$(y_3)_0 = (y_3)_{x=0} = f'''(0),$

and in general $\quad (y_n)_0 = (y_n)_{x=0} = f^n(0),$

4. If $y = (x^2 - 1)^n$, prove that $(x^2 - 1)y_{n+2} + 2x\, y_{n+1} - n(n+1)\, y_n = 0$

5. If $y = \sin(m \sin^{-1} x)$, prove that $(1 - x^2)y_{n+2} - (2n+1)x\, y_{n+1} - (n^2 - m^2)\, y_n = 0$

Also find the value of y_n when $x = 0$

6. If $f(x) = e^{m\cos^{-1} x}$, find $f^n(0)$ when n is even.

7. If $f(x) = \{\log(x + \sqrt{1+x^2})\}^2$, show that $f^{n+2}(0) = -n^2 f^n(0)$

8. If $x + y = 1$, prove that $D^n(x^n y^n) = n![y^n - ({}^nC_1)^2 y^{n-1}x + ({}^nC_2)^2 y^{n-2}.x^2 ... + (-1)^n x^n]$.

9. If $y = \log(x + \sqrt{x^2 + a^2})$, show that (*i*) $(a^2 + x^2)y_2 + xy_1 = 0$, (*ii*) $\lim\limits_{x \to 0}\left(\dfrac{y_{n+2}}{y_n}\right) = -\dfrac{n^2}{a^2}$.

ANSWERS

1. (*a*) $[9x^2 + 6nx + n(n-1)]e^{3x} \,.\, 3^{n-2}$

(*b*) $x^4 . 3^n \sin\left(3x + \dfrac{n\pi}{2}\right) + 4nx^3 \,.\, 3^{n-1} \sin\left\{3x + (n-1)\dfrac{\pi}{2}\right\}$

$+ ... + n(n-1)(n-2)(n-3)3x^{-4} \sin\left\{3x + (n-4)\dfrac{\pi}{2}\right\}$

(*c*) $e^x[x^n + {}^nC_1 \,.\, nx^{n-1} + {}^nC_2 \,.\, n(n-1)x^{n-2} + ... + n!]$

(*d*) $e^x\left[\log x + \dfrac{{}^nC_1}{x} - \dfrac{{}^nC_2}{x^2} + \dfrac{{}^nC_3 2!}{x^3} + ... + \dfrac{(-1)^{n-1}(n-1)!}{x^n}\right]$

5. $y_n(0) = 0$, when n is even $y_n(0) = m(1 - m^2)(3^2 - m^2)....\left[(n-2)^2 - m^2\right]$, when n is odd.

6. $f^n(0) = m^2(2^2 + m^2)(4^2 - m^2)....\left[(n-2)^2 - m^2\right]e^{\frac{m\pi}{2}}\, n$, is even

$= -m(1 + m^2)(3^2 - m^2)....\left[(n-2)^2 + m^2\right]e^{\frac{m\pi}{2}}\, n$ is odd

1.5. APPLICATION

1.5.1. Maxima and Minima

Differentiation is use to find the maximum and minimum values of a function. Because the provides informaction about the gradient of slope of the graph a function to locate points on a graph where the gradient is zero. Now, at least in their immediate locality as such points are associated by maximum and minimum value.

(*i*) Stationary points

Functions are used to describe the ways in variables change. Function has, its maximum and minimum values, its turning points. Drawing a graph of a function using a graphical plotting the precise location of such points we need to turn to algebra and differential calculus. In this section we look at how we can find maximum and minimum points in this way.

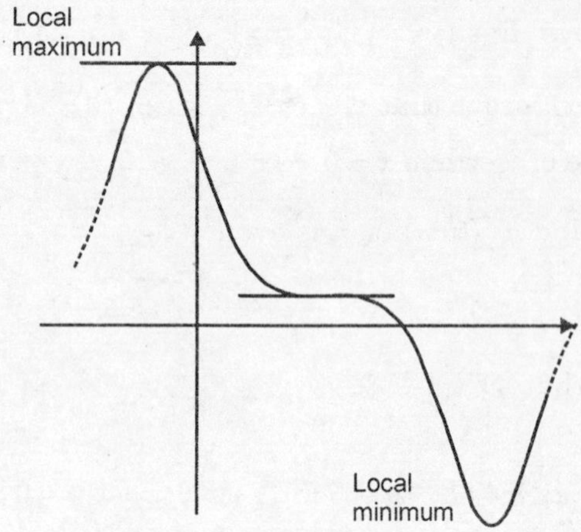

Fig. 1.

Let the graph of the function, $y(x)$. Shown in Figure 1. if at the points marked A, B and C, we draw tangents to the graph, note that these are parallel to the x axis. They are horizantal this means that at each of the points A, B and C the gradient of the graph is zero.

Fig 1. The gradient of this graph is zero at each of the points A, B and C gradient of a graph is given by $\frac{dy}{dx}$ and consequently, $\frac{dy}{dx}$ = 0 at points A, B and C. All of these points are known as **stationary points**.

Do you know

Any point at which the tangent to the graph is horizontal is called a **stationary point**. We can locate stationary points by looking for points at which $\frac{dy}{dx} = 0$.

(*ii*) Turning Points

Hence we see that points A and B the curve actually turns. These two stationary points are referred to as **turning points**. Point C is not a turining point because, although the graph is flat for a short time, the curve continues to go down from left to right.

But not all stationary points are turning points (*e.g.* point C).

Point A in Figure 1 is called a **local maximum** because in its immediate area it is the lowest point and so represent the least, or minimum, value of the function. We called local maximum and minimum. Similarly, a local minimum is often just called a minimum

(*iii*) Distinguishing maximum points from minimum points

Gradient of the graph as we travel through the minimum turning point, from left to right, that is as x increases as shown the Figure.

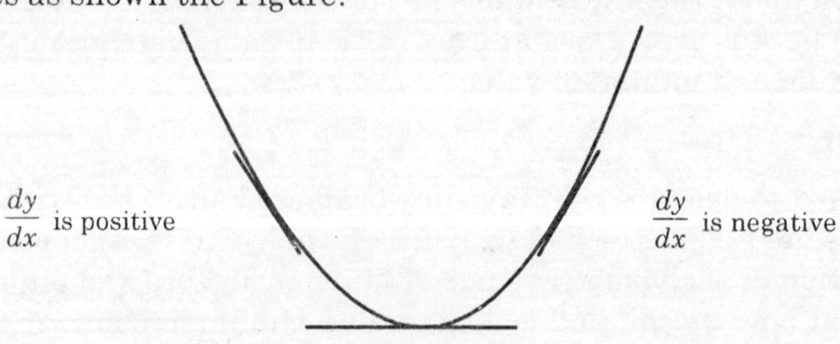

Fig. 2.

Figure 2. $\frac{dy}{dx}$ goes from negative through zero to positive as x increases. the left of the minimum point, $\frac{dy}{dx}$ is negative because the tangent has negative gradient. At the minimum point, $\frac{dy}{dx} = 0$.

1. To the right of the minimum point $\frac{dy}{dx}$ is positive, because here the tangent has a positive gradient. So, $\frac{dy}{dx}$ goes from negative to zero to positive as x increases. In other words, $\frac{dy}{dx}$ must be increasing as x increases.

2. We have found a stationary point to check if the point is a minimum. If $\frac{dy}{dx}$ is increasing near the stationary point then that point must be minimum. Now, if the derivative of $\frac{dy}{dx}$ is positive then we will know that $\frac{dy}{dx}$, is increasing so stationary point is minimum, so darivative of $\frac{dy}{dx}$ called the **second derivative** is written $\frac{d^2y}{dx^2}$. We conclude that if $\frac{d^2y}{dx^2}$ is positive at a stationary point, then that point must be a minimum turning point.

3. It is possible for a stationary point to be a minimum even if $\frac{d^2y}{dx^2}$ equals 0, although we cannot be certain other types of behaviour are possible. (However, we cannot have a minimum if $\frac{d^2y}{dx^2}$ is negative.)

Example 55. Discuss the maximum and minimum point of $y = x^4$

Sol. Maximum and minimum points of graph of this function is shown in Figure 3. There is clearly a minimum point when $x = 0$. But $\frac{dy}{dx} = 4x^3$ and this is clearly zero when x = 0. Differentiating again $\frac{d^2y}{dx^2} = 12x^2$ which is also zero when $x = 0$.

Figure 3. The function $y = x^4$ has a minimum at the origin where $x = 0$, but $\frac{d^2y}{dx^2} = 0$ and so is not greater than 0.

Now the gradient of the graph as shown maximum turing point, from left to right, that is as x increases.

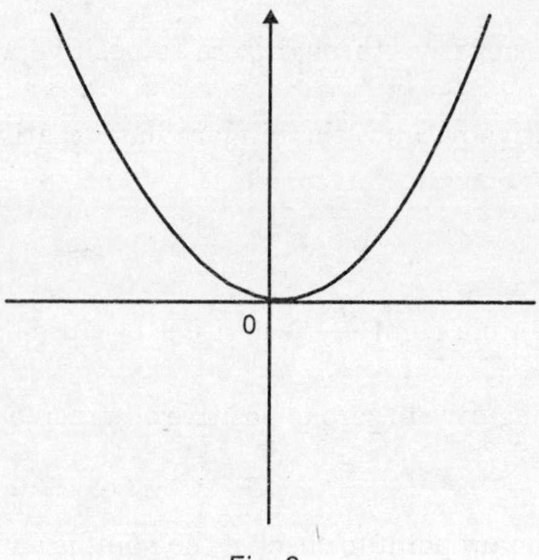

Fig. 3.

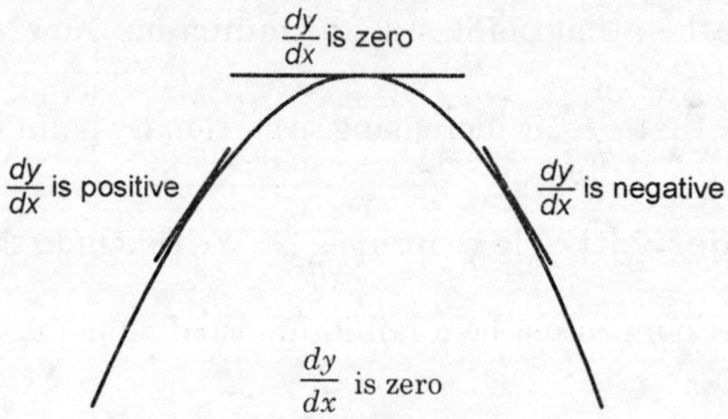

Fig. 4.

Fig. 4. $\frac{dy}{dx}$ goes from positive through zero to negative as x increases

If left of the maximum point, $\frac{dy}{dx}$ is positive because the tangent has positive gradient. At the maximum point, $\frac{dy}{dx} = 0$. To the right of the maximum point $\frac{dy}{dx}$ is negative because here the tangent has a negative gradient. So, $\frac{dy}{dx}$ goes from positive, to zero, to negative as x increases.

Now, we can check that stationary point is a maximum. If $\frac{dy}{dx}$ is **decreasing** near a stationary point then that point must **be maximum**. Now, if the derivative of $\frac{dy}{dx}$ is negative then we will know that $\frac{dy}{dx}$ is decreasing; so we will know that the stationary point is a maximum. As before, the derivative of $\frac{dy}{dx}$, the **second derivative** is $\frac{d^2y}{dx^2}$. We conclude that if $\frac{d^2y}{dx^2}$ is **negative** at a stationary point, then that point must be a **maximum** turning point.

It is important to realise that this test for a maximum is not conclusive. It is possible for a stationary point to be a maximum even if $\frac{d^2y}{dx^2} = 0$, although we cannot be certain other types of behaviour are possible. But we cannot have a maximum if $\frac{d^2y}{dx^2} > 0$, because as we have already seen the point would be a minimum.

Example 55A. *In functon $y = x^3 - 3x + 2$ at which point maximum or minimum*

Sol. We need to find where the turning points are, and whether we have maximum or minimum points First of all we carry out the differentiation and set $\frac{dy}{dx}$ equal to zero. This will enable us to look for any stationary points, including any turning points.

$$y = x^3 - 3x + 2$$

$$\frac{dy}{dx} = 3x^2 - 3$$

At stationary points, $\frac{dy}{dx} = 0$ and so

$$3x^2 - 3 = 0$$

$$3(x^2 - 1) = 0$$

(factorising)

$$3(x - 1)(x + 1) = 0$$

(factorising the difference of two squares)

Do you know

If $\frac{dy}{dx} = 0$ at a point, and if $\frac{d^2y}{dx^2} < 0$ there, then that point must be a maximum.

The second derivative test:

We can locate the position of stationary points by looking for points where $\frac{dy}{dx} = 0$.

As we have seen, it is possible that some such points will not be turning points.

We can calculate $\frac{d^2y}{dx^2}$ at each point we find.

If $\frac{d^2y}{dx^2}$ is **positive** then the stationary point is a **minimum** turning point.

If $\frac{d^2y}{dx^2}$ is **negative**, then the point is a **maximum** turning point.

If $\frac{d^2y}{dx^2} = 0$, it is possible that we have a **maximum**, or a **minimum**, or **indeed other** sorts of behaviour. So if $\frac{d^2y}{dx^2} = 0$ this second derivative test does not give us useful information and we must seek an alternative method.

It follows that either $x - 1 = 0$ or $x + 1 = 0$ and so either $x = 1$ or $x = -1$.

We have found the x coordinates of the points on the graph where $\frac{dy}{dx} = 0$, that is the stationary points. We need the y coordinates which are found by substituting the x values in the original function $y = x^3 - 3x + 2$.

when $x = 1$: $\quad y = 1^3 - 3(1) + 2 = 0.$

when $x = -1$ $\quad y = (-1)^3 - 3(-1) + 2 = 4.$

We have seen that the first derivative $\frac{dy}{dx} = 3x^2 - 3$. Differentiating this we can find the second derivative:

$$\frac{d^2y}{dx^2} = 6x$$

We now take each point in turn and use our test.

When $x = 1$: $\dfrac{d^2y}{dx^2} = 6x = 6(1) = 6$. We are not really interested in this value,

When $x = -1$: $\dfrac{d^2y}{dx^2} = 6x = 6(-1) = -6$. Again, what is important is its sign. Because it is negative we have a maximum point.

Finally, to finish this off we produce a quick sketch of the function now that we know the precise locations of its two turning points (See Figure 5).

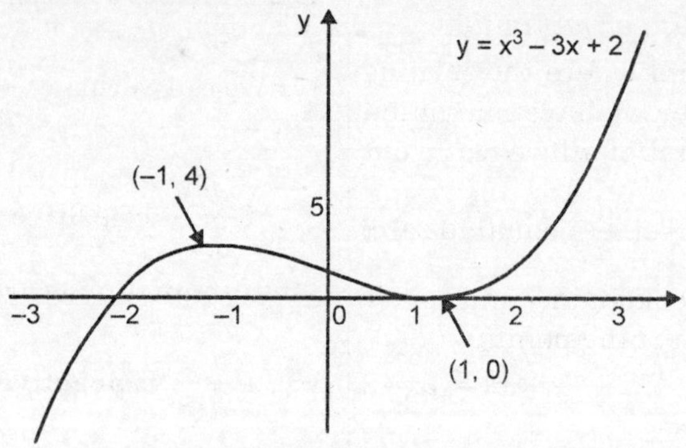

Fig. 5

EXERCISE 1.5.8.

Locate the position and nature of any turning points of the following functions.

1. $y = \dfrac{1}{2}x^2 - 2x$ **2.** $y = x^2 + 4x + 1$,

3. $y = 12x - 2x^2$, **4.** $y = -3x^2 + 3x + 1$

5. $y = x^4 + 2$, **6.** $y = 7 - 2x^4$,

7. $y = 2x^3 - 9x^2 + 12x$, **8.** $y = 4x^3 - 6x^2 - 72x + 1$

9. $y = -4x^3 + 30x^2 - 48x - 1$, **10.** $y = \dfrac{(x+1)^2}{x-1}$,

ANSWER

1. Minimum at (2–2) 2. Minimum at (–2–3),

3. Maximum at (–3,–54), 4. Maximum at $\left(\dfrac{1}{2}, \dfrac{7}{4}\right)$,

5. Minimum at (0,2), 6. Maximum at (0,7), 7. Maximum at (1,5), Minimum at (2,4),

8. Maximum at (–2,89), minimum at (3,–161), 9. Maximum at (4,31), minimum at (1,–23),

10. Maximum at (–1,0), minimum at (3,8),

$$= \frac{1}{2}\left(\frac{h^2 - 2x^2}{\sqrt{h^2 - x^2}}\right)$$

and,
$$\frac{d^2A}{dx^2} = \frac{(\sqrt{h^2 - 2x^2})(-4x) - (h^2 - 2x^2)\frac{1}{2}(h^2 - x^2)^{-1/2}(-2x)}{2(h^2 - x^2)}$$

$$= \frac{(2x^3 - 3xh^2)}{2(h^2 - x^2)^{3/2}}$$

Now,
$$\frac{dA}{dx} = 0 = (h^2 - 2x^2) = 0 = x = \frac{h}{\sqrt{2}}$$

$$\left[\frac{d^2A}{dx^2}\right]_{x=\frac{h}{\sqrt{2}}} = -2 < 0.$$

Thus, A is maximum at $x = (h/\sqrt{2})$

$$\text{base} = \sqrt{h^2 - x^2} = \sqrt{\left(h^2 - \frac{h^2}{2}\right)} = \frac{h}{\sqrt{2}} = \text{altitude}.$$

Hence, the triangle is isosceles.

Example 58. *If the sum of the lengths of the hypotenuse and a side of a right-angled triangle is given, show that the area of the triangle is maximum when the angle between them us (π/3).*

Sol. Let us consider a right-angled triangle with base = x and hypotenuse = y. Let $x + y = k$, where k is a constant. Let θ be the angle between the base and the hypotenuse. Let A be the area of the triangle. Then,

$$A = \frac{1}{2} \times BC \times AC = \frac{1}{2}\sqrt{y^2 - x^2}$$

$$A^2 = \frac{x^2(y^2 - x^2)}{4} = \frac{x^2[(k - x)^2 - x^2]}{4} \qquad [\because y = (k - x)]$$

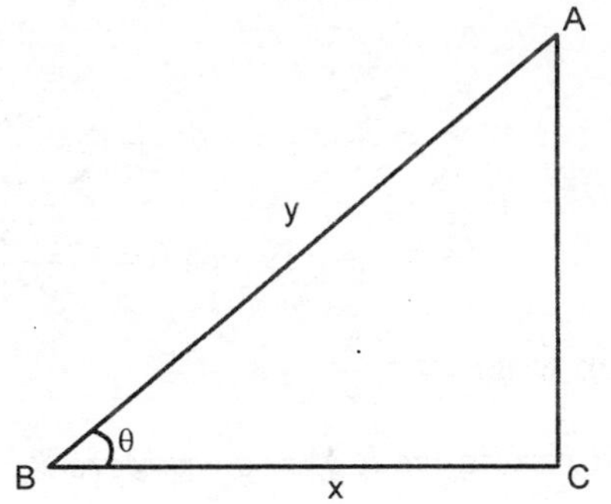

Fig. 7.

or
$$A^2 = \frac{k^2x^2 - 2kx^3}{4} \qquad ...(i)$$

Thus, A is maximum at $x = (h/\sqrt{2})$

$$\text{base} = \sqrt{h^2 - x^2} = \sqrt{\left(h^2 - \frac{h^2}{2}\right)} = \frac{h}{\sqrt{2}} = \text{altitude.}$$

Hence, the triangle is isosceles.

Example 58. *If the sum of the lengths of the hypotenuse and a side of a right-angled triangle is given, show that the area of the triangle is maximum when the angle between them us (π/3).*

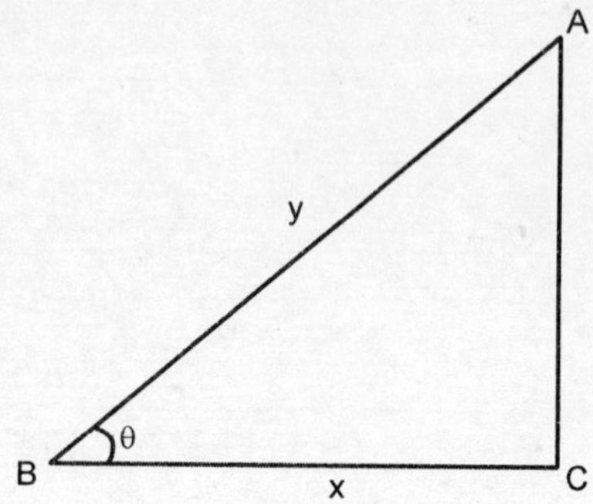

Fig. 7.

Sol. Let us consider a right-angled triangle with base = x and hypotenuse = y. Let $x + y = k$, where k is a constant. Let θ be the angle between the base and the hypotenuse. Let A be the area of the triangle. Then,

$$A = \frac{1}{2} \times BC \times AC = \frac{1}{2}\sqrt{y^2 - x^2}$$

$$A^2 = \frac{x^2(y^2 - x^2)}{4} = \frac{x^2[(k-x)^2 - x^2]}{4} \quad [\because y = (k - x)]$$

or
$$A^2 = \frac{k^2x^2 - 2kx^3}{4} \quad \ldots(i)$$

On differentiating (i), we get $2A \cdot \frac{dA}{dx} = \frac{2k^2x - 6kx^2}{4}$ $\quad \ldots(ii)$

or
$$\frac{dA}{dx} = \frac{k^2x - 3kx^2}{4A}$$

Now,
$$\frac{dA}{dx} = 0 = (k^2x - 3kx^2) = 0 = x = \frac{k}{3} \text{ [neglecting } x = 0].$$

Now, differentiating (ii), we get

$$2\left(\frac{dA}{dx}\right)^2 + 2A\frac{d^2A}{dx^2} = \frac{2k^2 - 12kx}{4} \quad \ldots(iii)$$

Putting $\frac{dA}{dx} = 0$ and $x = \frac{k}{3}$ on (iii), we get $\frac{d^2A}{dx^2} = \frac{-k^2}{4A} < 0$

Thus, A is maximum when $x = (k/3)$.

Now,
$$x = \frac{k}{3} = y = \left(k - \frac{k}{3}\right) = \frac{2k}{3}$$

$\therefore$
$$\frac{x}{y} = \cos\theta \Rightarrow \cos\theta = \frac{(k/3)}{(2k/3)} = \frac{1}{2} = = \frac{\pi}{3}$$

Hence, the area is maximum when $\theta = \frac{\pi}{3}$

Example 59. *The combined resistance R of two resistors R_1 and R_2 where $R_1, R_2 > 0$ is given by $\frac{1}{R} = \frac{1}{R_1} + \frac{1}{R_2}$. If $R_1 + R_2 = C$ (constant), show that the maximum resistance R is obtained by choosing $R_1 = R_2$.*

Sol. We have $R = \dfrac{R_1R_2}{(R_1 + R_2)}$ and $R_1 + R_2 = C$

$$R = \frac{R_1(C - R_1)}{C} \qquad [\because R_2 = (C - R_1)]$$

So, $$\frac{dR}{dR_1} = \frac{C - 2R_1}{C} \text{ and } \frac{d^2R}{dR_1^2} = -\frac{2}{C}$$

For a maxima or minima, we have $\dfrac{dR}{dR_1} = 0$

Now, $$\frac{dR}{dR_1} = 0 \quad \Rightarrow \frac{C - 2R_1}{C} = 0 \; R_1 = \frac{C}{2}$$

And, $$\left[\frac{d^2R}{dR_1^2}\right]_{R_1=(C/2)} = \frac{-2}{C} < 0$$

Thus, $R_1 = \dfrac{C}{2}$ is a point of maximum.

When, $R_1 = \dfrac{C}{2}$, we have $R_2 = \left(C - \dfrac{C}{2}\right) = \dfrac{C}{2}$. Hence, $R_1 = R_2$

Example 60. *A beam of length l is supported at one end. If W is the uniform load per unit length, the bending moment M at a distance x from the end is given by* $M = \left(\dfrac{lx}{2} - \dfrac{Wx^2}{2}\right)$. *Find the point on the beam at which the bending moment has the maximum* value.

Sol. $$M = \left(\frac{lx}{2} - \frac{Wx^2}{2}\right) = \frac{dM}{dx^2} = \left(\frac{1}{2} - Wx\right) \text{ and } \frac{d^2M}{dx^2} = -W.$$

For a maxima or minima, we have $\dfrac{dM}{dx} = 0.$

Now, $$\frac{dM}{dx} = 0 \Rightarrow \frac{1}{2} - Wx = 0 = x = \frac{1}{2W}$$

Also, $$\frac{d^2M}{dx^2} = -W < 0 \text{ for all values of } x.$$

$$x = \frac{1}{2W} \text{ is a point of maxima.}$$

So , the required point is at a distance of $\left(\dfrac{1}{2W}\right)$ from the supporting end.

Example 61. *A wire of length 25 m is to be cut into two peices. One of the wires is to be made into a square and the other into circle. What should be the lengths of the two pieces so that the combined area of the square and the circle is minimum?*

Sol. Let x metres and $(25 - x)$ metres be the required lengths.
Let a be the side od the square formed and r be the radius of the circle formed. Then,

$$4a = x \text{ and } 2\pi r = 25 - x \text{ or } a = \frac{x}{4} \text{ and } r = \frac{25-x}{2\pi}.$$

∴ Area of the square $= \left(\frac{x^2}{16}\right)$ sq metres.

Area of the circle $= \pi\left(\frac{25-x}{2\pi}\right)^2 = \frac{(25-x)^2}{4\pi}$

Now, the combined area $A = \frac{x^2}{16} + \frac{(25-x)^2}{4\pi}$

$$\frac{dA}{dx} = \frac{x}{8} - \frac{(25-x)}{2\pi} = \frac{(\pi+4)x-100}{8\pi}, \text{ and } \frac{d^2A}{dx^2} = \frac{(\pi+4)}{8\pi}$$

For maxima or minima, we have $\frac{dA}{dx} = 0$.

Now, $\frac{dA}{dx} = 0 \Rightarrow (\pi + 4)x - 100 = 0 \Rightarrow x = \frac{100}{(\pi+4)}$

And $\frac{d^2A}{dx^2} = \frac{(\pi+4)}{8\pi} > 0$ For all values of x.

$$x = \frac{100}{(\pi+4)} \text{ is a point of minimum.}$$

Lengths of the pieces are $\left(\frac{100}{\pi+4}\right)$ metres and $\left(\frac{25\pi}{\pi+4}\right)$ metres.

Example 62. *Show that a cylinder of a given volume which is open at the top has minimum total surface area, provided its height is equal to the radius of its base.*

Sol. Let r be the radius and h the height of the cylinder of given volume V.

Then, $V = \pi r^2 h = h = \frac{V}{\pi r^2}$

Let the total surface area be s. Then,

$$S = \pi r^2 + 2\pi r h = \left(\pi r^2 + 2\pi r \frac{V}{\pi r^2}\right) \quad \text{[using (}i\text{)].}$$

$$S = \left(\pi r^2 + \frac{2V}{r}\right)$$

So, $\frac{dS}{dr} = \left(2\pi r - \frac{2V}{r^2}\right)$ and $\frac{d^2S}{dr^2} = \left(2\pi + \frac{4V}{r^3}\right)$

For a maxima or minima, we have $\frac{dS}{dr} = 0$

Now, $$\frac{dS}{dr} = 0 \Rightarrow 2\pi r - \frac{2V}{r^2} = 0 \Rightarrow 2\pi r^3 = 2V = 2\pi^2 h \Rightarrow r = h$$

$$\left(\frac{d^2S}{dr^2}\right)_{r=h} = \left(2\pi + \frac{4V}{h^3}\right) > 0$$

Thus $r = h$ is a point of minimum

This shows that the total surface area of the cylinder is minimum where $h = r$, *i.e.*, when the height of the cylinder is equal to the radius of the base.

Example 63. *Show that the height of a cylinder which is open at the top, having a given surface and the greatest volume is equal to the radius of its base.*

Sol. Let r be the radius, h the height and S the given surface area of an open cylinder Then,

$$S = (\pi r^2 + 2\pi rh) \text{ or } h = \left(\frac{S - \pi r^2}{2\pi r}\right) \quad ...(i)$$

Let V be the volume of the cylinder

$$V = \pi r^2 h = \pi r^2 \left(\frac{S - \pi r^2}{2\pi r}\right) = \left(\frac{rS - \pi r^3}{2}\right) \quad \text{Using } (i)$$

$$\therefore \quad \frac{dv}{dr} = \left(\frac{s - 3\pi r^2}{2}\right) \text{ and } \frac{d^2v}{dr^2} = -3\pi r$$

Now, $$\frac{dV}{dr} = 0 \quad \Rightarrow s - 3\pi r^2 = 0 \quad \Rightarrow \pi r^2 + 2\pi rh - 3\pi r^2 = 0 = h = r$$

And, $$\left(\frac{d^2V}{dr^2}\right)_{r=h} = -3\pi h < 0. \text{ Thus, } r = h \text{ is a point of maximum.}$$

So, V is maximum when $r = h$, *i.e.*, when the height of the cylinder is equal to the radius of its base.

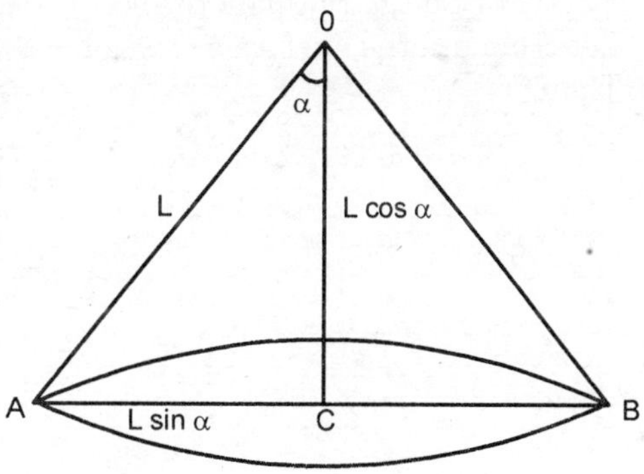

Example 64. *Show that the semivertical angle of a cone of maximum volume and of given slant height is* $tan^{-1}\sqrt{2}$.

Sol. Let α be the semivertical angle of a cone of given slant height l. Then, height of the cone $= l \cos \alpha$, radius of the base $= l \sin \alpha$.

Let V be the volume of the cone. Then,

$$V = \frac{1}{3}\pi r^2 h = \frac{1}{3}\pi (l^2 \sin^2 \alpha)(l \cos \alpha)$$

or $$V = \frac{1}{3}\pi l^3 \sin^2 \alpha \cos \alpha$$

$$\frac{dV}{d\alpha} = \frac{1}{3}\pi l^3 (-\sin^2 \alpha + 2 \sin \alpha \cos^2 \alpha)$$

$$= \frac{1}{3}\pi l^3 \sin\alpha(2\cos^2\alpha - \sin^2\alpha)$$

And $$\frac{d^2V}{d\alpha^2} = \frac{1}{3}\pi l^3 \cos\alpha(2\cos^2\alpha - \sin^2\alpha) + \frac{1}{3}\pi l^3 \sin\alpha(-4\sin\alpha\cos\alpha - 2\sin\alpha\cos\alpha)$$

$$= \frac{1}{3}\pi l^3 \cos^3\alpha(2 - 7\tan^2\alpha)$$

Now, for a maxima or minima, we have $\frac{dV}{d\alpha} = 0$.

Now, $$\frac{dV}{d\alpha} = 0 = \frac{1}{3}\pi l^3 \sin\alpha(2\cos^2\alpha - \sin^2\alpha) = 0$$

$\Rightarrow$ $$\sin\alpha = 0 \quad \text{or} \quad 2\cos^2\alpha - \sin^2\alpha = 0$$

$\Rightarrow$ $$\alpha = 0 \quad \text{or} \quad \alpha = \tan-1\sqrt{2}$$

Neglecting $\alpha = 0$, we have $\alpha = \tan^{-1}\sqrt{2}$, i.e., $\tan\alpha\ \sqrt{2}$

Now, $\tan\alpha = \sqrt{2} = \sec^2\alpha = 3 = \cos\alpha\ \frac{1}{\sqrt{3}}$

$$\left[\frac{d^2V}{d\alpha^2}\right]_{\alpha=\tan^{-1}\sqrt{2}} = \frac{1}{3}\pi l^3\left(\frac{1}{\sqrt{3}}\right)^3(2-14) = -\frac{4\pi l^3}{3\sqrt{3}} < 0$$

Thus, V is maximum when $\alpha = \tan^{-1}\sqrt{2}$, *i.e.*, when the semivertical angle of the cone is $\tan^{-1}\sqrt{2}$

Example 65. *Show that the semivertical angle of a right circular cone of given surface area and maximum volume is $\sin^{-1}(1/3)$.*

Sol. Let r be the radius, l the slant height and h the height of the cone.

Let S denote the surface area and V the volume of the cone.

Then, $$S = (\pi r^2 + \pi r l) = \text{constant.}$$

$$l = \left(\frac{S}{\pi r} - r\right) \qquad ...(i)$$

Now, $$V = \frac{1}{3}\pi r^2 h = \frac{1}{3}\pi r^2\sqrt{l^2 - r^2}$$

$$V^2 = \frac{1}{9}\pi^2 r^4(l^2 - r^2) = \frac{1}{9}\pi^2 r^4\left\{\left(\frac{S}{\pi r} - r\right)^2 - r^2\right\} \qquad \text{[using (i)]}$$

$$= \frac{1}{9}S\left(Sr^2 - \frac{2\pi}{9}Sr^4\right)$$

This, $$V^2 = \left(\frac{1}{9}S^2 r^2 - \frac{2\pi}{9}Sr^4\right)$$

$$2V\frac{dV}{dr} = \left(\frac{2}{9}S^2 r \frac{8\pi S}{9} r^3\right) = \frac{2rS}{9}(S - 4\pi r^2) \qquad ...(ii)$$

$$\frac{dV}{dr} = 0 \Rightarrow r = 0 \text{ or } (S–4\pi r2) = 0 \Rightarrow r^2 = \frac{S}{4\pi} \text{ (neglecting } r = 0)$$

On differentiating (*ii*), we get $2\left(\frac{dV}{dr}\right)^2 + 2V\frac{d^2V}{dr^2} = \frac{1}{9}S(2S - 24\pi r^2)$

Putting $\frac{dV}{dr} = 0$ and $r^2 = \frac{S}{4\pi}$, we get

$$2V\frac{d^2V}{dr^2} = \frac{1}{9}S(2s - 6s) = -\frac{4}{9}S^2 < 0$$

When the volume is maximum, we have

$$r^2 = \frac{S}{4\pi} = \frac{(\pi r^2 + \pi r l)}{4\pi} \Rightarrow l = 3r$$

Now, if α is the semivertical angle of the cone then

$$\frac{r}{l} = \sin\alpha \Rightarrow \frac{r}{3r} = \sin\alpha \Rightarrow \frac{1}{3} \Rightarrow \alpha = \sin^{-1}\left(\frac{1}{3}\right)$$

Hence, the semivertical angle of a right cone of a given surface and maximum volume is $\sin^{-1}(1/3)$.

Example 66. Show that the surface area of a closed cuboid with square base and given volume is minimum when it is a tube.

Sol. Let *V* be volume of a closed with angle *a*, breadth *a*, height *h*. Lets *S* be its surface area.

Then, $V = (a \times a \times h) \text{ or } h = \frac{V}{a^2}$ ...(*i*)

Now, $S = 2(a^2 + ah + ah) = 2(a^2 + 2ah) = 2\left[a^2 + \frac{2V}{a}\right]$ [using (*i*)]

i.e., $S = 2\left(a^2 + \frac{2V}{a}\right)$

$\therefore$ $\frac{dS}{da} = 2\left(2a - \frac{2V}{a^2}\right)$ and $\frac{d^2S}{da^2} = \left(4 + \frac{8V}{a^3}\right)$

Now, $\frac{dS}{da} = 0 \quad \Rightarrow V = a^3 \quad \Rightarrow a \times a \times h = a^3 \quad \Rightarrow h = a.$

Now , when $h = a$, we have $V = a^3$.

$$\left[\frac{d^2S}{da^2}\right]_{h=a} = \left(4 + \frac{8a^3}{a^3}\right) = 12 > 0.$$

So, *S* is minimum when length = *a*, breadth = *a* and height = *a*, *i.e.*, when it is a cube.

Examaple 67. *Show that the height of a closed cylinder of given surface and maximum volume is equal to the diameter of its base.*

Sol. Let *S* be the fixed surface area of a closed cylinder of radius *r* and height *h*. Let *V* be its volume.

Then, $S = (2\pi r^2 + 2\pi rh) \Rightarrow h = \left(\dfrac{S - 2\pi r^2}{2\pi r}\right)$ ...(*i*)

And, $V = \pi r^2 h = \pi r^2 \left(\dfrac{S - 2\pi r^2}{2\pi r}\right) = \left(\dfrac{rS - 2\pi r^3}{2}\right)$ [using (*i*)]

$$\frac{dV}{dr} = \left(\frac{S - 6\pi r^2}{2}\right) \text{ and } \frac{d^2V}{dr^2} = -6\pi r$$

Now, $\dfrac{dV}{dr} = 0 \Rightarrow (S - 6\pi r^2) = 0 \Rightarrow S = 6\pi r^2$

$\Rightarrow$ $(2\pi r^2 + 2\pi rh) = 6\pi r^2 \Rightarrow h = 2r$, *i.e.*, $r = \dfrac{1}{2}h$.

And, $\left[\dfrac{d^2V}{dr^2}\right]_{r=(h/2)} = -6\pi \times \dfrac{1}{2}h = -3\pi h < 0$

So, V is maximum when $h = 2\pi$.

EXCERCISE 1.5.9.

1. Find two positive numbers whose sum is 16 and the sum of whose squares is minimim. **Ans.** (7,7)
2. Find two positive numbers whose product is 49 and the sum is minimum. **Ans.** (8,8)
3. Divide 15 into two parts such that the square of one number multiplied with the cube of the other number maximum. **Ans.** (6,9)
4. Divide a into two parts such that the product of the *p*th power of one part and the *q*th power of the second part may be maximum. **Ans.** $\dfrac{ap}{p+q}, \dfrac{aq}{p+q}$
5. The rate of working of an engine is given by $R = 15v + \dfrac{6000}{v}$, where $0 < v < 30$ and v is the speed of the engine. Show that R is the least when $v = 20$.
6. Find the dimensional of the reactangle of area 96 cm^2 whose perimeter is the least. Also, find the perimeter of the reactangle.

   Ans. [Length $= 4\sqrt{6}$ cm, breadth $= \left(\dfrac{20}{\pi+4}\right) m$, height $= \left(\dfrac{10}{\pi+4}\right)$ m.]
7. Prove that the largest reactangle with a given perimeter is a square.
8. Given the perimeter of a reactangle, show that its diagonal is minimum when it is a square.
9. Show that a reactangle of maximum perimeter which can be inscribed in a circle of radius a is a square of side $\sqrt{2}\,a$.

10. The sum of the perimeter of a square and a circle is given. show that the sum of their areas is least when the side of the square is equal to the diameter of the circle.

11. Show that the right triangle of maximum area that can be inscribed in a circle ia an isosceles triangle.

12. Prove that the perimeter of a right-angled triangle of given hypotenuse is maximum when the triangle is isosceles.

13. An open box with a square base is to be made out of given cardboard of area c^2 square units. show that the maxiumum volume of the box is $\frac{c^3}{6\sqrt{3}}$ cubic units.

14. A cylindrical can is to be made hold 1 litre of oil. Find the dimensions which will minimize the cost of the metal to make the can.

Ans. $r = \left(\frac{500}{\pi}\right)^{1/3}$ cm, $h = \frac{1000}{\pi^{1/3}}(500)^{2/3}$ cm.

15. Show that the right circular cone of least curved surface and given volume has an altitude equal to $\sqrt{2}$ times the radius of the base.

16. Show that the height of a closed cylinder of given volume and the least surface area is equal to its diameter.

17. Prove that the volume the largest cone that can be inscribed in a sphere is $\frac{8}{27}$ of the volume of the sphere.

18. Show that the surface area of a closed cuboid with square base and given volume is minimum when it is a cube.

19. A reactangle is inscribed in a semicircle of radius r with one of its sides on the diameter of the semicircle. Find the dimensions of the reactangle so that its area is maximum. Find also this area.

Ans. Length $= r\sqrt{2}$, breadth $= r\sqrt{2}/2$, area πr^2

20. Two sides of a triangle have length a and b and the angle between them is θ. What value of θ will maximize the area of the triangle? **Ans.** $= \frac{\pi}{2}$

21. Show that the maximum volume of the cylinder which can be inscribed in a sphere of radius $5\sqrt{3}$ cm is $(500\pi)\text{cm}^3$.

22. A square tank of capacity 250 cubic meters has to be dug out. The cost of the land is Rs 50 per square metre. The cost of digging increases with the depth and for the whole tank, it is Rs $(400 \times h^2)$, where h meters is the depth of the tank. what should be the dimensions of the tank so that the cost is minimum?

Ans. Side $= 10$ cm, depth $= 2.5$ m,

23. A square piece of tin of side 18 cm is to be made into box without the top, by cutting a square piece from each cornor and folding up the flaps. what should be the side of the square to be cut off so that the volume of the box is maximum? Also, find the maximum volume of the box. **Ans.** 3 cm, 432 cm^2

24. An open tank with a square base and vertical sides is to be constructed from a metal sheet so as to hold a given quatity of water. show that the cost of the material will be least when the depth of the tank is half of its.

25. A wire of length 36 cm is cut into two pieces. One of the pieces is turned in the form of a square and the other in the form of an equilateral triangle. Find the length of each piece so that the sum of the area of the two be minimum.

Ans. $\dfrac{144\sqrt{3}}{4\sqrt{3}+9}$ cm, $\dfrac{324}{4\sqrt{3}+9}$

1.5.2. Equation of Tangents and Normals to a Curve

1. Definition of Tangent

Let P and Q be any two point on a curve $y = f(x)$, Now as point Q approaches P along the curve, the chord PQ, in general, tends to a straight line PT, which is defined as the tangent to the curve at the point P.

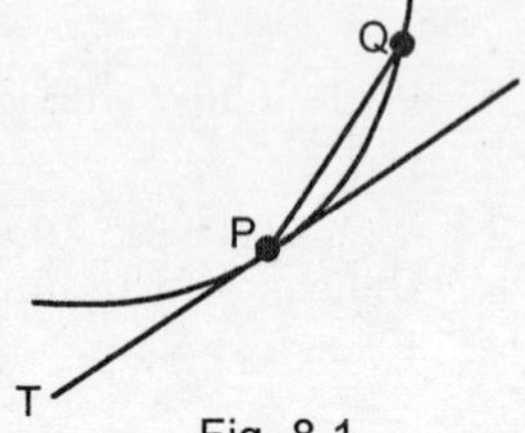

Fig. 8.1

2. Geometrical Interpretation of the Derivative

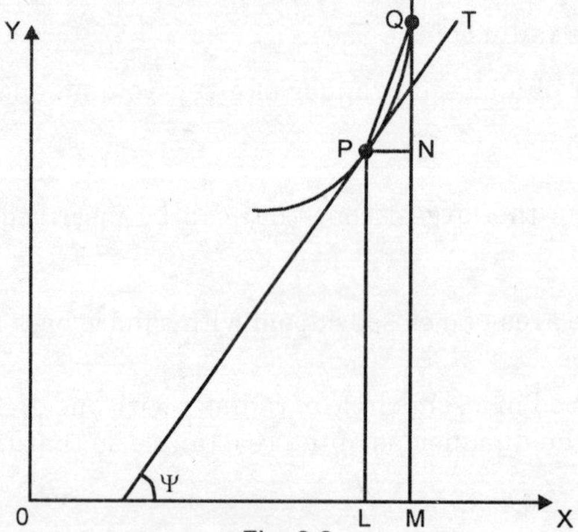

Fig. 8.2

Let $P(x, y)$ and $Q(x + \delta x, y + \delta y)$ be the points on a curve $y = f(x)$. Draw LP and MQ perpendiculars on the axis and PN perpendicular to MQ.

We have $LP = MV = y,$

$MQ = y + \delta y$

and $PN = \delta x$

Now $NQ = MQ - MN = MQ - LP$

$= (y + \delta y) - y = \delta y.$

From $\Delta\, PQN$, we have, $\tan \angle QPN = \dfrac{NQ}{PN} = \dfrac{\delta y}{\delta x}$ = Slop of chord PQ.

Now as the point Q approaches the points P along the curve, the chord PQ tends to tangents PT and $\angle QPN$ approaches ψ, the angle which the tangent PT makes with the x-axis .

$$\therefore \quad \tan \psi \ \underset{Q\to P}{Lt} \frac{\delta y}{\delta x} = \underset{\delta x\to Po}{Lt} - \frac{\delta y}{\delta x} = \frac{dy}{dx}.$$

$$\therefore \quad \tan \psi = \frac{dy}{dx}.$$

Thus the derivative of a function at a point denotes the slop of the tangent to the curve at that point.

1.5.2.1. Equation of Tangents to a Curve.

Let us find the equation of a tangent to the curve $y = f(x)$ at a point $P(x,y)$ on the curve. As already seen in previous article, the slop of tangent to the curve at (x,y) is $\frac{dy}{dx}$. Let (X, Y) be current co-ordinates of a points on the tangents, then its equation is given by,

$$Y - y = \frac{dy}{dx}(X - x).$$

Example 68. *Find the points on the curve $ax^2 + 2hxy + by^2 = 1$ at which tangents is parallel to axis.*

Sol. Here $ax^3 + 2hxy + by^2 = 1$ and let (x_1,y_1) be the required point. Differentiating both sides with respect to x, we have

$$2ax+2h\left(x\frac{dy}{dx}+y\right)+2by\frac{dy}{dx}=0$$

or
$$\frac{dy}{dx} = -\frac{ax+hy}{hx+by}$$

$$\left(\frac{dy}{dx}\right)_{(x_1,y_1)} = -\frac{ax_1+hy_1}{hx_1+by_1} \quad ...(1)$$

Tangent is parallel to x-axis, if

$$\left(\frac{dy}{dx}\right)_{(x_1,y_1)} = 0.$$

From (1), we have
$$ax_1 + hy_1 = 0$$

or
$$x_1 = -\frac{h}{a}y_1. \quad ...(2)$$

From equation of the curve, we have

$$ax_1^2 + 2hx_1y_1 + by_1^2 = 1$$

$$a\frac{h^2}{a^2}y_1^2 - \frac{2h^2}{a}y_1^2 + by_1^2 = 1$$

$\therefore$
$$(h^2 - 2h^2 + ab)y_1^2 = 1$$

or
$$y_1^2 = \frac{1}{ab-h^2}$$

$\Rightarrow$
$$y_1 = \pm\frac{1}{\sqrt{ab-h^2}}$$

Therefore, from (2),
$$x_1 = \mp\frac{h}{\sqrt{ab-h^2}}$$

Hence the required point is
$$\left(\mp\frac{h}{a\sqrt{ab-h^2}}, \mp\frac{1}{\sqrt{ab-h^2}}\right)$$

Example 69. *Find the equation of the tangent at t =* $\frac{\pi}{2}$ *to the cycloid*

$$x = a\,(t - \sin t),\; y = a(1 - \cos t).$$

Sol. Here $x = a(t - \sin t),\; y = a(1 - \cos t)$

$$\frac{dy}{dx} = \frac{dy}{dt}\Big|\frac{dx}{dt} = \frac{a\sin t}{a(1-\cos t)} = \frac{2a\sin\frac{t}{2}\cos\frac{t}{2}}{2a\sin^2\frac{t}{2}} = \cot t/2$$

$$\frac{dy}{dx} = \cot\frac{\pi}{4} = 1 \text{ at } t = \frac{\pi}{2}$$

Also, when $t = \frac{\pi}{2}$,

$$x = a\left(\frac{\pi}{2} - 1\right)\; y = a$$

Therefore, the required equation of the tangent is

$$y - a = 1\left[x - a\left(\frac{\pi}{2} - 1\right)\right]$$

or $$Y - x = 2a - \frac{\pi a}{2}$$

Example 70. *Prove that the curve* $\left(\frac{x}{a}\right)^n + \left(\frac{y}{b}\right)^n = 2$*, touches the straight line* $\frac{x}{a} + \frac{y}{b} = 2$ *at the point (a,b), whatever be the value of n.*

Sol. Here $$\left(\frac{x}{a}\right)^n + \left(\frac{y}{b}\right)^n = 2$$

$\therefore$ $$\frac{n}{a^n}x^{n-1} + \frac{n}{b^n}\cdot y^{n-1}\frac{dy}{dx} = 0$$

or $$\frac{dy}{dx} = -\frac{b^n}{a^n}\cdot\frac{x^{n-1}}{y^{n-1}}$$

or $$\left(\frac{dy}{dx}\right)_{(a,b)} = -\frac{b}{a}$$

Therefore, the equation of the tangent to the given curve at (a,b) is

$$y - b = -\frac{b}{a}(x - a) \Rightarrow \frac{x}{a} + \frac{y}{b} = 2.$$

Hence the line $\frac{x}{a} + \frac{y}{b} = 2$, always touches the given curve at the point (a,b), whatever may be value of n.

1.5.2.2. Equation of Normal

The Normal to a curve $y = f(x)$ at a point $P(x, y)$ is a stright line passing through P and perpendicular to the tangent there at.

Since the normal is at right angle to the tangent its slop is negative reciprocal of the slop of the tangent at $P(x, y)$.

Slop of normal at P $= -\dfrac{1}{\dfrac{dy}{dx}}$

Hence equation of normal at $P(x, y)$ is given by

$$Y - y = -\frac{1}{\dfrac{dy}{dx}}(X - x),$$

or $$(X - x) + \frac{dy}{dx}(Y - y) = 0$$

Example 71. *Find the equation of the normal to the curve $4x^2 + 9y^2 = 25$ at (2, 1).*

Sol. Here $$4x^2 + 9y^2 = 25$$

$$8x + 18y\frac{dy}{dx} = 0$$

or $$\frac{dy}{dx} = -\frac{8x}{18y} = -\frac{4x}{9y}$$

or $$\left(\frac{dy}{dx}\right)_{at(2,1)} = -\frac{8}{9} = m_1 \text{ (say)},$$

Therefore the slop of the normal $= -\dfrac{1}{m_1} = \dfrac{9}{8}$

Required equation of normal is $= y - 1 = \dfrac{9}{8}(x - 2)$

or $$9x - 8y = 10$$

Example 72. *Find the equation of normal to the parabola $y^2 = 4ax$ in the form $y = mx - 2am - am^3$, where m denotes the slope of the normal.*

Sol. Here $$y^2 = 4ax$$

$$2y\frac{dy}{dx} = 4a$$

or $$\frac{dy}{dx} = \frac{2a}{y}$$

Now, slope of the normal $= -\dfrac{y}{2a} = m$ (given)

$\therefore$ $$y = -2am$$

From the equation of the parabola

$$x = am^2$$

Thus the point of contact is $(am^2, -2am)$.

Hence the equation of the normal is,

$$y + 2\,am = m\,(x - am^2)$$

$$y = mx - 2\,am - am^3.$$

1.5.2.3. Some Important Theorems

Theorem 1. *Prove that the equation of a tangent to a curve* $y = f(x)$ *at a point* $P(x_1, y_1)$ *is given by* $\frac{y - y_1}{x - x_1} = \left(\frac{dy}{dx}\right)_{(x_1,y_1)}$, *where* $\left(\frac{dy}{dx}\right)_{(x_1,y_1)}$ *denotes the value of* $\frac{dy}{dx}$ *at* $x = x_1$ *and* $y = y_1$, *and the equation of the normal at* $P(x_1,y_1)$ *is given by* $\frac{y - y_1}{x - x_1} = \frac{-1}{\left(\frac{dy}{dx}\right)_{(x_1,y_1)}}$

Proof. we know that the slope of the tangent to the curve $y = f(x)$ at a point $p(x_1, y_1)$ is $\left(\frac{dy}{dx}\right)_{(x_1,y_1)}$. Thus, the tangent to the given curve at a point P is line passing through $(x_1,y_1)$with a slope of $\left(\frac{dy}{dx}\right)_{(x_1,y_1)}$. So, the equation of the tangent at P is $\frac{y - y_1}{x - x_1} = \left(\frac{dy}{dx}\right)_{(x_1,y_1)}$

Again, the normal to the given curve at P is a line perpendicular to the tangent at P.

So, the slope of the normal is $\frac{-1}{\left(\frac{dy}{dx}\right)_{(x_1,y_1)}}$

Also, the normal at P passes through (x_1,y_1).

So, the equation of the normal at P is $\frac{y - y_1}{x - x_1} = \frac{-1}{\left(\frac{dy}{dx}\right)_{(x_1,y_1)}}$

Theorem 2. The tangents to a curve $y = f(x)$ at a point $P(x_1, y_1)$ is parallel to the x-axis if and only if $\left(\frac{dy}{dx}\right)_{(x_1,y_1)} = 0$

Proof. The tangent is parallel to the x-axis $\Leftrightarrow$ its slope is 0 $\Leftrightarrow$ $\left(\frac{dy}{dx}\right)_{(x_1,y_1)} = 0$

Theorem 3. The tangent to a curve $y = f(x)$ at a point $P(x_1 - y_1)$ is parallel to the y–axis if and only if $\left(\frac{dx}{dy}\right)_{(x_1,y_1)} = 0$

Proof. The tangent is parallel to the y-axis

$$\Leftrightarrow \text{its slope is } 90° \Leftrightarrow \left(\frac{dy}{dx}\right)_{(x_1,y_1)} = \tan 90° = \infty \Leftrightarrow \left(\frac{dx}{dy}\right)_{(x_1,y_1)} = 0$$

Remark. If $x = f(t)$ and $y = g(t)$, where $f'(t) = 0$ then (*i*) equation of the tangent at '*t*' is $\dfrac{y-g(t)}{x-f(t)} = \dfrac{g'(t)}{f'(t)}$

and,(*ii*) equation of the normal at '*t*' is $\dfrac{y-g(t)}{x-f(t)} = -\dfrac{f'(t)}{g'(t)}$

Example 73. *Find the equations of the tangent and the normal to the curve $y = x^4 - 6x^3 + 13x^2 - 10x + 5$ at the point (1,3).*

Sol. The equation of the given curve is $y = x^4 - 6x^3 + 13x^2 - 10x + 5$.

$$\therefore \quad \frac{dy}{dx} = 4x^3 - 18x^2 + 26x - 10.$$

So,
$$\left(\frac{dy}{dx}\right)_{(1,3)} = (4 \times 1^3 - 18 \times 1^2 + 26 \times 1 - 10) = 2$$

∴ The required equation of the tangent is

$$\frac{y-3}{x-1} = 2 \text{ or } 2x - y + 1 = 0$$

And, the required equation of the normal is

$$\frac{y-3}{x-1} = \frac{-1}{2} \text{ or } x + 2y - 7 = 0.$$

Example 74. *Find the equation of the tangent and the normal to the curve $y = x^3 + 4x + 1$ at the point where $x = 3$.*

Sol. When $x = 3$, we have $y = (3^2 + 4 \times 3 + 1) = 22$.

So, the point of contact is (3, 22).

Now $y = x^2 + 4x + 1 \quad \Rightarrow \dfrac{dy}{dx} = 2x + 4 \quad \Rightarrow \left(\dfrac{dy}{dx}\right)_{(3,22)} = (2 \times 3 + 4) = 10.$

Equation of the tangent is $\dfrac{y-22}{x-30} \quad \Rightarrow 10x - y - 278 = 0.$

And, equation of the normal is $\dfrac{y-22}{x-3} = \dfrac{-1}{10} \Rightarrow x + 10y - 223 = 0.$

Example 75. *Show that the equation of the tangent to the ellipse*

$$\frac{x^2}{a^2} + \frac{y^2}{b^2} = 1 \text{ at } (x_1, y_1) \text{ is } \frac{xx_1}{a^2} + \frac{yy_1}{b^2} = 1$$

Sol. $\dfrac{x^2}{a^2} + \dfrac{y^2}{b^2} = 1 \Rightarrow \dfrac{2x}{a^2} + \dfrac{2y}{b^2} \cdot \dfrac{dy}{dx} = 0$ [no differentiating *w.r.t.x*]

$$\Rightarrow \quad \frac{dy}{dx} = \frac{-b^2 x}{a^2 y} \Rightarrow \left(\frac{dy}{dx}\right)_{(x_1, y_1)} = \frac{-b^2 x_1}{a^2 y_1}$$

So the equation of tangent at (x_1, y_1) is

$$\frac{y - y_1}{x - x_1} = \frac{-b^2 x_1}{a^2 y_1} \Rightarrow b^2 x x_1 + a^2 y y_1 = b^2 x^1{}_2 + a^2 y^2{}_1.$$

On diving throughout by a^2b^2, we get $\frac{xx_1}{a^2}+\frac{yy_1}{b^2} = \frac{x_1^2}{a^2}+\frac{y_1^2}{b^2}$

i.e., $$\frac{xx_1}{a^2}+\frac{yy_1}{b^2}=1\left[\because (x_1,y_1) \text{ lies on } \frac{x^2}{a^2}+\frac{y^2}{b^2}=1 \Rightarrow \frac{x_1^2}{a^2}+\frac{y_1^2}{b^2}=1\right]$$

Example 76. *Find the equation of the tangent to the curve* $y = \sqrt{5x-3}-2$, *which is parallel to the line* $4x - 2y + 3 = 0$

Sol. The given line is $4x - 2y + 3 = 0$ or $y = 2x + \frac{3}{2}$.

$\therefore$ Slope of the given line – 2.
So, the slope of the tangent = 2.
Let the point of contact be (x_1, y_1)

Now, $$y = \sqrt{5x-3}-2 \Rightarrow \frac{dy}{dx} = \frac{5}{2.\sqrt{5x-3}}$$

$$\Rightarrow \left(\frac{dy}{dx}\right)_{(x_1,y_1)} = \frac{5}{2\sqrt{5x_1-3}}$$

$$\therefore \frac{5}{2\sqrt{5x_1-3}} = 2 = \frac{25}{4(5x_1-3)} = 4, \text{ i.e., } x_1 = \frac{73}{80}$$

$$\therefore y_1 = \sqrt{5x_1-3}-2 = \sqrt{\left(5\times\frac{73}{80}-3\right)}-2 = \frac{-3}{4}$$

So, the point of contact is $\left(\frac{73}{80},\frac{-3}{4}\right)$

Hence the required equation of the tangent is $\dfrac{\left(y+\frac{3}{4}\right)}{\left(x-\frac{73}{80}\right)} = 2,$

i.e., $80x - 40y - 103 = 0$

Example 77. *Find the equations of the normals to the curve* $3x^2 - y^2 = 8$, *parallel to the line* $x + 3y = 4$.

Sol. The given line $x + 3y = 4$ or $y = -\frac{1}{3}x + \frac{4}{3}$

$\therefore$ Slope of the given line $= -\frac{1}{3}$

So, the slope of the required normal $= -\frac{1}{3}$ ...(*i*)

Let the point of contact be (x_1, y_1).

Now, $3x^2 - y^2 = 8 \Rightarrow 6x - 2y\frac{dy}{dx} = 0$ [on differentiating *w.r.t.x*]

$$= \frac{dy}{dx} = \frac{3x}{y} = \left(\frac{dy}{dx}\right)_{(x_1, y_1)} = \frac{3x_1}{y_1}$$

$\therefore$ Slope of the normal $= \dfrac{-1}{\left(\dfrac{dy}{dx}\right)_{(x_1,y_1)}} = \dfrac{-y_1}{3x_1}$ ...(*ii*)

Thus, from (*i*) and (*ii*) we get $\dfrac{-y_1}{3x_1} = \dfrac{1}{3} = y_1 = x_1$.

Also, since (x_1, y_1) lies on the given curve, we have

$$3x_1^2 - y_1^2 = 8 \text{ or } 3x_1^2 - x_1^2 = 8 \; [\because y_1 = x_1]$$

or $\quad x_1^2 = 4$ or $x_1 = \pm 2$.

$\therefore \quad y_1 = \pm 2 \; [\therefore y_1 = x_1]$.

Thus, the points of contact are (2, 2) and (–2, –2).

So, the equation of the required normal ar (2, 2) is $\dfrac{y-2}{x-2} = \dfrac{-1}{3}$

i.e., $\quad x + 3y - 8 = 0$

The equation of the required normal at (–2, –2) is $\dfrac{y+2}{x+2} = \dfrac{-1}{3}$

i.e., $\quad x + 3y + 8 = 0$

Example 78. *Prove that curve* $\left(\dfrac{x}{a}\right)^n + \left(\dfrac{y}{b}\right)^n = 2$ *touches the straight line* $\dfrac{x}{a} + \dfrac{y}{b} = 2$, *at the point (a, b) whatever be the value of n.*

Sol. The equation of the curve is $\left(\dfrac{x}{a}\right)^n + \left(\dfrac{y}{b}\right)^n = 2$.

On differentiating w.r.t. *x*, we get $n\left(\dfrac{x}{a}\right)^{n-1} . \dfrac{1}{a} + n . \left(\dfrac{y}{b}\right)^{n-1} . \dfrac{1}{b} . \dfrac{dy}{dx} = 0$

$$= \frac{nx^{n-1}}{a^n} + \frac{ny^{n-1}}{b^n} . \frac{dy}{dx} = 0 \Rightarrow \frac{dy}{dx} = -\frac{b^n x^{n-1}}{a^n y^{n-1}}$$

$$\Rightarrow \quad \left(\frac{dy}{dx}\right)_{(a,b)} = \frac{-b^n a^{n-1}}{a^n b^{n-1}} = -\frac{b}{a}$$

So, the equation of the tangent at (*a*, *b*) is $\dfrac{y-b}{x-a} = \dfrac{-b}{a}$

i.e., $\quad ay + bx = 2ab$

$\therefore \quad \dfrac{x}{a} + \dfrac{y}{b} = 2$, which is independent of *n*.

Example 79. *At what points will the tangent to the curve* $y = 2x^3 - 15x^2 + 36x - 21$ *be parallel to the axis ? Also, find the equations of tangents to the curve at these points.*

Sol. *Let the required point be $P(x_1, y_1)$.*

Then $$\left[\frac{dy}{dx}\right]_{(x_1,y_1)} = 0$$

Now, $$y = 2x^3 - 15x^2 + 36x - 21$$

$$\Rightarrow \quad \frac{dy}{dx} = 6x^2 - 30x + 36 = 6(x^2 - 5x + 6)$$

$$\Rightarrow \quad \left[\frac{dy}{dx}\right]_{(x_1,y_1)} = 6(x_1^2 - 5x_1 + 6).$$

$$\therefore \quad \left[\frac{dy}{dx}\right]_{(x_1,y_1)} = 0 \quad \Rightarrow 6(x_1^2 - 5x_1 + 6) = 0$$

1.5.2.4. Tangents at the Origin

Sometimes it is not possible to evaluate $\frac{dy}{dx}$ at the origin, as it takes the form $\left(\frac{0}{0}\right)$. To find the equations of tangents in such cases, a different method is adopted. We know equation of any straight line through the origin is of the form $y = mx$ (exception being y-axis). So we substitute. $y = mx$ in the equation of the curve and evaluate m as x approaches zero. The following example illustrates the use of this method.

Example 80. *Find the equations of the tangents at the origin to the curves,*
(a) $y^2 = x^2(1 + x)$ *(b)* $x^2 = 4ay$.

Sol. *(a)* Putting $y = mx$ in the equation of the curve, we have $m^2x^2 = x^2(1 + x)$

$\therefore$ $$(m^2 - 1)x^2 = x^3$$

or $$(m^2 - 1) = x$$

Now as $x \to 0$ $$m^2 - 1 = 0 \text{ or } m = \pm 1.$$

Therefore, the equations of tangents are $y = x$ and $y = -x$.

(b) Putting $y = mx$ in the equation of the curve, we get

$$x^2 = 4\,am\,x$$

or $$x = 4\,am$$

Now as $x \to 0$, $$m = 0;$$

Therefore the equation of tangent is $y = 0$, *i.e.* the x-axis.

The above example shows that equation of the tangent at the origin can be obtained by equating the lowest degree terms, in the equation of the curve to zero.

1.5.2.5. Angle of Intersection of two Curves

The angle of intersection of two curves is defined as the angle between the tangents at their common point of intersection.

Let m_1 and m_2 be the slopes of tangents to the curves at their point of intersection and if θ be the angle between them, then

$$\tan\theta = \frac{m_1 \sim m_2}{1 + m_1 m_2}$$

where $m_1 \sim m_2$ means positive difference between m_1 and m_2.

The acute angle θ between the tangents is taken to be the angle of intersection of the two curves.

When $m_1 m_2 = -1$, $\theta = 90^\circ$, the curves intersect each other at right angles and are called orthogonal.

Example 81. *Find the angle of intersection of the circle $x^2 + y^2 = 2a^2$ and the rectangular hyperbola $x^2 - y^2 = a^2$.*

Sol. Let the two curves intersect at (x_1, y_1).

$$\therefore \quad x_1^2 + y_1^2 = 2a^2 \quad ...(i)$$

$$\text{and} \quad x_1^2 - y_1^2 = a^2 \quad ...(ii)$$

From (*i*) and (*ii*), we get
$$x_1^2 = \frac{3a^2}{2} \quad ...(iii)$$

and
$$y_1^2 = \frac{a^2}{2} \quad ...(iv)$$

From the first curve
$$\frac{dy}{dx_{(x_1,y_1)}} = -\frac{x_1}{y_1} = m_1 \text{ (say)},$$

From the second curve
$$\frac{dy}{dx_{(x_1,y_1)}} = -\frac{x_1}{y_1} = m_2 \text{ (say)}.$$

Let θ be the angle between two curves then,

$$\tan\theta = \frac{m_1 \sim m_2}{1 + m_1 m_2} = \frac{\frac{x_1}{y_1} + \frac{x_1}{y_1}}{1 - \frac{x_1^2}{y_1^2}}$$

$$\tan\theta = \frac{2x_1 y_1}{x_1^2 - y_1^2}$$

$$\tan\theta = \frac{2\sqrt{\frac{3}{2}}\,.\,a\,.\,\frac{1}{\sqrt{2}}\,.\,a}{\frac{3a^2}{2} - \frac{a^2}{2}} = \sqrt{3}$$

$$\theta = \frac{\pi}{6} \text{ is the required angle.}$$

EXERCISE 1.5.10.

1. Find the equations of the tangent and normal at the point (x, y) on each of the following curves.

(*i*) $\frac{x^2}{a^2} + \frac{y^2}{b^2} = 1.$ (*ii*) $x^2 - y^2 = 7.$

(*iii*) $y^2 = 4px.$ (*iv*) $y = c \cos h\frac{x}{c}.$

2. Find the equations of the tangents to the curve $9x^2 + 16y^2 = 52$
(*i*) parallel to the line $9x - 8y = 10$.
(*ii*) perpendicular to this line.

3. (*a*) Find the points at which the tangent to the curve $y = x^3 + 5$ is
(*i*) parallel to the line $12x - y = 17$,
(*ii*) perpendicular to the line $x + 3y = 2$.

(*b*) Find the points on the curve $b^2y = \frac{x^3}{3} - ax^2$ at which tangent is parallel to the *x*-axis.

4. Find the equations of tangent and normal at the point t on each of the following curves
(*i*) $x = a(t + \sin t)$, $y = a(1 - \cos t)$,
(*ii*) $x = a \cos^3 t$ $y = b \sin^3 t$.

5. Find where the tangent is parallel to the axis of x and where it is perpendicular to that axis for the curve $y^3 = x^2(2a - x)$

6. Prove $\frac{x}{a} + \frac{y}{b} = 1$, touches the curve $y = b\,e^{-x/a}$ at the point where the curve crosses the y axis

7. Find the equations of the tangent and normal to the curve $y = \frac{8a^3}{4a^2 + x^2}$ at the point where it cuts the *y*-axis.

8. Tangents are drawn from the origin to the curve $y = \sin x$. Prove that their points of contact lie on the curve $x^2y^2 = x^2 - y^2$.

9. Show that the condition for the line $x \cos\theta + y \sin\theta = p$ to touch the curve $x^m y^n = a^{m+n}$ is $p^{m+n} \cdot m^m \cdot n^n = (m + n)^{m+n}\, a^{m+n} \cos^m\theta \sin^n\theta$.

10. Find the equation to the tangent at the point (a, a^3) on the curve $y = x^3$. Show that the tangent meets the curve again at the point $(-2a, -8a^3)$, and that the area included between the tangent and the curve is $\frac{27a^4}{4}$.

11. If OP_1 and OP_2 are the perpendiculars from O on the tangent and normal respectively at any point t on the curve $x = a \cos^3 t$ and $y = a \sin^3 t$, prove that $4 \cdot OP_1^2 + OP_2^2 = A^2$.

12. Show that the tangents to the curve $x^3 + y^3 = 3axy$ at the points other than origin where it meets the parabola $y^2 = ax$ are parallel to the *y*-axis.

13. In the catenary $y = a \cosh x/a$, prove that the length of the portion of the normal intercepted between the curve and the *x*-axis is y/a.

14. Find the angle of intersection of the following curves
(*i*) $y^2 = 4ax$ and $x^2 = 4by$.
(*ii*) $x^2 - 4x + y^2 = 0$, $x^2 + y^2 = 8$.
(*iii*) $y = x^2 + 2$ and $y = 2x^2 + 2$.
(*iv*) $x^2 + y^2 = \sqrt{2}a^2$ and $x^2 - y^2 = a^2$.

15. Show that the curves $\frac{x^2}{a^2 + \lambda_1} + \frac{y^2}{b^2 + \lambda_1} = 1$, and $\frac{x^2}{a^2 + \lambda_2} + \frac{y^2}{b^2 + \lambda_2} = 1$, intersect orthogonally.

16. Show that the angle between the tangent at any point p and the line joining p to the origin is the same at all points of the curve $\log(x^2 + y^2) = k \tan^{-1}\left(\frac{y}{x}\right)$

17. If α and β be the intercepts on the axes of x and y cut off by the tangent to the curve

$$\left(\frac{x}{a}\right)^n + \left(\frac{y}{b}\right)^n = 1, \text{ then}$$

$$\left(\frac{a}{\alpha}\right)^{\frac{n}{n-1}} + \left(\frac{b}{\beta}\right)^{\frac{n}{n-1}} = 1$$

18. The tangents at any point of the curve $x^3 + y^3 = 2a^3$ cut off lengths p and q on the co-ordinate axes. Show that $p^{-3/2} + q^{-3/2} = 2^{-1/2}\, a^{-3/2}$.

19. If x_1 and y_1 be the parts of the axes of x and y intercepted by the tangent at any point (x, y) on the curve $\left(\frac{x}{a}\right)^{3/2} + \left(\frac{y}{b}\right)^{3/2} = 1$, show that. $\frac{x_1^2}{a^2} + \frac{y_1^2}{b^2} = 1.$

20. Show that the ellipse $x^2 + 4y^2 = 8$ and the hyperbola $x^2 - 2y^2 = 2$, intersect orthogonally.

ANSWERS

1. (*i*) $\frac{Xx}{a^2} + \frac{Y.y}{b^2} = 1$ (*ii*) $Xx - Y.y = 7$

(*iii*) $Yy = 2p\,(X + x)$ (*iv*) $Y - y = \left(\sin h\frac{x}{C}\right)(X - x)$

2. $9x - 8y \pm 26 = 0$ (*ii*) $8x + 9x \pm 7 = 0$

3. (*a*) (*i*) (2, 13), (2, – 3), (*ii*) (1, 6), (–1, 4)

(*b*) (0, 0), $\left(2a, \frac{2a^3}{b^2}\right)$

4. (*i*) $y = (x - at)\tan t/2;$ $y + (x - at)\cos(t/2) = 0$

(*ii*) $\frac{x}{a\cos t} + \frac{y}{b\sin t} = 1$; by $\sin t - ax \cos t = b\sin^4 t - a\cos^4 t$.

5. Parallel at (0, 0), $\left[\frac{4a}{3}, \frac{(32)^{1/3}a}{3}\right]$, perpendicular at (0, 0), $(2a, 0)$

7. $y = 2a$; $x = 0$

14. (*i*) $\frac{\pi}{2}$ at (0, 0) and $\tan^{-1}\frac{3a^{1/3}b^{1/3}}{2(a^{2/3} + b^{2/3})}$ at $(4a^{1/3}\, b^{2/3}, 4a^{2/3}\, b^{1/3})$

(*ii*) $\frac{\pi}{2}$ (*iii*) 0°

(*iv*) 45°

1.5.2.6. Length of Tangent, Normal, Subtangent and Subnormal

Let tangent PT and normal PN to a curve $y = f(x)$ at a point $p(x, y)$, meet the x-axis at the points T and N respectively. Now $PR = y$, is the ordinate of the point P.

Then TR is called *Subtangent* and RN the *subnormal* of the curve. Let the tangent PT make an angle ϕ with the positive direction of the x-axis, then

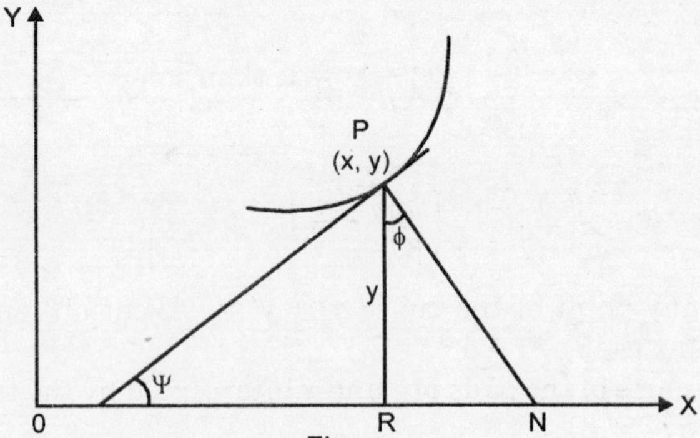

Fig.

$$\angle PTR = \phi = \angle RPN$$

and $$\tan\phi = \frac{dy}{dx} = y'.$$

(*i*) **Length of Tangent PT.**
From ΔTPR, we have

$$\frac{PT}{RP} = \text{cosec } \phi$$

$$\therefore \quad PT = RP \text{ cosec } \phi = y\sqrt{1+\cot^2\phi} = y\cdot\frac{\sqrt{1+\tan^2\phi}}{\tan\phi} \qquad [\because RP = y]$$

$$= y\cdot\frac{\sqrt{1+\left(\frac{dy}{dx}\right)^2}}{\frac{dy}{dx}} = \frac{y}{y'}\cdot\sqrt{1+y'^2} \qquad \left[\because \tan\phi = \frac{dy}{dx} = y\right]$$

Hence length of tangent $PT = \frac{y}{y'}.\sqrt{1+y'^2}$.

(*ii*) **Length of Normal *PN*.** From ΔRPN, we have

$$\frac{PN}{RP} = \sec\phi$$

$$\therefore \quad PN = RP\sec\phi = y\cdot\sqrt{1+\tan^2\phi} \qquad [\because RP = y]$$

$$= y\cdot\sqrt{1+\left(\frac{dy}{dx}\right)^2} = y\cdot\sqrt{1+y'^2} \qquad \left[\because \tan\phi = \frac{dy}{dx} = y'\right]$$

Hence length of normal $PN = y\sqrt{1+y'^2}$

(*iii*) **Length of subtangent *TR*.** From ΔTPR, we get

$$\frac{TR}{RP} = \cot\phi$$

$\therefore \qquad TR = RP \cot\phi = \dfrac{y}{\tan\phi} \qquad [\because RP = y]$

$$= \frac{y}{\frac{dy}{dx}} = \frac{y}{y'} \qquad \left[\because \tan\phi = \frac{dy}{dx} = y'\right]$$

Length of subtangent $TR = \dfrac{y}{y'}$.

(*iv*) **Length of subtangent RN.** From ΔRPN, we have

$$\frac{RN}{RP} = \tan\phi$$

$\therefore \qquad RN = RP\tan\phi = y\cdot\dfrac{dy}{dx} = yy' \qquad \left[\because RP = y \text{ and } \tan\phi = \dfrac{dy}{dx}\right]$

Hence length of Subnormal $RN = yy'$

Example 1. *Find the lengths of the tangent, subtangent, normal and subnormal at the point (1, 3) of the curve $y = x^2 + x + 1$.*

Sol. Here $\qquad y = x^2 + x + 1$

$$\frac{dy}{dx} = 2x + 1 \,;\; \frac{dy}{dx} = 3 \text{ at } (1, 3).$$

(*i*) Length of tangent $= \dfrac{y\sqrt{1+\left(\frac{dy}{dx}\right)^2}}{\frac{dy}{dx}} = \dfrac{3.\sqrt{1+3^2}}{3} = \sqrt{10}$

(*ii*) Length of subtangent $= \dfrac{y}{\frac{dy}{dx}} = \dfrac{3}{3} = 1.$

(*iii*) Length of normal $= y\sqrt{1+\left(\frac{dy}{dx}\right)^2} = 3\sqrt{1+3^2} = 3\sqrt{10}$

(*iv*) Length of subnormal $= y\cdot\dfrac{dy}{dx} = 3 \times 3 = 9.$

Example 82. *Find the lengths of tangent, normal and subnormal at the point $t = \dfrac{\pi}{2}$ on the cyclodid $x = a(t + \sin t)$, $y = a(1 - \cos t)$.*

Sol. Here $\qquad x = a(t + \sin t),\ y = a(1 - \cos t)$

$\therefore \qquad \dfrac{dx}{dt} = a(1 + \cos t),\ \dfrac{dy}{dt} = a \sin t.$

Now, $$y' = \frac{dy}{dx} = \frac{\frac{dy}{dt}}{\frac{dx}{dt}} = \frac{a\sin t}{a(1+\cos t)} = \frac{2a\sin t/2\cos t/2}{2a\cos^2 t/2} = \tan t/2.$$

$$y' = \frac{dy}{dx} = 1 \qquad \text{at } t = \frac{\pi}{2}$$

Also from the equation of the curve $y = a$ at $t = \frac{\pi}{2}$.

(*i*) Length of tangent $= \frac{y}{y'}\sqrt{1+y'^2} = \frac{a}{1}\cdot\sqrt{1+1^2} = a\sqrt{2}$.

(*ii*) Length of subtangent $= \frac{y}{y'} = a$.

(*iii*) Length of normal $= y\sqrt{1+y'^2} = a\sqrt{2}$

(*iv*) Length of subnormal $= yy' = a$.

Example 83. *Show that the subtangent at any point of the curve $x^m y^n = a^{m+n}$, varies as abscissa of the point of contact.*

Sol. Let the co-ordinates of the point of contact be (x, y).

We have to show that $\frac{\text{Subtangent}}{x} = \text{constant}$

Here $$x^m y^n = a^{m+n}$$

Taking logs on both sides, we have

$$m \log x + n \log y = (m + n)\log a$$

Differentiating both sides w.r.t. x, we get $\frac{m}{x} + \frac{n}{y}\frac{dy}{dx} = 0$,

$$\therefore \qquad \frac{dy}{dx} = -\frac{my}{nx}$$

Length of subtangent $$= \frac{y}{\frac{dy}{dx}} = -\frac{y.nx}{my} = -\frac{n}{m}x$$

$$\therefore \qquad \frac{\text{Subtangent}}{x} = -\frac{n}{m} = \text{constant}.$$

EXERCISE 1.5.11.

1. Show that for the parabola $y^2 = 4ax$, subtangent at any point is twice the abcissa and the subnormal is of constant length.
2. Show that in the exponential curve $y = be^{x/a}$, the subnormal varies as the square of the ordinate and subtangent is of constant length.
3. Find the subtangent, subnormal, normal and tangent at the point θ on the cycloid. $x = a(\theta - \sin\theta)$, $y = a(1 - \cos\theta)$

Ans. 3. S.T. $= 2a \sin^3 \theta/2 \sec \theta/2$; S.N. $= a \sin\theta$.
Normal $= 2a \sin \theta/2$, tangent $= 2a \sin \theta/2 \tan \theta/2$.

4. Prove that the length of the normal varies inversely as perpendicular from the origin on the tangent to the ellipse, $\frac{x^2}{a^2}+\frac{y^2}{b^2} = 1$.
5. Prove that in the curve $x^{m/n} = a^{m-n} y^{2n}$, m^{th} power of subtangent varies as n^{th} power of sub-normal.
6. Prove that for the curve $x = a(\cos t + \log \tan t/2)$, $y = a\sin t$. The portion of the tangent intercepted between the curve and the axis of x is of constant length.
7. Show that in the curve $y = a \log(x^2 - a^2)$ sum of the length of tangent and subtangent varies as the product of the coordinates of the point of contact.
8. If the coordinates of a point on the curve $x^{2/3} + y^{2/3} = a^{2/3}$ are represented as $x = a\sin^3 t$, $y = a \cos^3 t$, show that t is the angle which the perpendicular from the origin to the tangent makes with x-axis.

• Derivative of Arc (Cartesian)

Let $p(x, y)$ be any point on a curve $y = f(x)$, s the length of the arc AP, where A is a fixed point on the given curve and ϕ the angle that the tangent at p makes with x-axis. Further let $Q(x + \delta x, y + \delta y)$ be a neighbouring point to p on the curve and arc PQ and chord PQ be of length δs and δl respectively.

We shall assume that the arc RQ of the curve is concave to the chord PQ, then as $Q \to P$,

$$\frac{\text{arc } PQ}{\text{chord } PQ} = 1 \text{ or } \lim_{Q\to P} \cdot \frac{\widehat{PQ}}{PQ} = 1$$

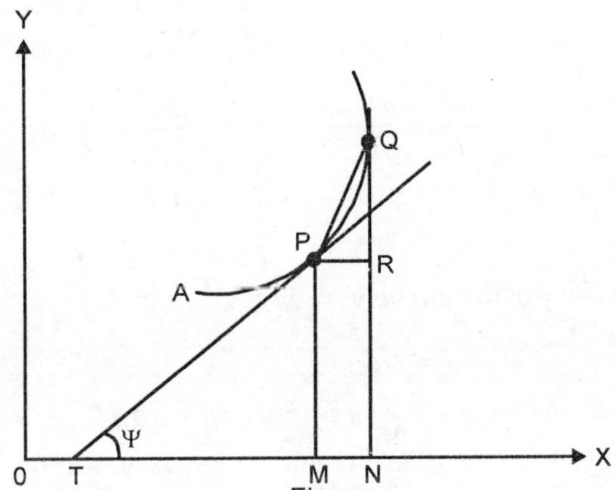

Fig.

Draw PM and QN perpendiculars on x-axis from the points P and Q respectively and PR be perpendicular to QN.

Now $\quad PR = MN = \delta x$ and $QR = QN - RN = (y + \delta y) - y = \delta y$.

Since $\quad RN = PM = y$

We have from ΔPQR, $\quad PQ^2 = PR^2 + QR^2$

or $\quad (\delta l)^2 = (\delta x)^2 + (\delta y)^2 \qquad (\because PQ = \delta l)$

or
$$\left(\frac{\delta l}{\delta x}\right)^2 = 1+\left(\frac{\delta y}{\delta x}\right)^2 \Rightarrow \frac{\delta l}{\delta x} = \sqrt{1+\left(\frac{\delta y}{\delta x}\right)^2}$$

Now
$$\frac{\delta s}{\delta x} = \frac{\delta s}{\delta l}\frac{\delta l}{\delta x} = \frac{\delta s}{\delta l}\cdot\sqrt{1+\left(\frac{\delta y}{\delta x}\right)^2} \qquad ...(1)$$

When $Q \to P$ *i.e.* $\delta x \to 0$, $\frac{\delta s}{\delta l} \to 1$, $\frac{\delta s}{\delta x}$ becomes $\frac{ds}{dx}$ and $\frac{\delta y}{\delta x}$ as $\frac{dy}{dx}$.

Then (1) becomes,

$$\frac{ds}{dx} = \sqrt{1+\left(\frac{dy}{dx}\right)^2}$$

Similarly we can show that $\frac{ds}{dy} = \sqrt{1+\left(\frac{dx}{dy}\right)^2}$

$$\frac{dx}{ds} = \frac{1}{\sqrt{1+\left(\frac{dy}{dx}\right)^2}} = \frac{1}{\sqrt{1+\tan^2\phi}} = \frac{1}{\sec\phi} = \cos\phi$$

and similarly $\frac{dy}{ds} = \sin\phi.$

When equation of the curve is in parametric form $x = f(t)$ and $y = \phi(t)$, we know

$$\frac{dy}{dx} = \frac{\frac{dy}{dt}}{\frac{dx}{dt}} \quad \text{and} \quad \frac{ds}{dx} = \sqrt{1+\left\{\frac{\left(\frac{dy}{dt}\right)^2}{\left(\frac{dx}{dt}\right)^2}\right\}}$$

$$\therefore \quad \frac{ds}{dt} = \frac{ds}{dx}\cdot\frac{dx}{dt} = \sqrt{1+\frac{\left(\frac{dy}{dt}\right)^2}{\left(\frac{dx}{dt}\right)^2}}\cdot\frac{dx}{dt} \quad \text{or} \quad \frac{ds}{dt} = \sqrt{\left(\frac{dx}{dt}\right)^2+\left(\frac{dy}{dt}\right)^2}.$$

Example 84. *Calculate* $\frac{ds}{dx}$ *for the curve* $y^2 = 4ax$

Sol. Here

$$y^2 = 4ax$$

$$\therefore \quad 2y\frac{dy}{dx} = 4a \quad \text{or} \quad \frac{dy}{dx} = \frac{2a}{y}$$

We know

$$\frac{ds}{dx} = \sqrt{1+\left(\frac{dy}{dx}\right)^2} = \sqrt{1+\frac{4a^2}{y^2}} = \sqrt{1+\frac{4a^2}{4ax}} \qquad (\because y^2 = 4ax)$$

or

$$\frac{ds}{dx} = \sqrt{1+\frac{a}{x}}.$$

Example 85. *Find* $\frac{ds}{dt}$ *for the astroid* $x = a\cos^3 t$, $y = a\sin^3 t$.

Sol. Here

$$x = a\cos^3 t,$$
$$y = a\sin^3 t$$

$$\therefore \qquad \frac{dx}{dt} = -\,3a\cos^2 t \sin t. \text{ and}$$

$$\frac{dy}{dt} = 3a\sin^2 t\cos t$$

We know
$$\frac{ds}{dt} = \sqrt{\left(\frac{dx}{dt}\right)^2 + \left(\frac{dy}{dt}\right)^2} = \sqrt{9a^2(\cos^4 t\sin^2 t + \sin^4 t\cos^2 t)}$$

$$= 3a\sin t\cos t\sqrt{\cos^2 t + \sin^2 t}$$

$$\frac{ds}{dt} = 3a\sin t\cos t.$$

EXERCISE 1.5.12.

1. Prove that for the catenary $y = c\cosh\frac{x}{c}$, $\frac{ds}{dx} = \frac{y}{a}$.
2. Prove that for the ellipse

   $x = a\cos t,\ y = b\sin t,\ \frac{ds}{dt} = a\sqrt{(1-e^2\cos^2 t)}$.

   Find $\frac{ds}{dx}$ for the following curves:

3. $3ay^2 = x^2(a - x)$ **Ans.** $\sqrt{1+\frac{(2a-3x)^2}{12a(x-a)}}$
4. $y = \log\frac{e^x - 1}{e^x + 1}$ **Ans.** $\frac{e^{2x}+1}{e^{2x}-1}$.
5. If $x = a(t + \sin t)$, $y = a(1 - \cos t)$, prove that $\frac{ds}{dy} = \sqrt{\frac{2a}{y}}$ and $\frac{ds}{dt} = 2a\cos\frac{t}{2}$

1.5.3. L'Hospital Rule for Solving Indeterminant form (0/0, ∞/∞)

Let,
$$\lim_{x\to 4}\frac{x^2-16}{x-4} \qquad \lim_{x\to\infty}\frac{4x^2-5x}{1-3x^2}$$

In the first limit if we put $x = 4$ we would get 0/0 and in the second limit if we "put" in infinity we would get $\infty/-\infty$ (recall that as x goes to infinity a polynomial will behave in the same fashion that its largest power). Both of these are called **indeterminate forms.**

In the case of 0/0 we typically think of a fraction that has a numerator of zero as being zero. So, we also tend to think of fractions in which the denominator is going to zero as infinity or might not exist at all.

In the case of $\infty/-\infty$ have a similar set of problems. If the numerator of a fraction is going to infinity we tend to whole fraction going to infinity. Also if the denominator is going to infinity we tend to whole fraction as going to zero. We also have the case of a fraction in which the numerator and denominator are the same (ignoring the minus sign).

This is the problem with indeterminate forms. It's just not clear what is happening in the limit. There are other types of indeterminate forms as well. Some other types are,

$(0)(\pm\infty)$ $\quad 1^{\infty} \quad 0^0 \quad \infty^0 \quad \infty - \infty$

Example 86. (*i*) $\lim\limits_{x\to 4} \dfrac{x^2-16}{x-4} = \lim\limits_{x\to 4}(x+4) = 8$

$$(ii)\ \lim_{x\to\infty} \frac{4x^2-5x}{1-3x^2} = \lim_{x\to\infty} \frac{4-\frac{5}{x}}{\frac{1}{x^2}-3} = -\frac{4}{3}$$

However what about the following two limits.

$$\lim_{x\to 0}\frac{\sin x}{x}; \qquad \lim_{x\to\infty}\frac{e^x}{x^2}$$

This first is a 0/0 indeterminate form, but we can't factor this one. The second is an ∞/∞ indeterminate form, but we can not just factor an x^2 out of the numerator. So, nothing that we've got in our bag of tricks will work with these two limits.

L'Hospital's Rule

Suppose that we have one of the following cases,

$$\lim_{x\to a}\frac{f(x)}{g(x)} = \frac{0}{0} \quad \text{or} \quad \lim_{x\to a}\frac{f(x)}{g(x)} = \frac{\pm\infty}{\pm\infty}$$

where a can be any real number, infinity or negative infinity. In these cases we have,

$$\lim_{x\to a}\frac{f(x)}{g(x)} = \lim_{x\to a}\frac{f'(x)}{g'(x)}$$

So, L'Hospital's rule tells us that if we have an indeterminate form 0/0 or ∞/∞ all we need to do is differentiate the numerator and differentiate the denominator and then take the limit.

Example 87. *Evaluate each of the following limits.*

(*a*) $\lim\limits_{x\to 0}\dfrac{\sin x}{x}$ $\qquad$ (*b*) $\lim\limits_{t\to 1}\dfrac{5t^4-4t^2-1}{10-t-9t^2}$ $\qquad$ (*c*) $\lim\limits_{x\to\infty}\dfrac{e^x}{x^2}$

Sol. (*a*) $\lim\limits_{x\to 0}\dfrac{\sin x}{x}$, so, we have already established that is a 0/0 indeterminate form so let's just apply L'Hospital's Rule. $\lim\limits_{x\to 0}\dfrac{\sin x}{x} = \lim\limits_{x\to 0}\dfrac{\cos x}{1} = \dfrac{1}{1} = 1$

(*b*) $\lim\limits_{t\to 1}\dfrac{5t^4-4t^2-1}{10-t-9t^3}$, in this case we also have *a* 0/0 indeterminate factor the numerator and denominator, simplify and take the limit $\lim\limits_{t\to 1}\dfrac{5t^4-4t^2-1}{10-t-9t^3}$ $= \lim\limits_{t\to 1}\dfrac{20t^3-8t}{-1-27t^2} = \dfrac{20-8}{-1-27} = -\dfrac{3}{7}$

(*c*) $\lim\limits_{x\to\infty}\dfrac{e^x}{x^2}$, This was the other limit that we started off looking at and we know that it's the indeterminate form ∞/∞ so let's apply L'Hospital's Rule.

$$\lim_{x\to\infty}\frac{e^x}{x^2} = \lim_{x\to\infty}\frac{e^x}{x^2} = \lim_{x\to\infty}\frac{e^x}{2x} = \lim_{x\to\infty}\frac{e^x}{2} = \infty$$

Now,

Example 88. *Evaluate the following limit.*

$$\lim_{x\to0^+} x \ln x$$

Sol. Using L'Hospital's Rule.

$$\lim_{x\to0^+} x \ln x = \lim_{x\to0^+}\frac{\ln x}{\frac{1}{x}}$$

Now, $-\infty/\infty$ and we can now use L'Hospital's Rule

$$\lim_{x\to0^+} x \ln x = \lim_{x\to0^+}\frac{\ln x}{\frac{1}{x}} = \lim_{x\to0^+}\frac{\frac{1}{x}}{-\frac{1}{x^2}} = \lim_{x\to0^+}\frac{\frac{1}{x}}{-\frac{1}{x^2}} = \lim_{x\to0^+}(-x) = 0$$

In the previous example we used the fact that we can always write a product of functions as a quotient by doing one of the following.

$$f(x)\,g(x) = \frac{g(x)}{\frac{1}{f(x)}} \qquad \text{OR} \qquad f(x)\,g(x) = \frac{f(x)}{\frac{1}{g(x)}}$$

Example 89. *Evaluate the following limit.*

$$\lim_{x\to-\infty} xe^x$$

Sol. So, it's in the form $(\infty)(0)$. This means that we'll need to write it as a quotient. Moving the x to the denominator worked in the previous example so let's try that with this problem.

$$\lim_{x\to-\infty} xe^x = \lim_{x\to-\infty}\frac{e^x}{1/x}$$

Writing the product in this way gives us a product that has the form 0/0 in the limit. So, let's use L'Hospital's Rule on the quotient.

$$\lim_{x\to-\infty} xe^x = \lim_{x\to-\infty}\frac{e^x}{\frac{1}{x}} = \lim_{x\to-\infty}\frac{e^x}{-\frac{1}{x^2}} = \lim_{x\to-\infty}\frac{e^x}{\frac{2}{x^3}} = \lim_{x\to-\infty}\frac{e^x}{-\frac{6}{x^4}} = \ldots$$

This does not mean however that the limit can't be done. It just means that we moved the wrong function to the denominator. Let's move the exponential function instead.

$$\lim_{x\to-\infty} xe^x = \lim_{x\to-\infty}\frac{x}{\frac{1}{e^x}} = \lim_{x\to-\infty}\frac{x}{e^{-x}}$$

L'Hospital's Rule.

$$\lim_{x\to-\infty} xe^x = \lim_{x\to-\infty}\frac{x}{e^{-x}} = \lim_{x\to-\infty}\frac{1}{-e^{-x}} = 0$$

Example 90. *Evaluate the following limit* $\lim_{x\to\infty} x^{1/x}$

Solution. Let $y = x^{1/x}$

by using L'Hospital's Rule $\quad \text{In}(y) = \text{In}\left(x^{1/x}\right) = \frac{1}{x}\text{In } x = \frac{\text{In } x}{x}$

$$\lim_{x\to\infty} \text{In}(y) = \lim_{x\to\infty} \frac{\text{In } x}{x} = \lim_{x\to\infty} \frac{\frac{1}{x}}{1} = 0$$

by using, L'Hospital's Rule $\quad e^{\text{In}(y)} = y$

$$\lim_{x\to\infty} x^{1/x} = \lim_{x\to\infty} y = \lim_{x\to\infty} e^{\text{In}(y)}$$

$$\lim_{x\to\infty} x^{1/x} = \lim_{x\to\infty} y = \lim_{x\to\infty} e^{\text{In}(y)} = e^{\lim_{x\to\infty} \text{In}(y)} = e^0 = 1$$

EXERCISE 1.5.13.

1. Find $\lim_{x\to 0} \frac{x - \sin x}{x^3}$ **[Ans.** 1/6]
2. Find $\lim_{x\to\infty} \frac{e^{2x} + x^2}{e^x + 4x}$ **[Ans.** 1]
3. Find $\lim_{x\to\infty} \frac{\sqrt{9x+1}}{\sqrt{x+1}}$ **[Ans.** 3]
4. Find $\lim_{x\to 1} \left(\frac{1}{x-1} - \frac{1}{\ln x}\right)$ **[Ans.** –1/2]
5. Find $\lim_{x\to\infty} \left(1 + \frac{1}{n}\right)^n$ **[Ans.** *e*]
6. Find $\lim_{x\to 0} x^x$ **[Ans.** 1]